METHODS IN MOLECULAR BIOLOGY

Series Editor
John M. Walker
School of Life and Medical Sciences
University of Hertfordshire
Hatfield, Hertfordshire, UK

For further volumes:
http://www.springer.com/series/7651

For over 35 years, biological scientists have come to rely on the research protocols and methodologies in the critically acclaimed *Methods in Molecular Biology* series. The series was the first to introduce the step-by-step protocols approach that has become the standard in all biomedical protocol publishing. Each protocol is provided in readily-reproducible step-by step fashion, opening with an introductory overview, a list of the materials and reagents needed to complete the experiment, and followed by a detailed procedure that is supported with a helpful notes section offering tips and tricks of the trade as well as troubleshooting advice. These hallmark features were introduced by series editor Dr. John Walker and constitute the key ingredient in each and every volume of the *Methods in Molecular Biology* series. Tested and trusted, comprehensive and reliable, all protocols from the series are indexed in PubMed.

Salamanders

Methods and Protocols

Edited by

Ashley W. Seifert

Department of Biology, University of Kentucky, Lexington, KY, USA

Joshua D. Currie

Department of Biology, Wake Forest University, Winston-Salem, NC, USA

Editors
Ashley W. Seifert
Department of Biology
University of Kentucky
Lexington, KY, USA

Joshua D. Currie
Department of Biology
Wake Forest University
Winston-Salem, NC, USA

ISSN 1064-3745 ISSN 1940-6029 (electronic)
Methods in Molecular Biology
ISBN 978-1-0716-2661-0 ISBN 978-1-0716-2659-7 (eBook)
https://doi.org/10.1007/978-1-0716-2659-7

This Humana imprint is published by the registered company Springer Science+Business Media, LLC, part of Springer Nature.
The registered company address is: 1 New York Plaza, New York, NY 10004, U.S.A.

Preface

For nearly three centuries, salamanders have remained key research organisms for developmental, evolutionary, and ecological studies. Building on advances in genetic model systems (e.g., *Drosophila*, *C. elegans*, and *Mus*), the last 20 years has seen a dizzying acceleration of molecular and genomic techniques that have made their way into amphibian research. This breadth and depth have made salamander and newt models ideal research organisms to study processes affecting biological scales from molecules to ecosystems.

Our goal in editing this new edition of *Methods in Salamander Research* was to assemble a definitive desk reference for researchers either new to using salamanders for research or expert labs that want to bring new techniques and approaches to bear on their scientific questions. The chapters contained in this volume highlight new voices in this field alongside the experience and perspective of veteran salamander scientists. We want to extend a heartfelt thanks to all this volume's authors for their efforts in the face of pandemic-related disruptions. We hope this reference will be a catalyst for researchers using amphibians to dialogue with the authors and the broader community in order to extend the boundaries of salamander research.

Lexington, KY, USA — *Ashley W. Seifert*
Winston-Salem, NC, USA — *Joshua D. Currie*

Contents

Part III Experimental Manipulations and Surgeries

Part IV Bioinformatics and Genomics

Part V Transgenics and Lineage-Tracing

Part VI Physiological and Organismal Techniques

Part VII Epilogue

Contributors

JENNIFER ÁLVAREZ • *Grupo Genética Regeneración y Cáncer, Institute of Biology, University of Antioquia, Medellin, Colombia*
ZACHARY BEAL • *Mount Desert Island Biological Laboratory (MDIBL), Kathryn W. Davis Center for Regenerative Biology and Aging, Bar Harbor, ME, USA*
BELFRAN CARBONELL • *Grupo Genética Regeneración y Cáncer, Institute of Biology, University of Antioquia, Medellin, Colombia*
WEIHAO CHEN • *Cellular Molecular and Structural Biology Program, Miami University, Oxford, OH, USA; Department of Chemical, Paper and Biomedical Engineering, Miami University, Oxford, OH, USA*
KATHARINE COURTEMANCHE • *Department of Stem Cell and Regenerative Biology, Harvard University, Cambridge, MA, USA*
MARCUS J. CRIM • *IDEXX BioAnalytics, Columbia, MO, USA*
JOSHUA D. CURRIE • *Department of Biology, Wake Forest University, Winston-Salem, NC, USA*
NADJIB DASTAGIR • *Mount Desert Island Biological Laboratory (MDIBL), Kathryn W. Davis Center for Regenerative Biology and Aging, Bar Harbor, ME, USA*
ANDREW K. DAVIS • *Odum School of Ecology, The University of Georgia, Athens, GA, USA; D. B. Warnell School of Forestry and Natural Resources, The University of Georgia, Athens, GA, USA*
KATIA DEL RIO-TSONIS • *Department of Biology, Miami University, Oxford, OH, USA; Center for Visual Sciences at Miami University, Oxford, OH, USA; Cellular Molecular and Structural Biology Program, Miami University, Oxford, OH, USA*
JEAN PAUL DELGADO • *Grupo Genética Regeneración y Cáncer, Institute of Biology, University of Antioquia, Medellin, Colombia*
TIMOTHY J. DUERR • *Department of Biology, Northeastern University, Boston, MA, USA*
GARRETT S. DUNLAP • *Biological and Biomedical Sciences, Harvard Medical School, Cambridge, MA, USA*
JI-FENG FEI • *Key Laboratory of Brain, Cognition and Education Sciences, Ministry of Education, Institute for Brain Research and Rehabilitation, South China Normal University, Guangzhou, China; Department of Pathology, Guangdong Provincial People's Hospital, Guangdong Academy of Medical Sciences, Guangzhou, China*
SAYA FURUKAWA • *Department of Biological Science Faculty of Science, Okayama University, Okayama, Japan*
JAMES GODWIN • *Mount Desert Island Biological Laboratory (MDIBL), Kathryn W. Davis Center for Regenerative Biology and Aging, Bar Harbor, ME, USA*
ANDREW HART • *Mount Desert Island Biological Laboratory (MDIBL), Kathryn W. Davis Center for Regenerative Biology and Aging, Bar Harbor, ME, USA*
MARCIA L. HART • *IDEXX BioAnalytics, Columbia, MO, USA*
WANSHENG JIANG • *Jishou University, Zhangjiajie, Hunan, China*
GABRIELA JOHNSON • *Mount Desert Island Biological Laboratory (MDIBL), Kathryn W. Davis Center for Regenerative Biology and Aging, Bar Harbor, ME, USA*
RENA KASHIMOTO • *Okayama University, Graduate school of Natural Science and Technology, Okayama, Japan*

AKANE KAWAGUCHI • *Vienna BioCenter (VBC), Research institute of Molecular Pathology (IMP), Wien, Austria*

MELISSA KEINATH • *Carnegie Institute of Science Department of Embryology (Lab 322), Baltimore, MD, USA*

RYAN KERNEY • *Department of Biology, Gettysburg College, Gettysburg, PA, USA*

MOSHE KHURGEL • *Department of Biology & Environmental Science, Bridgewater College, Bridgewater, VA, USA*

THOMAS KURTH • *Center for Molecular and Cellular Bioengineering (CMCB), Technology Platform, Electron Microscopy and Histology Facility, Technische Universität Dresden, Dresden, Germany*

NICHOLAS D. LEIGH • *Molecular Medicine and Gene Therapy, Wallenberg Centre for Molecular Medicine, Lund Stem Cell Center, Lund University, Lund, Sweden*

YANMEI LIU • *Key Laboratory of Brain, Cognition and Education Sciences, Ministry of Education, Institute for Brain Research and Rehabilitation, South China Normal University, Guangzhou, China*

ALEX M. LOVELY • *Department of Biology, Northeastern University, Boston, MA, USA; Northeastern University, Institute for Chemical Imaging of Living Systems, Boston, MA, USA*

MALCOLM MADEN • *Department of Biology & UF Genetics Institute, University of Florida, Gainesville, FL, USA*

JOHN C. MAERZ • *D.B. Warnell School of Forestry and Natural Resources, The University of Georgia, Athens, GA, USA*

ARIELLE MARTINEZ • *Department of Biology, Miami University, Oxford, OH, USA*

WOUTER MASSELINK • *Research Institute of Molecular Pathology (IMP), Vienna BioCenter (VBC), Wien, Austria*

CATHERINE D. MCCUSKER • *Department of Biology, University of Massachusetts, Boston, MA, USA*

ALYSSA MILLER • *Department of Biology, Miami University, Oxford, OH, USA*

JAMES R. MONAGHAN • *Department of Biology, Northeastern University, Boston, MA, USA; Northeastern University, Institute for Chemical Imaging of Living Systems, Boston, MA, USA*

EVAN T. MUN • *Department of Biology, Northeastern University, Boston, MA, USA*

PRAYAG MURAWALA • *Mount Desert Island Biological Laboratory (MDIBL), Salisbury Cove, ME, USA; Clinic for Kidney and Hypertension Diseases, Hannover Medical School, Hannover, Germany*

SERGEJ NOWOSHILOW • *Vienna BioCenter (VBC), Research institute of Molecular Pathology (IMP), Wien, Austria*

HELENA OKULSKI • *Research Institute of Molecular Pathology (IMP), Vienna Biocenter (VBC), Vienna, Austria*

CATARINA R. OLIVEIRA • *Center for Regenerative Therapies (CRTD), Technische Universität Dresden, Dresden, Germany; Graduate Program in Areas of Basic and Applied Biology (GABBA), University of Porto, Porto, Portugal*

JOSÉ RAÚL PÉREZ-ESTRADA • *Department of Biology, Miami University, Oxford, OH, USA; Center for Visual Sciences at Miami University, Oxford, OH, USA*

SRUTHI PURUSHOTHAMAN • *Department of Biology, University of Kentucky, Lexington, KY, USA*

MICHAEL RAYMOND • *Department of Biology, University of Massachusetts, Boston, MA, USA*

CAMILO RIQUELME-GUZMÁN • *Center for Regenerative Therapies Dresden (CRTD), Technische Universität Dresden, Dresden, Germany*
STÉPHANE ROY • *Department of Stomatology, Faculty of Dentistry, Université de Montréal, Montreal, Que, Canada*
ANTHONY SALLESE • *Department of Biology, Miami University, Oxford, OH, USA; Center for Visual Sciences at Miami University, Oxford, OH, USA*
TATIANA SANDOVAL-GUZMÁN • *Center for Regenerative Therapies Dresden (CRTD), Technische Universität Dresden, Dresden, Germany; Center for Healthy Aging, Universitätsklinikum Dresden, Dresden, Germany*
GLORIA A. SANTA-GONZÁLEZ • *Grupo Genética Regeneración y Cáncer, Institute of Biology, University of Antioquia, Medellin, Colombia; Biomedical Innovation and Research Group, Faculty of Applied and Exact Sciences, Instituto Tecnologico Metropolitano, Medellin, Colombia*
AKIRA SATOH • *Okayama University, Graduate school of Natural Science and Technology, Okayama, Japan; Okayama University, Research core for interdisciplinary sciences (RCIS), Okayama, Japan*
MARITTA SCHUEZ • *Center for Regenerative Therapies Dresden (CRTD), Technische Universität Dresden, Dresden, Germany*
ASHLEY W. SEIFERT • *Department of Biology, University of Kentucky, Lexington, KY, USA*
KONSTANTINOS SOUSOUNIS • *Department of Stem Cell and Regenerative Biology, Harvard University, Cambridge, MA, USA*
DAVID F. STEIN • *Department of Biology, Northeastern University, Boston, MA, USA*
ELLY M. TANAKA • *Research Institute of Molecular Pathology (IMP), Vienna BioCenter (VBC), Wien, Austria*
YUKA TANIGUCHI-SUGIURA • *Vienna BioCenter (VBC), Research institute of Molecular Pathology (IMP), Wien, Austria*
HAIFENG TIAN • *Yangtze River Fisheries Research Institute, Wuhan, Hubei, China*
VLADIMIR TIMOSHEVSKIY • *Department of Biology, University of Kentucky, Lexington, KY, USA*
GEORGIOS TSISSIOS • *Department of Biology, Miami University, Oxford, OH, USA; Center for Visual Sciences at Miami University, Oxford, OH, USA; Cellular Molecular and Structural Biology Program, Miami University, Oxford, OH, USA*
ELLI VICKERS • *Department of Biology, Gettysburg College, Gettysburg, PA, USA*
HANNAH E. WALTERS • *Technische Universität Dresden, CRTD/Center for Regenerative Therapies Dresden, Dresden, Germany*
HUI WANG • *Department of Chemical, Paper and Biomedical Engineering, Miami University, Oxford, OH, USA*
LIQUN WANG • *Key Laboratory of Brain, Cognition and Education Sciences, Ministry of Education, Institute for Brain Research and Rehabilitation, South China Normal University, Guangzhou, China*
JESSICA L. WHITED • *Department of Stem Cell and Regenerative Biology, Harvard University, Cambridge, MA, USA; The Harvard Stem Cell Institute, Cambridge, MA, USA; The Broad Institute of MIT and Harvard, Cambridge, MA, USA*
SAKIYA YAMAMOTO • *Department of Biological Science Faculty of Science, Okayama University, Okayama, Japan*
ANASTASIA S. YANDULSKAYA • *Department of Biology, Northeastern University, Boston, MA, USA*

MAXIMINA H. YUN • *Technische Universität Dresden, CRTD/Center for Regenerative Therapies Dresden, Dresden, Germany; Max Planck Institute for Molecular Cell Biology and Genetics, Dresden, Germany*
QINGHAO YU • *Technische Universität Dresden, CRTD/Center for Regenerative Therapies Dresden, Dresden, Germany*
YAN-YUN ZENG • *Key Laboratory of Brain, Cognition and Education Sciences, Ministry of Education, Institute for Brain Research and Rehabilitation, South China Normal University, Guangzhou, China*
LU ZHANG • *Sun Yat-sen University, Guangzhou, Guangdong, China*

Chapter 1

Salamanders as Key Models for Development and Regeneration Research

Malcolm Maden

Abstract

For 70 years from the very beginning of developmental biology, the salamander embryo was the pre-eminent model for these studies. Here I review the major discoveries that were made using salamander embryos including regionalization of the mesoderm; patterning of the neural plate; limb development, with the pinnacle being Spemann's Nobel Prize for the discovery of the organizer; and the phenomenon of induction. Salamanders have also been the major organism for elucidating discoveries in organ regeneration, and these are described here too beginning with Spallanzani's experiments in 1768. These include the neurotrophic hypothesis of regeneration, studies of aneurogenic limbs, the concept of dedifferentiation and transdifferentiation, and advances in understanding pattern formation via molecules located on the cell surface. Also described is the prodigious power of brain and spinal cord regeneration and discoveries from lens regeneration, all of which reveal how important salamanders have been as research models.

Key words Development, Regeneration, Salamanders, Axolotl, Newt, Embryology, Spemann, Spallanzani, Limb regeneration, Limb development

1 Early Studies of Salamanders

Few disciplines in science or introductions to model systems began with a treatise from a famous natural historian and polymath who was also educated in theology, languages, philosophy, logic, mathematics, metaphysics and Greek. And yet, such was the case with the publication of "Prodromo di un' opera da imprimersi sopra le riproduzioni animali" by Lazzaro Spallanzani in 1768. This was translated into English the following year by the secretary of the Royal Society of London to whom Spallanzani had sent his treatise as an example of his work. It was a successful ploy supported by letters from Charles Bonnet and Abraham Trembley, and as such, he was elected as a foreign correspondent of the society [1].

In this "Prodromo" Spallanzani described an astounding number of regeneration experiments he had performed on his local salamander, *Salamandra salamandra* (fire salamander). Having

Ashley W. Seifert and Joshua D. Currie (eds.), *Salamanders: Methods and Protocols*,
Methods in Molecular Biology, vol. 2562, https://doi.org/10.1007/978-1-0716-2659-7_1,

first shown that, despite its name, the fire salamander is no more resistant to the heat of hot water or "hot embers than other animals of the terrestrial, amphibious, and aquatic kind", he showed both limbs and tails regenerated; the jaws regenerated while replacing teeth, bones, cartilages, muscles, veins, and arteries; he showed that exactly what was cut off is replaced and that regeneration occurred after amputation at many different levels; he disarticulated the limbs at the girdle; he cut off all four limbs at once (a problem getting through today's regulatory authorities!); all different species he knew could regenerate; all different ages could regenerate; regeneration was faster at young stages; small adult species regenerated faster that large adult species; regeneration of toes was so slow that it took the same time as the regeneration of the whole leg; the digits both develop and regenerate in a particular sequence; regeneration still occurred if the animals were fasted; regeneration was not always perfect as there were often differences in the numbers of toes, both too few and too many (more often too many) or differences in the numbers of phalanges in the toes; he repeatedly amputated all four limbs and the tail a total of six times, and the last regeneration took the same time as the first regeneration; one of these salamanders regenerated 687 bones; amputating the leg was less injurious than breaking a bone which formed a callus and often resulted in poor use of the limb; and he described what we now call the blastema in this way: "the beginning of regeneration is the formation of a cone of gelatinous substance endued with the most exquisite feeling".

What an incredible treatise! 250 years later, many of us are still working on molecular explanations for the observations he made. Not content with this work, he also studied the blood circulation of the tails while they were regenerating and was especially interested in the bones which he stained with madder root. Madder (*Rubia tinctorum*) is a climbing vine whose root had been used since the ancient Egyptians as a textile dye and contains alizarin and purpurin so these studies must be one of the earliest examples of alizarin-stained whole-mounts with red bones! I wonder if they still exist in the museum at the University of Pavia where he spent the last 30 years of his productive academic career following the publication of his "Prodromo".

Spallanzani was a polymath. In addition to these regeneration studies, he published work on digestion, respiration, artificial inseminations, geology, and volcanology, and after having obtained a better microscope, he dismissed spontaneous generation a century before Louis Pasteur.

While the world of natural philosophy was wrestling with the concepts of preformation versus epigenesis which these regeneration experiments clearly had much to contribute, from out of the blue waters of the Mediterranean came the second work on salamanders from Tweedy John Todd [2]. Todd was in the British navy and stationed for several years in the British naval base in Naples. In this paper he first described the regenerative process of the limbs and tails of "the aquatic salamander" (presumably also *Salamandra salamandra*) as consisting of three phases, growth (blastema production), organization (differentiation), and increase (catch-up growth), all of which were preceded by inflammation. He noted that the digits regenerated in a specific sequence—the second and third digits—appeared first followed by the first, fourth and fifth and then went on to describe attempts to "induce any derangement in the process of reproduction". He tried angled cuts, spear-shaped cuts, repeated amputation, and season of the year or starvation, all of which had no deleterious effect. Then he cut the sciatic nerves and discovered that regeneration was dramatically inhibited. He attributed this not to the vascular derangement but to "something peculiar" in the influence of the nerve. Todd ended with the comment that he had not been able to get hold of a copy of Spallanzani's "Prodromo" so he had not been able to read it, despite having done many similar experiments!

2 Developmental Biology Begins: Entwicklungsmechanik

Forty years onwards from the beginnings of regeneration research using salamanders, the science of experimental embryology was founded by Wilhelm Roux in the 1880s. Again it was salamanders that played a dominant role in this new science for 70+ years. Without a doubt, the vast majority of our understanding of early vertebrate embryology, inductive processes, formation of the neural plate and neural tube, head-to-tail body patterning, limb development and three-dimensional axial organization was thanks to the use of salamander embryos. Of course, simple observation of the development of embryos had long been recorded by many naturalists including Aristotle who recorded the development of fish, bird, cephalopod and crustacean eggs around 340 BC. Spallanzani himself observed the eggs of salamanders noting that they are "of a dun colour and surrounded with a glutinous matter, not unlike that of frogs". But it was Roux along with von Baer, Driesch and His who moved embryology from a purely descriptive science to one which explored *causal* mechanisms (Entwicklungsmechanik).

In attempting to establish the causal nature of animal development, these experimental embryologists were drawn into a scientific debate centred around how new generations of animals arose, preformation and epigenesis, viewpoints whose origin lay in antiquity. The former considered that all structures of the organism are preformed in the egg (ovists) or in the sperm (animalculists) with development being the unfolding and growth of these preformed miniatures. The homunculus drawn in the sperm head by Hartsoeker in 1694 entered the egg at fertilization and then simply grew using the nutritive substances in the egg. It is probably the most widely known depiction of preformation. The latter view, epigenesis, considered that the egg is undifferentiated and during development structures emerges as a result of internal and external influences including the interactions between different parts of the developing embryo.

This was a dichotomy that Entwicklungsmechanik could discern with experimental analyses. Roux's experiments (Roux, 1888; translated in [3]) using *Rana esculenta* eggs to decide between these two alternatives became known as the "pricking experiment". He used a hot needle to kill one cell of a two-cell stage *Rana* embryo but did not remove the dead cell. The result was that the surviving blastomere formed half an embryo, a result which seemingly supported the preformation view since regulation to form a complete embryo did not occur. It was these experiments and the results that inclined Roux towards the preformationist view. In an attempt to provide further support for Roux's results, Driesch (1891; translated in [3]) used sea urchin embryos at the two-cell stage where after separating them by shaking, each formed a complete, albeit smaller, normal larva. This was the momentous discovery of regulation, and as a result regulation and interaction between parts became the hallmark of development.

According to Hamburger [4], "Roux and Driesch excelled as theorists but their experiments were rather artless"! It was salamander embryos to the rescue as two embryologists with outstanding experimental skills—Spemann in Germany and Harrison in the USA – led the charge of experimental embryology.

3 A Nobel Prize for a Revolutionary Discovery Using Salamander Embryos

In fact, salamander eggs turned out to be much more favourable for experimental procedures than frog's eggs, and they had already begun to be used for embryological studies following a staging paper published 10 years before Entwicklungsmechanik began, in 1878 ([5] Fig. 1). The first separation of two blastomeres using salamander eggs was done, unsuccessfully, by Hertwig in 1893 and then successfully by Endres in 1895 and later by Herlitzka in 1897 [4].

Hans Spemann took over from Hertwig and remained faithful to the salamander embryo for the rest of his life with only occasional experimental forays using frog embryos. He began a series of experiments which were much more sophisticated than simply killing cells or separating blastomeres, namely, constriction experiments (Spemann's papers of 1901, 1902 and 1903), using embryos of the common European salamander, *Triton* (subsequently *Triturus*) *taeniatus*. He performed these experiments "one spring after another" (i.e. at each breeding season) by constricting the eggs

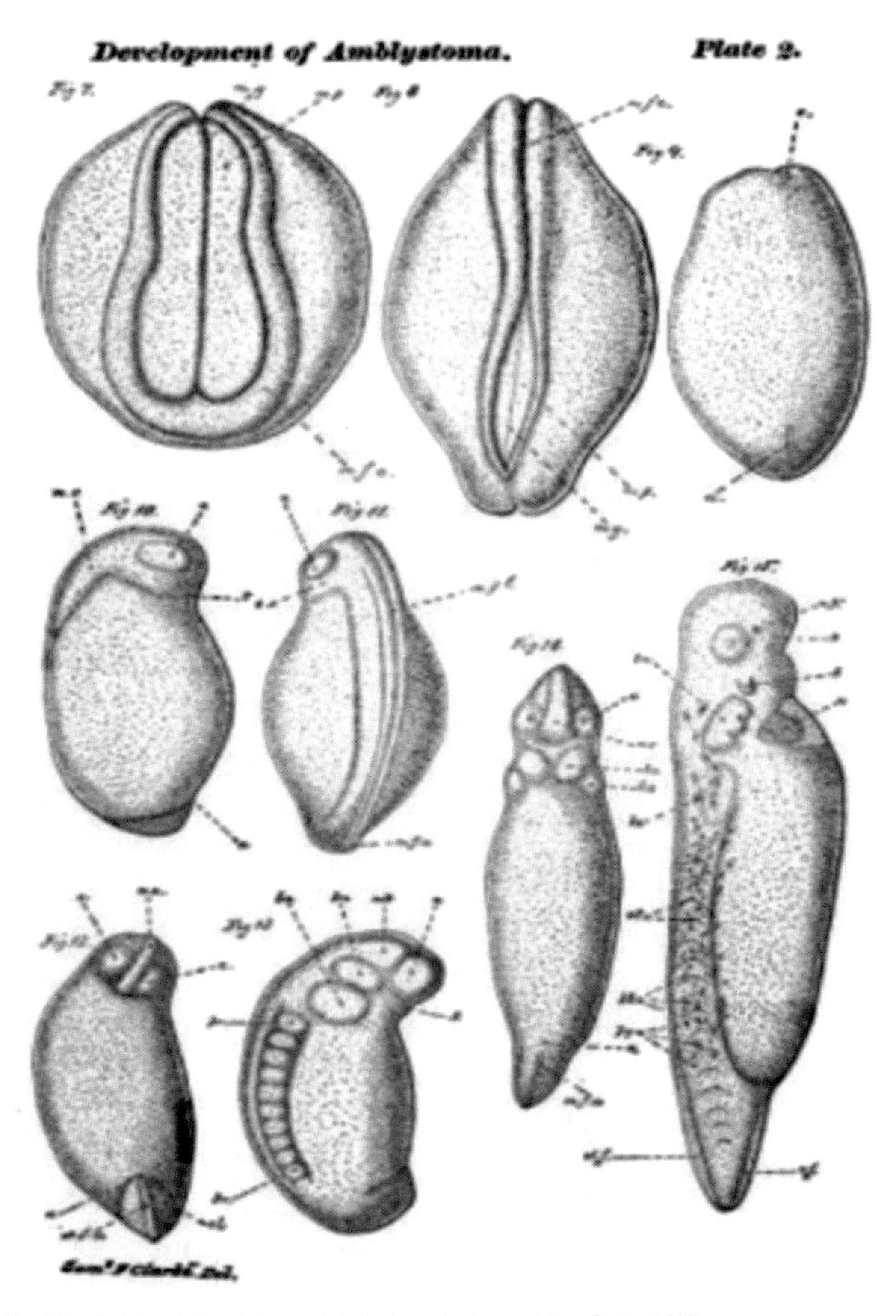

Fig. 1 The first description of stages of *Ambystoma* development from Clarke (1878)

with a fine baby hair from his own recently born daughter. He generated the classical results that we are familiar with today—anterior duplications from a partial constriction, twin embryos from a median constriction through the blastopore lip and a whole embryo and belly piece from frontal (dorsoventral) constrictions. These experiments demonstrated that the capacity for regulation decreased with the progression of gastrulation such that axial organs became irreversibly determined at the end of gastrulation [6].

Spemann's experiments using frog embryos took place at the same time as these salamander experiments, and he used the European frog, *Rana fusca*, to investigate eye and lens induction (Spemann's paper of 1901). Using his expertise in crafting microsurgical instruments from glass, he removed the eye vesicle or transplanted the optic vesicle under the flank or replaced head epidermis over the eye with flank epidermis. These experiments were also performed by Lewis using *Rana palustris* [7], and since a lens differentiated from flank epidermis, this was the first demonstration of embryonic induction, although the organizer experiments became the most famous.

The next series of salamander experiments by Spemann used interspecies grafting between *Triton cristatus* (nearly unpigmented) and *Triton taeniatus* (brownish). This was the most crucial use of salamander embryos which directly led to the discovery of induction by the organizer. In these initial interspecies grafting experiments, he placed presumptive neurectoderm from the graft species into the host's presumptive epidermis (Spemann's paper of 1921). He had previously performed the same experiment without heterospecific transplantation, so could not tell the respective contributions of graft and host (Spemann's paper of 1918). These experiments produced the result that prior to gastrulation and neurulation the grafted epidermis could clearly change fate. The corresponding figure from Spemann's paper shows an external view of the grafted (unpigmented) *T. cristatus* epidermis in the *T. taeniatus* neural plate (Fig. 2a), and a few days later, the incorporation of the *T. cristatus* cells into the eye and forebrain can clearly be seen at the cellular level which in addition the larger *T. cristatus* species produced a characteristically thicker forebrain neuroepithelium even within a foreign brain (Fig. 2b).

Having developed interspecific grafts in *Triturus* and perfected the hand-made tools and techniques for grafting small segments of the embryo, the stage was set for the discovery of the organizer (Spemann & Mangold, 1924 translated in [3]). Indeed, the discovery of the organizer is surely the most famous experiment in developmental biology and one that garnered the Nobel Prize in Physiology or Medicine to Hans Spemann in 1935 for "his discovery of the organizer effect in development". Unfortunately, his talented graduate student Hilde Mangold, who had worked on

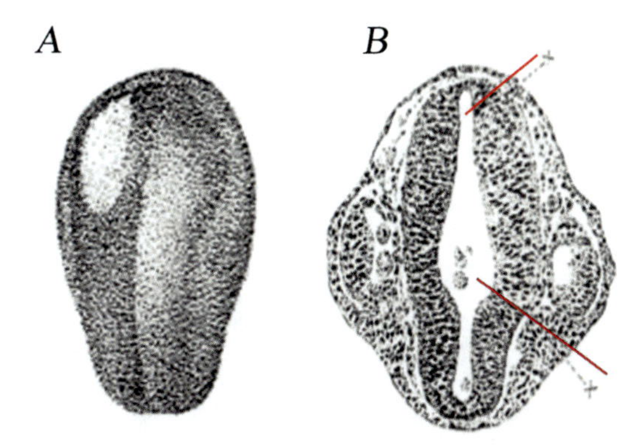

Fig. 2 (**a**) Figure from Spemann's 1918 paper showing the result of a graft of unpigmented epidermis from *T. cristatus* into the pigmented *T. taeniatus* neural plate. (**b**) A section through the later embryo showing that the *T. cristatus* epidermis has been fate changed into brain tissue and the origin of the graft can be determined from the lack of pigmentation of the *T. cristatus* cells between the two red lines

this problem with him for her PhD thesis, died in a tragic accident in 1924, an event which deprived her of sharing the prize since it is not awarded posthumously. However, she is fondly remembered in developmental biology lore, and her thesis still stands as a rare example of a PhD thesis leading directly to a Nobel Prize!

Hilde Mangold was tasked by Spemann to transplant the dorsal blastopore lip into the ventral region of another embryo. The grafts were between *T. cristatus* (nearly unpigmented), *T. taeniatus* (weakly pigmented), and *T. alpestris* (pigmented). Of the 259 grafts performed, only 42 survived and 18 of these showed secondary axis induction (43%). Interestingly, in the organizer paper, only six of these cases are described because these showed "all the basic facts" [8] that the secondary structures that had been induced by the dorsal lip graft were to differing degrees composed of host cells. Host cells had been diverted from their normal ventral fate to be induced to form dorsal structures instead (Fig. 3). Indeed, to emphasize the importance of this grafting regimen, it is often said [8] that Lewis actually discovered induction in his 1907 paper [9]. Using *Rana palustris* he transplanted blastopore lips into the otic region of other embryos and obtained the induction of nerve tube, notochord and somites. But because he had no means of distinguishing between graft and host cells, there was no proof of an inductive event. Salamander embryos won the Nobel Prize!

As an interesting side note, in all of the Spemann lab experiments on *Triturus* where early embryos were removed from the jelly coat, the survival of the embryos was horribly low, usually in the region of 10%. This means the vast majority of these tedious and complicated grafts did not survive—Hilde Mangold performed 259 grafts for her thesis work, only a proportion survived, leading

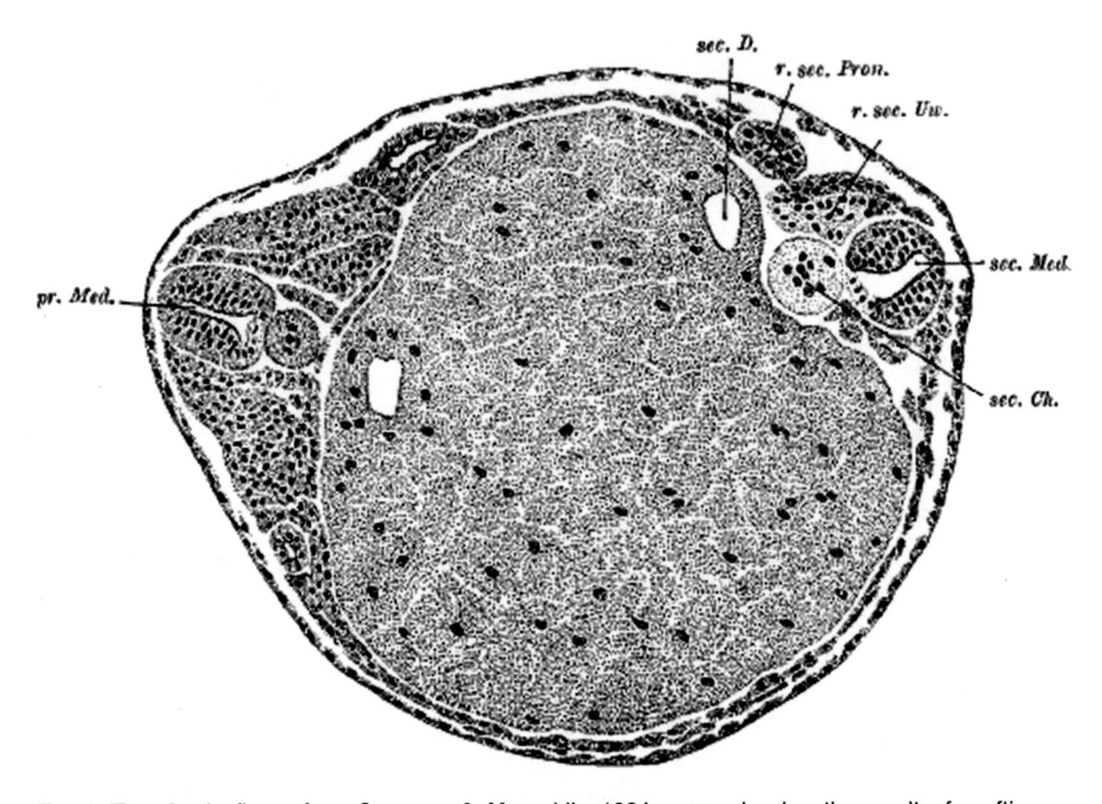

Fig. 3 The classic figure from Spemann & Mangold's 1924 paper showing the result of grafting an unpigmented *T. cristatus* dorsal lip into the ventral region of a *T. taeniatus* embryo. A new embryonic axis has been formed on the right side of the section consisting of *T. cristatus* cells (unpigmented) making up the notochord and some somite cells, and the majority of the remaining *T. taeniatus* cells in that region have been induced to change their fate from ventral to dorsal under the action of the dorsal lip graft

to the inclusion of only six cases in their Nobel Prize paper. This is in dramatic contrast to the papers of Harrison whose experiments were performed on later embryos allowing for much greater survival rates and the inclusion of 136 cases in his 1921 paper [10]. As Hamburger states, Spemann's technical acumen was not matched by a proficiency in chemistry, and all that was used as a culture medium for salamander embryos was filtered tap water. At one point when Spemann was in Berlin, he had water shipped in from Wurzburg whose magic ingredient was a high calcium content. This problem was finally solved by Holtfreter who devised the balanced salt solution still in use today by labs that work with salamanders.

4 Developmental Studies with *Ambystoma*

Soon after Entwicklungsmechanik began, developmental studies conducted in the USA mostly used *Ambystoma* embryos because

Fig. 4 An example of a beautiful lithograph from Harrison [12] showing the result of transplanting *Ambystoma* left limb bud mesoderm to the right flank resulting in the production of a supernumerary limb said to consist of a right limb developed from left limb mesoderm (labelled PRIM) and a duplication (DUPL)

these were the species available in local ponds; that and a staging series had been published very early on in this period of experimentation [5] (Fig. 4). Initially, experiments were performed on the developing limbs with the intention of discovering whether the myotomes which had been seen to generate "muscle buds" from their ventral edges made any contribution to the limb buds. Byrnes [11] examined this question in *Ambystoma punctatum* and several species of *Triton* (also in *Rana* and *Bufo*) both by observing development and by deleting the ventral myotomes with a hot needle. She concluded that they did not contribute to the development of the limb, but instead only contributed to the ventral muscles of the body wall. A similar conclusion was reached by Lewis [7] who removed one or several of the first few myotomes to examine the effect on the limb bud, concluding that "the anterior limb of *Ambystoma* develops in situ and is in no way derived from the myotomes or their ventral processes". It is notable that Lewis also featured in the experiments on early induction as described above.

Having begun studying nervous system development using frogs in the first decade of the twentieth century, Harrison also investigated fin development in salmon before he turned his attention to *Ambystoma* limb buds. He turned to the salamander embryo for his experiments to ask how the limb buds form and what factors were important in determining limb bud differentiation—the ectoderm, the somatopleure, the pronephros, the myotomes—and he transplanted the early limb field to different locations. As was typical for that period, Harrison collected *Ambystoma punctatum* eggs each spring from local ponds, performed many hundreds of experiments each year and wrote large papers summarizing several years of work. In his 1918 paper [12], he removed the limb bud, covered it with skin from another part of the body, split the limb bud, superimposed limb buds and removed the mesoderm. Approximately 600 cases are described in 8 tables and 45 figures (a dramatic contrast with Spemann and Mangold (1925) whose paper described six cases and who did not believe in tables to summarize the work!). Harrison concluded, as the title of the paper suggests, that the mesoderm of the limb bud is a self-differentiating equipotential system. In his 1921 paper [10], Harrison rotated limb buds, transplanted them left to right with and without rotation, transplanted half limb buds and superimposed limb buds. With 5 tables and 134 figures, he concluded by outlining the "rules of symmetry" which were that the asymmetry of the limb bud is determined *both* by the limb itself and its surroundings. This is because at the stages when these experiments were conducted the proximodistal and anteroposterior axes were already determined (by the limb itself) and could not be altered, but the dorsoventral axis could be altered and was determined by the surroundings. This is how he explained the production of left limbs after rotating right limb buds upside-down which seems to contradict the self-differentiating equipotential conclusion. In my opinion, this is a misinterpretation because the only marker he had for the dorsoventral axis was bending of the digits. In fact, these experiments should be repeated using GFP limb buds for transplantation to examine the musculature of the limb to see if they really are dorsoventrally reversed.

Although these are the dominant two papers of the time, many other developmental biologists were studying axial organization in *Ambystoma* limb buds such as Detwiler who examined the development of the shoulder girdle and the innervation of the limb, Swett; Schwind and Nicholas did similar rotation experiments to study axial organization, Stultz studied the hindlimb bud; Hollinshead studied the dorsoventral axis; and the innervation of the limb was studied by Detwiler, Piatt and Deck. In Europe, axial organization of the limb bud was studied in *Triturus* or *Pleurodeles* by Suzuki and Rotmann.

Of course, the whole range of developmental systems, not just limbs, was studied in *Ambystoma*. For example, development of the spinal cord and growth of nerves (Detwiler); development of the CNS (Burr); development of lateral line, taste buds, eye and lens (Stone); observations on living nerves (Speidel); development of cranial ganglia, ear induction and transplantation (Yntema, Richardson); musculature development in a range of species of salamanders (Piatt); development of the cerebellum and localization of function in the nervous system (Herrick); development of the digestive tract (Dorris); heart development (Lemanski); and the work of Frankhauser on fertilization led to the elaboration of the method of making triploid embryos which became one of the earliest methods for studying cell lineage in limb regeneration.

5 Einsteck and Neural Induction: Another Major Contribution from Salamanders

Continuing his studies on how the neural plate is induced by the organizer, together with Otto Mangold (the husband of Hilde Mangold), Spemann invented a technique in salamander embryos, again transplanting between *T. cristatus* and *T. taeniatus* to determine cellular contributions, which became known as the Einsteck method. This involved grafting differently staged blastopore lips (early or late) into the blastocoel of the embryo, and as a result, an ectopic neural plate was often induced which demonstrated the inductive capacity of the graft, as did the original transplants. However, this method was much easier to perform, and it also allowed xenotransplantation with the introduction of other tissues such as mesoderm. Studies using mesoderm in the Einsteck method led to an elaboration of the regional specificity of the mesoderm by Mangold [4]. When Mangold peeled back the neural plate to reveal the mesoderm of the archenteron roof, he divided it into four parts along the rostrocaudal axis. Then he grafted this mesoderm into the blastocoel of another embryo which induced structures that were positionally related to the mesoderm graft—rostral mesoderm-induced rostral structures (balancers, eyes, brain vesicles) and caudal mesoderm-induced spinal cord (pronephros and tails). The same result was obtained when the neural plate itself was used as a graft showing that both the mesoderm and the neural plate demonstrate regional specificity when they exhibit their inductive powers during development.

The Einsteck method and the sandwich technique, whereby the tissues to be tested were wrapped in ectoderm, formed the foundation for investigators searching for the biochemical nature of the inducer: a search which proved to be a 50-year dead end until molecular biology and cloning of the secreted factors expressed by the blastopore lip came to the rescue. By that time, salamander embryos had been replaced by *Xenopus* as the primary amphibian model used by experimental embryologists.

A huge variety of living and dead tissues were implanted into the *Triturus* blastulae—heat-killed tissues, adult tissues, chick embryo extract, mouse kidney (fresh and boiled), guinea pig thymus, salamander liver, perch liver and kidney, viper liver and kidney and ultimately HeLa cells to name but a few. To render living tissue dead, it was killed by boiling or placed in 70% alcohol for several weeks. These experiments were performed by Holtfreter, Chuang and Tiedemann in Germany, Toivonen and Saxen in Finland and

Yamada in Japan. The type of tissue (mesoderm, neural) and its rostrocaudal origin (brain, spinal cord, balancers) were recorded, and often more than one of these inducers were placed into the blastocoel. For example, Toivonen and Saxen implanted guinea pig liver (neural inducer) and guinea pig bone marrow (mesoderm inducer) and came up with a double-gradient theory of induction [13]. Tiedemann came closest to identifying the nature of these inducers, purifying them from chick embryos, and he identified them as proteins and found both a mesodermal factor called a vegetalizing factor (now known to be activin) and a neural factor [14]. Although they did not ultimately identify the organizer molecule(s), these salamander experiments refined the role of the blastopore lip during gastrulation, confirmed of the presence of a head organizer, a trunk organizer or a double-gradient theory and determined that the mesoderm-inducing agent is independent of the neural-inducing agents as originally proposed by Spemann.

6 Axolotls Appear and Are Replaced by *Xenopus*

By the time Nieuwkoop developed his modification to these neural induction ideas in the 1950s, axolotls (*Ambystoma mexicanum*) had become experimental organisms not only for regeneration research but also for embryological research. As a result of a series of grafting experiments involving transplanting folds of competent ectoderm to different locations on the neural plate in axolotls, he proposed the activation/transformation model. This model suggested that there were two inductive mesodermal events which patterned the early nervous system, the first one being a non-specific inducer of anterior neural character followed by a second one which gradually transforms the anterior neural tissues into more posterior fates [15]. This model, developed on axolotls and also *Triturus* embryos, has stood the test of time and been placed on a molecular footing by the discovery of several secreted molecules released from different regions of the embryo. Chordin, noggin and follistatin are secreted by Spemann's organizer which acts to inhibit BMP signalling and are the "activation factors", whereas Wnts, retinoic acid and FGFs have been identified as the "transformation factors" in Nieuwkoop's model.

However, all these molecular details were elaborated in *Xenopus* embryos which by the late 1950s had displaced salamander embryos as the laboratory amphibian of choice and which subsequently became amenable to molecular biology. *Xenopus* had become widely available as a tractable laboratory organism because Lancelot Hogben, a reproductive physiologist, had developed it for use in pregnancy testing. He had discovered that injecting just one dose of urine containing gonadotrophic hormone into the dorsal lymph sac would induce egg-laying 8–12 h later. This became known as the "Hogben test", and all these eggs could now be used for embryological research. From his lab in the basement of the London School of Economics, *Xenopus* spread throughout Europe and the USA displacing *Triturus* which had reigned supreme since the 1890s.

Similarly, Nieuwkoop who subsequently became Director of the Hubrecht Laboratory in Utrecht following this work on the development of the neural plate (*see* above) played a key role in making *Xenopus* the dominant force in embryological research by emphasizing the ease of decapsulation, favourable tissue consistency for microsurgery and excellent survival after surgery even though the eggs were smaller and had more rapid development and the eggs could be generated all year round with one injection of mammalian hormone preparations. In addition, an anucleolate mutant had been isolated, and when heterozygous tissue was grafted, it provided a cell marker which had done so much for salamander research, thus rang the death knell for salamanders, a tone that passed throughout embryological research labs [16].

Axolotls, however, continued to maintain a developmental presence because they were readily available from the Ambystoma Genetic Stock Center, which was initiated by Humphrey in the late 1950s at Indiana University and subsequently moved to the University of Kentucky. Eggs could be obtained all year round, thereby solving the seasonal egg laying of wild *Ambystoma* and *Triturus*, and in addition many laboratories had their own colonies of axolotls for the generation of embryos. Limb development was studied in a return to Harrison's work examining the development of axial polarity this time with an emphasis on a signal from posterior flank tissues being responsible for the control of anteroposterior polarity [17, 18]. Many gene expression patterns were studied by in situ hybridization both in the developing limb and other regions of the embryo. For example, *HoxA* and *HoxD* genes, bone development genes, retinoic acid signalling genes and the effect of RA; *Wnt* genes; *Msx* genes; *Tbx* genes; *Bmp Prod1*, *Shh*, *Sox* and collagen genes; and latterly *Fgf* genes emphasize how different signalling in axolotl limb development is from the classical limb development models of chick and mouse [19].

More recently, a huge boost to developmental studies using axolotl embryos has been provided by the development of transgenic lines expressing fluorescent tags such as GFP and RFP. These lines permit grafting of embryonic tissue between fluorescent embryos and white embryos, transplants that provide for lineage tracing of known embryonic tissues [20]. This technique has been used to study the origin of mesenchyme contributing to fin development; the development of fore- and hindlimb musculature; the neural crest contribution to a range of head structures including the teeth, hypobranchial apparatus and pectoral girdle; the origin of the posterior neural plate; and the location of haematopoiesis and

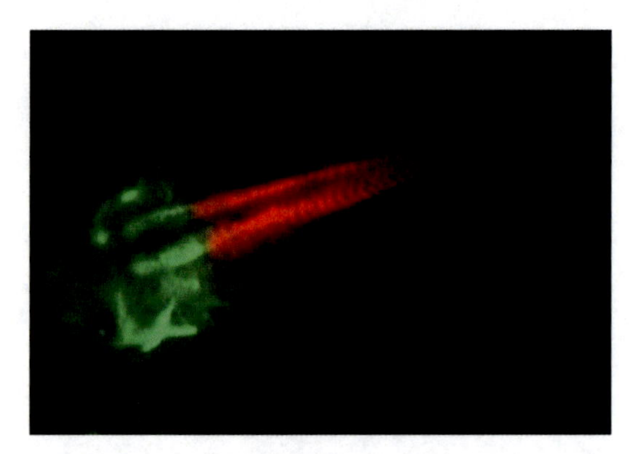

Fig. 5 A half GFP and half nuclear cherry red larva created to examine the site of long-term haematopoietic cells in axolotl from [21]

haematopoietic stem cells, works in which RFP embryos were also used to create half red/half green chimeric larva [21] (Fig. 5).

A final boost to axolotl developmental studies will surely be the application of CRISPR technology to examine the role of various genes in development which has already been performed for *Brachyury*, *Pax7* and *Sox2* [22–24]. Since salamanders have the ability to develop organs and then regenerate them (*see* below), these CRISPR studies will be uniquely valuable in being able to examine the role of individual genes in both development *and* regeneration and thus help disentangle the ever-fascinating relationship between these processes (for instance, with *Sox2*) [23].

7 Regeneration

Although salamanders are the only vertebrates that can regenerate their limbs after amputation, all of the other systems of their bodies seem to as well, and thus it is not surprising that the vast majority of discoveries in vertebrate organ regeneration have emerged from working with salamanders. This is not the place for an extensive review of regeneration or limb regeneration in particular, but below I identify several ground-breaking scientific concepts that emerged from working with salamanders as a model system for regeneration.

7.1 Neurotrophic Control

One novel concept was the neurotrophic control of regeneration which was completely different from the classical function of nerves—conduction of the action potential. Following Todd's report of the nervous control of limb regeneration in 1823 [2], there was a considerable hiatus before regeneration experiments began to be reported in the scientific literature again. It was from Paris in 1866 that Philippeaux reported that if the whole limb of *Triton cristatus* is extirpated including the scapula, then it does not regenerate, but if the limb is cut off flush with the body through the humerus, and then regeneration takes place rapidly [25]. The following year in 1867, he did the same experiment on axolotls (then called *Siron pisceformis*) given to him by A. Dumeril at the Le Museum National d'Histoire Naturelle in Paris who was the recipient of the original shipment of axolotls from Mexico in 1863. This was surely the first published axolotl limb regeneration experiment!

The subject of the nervous control of limb regeneration became the major topic of interest in the early 1900s from the laboratories of Wintrebert, Goldfarb, Barfurth, Godlewski and Rubin, most of whom kept controversy alive by concluding that the nervous system was *not* required for limb regeneration. These experiments were done on the hindlimbs of axolotls and *Triturus*. In the 1920s Locatelli and Schotte examined the type of nerve that is required for regeneration by severing sensory or motor nerves. Locatelli concluded that the sensory nerves are required for regeneration and made two other striking discoveries [26]. One was that if the brachial nerves of *T. cristatus* are deviated to the shoulder region, accessory limbs were produced so the nerves contained remarkable "morphogenetic", power and the second was that the nerves to the limb need not be connected to the spinal cord—connections from the ganglia to the limb were enough. Schotte used the forelimbs of larval *Triturus cristatus*, *T. alpestris* and *T. palmatus* to discover that transected nerves will rapidly reinnervate the limb, so delayed amputation will often result in regeneration following reinnervation; conversely if denervation is delayed after amputation by a certain stage, the regenerate becomes independent from the nervous supply. Furthermore, in those cases where regeneration was inhibited by denervation, they could be rescued if months later the limb was reamputated. However, Schotte came to the conclusion that it was the sympathetic nervous system which was the principle determining factor for regeneration.

Interestingly, Schotte was one of the few researchers who spanned both development and regeneration because after performing this work in the laboratory of Guyenot in Geneva, he went to Spemann's lab to do an experiment dreamed up by Spemann to transplant oral ectoderm between salamanders and frogs to answer the question of whether the larvae would develop teeth (salamanders) or horny jaws (frogs). Schotte succeeded in this transplant and obtained salamander larvae with frog-type horny jaws and suckers and frog tadpoles with balancers and teeth [4]. Schotte then went to Amherst, MA, and returned to the subject of the nervous control of limb regeneration. There he and Butler performed experiments on larval *Ambystoma maculatum* (now he was in the USA) and showed the differences between denervating adult and larval limbs—that larval limbs would completely regress in the absence of nerves and then regenerate when the nerves returned.

It was Singer's masterly series of studies on adult *Notophthalmus viridescens* performed for four decades from 1942 onwards [27] that really established the role of the nervous system on limb regeneration. All combinations of spinal nerves were removed, nerve fibre numbers were counted at all limb levels and it was found to be the quantity of nerve fibres that was important, not quality. This was elaborated into the quantitative theory of nervous control and subsequently the neurotrophic hypothesis which proposed that a trophic factor is released from the severed axons of the nerves. This was incredibly influential throughout the regeneration world, extending beyond salamanders to include invertebrates, fish and even mammals. This hypothesis also extended across all organs that regenerate because it established a principle that complex regenerating organs were under the control of the nervous system via a putative neurotrophic factor. It led to attempts to induce regeneration in other non-regenerating organisms by supplementing the nervous supply, in frogs (partially successful [28]), lizards and mammals (unsuccessful). The identity of the neurotrophic factor has had many potential candidates including fibroblast growth factor 2, glial growth factor, substance P, transferrin, insulin and nerve growth factor. Two more recent candidates which fulfil the required criteria (must be present, must disappear on denervation, antagonizing, it mimics denervation, must substitute for the nerves) are newt anterior gradient (nAG) and neuregulin. Transfection of a plasmid expressing nAG can rescue regeneration in a denervated newt limb [29] although the regenerate seems to lack muscle, and neuregulin supplied on bead implanted into the blastema replaces the nerves in denervated axolotl limbs [30]. Neuregulin is also involved in mammalian peripheral nervous system and in fish and mammalian cardiac regeneration, thereby fulfilling Singer's neurotrophic criteria of being involved in regenerating organs across all the vertebrates.

A remarkable and celebrated twist to the neurotrophic hypothesis arose from the finding that limbs that have never been innervated (innervation is not required for limb development) can regenerate normally. This exception to the neurotrophic theory was performed on *Ambystoma maculatum* where two embryos were joined in parabiosis, and in one of them the neural tube was excised. The nerveless embryo survives due to parabiosis, and when the limbs develop, they are either sparsely innervated or aneurogenic but surprisingly can regenerate only marginally slower than normal [31]. Interestingly, aneurogenic limbs are virtually devoid of muscle, just as the denervated nAG rescued newt limbs are (*see* above) [32].

Furthermore, when these aneurogenic limbs are transplanted onto normal hosts, they become innervated for the first time. By periodically denervating and amputating them, it was found that after 13 days of innervation they become dependent on the nervous system [33], and, interestingly, there was a suggestion that the regenerative ability can be assigned to a property of the skin [34]. This was explained in terms of the "addictive hypothesis" whereby the limb gradually became addicted to the presence of nerves as innervation progressed.

Recently the aneurogenic limb has been revived in studies of the mechanism of action of nAG [35]. Using an antibody to AG in the normally innervated regenerating limb, this protein is expressed in nerve fibres, in the Schwann cells present in the distal cut nerve fibres and then in the gland cells of the wound epidermis. In the aneurogenic limb AG is present throughout the epidermis in the secretory Leydig cells and as nerves proceed to innervated, the limb epidermal expression is downregulated. This confirms the suggestion that the regeneration of aneurogenic limbs is a property of the skin and shows a striking interaction between the nerves and epidermis in the "take-over" of the control of limb development/ regeneration by the nervous system. These unique findings on the nervous control of regeneration were entirely attributable to this research on salamander limbs.

7.2 Dedifferentiation

The second ground-breaking concept that arose from regeneration studies was that of dedifferentiation. Experiments on the effects of X-rays made a significant contribution to understanding the source of cells that would give rise to the blastema. Some had earlier suggested that the blood was the source of blastemal cells, but when it was concluded that X-rays could inhibit cell division and permanently prevent regeneration, localized irradiation was introduced by shielding part of the salamander body and limb with lead plates. After amputation through an irradiated region, such as the foot, no regeneration occurred, but if that same limb was subsequently amputated through the shielded region such as the knee, regeneration occurred [36, 37]. This led to the important finding that the source of blastemal cells was 1–2 mm from the amputation plane and not from a source of "reserve" cells somewhere else in the body. Detailed histological studies on the local cells that give rise to the blastema and their proliferation rates [38] led to the concept of "dedifferentiation" whereby stump tissues rapidly lose their differentiated characteristics, their nuclei expand and they begin to proliferate as the blastema forms. Dedifferentiation was a novel concept in biology which originated here in the salamander limb blastema.

The dedifferentiation studies of Hay, at the electron microscopic level on adult newt muscle cells showing loss of striations, generation of single cells and commencement of proliferation, showed that dedifferentiation as a developmental reversion was a true phenomenon in the creation of the blastema (at least in the muscle). A confirmation of dedifferentiation as the means of muscle regeneration has recently been observed in transgenic adult newts but with the surprising rider that the mechanism is different in larval newts where satellite stem cells perform this function instead [39]. However, the real argument to emerge was over the fate of the dedifferentiated cells—did they revert to what they were before, known as modulation, or did they change their fate and differentiate into something different—a far more contentious issue. Many attempts were made using grafts of radioactive tissue, triploid tissue, tissue from black or white axolotls into normal limbs and into X-rayed limb stump (to "force" the cells to do more than they might normally), e.g. [40], and in general the result was that while some flexibility in the connective tissues of the limb (cartilage, tendon, dermis) could be detected, there was little flexibility among cells in the epidermis or muscle. These lineage relationships have recently been thoroughly confirmed using grafts of GFP labelled cells [41]: the epidermis can only generate the epidermis in the regenerate, the muscle can only make muscle and the connective tissue fibroblasts can differentiate into cartilage, dermis, tendon and ligaments.

The concept of dedifferentiation followed by a change of fate (see also lens regeneration below) no longer surprises us today with the generation of induced pluripotent stem cells from fully differentiated cells by the introduction of only four factors, the Yamanaka factors, but at the time these revolutionary ideas from limb regeneration were highly controversial.

7.3 Pattern Formation

Another aspect of limb regeneration which generated a unique concept was the consideration of pattern formation and axial organization in the regeneration blastema. At the time, the dominant idea for how three-dimensional patterning was organized as it related to the three Cartesian limb axes (i.e. the proximodistal, anteroposterior and the dorsoventral axis) (see the experiments of Harrison, above) was based on how these axes arose during embryonic development from gradients of signalling molecules, usually extracellular. Thus, the position of a cell in any axis is calculated by assessing the concentration of these signalling molecules in their environment, a measure known as positional information.

Limb regeneration studies, however, did not follow these dogmas, and novel concepts of positional information were elaborated. Firstly, in experiments where blastemas were transplanted, for example, a distal blastema transplanted to proximal amputation site, the distal blastema did not integrate into the limb until regeneration to the distal level had occurred [42]. This demonstrated that blastemal cells can assess their positional information but do so by cell surface interactions and not measuring extracellular gradients. Proximal blastemal cells envelop distal blastemal cells [43] in a result resembling Steinberg's differential adhesion hypothesis. The measurement of positional information was thus considered to be via a molecule(s) located on the surface of the blastemal cell and not by extracellular gradients.

The identification of a potential positional molecule was enabled by the discovery of the striking effects of the signalling molecule retinoic acid on the regenerating limb [44]. When limb amputation is performed at distal levels, such as through the hand, and a precise concentration of RA is applied; then instead of regenerating a hand, a whole limb can be produced resulting in two limbs in tandem. Clearly the positional information of the distal blastema had been respecified in a proximal direction by RA and in the blastemal relocation experiments described above the surface assessment of position has changed too [45]. In a remarkable differential screen between normal and RA-treated distal blastemas, one GPI-linked cell surface molecule, named Prod-1, was identified [46]. When Prod-1 is overexpressed in distal cells, then, just like RA treatment, these cells move proximally and become incorporated into the regenerate at a new positional level [47]. These cell surface concepts, the identification of cell surface molecules and their pathways of which the most recent player is *Tig1* [48] show how limb regeneration studies in salamanders are making revolutionary discoveries in molecular and cell biology.

The movement of blastemas according to their surface interactions can also be seen when a blastema is cut off from the stump and rotated 180° and replaced on that stump. In a proportion of cases, the blastema slowly de-rotates and returns to its former position. These experiments were performed during investigations into the generation of supernumerary limbs which is a striking phenomenon also seen in the developing limb bud rotations of Harrison (*see* above) and characterized in many invertebrates [48]. When limb blastemas are exchanged between left and right limbs or rotated 180°, the complete extra limbs are produced usually at the places of maximum incongruity to typically generate three limbs instead of the normal one [49]. The similarity between these results and those seen in invertebrate limbs and also imaginal discs led to the development of a model unique to limb regeneration in vertebrates and thus to salamanders: the polar coordinate model [50]. This envisaged the limb not as a three-dimensional Cartesian coordinate system, but as a polar coordinate system with the AP and DV axes represented as the numbers on a clockface and the proximodistal axis represented as concentric circles inside each other. This provided a remarkably efficient explanation of firstly the location of the supernumerary limbs and secondly the local cell/cell interactions that occur in the blastema rather than three axial gradients of morphogens. As such this was a regeneration model but encountered great resistance among developmental biologists when it was applied to the developing limb bud where it had considerable difficulty in explaining the results of manipulations.

7.4 Brain Regeneration

Limb regeneration is not the only salamander system which has enlightened the world of regeneration mechanisms; there are many others which have provided ground-breaking data. For example, many regions of the brain of salamanders can regenerate following the removal of remarkably large segments [51]. The forebrain will regenerate after extensive damage including its entire removal, and the fibre tracts and molecularly distinct sub-types of neurons can be replaced although perhaps not all long-distance fibre tracts [52, 53]. In the midbrain up to 70% of the optic tectum and the retinotectal connections will regenerate [54], and dopaminergic neurons are completely regenerated after 6-hydroxydopamine-induced ablation, serving as a unique model for Parkinson's disease [55]. Although brain regeneration has also been investigated in lizards and zebrafish where more molecular progress has been made in zebrafish due to its vast molecular resources, the damage that can be regenerated is much more limited, and the typical injury is a cortical stab. Salamanders with their incredible ability to regenerate large segments of the brain and spinal cord will surely provide major insights into how to regenerate neuronal damage in the future.

7.5 Lens Regeneration

Another regenerating system which provided the definitive demonstration of dedifferentiation and redifferentiation is the regenerating lens. This phenomenon, first described in the late 1800s [56], surprisingly does not occur in all salamanders, but most of them. Of those tested, 18 US species, 9 European species and 2 Japanese species did, and 8 did not including all of the *Ambystomatidae* [57], although it was subsequently shown that axolotls can regenerate their lens during a narrowly defined larval stage [58]. Regeneration of the lens takes place from the dorsal iris (not the ventral iris), and during this process, cells of the iris lose their pigmentation, proliferate and then differentiate into lens cells eventually expressing crystallins. Because there is no "embryonic/stem cell" intermediate, this process is known as transdifferentiation and also occurs in culture where after a prolonged culture period ventral iris can be made to transdifferentiate [59]. This regeneration model has become a very significant system for the study of the molecular control of regeneration, cell determination, aging and reprogramming. For example, the normally non-regenerating ventral iris can be induced to regenerate the lens in vivo by the combination of inhibiting the BMP pathway, transfecting the iris with the *Six3* gene and treating it with retinoic acid—events which do not occur in the regenerating dorsal iris [60]. And in an 18-year experiment in which the lens was repeatedly removed up to 19 times by which time the animals were 30 years old, the last lens was structurally and transcriptionally similar to young lenses removed [61]. These advances in regeneration and aging have not been approached in any other regeneration system and are unique to salamanders.

8 Conclusions

From 1768 onwards, salamanders have been an incredible resource for understanding both development and regeneration all the way from its phenomenology to discovering the molecular and cellular controls. The vast majority of early developmental biology concepts from the role of the organizer to patterning the mesoderm and neural plate to patterning the limb were derived from experiments conducted from the beginnings of Entwicklungsmechanik to the late 1950s. These experiments were conducted on wild-caught *Ambystoma* in the USA or *Triturus* in Europe and latterly on laboratory stocks of axolotls. With the advent of transgenesis, axolotl development still has more discoveries to provide to the world of developmental biology.

Since limb regeneration is, uniquely among the vertebrates, a salamander phenomenon, almost all discoveries in this field were of major significance, and the regeneration of other organs also provided major insights. It is important to emphasize that other models of development such as *Xenopus*, *Drosophila*, zebrafish and mouse may have eclipsed salamanders in the field of developmental biology, but no other model can both develop and then regenerate such a wide range of organs at all stages of their life cycle. Salamanders are therefore a unique resource which will continue to surprise us in the years ahead.

References

1. Dinsmore CE (1991) A history of regeneration research: milestones in the evolution of a science. Cambridge University Press
2. Todd TJ (1823) On the process of reproduction of the members of the aquatic salamander. Q J Lit Sci Arts 16:84–96
3. Willier BH, Oppenheimer JM (1964) Foundations of experimental embryology. Prentice-Hall Inc., Englewood Cliffs
4. Hamburger V (1988) The heritage of experimental embryology. Hans Spemann and the organizer. Oxford University Press, Oxford
5. Clarke SF (1878) The development of Amblystoma punctatum – Baird. Part 1, External Biol Studies, vol 1. Johns Hopkins University
6. Fassler PE (1996) Hans Spemann (1869-1941) and the Freiburg school of embryology. Int J Dev Biol 40:49–57
7. Lewis WH (1910) The relation of the myotomes to the ventrolateral musculature and to the anterior limbs in Amblystoma. Anat Rec 4(5):183
8. Fassler PE, Sander K (1977) Hilde Mangold (1898–1924) and Spemann's organizer: achievement and tragedy. In: Landmarks in developmental biology 1883–1924. Springer-Verlag, Berlin Heidelberg
9. Lewis WH (1907) Transplantation of the lips of the blastopore in *Rana palustris*. Am J Anat 7:137–143
10. Harrison RG (1921) On relations of symmetry in transplanted limbs. J Exp Zool 32:1–136
11. Byrnes EF (1898) Experimental studies on the development of limb-muscles in Amphibia. J Morphol 24:105–140
12. Harrison RG (1918) Experiments on the development of the fore limb of Amblystoma, a self-differentiating equipotential system. J Exp Zool 25:413–461
13. Toivonen S, Saxen L (1955) The simultaneous inducing action of liver and bone-marrow of the guinea pig in implantation and explantation experiments with embryos of *Triturus*. Exp Cell Res Suppl 3:346–357

14. Grunz H (2001) Developmental biology of amphibians after Hans Spemann in Germany. Int J Dev Biol 45:39–50
15. Nieuwkoop PD, Nigtevecht GV (1952) Neural activation and transformation in explants of competent ectoderm under the influence of fragments of anterior notochord in urodeles. J Embryol Exp Morphol 2:175–193
16. Gurdon JB, Hopwood N (2000) The introduction of Xenopus laevis into developmental biology: of empire, pregnancy testing and ribosomal genes. Int J Dev Biol 44:43–50
17. Slack JMW (1976) Determination of polarity in the amphibian limb. Nature 261:44–46
18. Slack JMW (1977) Control of anteroposterior pattern in the axolotl forelimb by a smoothly graded signal. J Embryol Exp Morphol 39: 1690182
19. Purushothaman S, Elewa A, Seifert AW (2019) Fgf-signaling is compartmentalized within the mesenchyme and controls proliferation during salamander limb development. eLife 8:e48507. https://doi.org/10.7554/eLife.48507
20. Sobokow L, Epperlein H-H, Herklotz S, Straube WL, Tanaka EM (2006) A germline GFP transgenic axolotl and its use to track cell fate: dual origin of the fin mesenchyme during development and the fate of blood cells during regeneration. Dev Biol 290: 386–397
21. Lopez D, Monaghan JR, Cogle CR, Bova FJ, Maden M, Scott EW (2014) Mapping hematopoiesis in a fully regenerative vertebrate: the axolotl. Blood 124:1232–1241
22. Nowoshilow S et al (2018) The axolotl genome and the evolution of key tissue formation regulators. Nature 554:50–55
23. Fei JF et al (2014) CRISPR-mediated genomic deletion of Sox2 in the axolotl shows a requirement in spinal cord neural stem cell amplification during tail regeneration. Stem Cell Rep 3: 444–459
24. Flowers GP, Timberlake AT, Mclean KC, Monaghan JR, Crews CM (2014) Highly efficient targeted mutagenesis in axolotl using Cas9 RNA-guided nuclease. Development 141: 2165–2171
25. Philippeaux JM (1867) On the regeneration of the limbs in the Axolotl (Siren pisciformis). Ann Mag Nat Hist 20(116):149–149
26. Wallace H (1981) Vertebrate Limb Regeneration. Wiley & Sons, Hoboken
27. Singer M (1942) The nervous system and regeneration of the forelimb of adult Triturus. I. The role of the sympathetics. J Exp Zool 90:377–399
28. Singer M (1954) Induction of regeneration of the forelimb of the postmetamorphic frog by augmentation of the nerve supply. J Exp Zool 126:419–471
29. Kumar A, Godwin JW, Gates PB, Garza-Garcia AA, Brockes JP (2007) Molecular basis for the nerve dependence of limb regeneration in an adult vertebrate. Science 318:772–777
30. Farkas JE, Freitas PD, Bryant DM, Whited JL, Monaghan JR (2016) Neuregulin-1 signaling is essential for nerve-dependent axolotl limb regeneration. Development 143:2724–2731
31. Yntema CL (1959) Regeneration of sparsely innervated and aneurogenic forelimbs of *Amblystoma* larvae. J Exp Zool 140:101–123
32. Popiela H (1967) *In vivo* limb tissue development in the absence of nerves: a quantitative study. Exp Neurol 53:214–226
33. Thortnon CS, Thornton MT (1970) Recuperation of regeneration in denervated limbs of *Ambystoma* larvae. J Exp Zool 173:293–302
34. Steen TP, Thornton CS (1963) Tissues interaction in amputated aneurogenic limbs of Ambystoma larvae. J Exp Zool 154:207–221
35. Kumar A et al (2011) The aneurogenic limb identifies developmental cell interactions underlying vertebrate limb regeneration. PNAS USA 108:13588–13593
36. Scheremetieva EA, Brunst VV (1938) Preservation of the regeneration capacity in the middle part of the limb of newt and its simultaneous loss in the distal and proximal part of the same limb. Bull Biol Med Exp URSS 6:723–724
37. Butler EG, O'Brien JP (1942) Effects of localized X-irradiation in regeneration of the urodele limb. Anat Rec 84:407–413
38. Chalkley DT (1954) A quantitative histological analysis of forelimb regeneration in Triturus viridescens. J Morphol 94:21–70
39. Tanaka HV, Ng NY, Yu ZY, Casco-Robles MM, Maruo F, Tsonis PA, Chiba C (2016) A developmentally regulated switch from stem cells to dedifferentiation for limb muscle regeneration in newts. Nat Commun 7:11069
40. Namenwirth M (1974) The inheritance of cell differentiation during limb regeneration in the axolotl. Dev Biol 41:42–56
41. Kragl M, Knapp D, Nacu E, Khattak S, Maden M, Epperling HH, Tanaka EM (2009) Cells keep a memory of their tissue origin during axolotl limb regeneration. Nature 460: 60–65
42. Crawford K, Stocum DL (1988) Retinoic acid coordinately proximalizes regenerate pattern and blastema differential affinity in axolotl limbs. Development 102:687–698

43. Nardi JB, Stocum DL (1984) Surface properties of regenerating limb cells: evidence for gradation along the proximodistal axis. Differentiation 25:27–31
44. Maden M (1982) Vitamin a and pattern formation in the regenerating limb. Nature 295: 672–675
45. da Silva SM, Gates PB, Brockes JP (2002) The newt orthologue of CD59 is implicated in proximodistal identity during amphibian limb regeneration. Dev Cell 3:547–555
46. Echeverri K, Tanaka EM (2004) Proximodistal patterning during limb regeneration. Dev Biol 279:391–401
47. Oliveira CR, Knapp D, Elewa A, Malagon SGG, Gates PB, Petzhold A, Arce H, Cordoba RC, Chara O, Tanaka EM, Simon A, Yun MH (2021) Tig1 regulates proximo-distal identity during salamander limb regeneration. bioRxiv. https://doi.org/10.1101/2021.02.03.42834
48. Bateson W (1894) Material for the study of variation. Reprinted 1992. Johns Hopkins University Press, Baltimore
49. Maden M (1980) Structure of supernumerary limbs. Nature 286:803–805
50. French V, Bryant PJ, Bryant SV (1976) Pattern regulation in epimorphic fields. Science 193: 969–981
51. Kirsche W (1983) The significance of matrix zones for brain regeneration and brain transplantation with special consideration of lower vertebrates. In: Wallace R, Das GD (eds) Neural tissue transplantation research. Springer, New York, pp 65–104
52. Maden M, Manwell LA, Ormerod BK (2013) Proliferation zones in the axolotl brain and regeneration of the telencephalon. Neural Dev 8:1
53. Amamoto R, Lopex Huerta VG, Takahashi E, Dai G, Grant AK, Fu Z, Arlotta P (2016) Adult axolotls can regenerate original neuronal diversity in response to brain injury. elife 5:e13998
54. Okamoto M, Ohsawa H, Hayashi T, Owaribe K, Tsonis PA (2007) Regeneration of retinotectal projections after optic tectum removal in adult newts. Mol Vis 13: 2112–2118
55. Parish CL, Beljajeva A, Arenas E, Simon A (2007) Midbrain dopaminergic neurogenesis and behavioural recovery in a salamander lesion-induced regeneration model. Development 134:2882–2887
56. Tsonis PA, Mahadavan M, Tancous EE, Del Rio-Tsonis K (2004) A newt's eye view of lens regeneration. Int J Dev Biol 48:975–980
57. Stone LS (1967) An investigation recording all salamanders which can and cannot regenerate a lens from the dorsal iris. J Exp Zool 164: 87–103
58. Suetsugu-Maki R, Maki N, Nakamura K, Sumanas S, Zhu J, Del Rio-Tsonis K, Tsonis PA (2012) Lens regeneration in axolotl: new evidence of developmental plasticity. BMC Biol 10:103
59. Eguchi G (1988) Cellular and molecular background of Wolffian lens regeneration. In: Eguchi G, Okada TS, Saxen L (eds) Regulatory mechanisms in developmental processes. Elsevier, Limerick, pp 147–158
60. Grogg MW, Call MK, Okamoto M, Vergara MN, Del Rio-Tsonis K, Tsonis PA (2005) BMP inhibition-driven regulation of six-3 underlies induction of newt lens regeneration. Nature 438:858–862
61. Eguchi G, Eguchi Y, Nakamura K, Yadav MC, Millan JL, Tsonis PA (2011) Regenerative capacity in newts is not altered by repeated regeneration and ageing. Nat Commun 2:384

Part I

Laboratory Colony Husbandry

Chapter 2

Establishing a New Research Axolotl Colony

Anastasia S. Yandulskaya and James R. Monaghan

Abstract

The field of regenerative biology has taken a keen interest in the Mexican axolotl (*Ambystoma mexicanum*) over the past few decades, as this salamander successfully regenerates amputated limbs and injured body parts. Recent progress in research tool development has also made possible axolotl genetic manipulation and single-cell analysis, which will help understand the molecular mechanisms of complex tissue regeneration. To support the growing popularity of this model, we describe how to set up a new axolotl housing facility at a research laboratory. We also review husbandry practices for raising axolotls and using them in biological research, with a focus on diet, water quality, breeding, and anesthesia.

Key words Axolotl, Husbandry, Housing, Aquatic

1 Introduction

The Mexican axolotl (*Ambystoma mexicanum*) is an aquatic salamander used in biomedical research to study development, evolution, and regeneration. Axolotls were first collected from the Lake Xochimilco System, Mexico, in 1863, making them the oldest laboratory animal model [1, 2]. The use of axolotls has considerably grown in the past decades, partly due to an increased interest in stem cells and regeneration, as well as technical advances in transcriptomic and transgenic approaches. The axolotl is an appealing laboratory animal because it is relatively hardy, thrives in near room-temperature freshwater, and breeds year-round. Axolotls also have a striking ability to regenerate organs and appendages after injury. Organ regeneration occurs in the brain, spinal cord, retina, heart, ovaries, liver, and teeth [3–9]. One of the more striking examples of regeneration is that of the limb [10, 11], which has motivated several new groups to adopt the axolotl system.

Ashley W. Seifert and Joshua D. Currie (eds.), *Salamanders: Methods and Protocols*,
Methods in Molecular Biology, vol. 2562, https://doi.org/10.1007/978-1-0716-2659-7_2,

This chapter focuses on creating a new axolotl breeding colony for laboratory research. Also, we encourage readers to reference other reviews on the topic of axolotl housing and maintenance practices [12–15] and the Ambystoma Genetic Stock Center (https://ambystoma.uky.edu/genetic-stock-center/).

2 Materials

2.1 Housing

1. A large colony requires a recirculating aquatic system in a windowless or a blacked-out room. Our laboratory has two Aquatic Habitats® Z-Hab Systems (Pentair Aquatic Eco-systems, Inc.), but newer models such as the Iwaki LAb-REED systems will serve the same purpose. Each system houses ~100 adult axolotls, which are sufficient for establishing a breeding colony.
2. Food-grade plastic storage containers, at least 4.5 L in volume, for housing adult animals off the recirculating system.
3. Food-grade Choice 8 oz. clear deli containers (500/case) (Item # 127DM8BULK, Webstaurantstore.com), for housing young juveniles.
4. Food-grade Choice 2 oz. clear plastic portion cups (2500/case) (Item # 127P2C, Webstaurantstore.com), for housing larvae individually.
5. Cafeteria trays for storing dishes and cups with young animals.
6. Insulation tape for sticking onto bottoms of housing trays.
7. MetroMax I open grid cart with rubber casters 24″ × 48″ (Item # 461X556BGX3, Webstaurantstore.com), for storing containers and trays with animals.
8. Vented Brute® 32-gallon trash can, with a lid and a dolly, for storing housing water (SKU: FG263200GRAY, FG263100GRAY, FG263100GRAY, Rubbermaid Commercial Products).
9. Scotch-Brite® Little Handy scrubbers, for cleaning water barrels.
10. Stainless steel tube brushes for cleaning containers and dishes.
11. Fishnets (8″, 5″, and 3″).
12. Room temperature controller (Schneider Electric).
13. Janitorial bucket 26 quarts with a side press wringer (Lavex®).
14. A floor mop, such as Unger SmartColor™ RoughMop String Mops Heavy Duty, with a handle.
15. A squeegee broom, such as Unger AquaDozer® Heavy Duty, with a mop handle.

16. A dustpan with a broom.
17. A disinfectant cleaner.
18. A stainless sink (Aero Aerospec Sink 1-bowl 20″—2F1-2020-30LR).

2.2 Water Supply

1. In-house reverse osmosis deionized (RO/DI) water will provide the best source for clean water, which can be used to make salamander housing water at pH 7.0, 1000 μS/cm.
2. City water can be used depending upon local water quality. We triple filter Boston City water with one in-line 20″ sediment and two 20″ carbon filters.
3. Ninety-gallon storage tank and distribution system (Pentair).

2.3 Diet

1. A simple brine shrimp hatchery can be constructed out of a sizeable upright square cooler (>45 cm in height). A small fluorescent light bulb can be used to generate enough heat (~30 °C) and light to trigger egg hatching.
2. A hatchery cone with a stand for hatching brine shrimp (Brine Shrimp Direct).
3. A Tetra Whisper Aquarium Air Pump.
4. A rotifer sieve for collecting brine shrimp (Brine Shrimp Direct).
5. *Artemia* cysts (such as INVE Aquaculture).
6. California blackworms (Eastern Aquatics).
7. Rangen 3/16″ Soft Moist Salmon Pellets for feeding small axolotls (Aquatic Foods and Blackworm Company).
8. Rangen 1/8″ Soft Moist Salmon Pellets for feeding adult axolotls (Aquatic Foods and Blackworm Company).

2.4 Water Quality

1. Sea salt for making salamander housing solution (Instant Ocean, Pentair).
2. Baking soda for making salamander housing solution and cleaning tanks (such as Arm & Hammer™).
3. eXact® Eco-Check® Photometer Kit, for monitoring water quality (Industrial Test Systems, Inc.).
4. Kordon AmQuel Plus, for removing ammonia products from housing water (#AM75P, Pentair).
5. Kordon NovAqua Plus, for removing heavy metals from housing water (#NA5P, Pentair).
6. Fisherbrand™ polypropylene rectangular carboy with a spigot, 20 L, for storing conductivity solution (03-007-648, Fisher Scientific).

7. Carbon filters for the flow-through system (gpe50-20, Pentair).
8. Pad filters for the flow-through system (PF11P4-A, Pentair).

2.5 Anesthetic

1. Salamander housing water.
2. 95–100% EtOH.
3. Benzocaine powder.

3 Methods

3.1 Housing

Just a few adult male and female animals are needed to establish a new colony, which can be kept in free-standing tanks of at least 4.5-L capacity if the water is changed every other day. Axolotls should be housed in a temperature-controlled, windowless room on a 12: 12 light/dark cycle.

To grow the colony, mate adult animals and raise the resulting larvae in individual dishes to prevent overcrowding and cannibalism (*see* **Note 1**). Larvae and juveniles should be transferred to larger containers when they start outgrowing their dishes.

When the number of adult axolotls in free-standing tanks becomes challenging to manage, it is time to establish a housing system with automated water exchange. Recirculating systems for zebrafish (without heating element) are well suited to hold axolotl tanks because they control the number of contaminants in the water [16]. However, axolotls are sensitive to strong water flow, so only adult animals should be housed in a standard recirculating system. Even then, axolotls are most comfortable if new water is introduced to their tanks as a slow drip. Ten percent daily water exchange is optimal. A forceful stream can be turned on to flush out foul water but return to low flow as soon as possible. However, the best way to routinely clean system tanks is to remove waste and uneaten food with a serological pipette. Water flow can injure or upset smaller animals, so juveniles and larvae should reside in size-appropriate free-standing containers with salamander housing solution.

Stress and subpar water quality are the chief drivers of disease in axolotls (*see* **Note 2**). The best way to ensure the colony's health is to closely monitor the temperature in the animal facility, handle the animals with care, and check the water quality regularly. Axolotls thrive in water cooler than room temperature; the ideal water temperature is 16–18 °C (*see* **Note 3**). Installing chillers for the flow-through systems will help maintain the optimal temperature of the system's water. The ambient room temperature should be kept below 24 °C at all times.

If the number of staff is limited, such as during holidays or quarantine, the animals can be fed and their water changed twice a week. This excludes larvae, who still need to be fed brine shrimp and have their water changed daily or every other day. If operating on this reduced schedule, it is vital to continuously monitor animal health for signs of stress or disease. Any ill-looking animals should be removed from the recirculating system and housed in free-standing containers (*see* **Note 4**). Tanks should be cleaned if they are dirty, regardless of the schedule.

Often, washing filthy tanks with water is not enough to remove the slimy yellow residue. However, no soap or detergents should ever come into contact with any of the axolotl equipment or cleaning tools because they are toxic to the animals. Food-grade sodium bicarbonate (baking soda) is a suitable replacement for soap when scrubbing housing containers, and it also possesses mild antimicrobial properties [17].

In a recirculating system, tanks are filled with water from tubes that are connected to supply lines. Over time, the insides of those tubes accumulate dirt. A short burst of forceful water flow removes some of this debris. To remove some of the remaining debris, put the tubes in a container with a concentrated sodium bicarbonate solution and leave it on a magnetic stirrer with a stir bar for several hours. Rinse the tubes well before replacing them.

Axolotls of all sizes can jump out of their tanks or containers and desiccate to death if not discovered in time, so it is best to keep them covered with lids or trays. However, young animals may suffocate if their dishes are full to the brim and covered very tightly, such as with other trays in a stack. To prevent sealing of containers in a tray stack and ensure air exchange for animals, attach insulation tape to the bottom side of the trays (Fig. 1).

Axolotls' sudden bouts of agility are best kept in mind when designing the animal room and planning the placement of recirculating systems. Even large axolotls sometimes manage to escape their tanks. Cleaning is especially risky since it requires sliding off lids to allow in serological pipettes and animals are agitated by the commotion. If a salamander does disappear in the system's innards, partially disassemble the system to rescue. This is easiest if all sides of the system are readily accessible and not against a wall. To keep animals from jumping during cleaning, avoid touching them with the serological pipette and replace their lids the moment they start thrashing around their tanks. They will calm down in a minute and not object to resuming the cleaning.

Even if an agitated axolotl does not jump out, it may knock a water supply tube out of its tank. If not discovered within an hour or two, this tube will flood the room. In anticipation of an inevitable deluge, a well-designed animal room will have a drain in the floor, and the floor will be slanted toward it and away from any doors. A squeegee broom is also useful for quickly drying the floor.

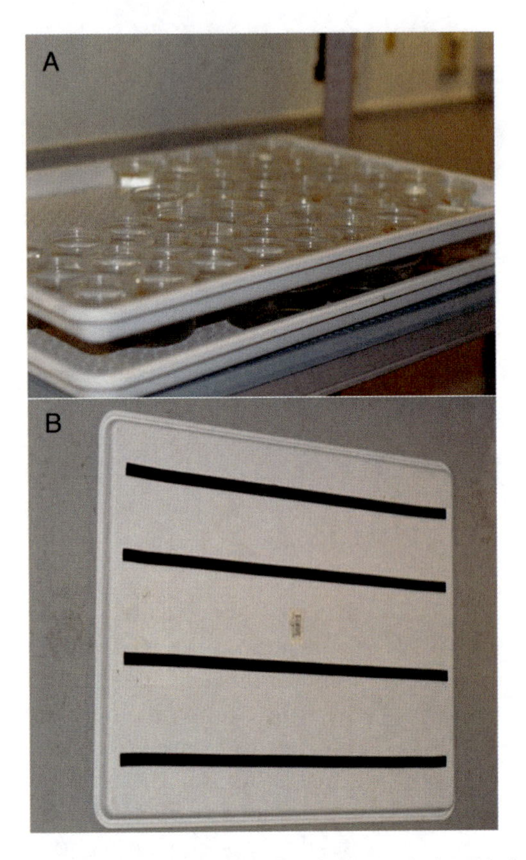

Fig. 1 Housing juvenile axolotls. (**a**) Young axolotls (up to 1 cm in length) can be housed in food-grade plastic dishes and stored on stacked cafeteria trays. (**b**) Insulation tape on the bottom of trays prevents animals on the tray below from suffocating

After a flood, monitor the pH and salinity of the system's water, as they may drop due to the large water loss. To prevent flooding, check for water drips from tanks and for puddles on the floor after cleaning and ensure that no water supply tubes are out of the tanks.

3.2 Water Quality

Water quality is measured by its pH, salinity, mineral levels, toxic contaminants like chlorine, and waste products. All these parameters must be monitored regularly, at least monthly. pH and conductivity should be checked daily.

Housing water should be at neutral pH; the acceptable range is 7.0–7.5. If the pH is too low and the recirculating housing system lacks a pH adjustment tool, the pH of the water can be increased by slowly introducing a solution of sodium bicarbonate (20 g to 1 L RO/DI water) into the recirculating system, such as by pouring it into an empty tank and setting the drip to low. Never add any solutions directly to occupied tanks because sudden water changes will distress axolotls.

Axolotls are content with water of varying conductivity, from 500 to 2300 microsiemens (μS)/cm [14, 15]. In our experience, maintaining adult housing water salinity between 850 and 1000 μS/cm works well. Flow-through system salinity is adjusted with a high salinity solution by filling a 20 L carboy with deionized water and dissolving 500 g of sea salt in it. Conductivity solution should be introduced either through a recirculating system-specific installed salinity adjustment bucket or an unoccupied tank with drip set to low.

Some of the biggest threats to axolotl wellbeing are ammonia and its products, nitrite and nitrate. Ammonia is especially toxic. It is released by animal waste and decomposing uneaten food, which is why any debris should be removed from tanks and dishes as soon as possible. Young animals are particularly sensitive to dirty water. AmQuel Plus and NovAqua Plus conditioners help control ammonia levels and its products. These can be added directly to the flow-through system according to water volume in the system. In free-standing water, add each conditioner according to manufacturer directions to every 32 gallons of salamander housing solution. Conditioning housing water may change its pH and conductivity, and regular monitoring of water quality is advised.

Recirculating systems also contain biological filters, where bacteria convert ammonia into nitrite and then relatively safe nitrate. However, those bacteria require a period of adjustment in a newly established recirculating system, and ammonia levels should be monitored daily in the beginning [14]. Target levels for ammonia and nitrite in the housing water are zero.

Recirculating systems can operate on municipal tap water depending upon water quality, which is usually treated with chlorine or chloramines. These should be removed from water before it reaches the system. We filter tap water in one 20″ sediment filter and two 20″ carbon filters before storing it in a 90-gallon recirculating storage tank. To dechlorinate free-standing salamander housing solution, it can be prepared from deionized or distilled water with chlorine, treated with AmQuel Plus, and allowed to rest for 24 h before use. We find that carbon filters can be changed once a month and pad filters switched out once a week. Systems relying on carbon or reverse osmosis filtration require regular monitoring of water for chlorine to ensure proper filtration.

3.3 Diet

Animals living in recirculating systems thrive on large salmon pellets. A serving of 5–7 pellets 2–3 times a week is sufficient for an adult axolotl. Juveniles that are almost large enough to move into the systems (about 10 cm long) can also be fed 3–4 large pellets at the same intervals. It is critical to avoid overfeeding animals with pellets because any uneaten ones will soon decay and contaminate the housing water. If leftover pellets are routine, reduce portions.

Smaller juveniles (about 5 cm long and larger) can eat small salmon pellets (5–7 at a time), and their diet can be supplemented with brine shrimp (*Artemia*). However, juvenile axolotls grow fastest on California blackworms (*L. variegatus*) [18]. For animals too small to swallow an entire worm, the worms can be cut up with scissors or a transfer pipette. Overfeeding with blackworms is acceptable, and any leftover worms can be reused for the next feeding.

California blackworms should be stored in shallow levels of salamander housing solution at 4 °C. There is no need to feed them. They crawl up the walls of their container, so it should be covered but not airtight. The worms should be washed daily with copious changes of salamander housing solution until the water runs clear and most floating dead worms are gone. White clumps of decayed dead worms should be removed with a transfer pipette. If there are any leeches attached to the container's bottom, they should be wiped away with a paper towel. Avoid feeding dead worms and leeches to axolotls.

When axolotl larvae lose their yolk, which is indicated by the loss of dark pigmentation, they can be introduced to brine shrimp. Add generous amounts of hatched brine shrimp into their dishes with a pipette and let the larvae feed for at least 1 h. When they are full of shrimp, their stomachs become round and orange. Very young larvae should not be left in the water with brine shrimp overnight, as the shrimp die in fresh water and can be toxic. Older larvae are more tolerant. Young axolotls can live on this diet until they are big enough to eat cut up blackworms. If the supply of blackworms runs out and some juveniles are still too small for salmon pellets, they can survive off brine shrimp or even crushed salmon pellets, although most young axolotls refuse them.

Brine shrimp (*Artemia*) should be hatched from cysts daily according to the manufacturer's instructions. A clear hatching cone works very well, as it allows to easily discard most unhatched cysts floating at the water surface and dead shrimp at the bottom. Because brine shrimp hatch in water with high salt content, they should be thoroughly washed with salamander housing solution after being harvested. Allow the collected shrimp to sit in a container full of salamander housing solution for at least 10 min before serving them to axolotl larvae so that any dead shrimp sink and any unhatched ones float. The layer of live shrimp will be in the middle, and they also swim toward light. Larvae should only be fed live shrimp.

Sick adult axolotls ignoring salmon pellets may instead accept blackworms. If an axolotl is so ill that it does not hunt its worms, dangling them in front of its face from a transfer pipette may entice it to eat. This deluxe diet may help the animal recover.

3.4 Anesthesia

It is imperative to properly sedate and anesthetize axolotls before carrying out any surgeries, injections, or other procedures. The best way to administer anesthesia is to take advantage of the axolotls' permeable skin and bathe them in an anesthetic solution.

Benzocaine is an efficient and economical choice of anesthetic. To prepare stock benzocaine solution, fully dissolve 1 g of benzocaine powder in 30 mL of absolute ethanol and add 970 mL of deionized water. For temporary anesthesia, prepare a 0.01% solution by diluting one part of stock benzocaine in nine parts of housing solution (by volume). For terminal anesthesia, prepare a 0.05% solution by combining stock benzocaine and housing solution in equal volumes. To ensure death after terminal anesthesia, animals must also undergo pithing. It is important to remember that prolonged exposure (several hours) to 0.01% benzocaine can also kill axolotls, so avoid leaving animals in anesthetic for longer than necessary.

Other possible anesthetics, with varying concentrations and delivery methods, are propofol, tricaine sulfonate, butorphanol, and buprenorphine [19–21].

The time for axolotls to become anesthetized depends on their size. Juveniles typically are fully sedated after 15–20 min in the anesthetic, while large adults can take 45 min to an hour. Monitoring animals is critical. An anesthetized animal will not react to a gentle poke or protest if flipped upside down. However, anesthesia will start wearing off after the animal is removed from the solution. To prolong sedation during a surgery, soak a thin, lint-free piece of tissue in the anesthetic solution and cover the animal, especially its gills. This precaution will also protect the animal's skin from desiccation.

After the surgery, return the animal to its housing tank or allow to regain mobility in shallow water under the covers of wet thin tissue.

3.5 Courtship and Mating

Axolotls reach sexual maturity around the age of 12 months. Males develop dark nails, cloacal bulges, and deep costal grooves; females grow broad and round bellies due to an internal egg supply [12, 15]. Prior to the onset of sexual maturity, a genotyping assay is the only way to distinguish males from females [22].

The axolotl breeding season begins in midwinter and ends in late spring. Axolotls are reluctant to procreate midsummer to late fall, although an unvarying artificial light schedule minimizes this seasonality. Males can be mated once a month, and females once every 2–3 months. To breed axolotls, put a male and a female together in a large tank that is lined with foil on the outside to protect them from light. Place terracotta clay saucers or flat natural rocks on the bottom of the tank so that the male can attach his spermatophores to a textured surface. Put several reusable ice packs in the tank as well, so that the water stays cool. Ensure a sufficient

Fig. 2 Remote surveillance of an egg-laying female. The female is preparing to spawn in a temperature-controlled incubator, and a tablet is being used to monitor the process during the night

supply of housing water and leave them together overnight. There is no need to provide food.

Soon after being introduced, the male deposits spermatophores and begins a courtship dance. He walks in front of the female, prods her with his snout, and waves his tail, secreting pheromones from his cloaca [23]. Responding to his odorants, the female follows him and takes up the spermatophores with her cloaca, storing them [24]. After the female takes up the spermatophores, remove the male, put one or two mesh nets into the breeding tank, change out the ice packs, and let the female rest in solitude.

The female will usually begin to spawn within 12–24 h. Eggs leave her oviducts and enter the cloaca, where they encounter the spermatozoa from the stored spermatophores and become fertilized [15]. The female coats the fertilized eggs in a thick, sticky layer of clear jelly. She then climbs into a mesh net and lays the eggs in strings. Avoid disturbing her as much as possible. When she takes a break, collect the eggs from the net with a cutoff transfer pipette or a razor blade. This process can be delayed by putting the female into a 4 °C refrigerator for several hours or overnight. Refrigeration will slow down the spawning and the development of eggs. If the female does not begin to lay on the same day but it is important to know exactly when the eggs are fertilized or to work on recently laid embryos, it may help to set up a tablet in front of the breeding tank and use baby-monitoring software to observe the female remotely overnight (Fig. 2).

Within one spawning, a single female can lay from 200 to 1000 eggs. Ensure that she is done laying before transferring her back to her housing tank. It may be useful to keep track of successful breeders to facilitate future spawnings.

Occasionally, breedings fail. Some potential couples ignore each other in the breeding tank and the male does not lay any spermatophores even over the course of several nights. If natural spawning is difficult to achieve, embryos can be generated via in vitro fertilization.

Adult axolotls can be induced to breed with hormonal treatment. Females lay eggs after being injected with follicle-stimulating hormone [25], and an injection of human chorionic gonadotrophin can induce both spermiation in males and ovulation in females. Gentle massage of the animals' abdomens can promote deposition of gametes. In vitro fertilization is carried out by mixing sperm with eggs and then adding fertilization saline solution. High pH of the solution improves the hatching rate [26]. Axolotl sperm can be refrigerated and remain viable for up to 28 days [27].

4 Notes

1. Largest juveniles nip limbs and tails of their smaller siblings.
2. Even though axolotls are usually resilient against bacterial infection, chronic stress (caused by poor water quality, high temperature, or a physical injury) may undermine this resistance. Axolotls are vulnerable to fungal and viral infections. Albino larvae may be especially at risk, possibly due to their weaker immune systems [28].
3. Axolotls living at a constantly elevated water temperature (greater than 18 °C) may be at risk of swelling of the body cavity. Exact causes of this common amphibian disorder are unknown, and it may also be a symptom of other diseases such as flavobacteriosis [29, 30].
4. Ill or stressed axolotls typically float at the water surface, refuse food, shed skin, and/or have curled tails. The bright red gill branches may grow paler and smaller. The animals may also become pale in color themselves (except for the dark wild-type axolotls).

Acknowledgments

We would like to thank the Ambystoma Genetic Stock Center for providing all the animals and methodologies these methods have been built upon. We also thank Jackson Griffiths for taking the pictures and Michelle Lim for lending the camera.

Competing Interests Authors declare no competing interests.

Grant Sponsor NSF grants 1558017 and 1656429 to JRM. NIH grant HD099174 to JRM. Material and information obtained from the Ambystoma Genetic Stock Center funded through NIH grant: P40-OD019794.

References

1. Reiß C, Olsson L, Hoßfeld U (2015) The history of the oldest self-sustaining laboratory animal: 150 years of axolotl research. J Exp Zool B Mol Dev Evol 324(5):393–404. https://doi.org/10.1002/jez.b.22617
2. Voss SR, Woodcock MR, Zambrano L (2015) A tale of two axolotls. Bioscience 65(12): 1134–1140. https://doi.org/10.1093/biosci/biv153
3. Amamoto R, Huerta VG, Takahashi E et al (2016) Adult axolotls can regenerate original neuronal diversity in response to brain injury. Elife 5. https://doi.org/10.7554/eLife.13998
4. Maden M, Manwell LA, Ormerod BK (2013) Proliferation zones in the axolotl brain and regeneration of the telencephalon. Nerual Dev 8(1):1–15
5. Clarke JD, Alexander R, Holder N (1988) Regeneration of descending axons in the spinal cord of the axolotl. Neurosci Lett 89(1):1–6
6. Svistunov SA, Mitashov VI (1983) Proliferative activity of the pigment epithelium and regenerating retinal cells in Ambystoma mexicanum. Ontogenez 14(6):597–606
7. Erler P, Sweeney A, Monaghan JR (2017) Regulation of injury-induced ovarian regeneration by activation of oogonial stem cells. Stem Cells 35(1):236–247. https://doi.org/10.1002/stem.2504
8. Ohashi A, Saito N, Kashimoto R, Furukawa S, Yamamoto S, Satoh A (2020) Axolotl liver regeneration is accomplished via compensatory congestion mechanisms regulated by ERK signaling after partial hepatectomy. Dev Dyn. https://doi.org/10.1002/dvdy.262
9. Makanae A, Tajika Y, Nishimura K, Saito N, Tanaka JI, Satoh A (2020) Neural regulation in tooth regeneration of Ambystoma mexicanum. Sci Rep 10(1):9323. https://doi.org/10.1038/s41598-020-66142-2
10. Haas BJ, Whited JL (2017) Advances in decoding axolotl limb regeneration. Trends Genet 33(8):553–565. https://doi.org/10.1016/j.tig.2017.05.006
11. Simon A, Tanaka EM (2013) Limb regeneration. Wiley Interdiscip Rev Dev Biol 2(2): 291–300. https://doi.org/10.1002/wdev.73
12. Björklund NK, Duhon ST (1997) The Mexican axolotl as a pet and a laboratory animal. In: Biology, husbandry and health care of reptiles and amphibians. Tropical Fish Hobbyist, Jersey City
13. Gresens J (2004) An introduction to the Mexican axolotl (Ambystoma mexicanum). Lab Anim (NY) 33(9):41–47. https://doi.org/10.1038/laban1004-41
14. Khattak S, Murawala P, Andreas H et al (2014) Optimized axolotl (Ambystoma mexicanum) husbandry, breeding, metamorphosis, transgenesis and tamoxifen-mediated recombination. Nat Protoc 9(3):529–540. https://doi.org/10.1038/nprot.2014.040
15. Farkas JE, Monaghan JR (2015) Housing and maintenance of Ambystoma mexicanum, the Mexican axolotl. Methods Mol Biol 1290: 27–46. https://doi.org/10.1007/978-1-4939-2495-0_3
16. Lawrence C, Mason T (2012) Zebrafish housing systems: a review of basic operating principles and considerations for design and functionality. ILAR J 53(2):179–191. https://doi.org/10.1093/ilar.53.2.179
17. Corral LG, Post LS, Montville TJ (1988) Antimicrobial activity of sodium bicarbonate. J Food Sci 53(3):981–982. https://doi.org/10.1111/j.1365-2621.1988.tb09005.x
18. Slight DJ, Nichols HJ, Arbuckle K (2015) Are mixed diets beneficial for the welfare of captive axolotls (Ambystoma mexicanum)? Effects of feeding regimes on growth and behavior. J Vet Behav 10(2):185–190. https://doi.org/10.1016/j.jveb.2014.09.004
19. Thygesen MM, Rasmussen MM, Madsen JG, Pedersen M, Lauridsen H (2017) Propofol (2,6-diisopropylphenol) is an applicable immersion anesthetic in the axolotl with potential uses in hemodynamic and neurophysiological experiments. Regeneration (Oxf) 4(3): 124–131. https://doi.org/10.1002/reg2.80

20. Zullian C, Dodelet-Devillers A, Roy S, Vachon P (2016) Evaluation of the anesthetic effects of MS222 in the adult Mexican axolotl (Ambystoma mexicanum). Vet Med (Auckl) 7:1–7. https://doi.org/10.2147/vmrr.S96761
21. Burns PM, Langlois I, Dunn M (2019) Endoscopic removal of a foreign body in a Mexican axolotl (Ambystoma mexicanum) with the use of Ms222-induced immobilization. J Zoo Wildl Med 50(1):282–286. https://doi.org/10.1638/2012-0118
22. Keinath MC, Timoshevskaya NY, Hardy DL, Muzinic L, Voss SR, Smith JJ (2017) A PCR based assay to efficiently determine the sex of axolotls. Axolotl Newsl 2:5–7
23. Maex M, Van Bocxlaer I, Mortier A, Proost P, Bossuyt F (2016) courtship pheromone use in a model urodele, the Mexican axolotl (Ambystoma mexicanum). Sci Rep 6:20184. https://doi.org/10.1038/srep20184
24. Park D, McGuire JM, Majchrzak AL, Ziobro JM, Eisthen HL (2004) Discrimination of conspecific sex and reproductive condition using chemical cues in axolotls (Ambystoma mexicanum). J Comp Physiol A Neuroethol Sens Neural Behav Physiol 190(5):415–427. https://doi.org/10.1007/s00359-004-0510-y
25. Khattak S, Richter T, Tanaka EM (2009) Generation of transgenic axolotls (Ambystoma mexicanum). Cold Spring Harb Protoc 2009(8). https://doi.org/10.1101/pdb.prot5264
26. Mansour N, Lahnsteiner F, Patzner RA (2011) Collection of gametes from live axolotl, Ambystoma mexicanum, and standardization of in vitro fertilization. Theriogenology 75(2):354–361. https://doi.org/10.1016/j.theriogenology.2010.09.006
27. Figiel CR (2020) Cold storage of sperm from the axolotl, Ambystoma mexicanum. Herpetol Conserv Biol 15(2):367–371
28. Zullian C, Dodelet-Devillers A, Roy S, Hélie P, Vachon P (2015) Abdominal distension associated with luminal fungi in the intestines of axolotl larvae. Case Rep Vet Med 2015:851689. https://doi.org/10.1155/2015/851689
29. Takami Y, Une Y (2017) A retrospective study of diseases in Ambystoma mexicanum: a report of 97 cases. J Vet Med Sci 79(6):1068–1071. https://doi.org/10.1292/jvms.17-0066
30. Densmore CL, Green DE (2007) Diseases of amphibians. ILAR J 48(3):235–254. https://doi.org/10.1093/ilar.48.3.235

Chapter 3

Health Monitoring for Laboratory Salamanders

Marcus J. Crim and Marcia L. Hart

Abstract

Laboratory animal health monitoring programs are necessary to protect animal health and welfare, the validity of experimental data, and human health against zoonotic infections. Health monitoring programs should be designed based on a risk assessment and knowledge about the biology and transmission of salamander pathogens. Both traditional and molecular diagnostic platforms are available for salamanders, and they provide complementary information. A comprehensive approach to health monitoring leverages the advantages of multiple platforms to provide a more complete picture of colony health and pathogen status. This chapter presents key considerations in the design and implementation of a colony health monitoring program for laboratory salamanders, including protocols for necropsy and sample collection.

Key words Salamander, Axolotl, Health monitoring, Biosecurity, Quarantine, Necropsy, Pathology, Pathogen, Parasite, Environmental monitoring

1 Introduction

1.1 Biosecurity and Health Monitoring Program Design

Key objectives for laboratory animal colony health monitoring programs include the protection of animal health and welfare, ensuring the validity of experimental data, preventing the spread of infectious agents among colonies and institutions, and protecting human health against zoonotic infections. Both clinical and subclinical naturally occurring infections can introduce variability that can lead to invalid or misinterpreted experimental results in animal models [1]. For example, subclinical infections in salamanders can result in altered gene expression [2], immune function [3], and susceptibility to coinfections [4, 5].

Laboratory animal biosecurity comprises all measures used to prevent, detect, contain, control, and eradicate known or unknown infectious agents in laboratory animals [6]. Colony health monitoring is a program of routine surveillance to assess the pathogen status of research animals. Routine health surveillance programs should include a list of agents of concern, testing methodology, and testing frequency for each agent. Health monitoring programs

Ashley W. Seifert and Joshua D. Currie (eds.), *Salamanders: Methods and Protocols*,
Methods in Molecular Biology, vol. 2562, https://doi.org/10.1007/978-1-0716-2659-7_3,

should be periodically reviewed and adjusted as new information becomes available about salamander pathogens, suppliers, and researcher needs.

Animal import procedures and colony health monitoring programs should be designed based on a risk assessment. Many factors contribute to the risk of introducing an infectious agent into a laboratory research colony. For example, the risk is minimal in closed colonies, and facilities that only occasionally import salamanders or embryos compared to colonies that import frequently. Colonies that include wild-caught salamanders or salamander species for which purpose-bred and laboratory-reared animals are not available are subject to much greater risks. Similarly, facilities that import salamanders from multiple sources or sources with no available health history have greater risk exposure than facilities that import embryos or salamanders from a single supplier with a known health history. It is also important to consider whether colonies are housed in close proximity to other laboratory species or share husbandry staff, equipment, or supplies.

A variety of tools can be applied to facilitate health management for research colonies, including approved vendor lists, quarantine and other import procedures, colony-specific exclusion lists, routine surveillance using multiple diagnostic platforms, environmental monitoring, and work and material flows designed to preserve colony health status. While the application of these tools is standard in the health management of laboratory rodents, many of these tools have been applied in a very limited way in salamander colonies at most institutions. The implementation of health monitoring programs for salamander colonies is limited by a lack of information about the species susceptibility, prevalence, transmission, and treatment of many infectious agents, as well as a lack of awareness of the adverse impacts of undetected infectious diseases on research. The selection of infectious agents to monitor should consider the type of research conducted, colony health history, and budget.

An approved vendors list is a register of authorized suppliers that are able to supply animals or other materials that meet specific quality control criteria. Obtaining research animals that are exclusively purpose-bred and laboratory-raised from colonies with known health status and adequate biosecurity practices substantially reduces the risk of epizootics as well as other adverse consequences of introducing infectious agents into a laboratory colony. There have been substantial efforts by laboratory animal organizations to make recommendations that improve the consistency of exclusion lists among institutions for the pathogens of laboratory rodents and rabbits [7, 8]. However, health monitoring and reporting for laboratory salamanders are sporadic with relatively little coordination among institutions. Obtaining the health status of salamanders from other institutions is often challenging.

Nevertheless, laboratory salamanders are susceptible to infection with zoonotic agents as well as significant amphibian pathogens. When designing a biosecurity plan for an existing salamander colony, it is advisable to screen the existing population for infectious agents of interest prior to including them on the exclusion list in order to avoid using resources in an effort to exclude agents that are already enzootic in the colony. Biosecurity programs should minimize the risk to salamander colonies while simultaneously facilitating collaboration among investigators, including sharing of salamanders among institutions.

Quarantine of newly acquired salamanders prior to introduction to the main colony allows them to recover from shipping stress and acclimate to a new laboratory environment. Quarantine allows the importation of animals from colonies that have an undefined health status, which is important for salamander research due to the lack of commercial sources providing salamanders that are guaranteed to be free of specific pathogens. While segregated in quarantine, new arrivals can be observed for clinical signs and tested for infectious agents. If undesirable agents are detected in quarantine, they can be eliminated from the smaller population while they are still in quarantine without exposing other salamanders. Elimination refers to the removal of an infectious agent from a group of imported animals or a small population of animals, whereas eradication refers to the removal of an infectious agent from an entire colony or facility. To date, rederivation procedures to eliminate enzootic pathogens have not been developed for salamanders. Surface disinfection of fish embryos is widely practiced in zebrafish facilities and commercial aquaculture, and the toxicity of several potential protocols for the disinfection of anuran egg masses has been recently investigated [9]. However, there are no published reports of salamander embryo surface disinfection. Alternatives to rederivation include test-and-cull approaches, saline bath treatment, husbandry changes, probiotic therapy, and antimicrobial chemotherapy.

As with other laboratory species, the exclusion of pathogens from the colony is the most effective tool for protecting colony health from infectious diseases. However, the purpose of biosecurity programs is not to exclude all infectious agents from the colony, but to exclude those agents that pose a significant risk to the animals, personnel, or the research being performed. Routine colony health monitoring consists of a regular schedule of surveillance to detect changes in pathogen status resulting from breaks in biosecurity. Thus, the agents monitored typically correspond to the agents on the exclusion list. In some cases, facilities may monitor agents that are not on the exclusion list for a colony, for example, to track the impact on the positivity rate when husbandry changes are made, or in different groups of animals over time. The selection of infectious agents for routine monitoring of salamander colonies should reflect the needs of the research being conducted, and there must be flexibility to accommodate the changing needs of investigators. In general, agents that pose a greater risk to the research or the colony should be surveyed more frequently.

Sentinel health monitoring is an approach that maximizes pathogen transmission to a small number of sentinel animals, from which samples are collected to represent the colony. Sentinels are a major component of health monitoring for rodent and zebrafish colonies but are only rarely used in amphibian colonies. A more common approach is to routinely monitor antemortem and environmental samples via real-time PCR with a more complete diagnostic investigation of significant clinical disease, including necropsy and multiple diagnostic platforms.

1.2 Diagnostic Platforms

Several diagnostic methodologies are available for the assessment of salamander health and pathogen status. There are advantages and disadvantages to each diagnostic platform, and they provide slightly different but often complementary information. Important differences among platforms include sensitivity, specificity, speed, the amount of expertise required for interpretation, the range of possible sample types, and the perishability of samples. Therefore, comprehensive approaches to colony health monitoring leverage the different advantages of multiple methodologies in order to provide a more complete picture of animal health.

1.2.1 Simple Direct Examinations

Although used less frequently for routine surveillance, simple direct examination of samples is a very useful methodology providing preliminary disease diagnosis and allowing rapid initiation of treatment or quarantine strategies [10]. For example, fecal samples collected from live colony animals and examined under a light microscope can be used to determine the presence of parasites allowing antiparasitic treatment as necessary. Simple direct examinations are particularly valuable at necropsy in the diagnostic investigation of an epizootic. Examples include the microscopic evaluation of skin scrapings, gill clippings, or intestinal contents at necropsy when animals have easily discernable lesions such as on the skin, gills, or internal organs. Biopsy samples can be collected from either valuable live anesthetized animals or freshly euthanized animals, following Institutional Animal Care and Use Committee (IACUC) protocols. Tissue biopsies should be performed and evaluated by trained individuals. Simple direct exams are advantageous because they are extremely rapid compared to other diagnostic methods; however, this approach involves perishable samples, can be insensitive, and requires expertise. Identification of parasites on simple direct examination is typically based on the recognition of specific morphologic features as well as characteristic motility, or in some cases a lack of motility, e.g., *Piscinoodinium pillulare*. Several resources with descriptions including measurements and images of salamander parasites are available [11–13]. In addition, many of the organisms that parasitize aquatic salamander life stages also parasitize fishes, for which extensive resources including pictorial guides are available [14–22]. It is important to contact veterinary staff as soon as possible when unexpected lesions, increased morbidity, or increased mortality is observed in colony animals in order to facilitate rapid disease diagnosis and implementation of treatment.

1.2.2 Histopathology

Histopathology is a commonly used diagnostic method that allows for visualization and evaluation of biological tissues at the cellular level to determine the presence of disease or tissue changes due to infection, noninfectious conditions, or experimental manipulation. Tissues are typically collected during necropsy and preserved in a fixative solution such as buffered formalin or ethanol. Once samples have been fixed for an adequate period of time, samples undergo slide preparation using a five-step process: trimming to create a smaller sample, processing to dehydrate the sample, embedding in paraffin, sectioning into thin slices mounted onto a glass slide, and staining with dyes that allow visualization of cellular detail using light microscopy.

Samples can be evaluated for the presence of tissue changes at the cellular level using routine stains such as hematoxylin and eosin (H&E) that allow for visualization of nuclear and cytoplasmic cellular detail as well as the cellular arrangement and tissue architecture. Special stains such as Gram stain, Ziehl–Neelsen, Grocott–Gomori's methenamine silver stain (GMS), or periodic acid-Schiff (PAS) can be used to evaluate the presence of bacteria or fungi within the tissue that may be associated with disease processes. The accuracy of histopathology is dependent on the appropriate tissue collection and preservation at the time of necropsy as well as the ability to capture the lesion (and any associated organisms) within the plane of sectioning. Histopathology can be used as a stand-alone test for animal health monitoring or can be an adjunctive method for other platforms.

1.2.3 Cytology

Cytology is another methodology that can be used to evaluate pathologic changes at the cellular level by examination of single cells or clusters of cells placed on a glass slide and evaluated using light microscopy. Samples often include bodily fluids such as blood, urine, and peritoneal fluid or fine needle aspirates (biopsy) of masses or abnormal tissues.

1.2.4 Immunology

The immune system of larval and adult amphibians is similar to other vertebrates, with distinct innate and adaptive immune components [23]. Because of this, there is potential to monitor the health of amphibians using traditional immunologic assay methods utilized for other species. However, routine monitoring of amphibian immune response to disease remains in its infancy, largely due to limited amphibian-specific reagent availability.

1.2.5 Parasitology

Amphibians can be hosts to a variety of parasites and may or may not display clinical signs [24, 25]. Clinical signs, when present, can vary depending on the species of parasite, burden, host age, host immune status, and whether the animals were wild-caught or captive-bred. Clinical signs associated with parasite infections can include dehydration, anorexia, emaciation, change in fecal consistency, changes in skin appearance, petechia, coelomitis, and death. Wet mounts of gill tissue, skin scrapings, or gastrointestinal contents can also provide the opportunity to observe motility, which is often very useful in the identification of protozoal parasites and commensals.

1.2.6 Microbiology

Diagnostic accuracy depends on the collection of samples that are appropriate for the platform and representative of the animal's pathogen status, in addition to the sensitivity and specificity of individual diagnostic assays [26]. Microbiology offers several advantages as a diagnostic platform, including the ability to test a wide variety of specimen and sample types, species-level identification, bacterial phenotyping, and the ability to do antibiotic sensitivity testing. Disadvantages of microbiology as a diagnostic platform include wide variability in sensitivity, a high level of necessary expertise, specialized incubation conditions for many organisms, and variable turnaround times.

The growth requirements of aquatic pathogens and environmental opportunists can sometimes delay or complicate identification in diagnostic laboratories using protocols that are optimized for the identification of mammalian pathogens [27]. Salamanders are ectothermic and are often housed at cooler temperatures, and some pathogenic bacterial species grow poorly or not at all at the incubation temperatures commonly used in most diagnostic microbiology laboratories (35–37 °C). Microbiological cultures from salamanders should often be incubated at a cooler temperature and for much longer incubation periods [28].

The bacterial pathogens of laboratory rodents and other mammalian species are generally well characterized and readily identifiable when cultured in diagnostic laboratories [27]. However, the identification of salamander pathogens, opportunists, commensals, and environmental organisms can be very difficult using traditional biochemical and phenotypic tests. The phenotypes of some microorganisms are poorly characterized, and the variability in biochemical traits among different isolates within the same microbial taxa can make interpretation challenging. Automated systems for microbial identification based on biochemical tests are widely used, but can only identify a limited number of organisms accurately due to the limited number of biochemical tests available and the limited database of phenotypes compared to the vast diversity of microorganisms that can be cultured from salamanders and their environments.

Matrix-assisted laser desorption/ionization time-of-flight mass spectrometry (MALDI-TOF MS) has emerged as an extremely powerful and rapid spectrophotometric method for microbial identification. Bacteria are identified by their unique molecular fingerprints of highly abundant 2–20 kDa proteins. MALDI-TOF MS can substantially reduce the time required to identify organisms that have historically been difficult to identify [29]. Some salamander pathogens may be underrepresented in spectral databases. However, spectra for many potential pathogens of salamanders are well represented in commercially available reference databases, often because they have been included as organisms of importance for human or domestic animal health, as aquaculture pathogens [30], or as organisms associated with seafood spoilage [31–33]. The breadth of organisms represented in commercially available spectral databases is steadily increasing.

1.2.7 Molecular Diagnostics

Several different molecular platforms can be used for diagnostics, including conventional PCR, real-time PCR [34, 35], droplet digital PCR [36], loop-mediated isothermal amplification (LAMP) [37], and CRISPR [38, 39], among others. However, real-time PCR has been the most widely adopted for veterinary diagnostics for several reasons, including excellent sensitivity and specificity, short turnaround times, scalability, relatively low cost, and the ability to pool samples [35].

Compared to traditional diagnostic platforms such as histopathology and microbiology, real-time PCR offers several important advantages, including increased sensitivity and increased specificity. Histopathology allows surveillance for noninfectious lesions, which may provide insight into genetic conditions or husbandry issues. While histopathology provides substantial information about tissue architecture and the host response to a pathogen that cannot be captured using other methodologies, histopathology lacks sensitivity, particularly for focal or rare multifocal lesions that may not be present in the planes of section examined by a pathologist. The specificity of histopathology can be limited. Organisms that are observed can be classified into categories such as "gram-negative rod-shaped bacteria" or "septate branching filamentous fungal hyphae," but are not identified to the species level. For some salamander pathogens, such as *Mycobacterium* spp., species-level identification is important because related species differ in pathogenicity and zoonotic potential. Another key advantage of real-time PCR compared to histopathology is the ability to test a wide variety of specimens and sample types, including antemortem and environmental samples in addition to tissues. Example specimens and sample types include skin, oral swabs, cloacal swabs, feces, live feed cultures, biofilms, environmental detritus, concentrated water samples, and eggs and sperm.

Real-time PCR is more sensitive than microbiology for most infectious agents. Some pathogens have not yet been cultivated in vitro but can be detected using real-time PCR. However, the sensitivity of real-time PCR can be limited compared to culture-based methods by the amount of sample that can be processed, extracted, and tested using commercially available instruments. Thus, in some situations, an organism present in low numbers can be detected by microbiology simply because a much larger inoculum can be tested, or when a series of enrichment steps are used to improve culture sensitivity. There are also other limitations of real-time PCR compared to other diagnostic platforms. Advance knowledge of the genetic sequence for a pathogen of interest is required for the design and validation of diagnostic real-time PCR assays. Real-time PCR provides information about the presence or absence of a targeted infectious agent; however, it cannot differentiate between live and dead organisms, or provide any information about the host's response to colonization or infection.

Environmental monitoring, which can include real-time PCR and microbiology, is a useful adjunct method to complement routine surveillance of animals. Environmental detritus often accumulates on the floor of aquaria or enclosures housing salamanders, which may include pieces of uneaten food, feces, dust, sediment, sloughed skin, and other debris. Detritus is an attractive sample because it can usually be collected with minimal disturbance to the animals. In addition, aquatic detritus contains particulates coated in biofilms that often include facultative pathogens such as mycobacteria [40]. Dust and sediment have been used as a real-time PCR sample type to detect infectious agents in a wide range of terrestrial and aquatic laboratory animal species [40–45]. Other environmental sample types include environmental swabs and water samples that have been concentrated by filtration [40]. The utility of real-time PCR to test environmental samples facilitates more efficient detection of low-prevalence pathogens, depending on the pathogen's life cycle [40].

2 Materials

2.1 General Biosecurity

1. Personal protective equipment (PPE) that is appropriate for the species housed and the research being performed.
2. An additional set of husbandry supplies for quarantine.
3. Disposable containers to transport solutions or live feeds prepared in a central location to quarantine.
4. Disinfectant solutions adequate for inactivation of the most resistant excluded or monitored infectious agents.
5. Large bags for the removal of supplies from quarantine.

2.2 Necropsy and Postmortem Sample Collection

1. Down draft table and fume hood.
2. Latex or nitrile gloves.
3. Laboratory coat.
4. Eye protection.
5. Plastic dissection board that can be easily sanitized and disinfected.
6. Blunt end forceps.
7. Small dissecting scissors.
8. Surgical scissors.
9. Syringes (1 and 3 mL most commonly used) with a 21-gauge needle.
10. Scalpel handle and scalpel blades.
11. Small jars or plastic containers for tissue collection.
12. Adhesive labels for sample containers.
13. Sterile dry, flocked swabs with a synthetic (plastic) handle.
14. Sterile culture swabs.
15. Sterile bacterial culture loops.
16. Bacterial culture medium as necessary (Table 1).
17. Fungal culture medium as necessary (Table 1).
18. Conical and microcentrifuge tubes for additional sample types (fecal contents for parasitology, blood, urine, and other body fluids).
19. Sterile phosphate-buffered saline (PBS) solution [47].
20. Glass slides.
21. Pencil for labeling glass slides.
22. Euthanasia solution as approved by Institutional Animal Care and Use Committee and in compliance with guidelines from the American Veterinary Medical Association (AVMA) [48] or the appropriate regulatory organization(s) (*see* **Note 1**).
23. 10% neutral-buffered formalin (also denoted as 4% (v/v) formaldehyde solution) such as Sigma-Aldrich® HT501128. Published formulations are also available [49, 50].
24. Instrument holder with 70% ethanol.
25. Instrument holder with 10% bleach (used when collecting tissues for molecular diagnostics).
26. Handheld camera for taking images of gross lesions (as needed).
27. Tablet or pen and notebook to record lesions.
28. Ruler that can be easily disinfected for measuring observed gross lesions (as needed).

Table 1
Selected microbiological culture media

Culture medium	Type	Primary utility	Notes	Catalog number or Reference
Cetrimide agar	Selective	Bacteria	Selects for *Pseudomonas aeruginosa* at 42 °C	BD Difco™ 285420
Chocolate agar	Nonselective, enriched	Bacteria and fungi	Contains lysed blood cells	BD BBL™ 221267
Hektoen enteric (HE) agar	Selective and differential	Bacteria	Useful for *Salmonella enterica*	BD Difco™ 221366
MacConkey agar	Selective and differential	Bacteria	Gram-negative and enteric bacteria	BD Difco™ 221270
Potato dextrose agar (PDA)	Nonselective	Fungi	Chloramphenicol can be added to suppress bacterial growth	Oxoid CM0139
Sabouraud dextrose agar	Nonselective	Fungi	Low pH; antibiotics can be added to suppress bacterial growth	BD Difco™ 211661
Trypticase™ soy agar (TSA) with 5% sheep blood	Nonselective, enriched	Bacteria and fungi	Detection of hemolysis	BD BBL™ 221261
Tryptone yeast extract salts (TYES) agar	Nonselective, minimal	Bacteria and fungi	Useful for isolating *Flavobacterium columnare*	[46]
Xylose lysine deoxychlate (XLD) agar	Selective and differential	Bacteria	Useful for *Salmonella enterica*	BD Difco™ 278850

29. Clean paper towels.
30. Biohazard bag for carcass disposal.

2.3 Antemortem and Environmental Sample Collection

1. Conical tubes with screw top lids.
2. Microcentrifuge tubes with screw top lids.
3. Dry, synthetic flocked swabs with synthetic (plastic) handles.
4. Culturettes with transport medium, e.g., Amies transport medium without charcoal.
5. Parafilm® (Bemis Company, Inc., Neenah, WI).
6. Polystyrene foam boxes.
7. Cold packs.

3 Methods

3.1 Import Procedures

3.1.1 Approved Vendors List

1. Define an approved vendors list for the salamander colony (*see* **Note 2**). However, it is often impractical to set rigid institutional requirements for the importation of salamanders, in general, because many species are used in research for which there are no genetic stock centers, and investigators using axolotls may need to import animals from collaborating institutions that do not track colony health.
2. Define two sets of import procedures, including a simpler set of import procedures for animals or embryos received from preferred suppliers, and a more extensive set of import procedures to mitigate the risk of importing from acceptable, but not preferred vendors (*see* **Note 3**).

3.1.2 Exclusion Lists

1. Define an exclusion list for the colony. An exclusion list is a defined list of infectious agents that are absent in a population of laboratory animals according to laboratory testing, and for which measures are taken to prevent introduction to the colony or vivarium. A partial list of salamander infectious agents is presented in Table 2.
2. Consider the key factors when deciding on agents that must be excluded from the colony, including the type of research for which the salamanders are being used, whether each infectious agent is likely to cause morbidity and/or mortality in the colony, the overall prevalence of the agent, the difficulty of excluding the agent, and the known impacts of subclinical infection on physiology, immune function, and other parameters.

3.1.3 Quarantine

1. Design a quarantine program for the colony, considering the sources of animals, species, intended use, life stages, health history, and the needs of the researchers. Quarantine newly imported animals in a location that is physically separated from the established colony for a defined period of time (*see* **Note 4**).
2. Receipt procedures will vary according to the salamander species, life stage, and husbandry requirements. Minimize handling stress and provide newly arrived salamanders with a stable environment to begin the acclimation process as soon as possible.
3. Immediately upon receipt, carefully inspect salamander or embryo shipping containers for damage.
4. Thoroughly inspect the embryos or salamanders for injury, illness, or signs of distress as soon as possible after arrival.

Table 2
Selected infectious agents of laboratory salamanders

Agent	Category	Zoonotic	Susceptible[a] laboratory species	Known modes of transmission	Associated clinical signs	Molecular and adjunct diagnostic methods[b]	References
Acinetobacter haemolyticus/ Acinetobacter calcoaceticus	Gram-negative bacterium	Possible	At	Waterborne	Hemorrhage, edema, ventral hyperemia including pelvic limbs, death	BC, H	[51]
Aeromonas hydrophila[c]	Gram-negative bacterium	Yes	Am, At	Direct contact, open wound, waterborne	Anorexia, edema, hydroceolom, lethargy, petechiae or ecchymoses, ventral hyperemia including pelvic limbs, death	PCR, BC, H	[52–54]
Amphibiocystidium spp.	Mesomycetozoea	No	Am, Nv	Unknown	Nodular dermatitis, subclinical, subcutaneous cysts, skin ulceration, death	H	[12, 55]
Batrachochytrium dendrobatidis	Chytrid fungus	No	Am, At, Nv, Pw	Direct contact, waterborne	Subclinical, black spots on ventrum and/or extremities, discoloration of the skin, sloughing skin, death	PCR, H	[56–61]
Batrachochytrium salamandrivorans	Chytrid fungus	No	Nv, Pw	Direct contact, waterborne	Subclinical, anorexia, lethargy, ataxia, skin lesions including the appearance of black spots and ulceration, death	PCR, H	[5, 62–65]
Chilodonella sp.	Ciliate ectoparasite	No	Am	Direct contact	Discoloration of the skin, gill lesions	SDE, H	[12, 66]

Chilomastix caulleryi	Flagellate endocommensal	No	Am	Fecal-oral	Typically subclinical or commensal	PCR, SDE, H	[11]
Flavobacterium columnare	Gram-negative bacterium	No	Am, At	Direct contact, waterborne	Edema, gill lesions, hydroceolom, petechiae or ecchymoses, respiratory difficulty, discoloration of the skin	PCR, BC, H	[67–69]
Hexamita spp.	Flagellate endoparasite	No	Am	Fecal-oral	Subclinical, diarrhea	SDE, H	[11, 70, 71]
Ichthyobodo sp.	Flagellate ectoparasite	No	Am	Direct contact	Gill lesions, discoloration of the skin	PCR, SDE, H	[12, 13, 66, 70, 72]
Ichthyophonus sp.	Mesomycetozoea	No	Nv	Feeding on infected tissues, parasite vectors, waterborne	Skin ulceration, edema	PCR, H	[73–76]
Mycobacterium marinum	Acid-fast gram-positive bacterium	Yes	Am	Feeding on infected tissues, open wound, waterborne	Anorexia, edema, lethargy, nodular dermatitis, skin ulceration, death	PCR, H, BC (NR)	[77, 78]
Mycobacterium spp.	Acid-fast gram-positive bacterium	Yes	Am, Pw	Feeding on infected tissues, open wound, waterborne	Anorexia, edema, lethargy, nodular dermatitis, skin ulceration, death	PCR, H, BC (NR)	[79–81]
Opalina spp.	Ciliate endocommensal	No	Am	Fecal-oral	Typically subclinical or commensal	SDE, H	[70, 72]
Piscinoodinium pillulare	Dinoflagellate ectoparasite	No	Am	Direct contact, waterborne	Subclinical, gill lesions, discoloration of the skin	PCR, SDE, H	[11, 82–84]
Pseudomonas aeruginosa[cb]	Gram-negative bacterium	Possible	At, Nv	Feeding on infected tissues, open wound, waterborne	Hemorrhage, edema, ventral hyperemia including pelvic limbs	PCR, BC, H	[51, 55, 56]

(continued)

Table 2 (continued)

Agent	Category	Zoonotic	Susceptible[a] laboratory species	Known modes of transmission	Associated clinical signs	Molecular and adjunct diagnostic methods[b]	References
Ranavirus, e.g., Ambystoma tigrinum virus (ATV)	DNA virus	No	Am, At, Nv, Pw	Direct contact, feeding on infected tissues, Fecal-oral, waterborne	Edema, lethargy, hemorrhage, subclinical, skin ulceration, death	PCR, H, VC (NR)	[85–89]
Salmonella enterica	Gram-negative bacterium	Yes	Nv	Fecal-oral	Typically commensal	PCR, BC, H	[90]
Saprolegnia spp.	Oomycete	No	Am, At	Waterborne	White or greyish hyphal growth, discoloration of the skin	FC, H	[12, 91–94]
Serratia marcescens[c]	Gram-negative bacterium	Possible	Am	Unknown	Anorexia, hemorrhage, hydroceolom, lethargy, death	PCR, BC, H	[95]
Trichodina sp.	Ciliate ectoparasite	No	Am	Direct contact	Discoloration of the skin, gill lesions	SDE, H	[12, 66, 70, 72]

[a] *Ambystoma mexicanum* (Am), *Ambystoma tigrinum* (At), *Notophthalmus viridescens* (Nv), *Pleurodeles waltl* (Pw); Susceptibility of many salamander species to various pathogens is unknown and the available scientific literature is limited; therefore, salamander species not listed may also be susceptible to infection

[b] Bacterial culture (BC), fungal culture (FC) histopathology (H), not recommended (NR), polymerase chain reaction (PCR), simple direct examination (SDE) viral culture (VC)

[c] Organisms that are part of the normal microbiota of many salamander species that can also opportunistically cause bacterial dermatosepticemia

5. Supply the quarantine room with all necessary equipment, PPE, and other supplies to prevent or minimize the movement of materials in and out of quarantine.
6. Any reagents, solutions, feed, or live feeds for use in quarantine that are prepared daily or routinely outside of quarantine should ideally be transported to quarantine in disposable containers that are not returned to the preparation area.
7. Allow newly arrived salamanders to acclimate to laboratory conditions for 2 weeks prior to any attempts to breed or other manipulations while in quarantine. Carefully observe them for clinical and behavioral signs throughout this period (*see* **Note 5**).
8. Perform regular health assessments at least daily on salamanders while in quarantine.
9. Screen imported salamanders for adventitious agents using antemortem and/or environmental samples.
10. If agents on the main colony exclusion list are identified in quarantined animals, formulate a plan to eliminate the agent (*see* Subheading 3.2.1) from the imported population while they are still held in quarantine, avoiding exposure to main colony animals.

3.2 Biosecurity

3.2.1 Exclusion

1. Review the ecological niche (i.e., reservoir), potential source (s) of infection, and the modes of transmission for agents included on the exclusion list (*see* **Note 6**).
2. Reduce horizontal transmission by physical separation of different salamander species, or subpopulations with different health statues or susceptibility to infectious agents. Many pathogens spread most efficiently by direct contact.
3. Reduce the risk presented by other potential source(s) of infection. Pathogens can also spread indirectly via fomites. Examples of potential sources for salamander exposure to pathogens could include live feeds, water, nets, environmental enrichment, equipment and supplies, and substrate.
4. Maintain pest control to (1) prevent the introduction of zoonotic agents such as *Salmonella enterica* that can be carried by feral rodents and also colonize salamanders [90] and (2) prevent transmission of pathogens and opportunistic pathogens by potential vectors, such as arthropods or other invertebrates (*see* **Note 7**).

3.2.2 Work and Material Flow

1. Large salamander colonies should be physically separated into different areas or rooms based on quarantine status, health status, and life stage in order to reduce pathogen transmission.
2. Provide designated supplies for each housing area or room.

3. Personnel should work with the salamanders with the highest health status first, followed by healthy salamanders of lower health status, then animals in quarantine, and finally isolated or diseased animals.

3.2.3 Disinfection

1. Develop a disinfection plan. While sterilization refers to the complete inactivation of all microorganisms, disinfection refers to the use of physical or chemical means to greatly reduce the presence of viable microorganisms. Disinfection of equipment, supplies, and surfaces reduces the risk of pathogen transmission via fomites. Disinfection is particularly important in quarantine and when personnel, supplies, or equipment are shared between animals of different health status.
2. In some cases, physical means of disinfection such as heat exposure are appropriate and avoid the toxicity associated with chemical disinfectants.
3. Select appropriate chemical disinfectants considering both the toxicity profile and efficacy. Ideally, the selected chemical disinfectant(s) should be effective against the most environmentally stable pathogens on the colony exclusion list [26]. Chemical disinfectants vary considerably in both toxicity and efficacy for inactivating different pathogens [96–100].
4. Clean equipment or surfaces so that they are completely free of gross organic debris prior to the application of chemical disinfectants.
5. Use a freshly prepared or diluted disinfection solution if required.
6. Ensure adequate contact time for efficacy, which varies by disinfectant.
7. Rinse the disinfected surfaces to help to remove any toxic residues and prevent damage to equipment or surfaces by disinfectants that can be corrosive.

3.2.4 Elimination

1. When an excluded agent is identified, repeat the testing to verify the initial test results before proceeding with procedures to eliminate the agent that may be disruptive to research.
2. Consider a test-and-cull approach for elimination. Test-and-cull approaches can be used to eliminate infectious agents from small populations, but take longer and are more difficult for larger populations and for pathogens that are environmentally persistent and resistant to disinfection [101].
3. Consider saline bath treatment for skin infections and ectoparasites [10, 67, 102]. Salt bath treatments can be used for individual animals or small numbers of animals that are co-housed. Recommendations for salt bath treatment are

anecdotal with recommendations ranging from 5 to 20 g/L of sodium chloride concentrations and exposure for less than 10 min once to twice daily. Alternatively, 100% concentration baths of modified Holtfreter's [103] or Steinberg's solution [104] can be used. Prolonged (72-h) immersion in 4–6 g/L saline baths has also been described as a treatment option for *Piscinoodinium pillulare* and *Trichodina* spp. [105]. When using saltwater baths to treat disease, it is important to determine species susceptibility to salt toxicosis at high salt concentrations as well as determine effects of hyperchloremia on research parameters.

4. Consider altering temperature or other husbandry parameters to reduce pathogen replication, reduce transmission, or aid pathogen clearance. Temporary housing of singly housed or co-housed infected animals at a lowered temperature (4–8 °C) can be used to treat some amphibians, such as the axolotl, with bacterial or fungal infections [67, 106]. A reduction in body temperature can slow amphibian metabolism and also slows the growth of some infectious agents, allowing healing to occur [67, 106]. Temporary housing of infected salamanders at an increased temperature can also help to clear certain infectious agents such as *Batrachochytrium salamandrivorans* and *B. dendrobatidis,* sometimes as an adjunct therapy [107, 108]. However, some species including axolotls tolerate increased temperatures poorly [68, 70, 109]. Thus, tolerance of altered temperatures is species-specific, and can also depend on the developmental stage [109, 110]. Isolation of infected animals and increasing water changes can reduce pathogen transmission and decrease spread throughout the colony.
5. Although there is only very limited information available at the time of this writing, there is some evidence that administration of probiotics may be protective against pathogens in some cases. For example, several species of bacteria that have been isolated from salamander skin are protective against *Batrachochytrium dendrobatidis* or *B. salamandrivorans* [111–114].
6. Consider chemotherapy using antimicrobial drugs. There are a few reports describing chemotherapeutic approaches for eliminating or managing infectious diseases in laboratory salamanders [56, 66, 67], as well as other captive or ex situ salamander populations [57, 107, 115], and published formularies of drugs and dosages for amphibians [10, 116]. Antimicrobial treatment can eliminate infections in some cases; however, in other situations, the pathogen is only suppressed (*see* **Notes 8** and **9**).

3.2.5 Containment

1. In the event of an epizootic or introduction of a significant pathogen into a salamander room, inform all affected personnel, including investigators, veterinarians, laboratory managers, postdoctoral fellows, technicians, students, and animal husbandry staff.
2. Schedule a meeting to discuss the issue with affected parties as soon as possible. Review the details of the outbreak and the proposed plan for management of the outbreak.
3. Quarantine the affected salamander population. Limiting the number of personnel who access the room and halting any movement of animals or embryos in or out of an affected room will reduce the likelihood of accidental transmission to salamanders housed in other areas or other amphibians housed in the vivarium.
4. Avoid using organs or tissues collected from quarantined animals for transplantation into other salamanders.
5. Place supplies removed from the quarantined room into plastic bags and autoclave them.
6. Review colony records and investigate potential sources of contamination in order to identify gaps in biosecurity that need to be corrected.
7. Contact any receiving laboratories or institutions if embryos or salamanders have been recently exported from the quarantined room and inform them of the outbreak.
8. Determine whether embryos or salamanders have been recently moved from the quarantined room to other housing or holding areas in the vivarium or the laboratory.
9. Screen other groups of salamanders that may have been exposed to the pathogen.

3.2.6 Eradication

1. Repeat the testing as needed to confirm positive test results before proceeding with eradication procedures, which are costly and disruptive to research.
2. Consider test-and-cull approaches as an alternative to depopulation. Eradication by test-and-cull is a prolonged process and is not always successful. Test-and-cull approaches are more likely to be effective for pathogens that are not environmentally persistent because of indirect transmission [101].
3. The most reliable approach to eradication is depopulation and disinfection, followed by repopulation of the facility with newly acquired animals. Depopulation procedures for salamanders are similar to procedures that have been described for other laboratory species [26].
4. Euthanize all animals in the room or facility.

5. Discard supplies from the room that can be easily replaced.
6. Double-bag and autoclave reusable equipment or supplies that are removed from the room.
7. Thoroughly clean all surfaces in the room, including the floor and walls with a detergent solution and rinse away the detergent with water.
8. Sterilize the room using chlorine dioxide or vapor-phase hydrogen peroxide.
9. If sterilants are not available, thoroughly disinfect the room at least twice using disinfectants with a different mode of action each time.

3.3 Necropsy and Postmortem Sample Collection

3.3.1 General Considerations

Necropsy and histopathology can be the key procedures that complement other diagnostic methods in the evaluation of health and disease status. During necropsy, all body organs and cavities can be examined for macroscopic tissue changes that may be associated with disease or experimental manipulation. Tissues can be collected at the time of necropsy, preserved in a liquid fixative, trimmed for size and embedded in paraffin, mounted on glass slides, and later evaluated by a pathologist. Necropsy techniques developed for a variety of small mammalian species can be easily adapted for amphibians. The overall process for necropsy and postmortem sample collection is very similar to what is used for rodents with (1) examination of the live animal, (2) euthanasia, (3) exsanguination (if necessary), (4) opening of abdominal cavity, and (5) opening of the thoracic cavity.

1. When evaluating the live animal prior to euthanasia, pay careful attention to details such as animal behavior, body position, righting reflex, response to external stimuli, and buoyancy.
2. Once the animal is euthanized, perform a thorough physical examination. Include body condition score and external lesions such as skin discoloration or ulceration; palpable subcutaneous or abdominal masses; abdominal distention with fluid; and abnormalities of eyes, mouth, nasal, and urogenital openings. Describe and record any abnormalities.
3. Before starting the necropsy, bring all necessary necropsy materials (*see* Subheading 2.1) to the workspace.
4. Label all containers and media with appropriate animal/sample identification.
5. Place blunt-end forceps, small dissecting scissors, surgical scissors, and scalpel handle with attached blade in container with 70% alcohol.
6. For collection of tissues for PCR, place a set of blunt-end forceps and small dissecting scissors in a container with a 10% bleach solution.

7. Set to the side biohazard bag for carcass disposal.
8. Perform the necropsies in a systematic and consistent order to prevent inadvertent contamination of samples and overlooking tissues or lesions. Collect samples from the exterior surface of the salamander first and move inward as the necropsy progresses.
9. Describe lesions using correct pathologic terminology so that the notes and reports can be interpreted correctly by others (*see* **Note 10**).

3.3.2 Skin and Gills

1. To test for external parasites, fungi, or bacteria by real-time PCR; a sterile dry, flocked swab with a synthetic (plastic) handle can be used to swab the skin, including any lesions and the ventrum. After swabbing, place the swab into a small sterile tube (such as a microcentrifuge tube).
2. A sterile scalpel blade can be used to gently scrape the skin. Place the resultant skin debris on a clean glass slide with a drop of water or normal saline for microscopic evaluation. Examine the slide for ectoparasites and for oomycete or fungal hyphae.
3. To test by bacterial or fungal culture, use a sterile bacterial sampling loop to gently scrape the skin. Use the loop to inoculate solid or liquid culture media.
4. Use small dissection scissors to remove a small piece of skin. Place skin (subcutis side down) onto a paper index card. Cut the card into a smaller segment and place in a formalin-containing organ collection container (*see* **Note 11**).
5. Gills (if present) can also be swabbed for culture using a similar strategy as described for the skin.
6. Make wet mounts of the gills using small dissection scissors to remove a small section of the gill which is then gently teased apart, mounted in a small volume of sterile PBS, and the tissue is placed on a glass slide (with or without a cover slip) for microscopic evaluation.
7. Alternatively, the gill section can be placed in an organ collection container filled with formalin to be fixed for histologic evaluation.

3.3.3 Fluids

1. Coelomic (abdominal) fluid can be collected by carefully puncturing the abdominal wall with a 1 or 3 mL syringe with attached 21-gauge needle. Carefully withdraw fluid, making note of consistency and color. Fluid can be placed in a sterile tube for culture, cytology, or chemistry evaluation.
2. Blood can be collected by carefully puncturing the heart or abdominal vein using a 1 mL syringe and a 21-guage needle.

3.3.4 Internal Organs

1. Remove blunt-end forceps and scalpel (or surgical scissors) from 70% ethanol bath and incise skin along the midline from the mandibles to the pubis.
2. Reflect the skin on both sides of the incision.
3. Starting at the mouth, examine the oral cavity, tongue, esophagus, trachea, lungs, and heart, and evaluate the tissues for correct anatomical position, relative size, color, and shape recording any observed abnormalities or pathologic lesions.
4. If lesions are identified, use a sterile bacterial sampling loop to gently scrape the lesion surface. For discrete lesions, scrape the margins of the lesion. Use the loop to inoculate solid or liquid culture media for bacterial or fungal culture.
5. For molecular diagnostics, remove dissection scissors and blunt-end forceps from 10% bleach solution. Blot instruments on a clean paper towel to remove bleach. Carefully remove a small section of the lesion area and place in sterile tube (such as a microcentrifuge tube).
6. Once samples have been collected for culture and/or molecular diagnostics, tissues can then be carefully excised and placed in an organ collection container containing formalin for histologic evaluation (*see* **Note 12**).
7. Locate the kidneys, spleen, and liver. Evaluate the tissues for correct anatomical position, relative size, color, and shape recording any observed abnormalities or lesions.
8. For each tissue, if lesions are evident, use a sterile bacterial sampling loop to gently scrape the lesion surface. Use the loop to inoculate solid or liquid culture media for bacterial or fungal culture.
9. For molecular diagnostics, remove dissection scissors and blunt-end forceps from 10% bleach solution. Blot instruments on a clean paper towel to remove bleach. Carefully remove a small section of the lesion area and place in sterile tube (such as a microcentrifuge tube).
10. Once samples have been collected for culture and/or PCR, carefully excise tissues and place them in a formalin-containing organ collection container for histologic evaluation.
11. Locate the urinary bladder. Evaluate the tissues for correct anatomical position, relative size, color, and shape recording any observed abnormalities or pathologic lesions.
12. If urinary bladder is distended, a urine sample can be collected by carefully puncturing the organ with a 1 mL syringe and 21-gauge needle.

13. Carefully withdraw fluid, making note of consistency and color. Fluid can be placed in a sterile tube for culture, cytology, or chemistry evaluation.
14. Locate the reproductive organs. Evaluate the tissues for correct anatomical position, relative size, color, and shape, recording any observed abnormalities or lesions.
15. For bacterial or fungal culture, use a sterile bacterial sampling loop to gently scrape the lesion surface. Use the loop to inoculate solid or liquid culture media.
16. For molecular diagnostics, remove dissection scissors and blunt-end forceps from 10% bleach solution. Blot instruments on a clean paper towel to remove bleach. Carefully remove a small section of the lesion area and place in sterile tube (such as a microcentrifuge tube).
17. Beginning at the cloaca, follow the length of the intestine upward to the stomach to evaluate the large intestine, small intestine, and stomach. Record any abnormalities or lesions.
18. Collect intestinal contents for evaluation by parasitology (wet mount), real-time PCR, or microbial culture.
19. For parasitology, remove clean forceps and dissection scissors and make a small incision in the desired intestinal section (colon or jejunum/ileum). Using a sterile bacterial loop, remove a small amount of intestinal contents, mount in a small volume of sterile PBS on a glass slide.
20. For bacterial or fungal culture, use a sterile bacterial sampling loop to gently remove intestinal contents. Use the loop to inoculate solid or liquid culture media.
21. For molecular diagnostics, use a sterile bacterial sampling loop to remove a small amount of ingesta and place in sterile tube (such as a microcentrifuge tube).
22. For histopathology, remove small 1–2 cm sections of each section of the gastrointestinal tract and place in a formalin-containing organ collection container for histologic evaluation.

3.4 Routine Colony Health Monitoring

3.4.1 Selection of Agents to Monitor

1. Consider zoonotic agents. Zoonoses are infectious diseases that can be transmitted from animals to humans. Mycobacteriosis has been described in several species of salamanders [77, 117–119]. Axolotls and other salamander species are susceptible to infection with *M. marinum*, which exhibits greater zoonotic potential than many other *Mycobacterium* spp. [77–79]. *Salmonella enterica* a zoonotic agent that has been isolated from several species of salamanders [90]. *S. enterica* is not normally associated with clinical disease in salamanders.
2. Consider primary pathogens. Primary pathogens cause infections in otherwise healthy, immunocompetent animals, and may include viruses, bacteria, oomycetes, true fungi,

ectoparasites, and endoparasites. Obligate pathogens, including viruses and many parasites, cannot reproduce without a suitable host, and are similarly phylogenetically diverse. Facultative pathogens, including various protozoa and several *Mycobacterium* spp. can be free-living in the environment of aquatic systems and infect otherwise healthy, immunocompetent hosts [40, 120].

3. Consider opportunistic pathogens. Opportunistic pathogens only infect hosts that are immunocompromised, injured, or stressed. Sources of stress in laboratory salamanders can include husbandry concerns, such as incorrect temperature, vibration, inadequate nutrition, and poor water quality, but can also include other factors like intraspecific aggression. In most situations it is not practical to exclude opportunistic pathogens.
4. Consider whether infectious agents are likely to have the potential to alter experimental outcomes despite producing only subclinical infections. Adverse effects of subclinical infections have been demonstrated in a wide range of animal models and may include a wide array of changes, including altered cytokine levels, gene transcription, susceptibility to carcinogenesis, nutrition or gastrointestinal function, or future immune responses [121–127].
5. Consider agents that pose a significant health risk to other species. In some cases, one species can carry a pathogen subclinically (the reservoir host) but pass it to another, more susceptible host species where it causes clinical disease. The reservoir host and susceptible host do not necessarily need to both be salamander species. Some salamander infectious agents have a broad host range that includes fish, anurans, or other animals. Subclinical infections in reservoir host species are most problematic in the laboratory when a susceptible species is housed nearby or there are shared facilities, equipment, personnel, or supplies.
6. Review the life cycle and the biology of each pathogen to understand transmission, host range, pathogenicity, persistent environmental stages, the importance of alternate, intermediate, or paratenic hosts, and therapeutic options. A thorough understanding of pathogen life cycle informs the decision of whether the agent should be monitored by providing insight into how easily it can be excluded and what options are available for control.

3.4.2 Diagnostic Platforms

1. Make a list of agents to be routinely monitored that are best detected by real-time PCR. A list of selected salamander infectious agents and useful diagnostic approaches are presented in Table 2.

2. Determine which sample types would be necessary to monitor for the list of agents. Consider ease of collection.
3. Make a list of agents for which microbiology would be useful, for example, bacterial agents for which antibiotic sensitivity testing would be desirable (*see* **Note 13**).
4. Make a list of agents to be monitored by traditional coprological methods, e.g., direct smear, fecal flotation, or fecal sedimentation.

3.4.3 Sample Sizes

A statistical formula can be used to calculate the number of animals that must be randomly tested, based on the expected prevalence of the infectious agent in the colony, to detect the agent with 95% confidence [128–130]. Prevalence for an agent may differ substantially among facilities with different husbandry practices. Assumptions about transmission among colony animals may be invalidated by husbandry practices, and sampling for agents at low prevalence according to this approach may be cost-prohibitive (*see* **Note 14**). In practice, the number of animals tested is limited by budget, as with both rodents and zebrafish [120]. Pooling salamander samples for real-time PCR can be an economical approach, although few data are available for pooling salamander samples. In the absence of experimental pooling data, pooling very small numbers of samples (e.g., 2–3 samples) is unlikely to result in a substantial loss of diagnostic sensitivity.

Although rarely used for amphibians, sentinel animals can be placed for routine health monitoring with the intent to maximize pathogen detection using fewer animals. Unfortunately, there is a dearth of information about the efficacy of sentinel health monitoring for salamanders. It is well established for other laboratory species that some pathogens do not transmit to sentinels efficiently. Sentinels should be placed for at least several months and receive effluent water or soiled bedding from all monitored salamander enclosures. In addition to sentinel animals, clinically diseased salamanders should be submitted for postmortem evaluation.

Infectious agents with environmental life stages or that are shed into the environment at high levels may be more efficiently detected using environmental samples. Environmental positives do not indicate that any particular animal is infected, but that the agent has not been excluded from a population (*see* Subheading 3.4.6).

3.4.4 Testing Interval

1. Consider the risk that each agent poses to the colony. Some programs will test for agents that pose a greater risk, for example, due to higher prevalence or greater pathogenicity, more frequently while testing for uncommon or nonpathogenic agents infrequently.

2. Determine the appropriate monitoring interval for each agent or group of agents. Testing more frequently improves the likelihood that an infectious agent will be detected soon after it is introduced into a colony. When an agent is identified soon after introduction, biocontainment of the pathogen and treatment of infected animals are easier and more likely to succeed.

3.4.5 Antemortem Sample Collection

1. Collect feces into a rigid, plastic container, such as a conical or microcentrifuge tube with a screw top lid.
2. Collect skin swabs using dry, synthetic flocked swabs with synthetic (plastic) handles. The swab tip can be cut off using scissors into a microcentrifuge tube with a screw top lid to prevent leaking during transit.
3. Store samples for real-time PCR analysis frozen at −20 °C. Long-term storage at or below −70 °C may extend the useful life of the samples for detection of RNA viruses.
4. Ship samples in for real-time PCR analysis in rigid primary containers with screw top lids to prevent leaking during transit. Wrap the lids in Parafilm® (Bemis Company, Inc., Neenah, WI) for additional protection against leakage. The primary containers should be shipped in a rigid, insulated container, such as a polystyrene foam box with adequate cold packs for overnight delivery. Dry ice is usually not necessary for domestic shipments but may be necessary for international shipments.

3.4.6 Environmental Sample Collection

1. Environmental detritus can be collected into a rigid, plastic container, such as a conical or microcentrifuge tube with a screw top lid. Submission of up to 1 mL or more is helpful to allow the diagnostic laboratory to re-extract nucleic acids from the original sample if follow-up testing is needed.
2. Dry, synthetic flocked swabs with a synthetic (plastic) handle should be used for swabbing the environment. Environmental swabs can be collected from multiple locations including substrates, the edge of the air–water interface, and in gutters and sumps in the case of recirculating systems. Collect as much material as possible on the swab. Swabbing biofilms on plastic or glass is more effective if the biofilm is swabbed while it is not submerged in water. The swab tip can be cut off using scissors into a microcentrifuge tube with a screw top lid to prevent leaking during transit.
3. Store samples for real-time PCR analysis frozen at or below −20 °C. Long-term storage at or below −70° C may extend the useful life of the samples for the detection of RNA viruses.
4. Ship samples in for real-time PCR analysis in rigid primary containers with screw top lids to prevent leaking during transit. Wrap the lids in Parafilm® (Bemis Company, Inc., Neenah,

WI) for additional protection against leakage. The primary containers should be shipped in a rigid, insulated container, such as a polystyrene foam box with adequate cold packs for overnight delivery. Dry ice is usually not necessary for domestic shipments but may be necessary for international shipments.

4 Notes

1. AVMA acceptable methods [48] for amphibian euthanasia include (a) injectable sodium pentobarbital, typically 60–100 mg/kg of body weight depending on species, which can be delivered intracoelomically; (b) "dissociative agents such as ketamine hydrochloride or combinations such as tiletamine and zolazepam; inhaled agents; and IV administered anesthetics, such as propofol, or other ultra–short-acting barbiturates, may be used for poikilotherms to induce rapid general anesthesia and subsequent euthanasia, although application of an adjunctive method to ensure death is recommended"; (c) buffered tricaine methanesulfonate (MS-222) administered by water bath or injected into the coelomic cavity; and (d) benzocaine hydrochloride administered as a bath or in a recirculation system at concentrations $\geq$250 mg/L or applied topically to the ventrum as a 7.5% or 20% gel. Other methods are considered acceptable with conditions [48].
2. A list of vendors with established animal health monitoring programs can be compiled and designated as approved vendors. As animals from approved vendors typically enter an existing colony with a lowered or minimal quarantine screening process, vendors must be chosen that have evidence of strong animal monitoring procedures in place. Factors to consider are pathogens detected, frequency of pathogen screening, methods used to detect pathogens, and number of animals tested.
3. For example, an academic laboratory might primarily obtain axolotls and embryos from the Ambystoma Genetic Stock Center (AGSC) at the University of Kentucky, but occasionally imports axolotls from collaborators at other institutions that pose a greater risk to the colony. One set of import procedures could apply only to the AGSC, while a second set of more extensive import and quarantine procedures can be defined for animals received from investigators at other institutions or higher risk sources. Higher risk sources may include wild-caught salamanders or animals obtained from the pet trade, wholesalers, multispecies facilities, biological supply companies, or other laboratories that obtain animals from these sources [131].

4. Quarantine periods for laboratory animals in general are often 3–8 weeks, although some quarantine protocols used for zebrafish that involve raising embryos to adulthood while still in quarantine in order to disinfect their eggs may be as long as 3–6 months.
5. Clinical signs that may be observed in salamanders include anorexia, lethargy, deterioration of the gills, dropsy, cutaneous erythema, discolored skin, sloughing skin, petechial hemorrhage, anemia, icterus, buoyancy problems, cloudy eyes, exophthalmia, behavioral changes, masses, ulcers, and death [56, 70, 95, 132, 133].
6. The reservoir for a pathogen or opportunist can be the environment, one or more animal species, or some combination. Obligate animal pathogens have animal reservoirs, which may include a wide range of species or only a single species or group of closely related species if the host range is narrow. Animal reservoirs for microorganisms do not necessarily exhibit clinical disease; and it is possible for a resistant host to serve as an important reservoir for a more susceptible host. Facultative and opportunistic pathogens may have both environmental and animal reservoirs.
7. The use of chemical insecticides in salamander rooms is highly discouraged, because many preparations are highly toxic to salamanders and salamander larvae. If chemical insecticides must be used, they should be applied cautiously in very limited amounts and specified areas to prevent salamander exposure.
8. Antimicrobial therapy may be most effective in eliminating infectious agents that survive only a short time ex vivo, reducing the incidence of reinfection post-treatment.
9. There are also risks associated with antimicrobial treatment that should be considered, including potential toxicity at therapeutic doses and alteration of the composition of the salamander microbiota, which in some cases can be protective against infectious disease [112, 134].
10. Terms denoting location that are commonly used include: dorsal (back of the animal), ventral (abdominal surface of the animal), cranial (head region), caudal (tail region), rostral (organs or tissues closer to the nose region of the head), lateral (each side of the animal), medial (closer to the midline or middle of the body), proximal (close to the region/point of reference), and distal (farther away from the region/point of reference). Additionally, abnormalities can be described by size, shape, color, and consistency (fluid, soft, firm, hard, etc.).
11. Placing skin (subcutis side down) on a 3″ × 5″ index card can ensure that the skin remains flat during fixation. This helps facilitate embedding in paraffin such that a full-thickness cross-section of the skin can be easily obtained.

12. A 3 mL syringe with attached 21-gauge needle can be filled with 10% buffered formalin and be used to gently inflate the lungs prior to removal at necropsy. Use blunt-ended forceps to carefully grasp the trachea. Carefully pierce the trachea with syringe (caudal to forceps placement) and depress plunger slowly to release formalin. The lungs should start to expand. Once the lungs have been inflated, the syringe can be removed, and the lungs can be collected and placed in formalin-filled tissue collection container.
13. Although microbiology can be used for routine screening, it is particularly valuable for diagnostic investigations of clinical disease because of the ability to culture and identify a broad array of microorganisms without advance knowledge of the likely cause. Histopathology can support microbiological findings by demonstrating intralesional organisms that exhibit the same morphology and staining characteristics using special stains.
14. The sample size required to detect an infectious agent in a population of a given size depends on an estimate of the expected prevalence of the pathogen in that population. Pathogens that are highly prevalent are easier to detect and therefore require fewer samples. In contrast, a high level of confidence in detecting uncommon or rare pathogens requires much larger sample sizes, which are often cost-prohibitive. In contrast to natural populations, laboratory animals are typically housed in a way that minimizes the spread of pathogens in a colony which makes pathogens more difficult to detect in the laboratory setting.

Competing Interests

Marcus J. Crim and Marcia L. Hart are employees of IDEXX BioAnalytics, a division of IDEXX Laboratories, Inc., a company that provides veterinary diagnostics.

References

1. Baker DG (1998) Natural pathogens of laboratory mice, rats, and rabbits and their effects on research. Clin Microbiol Rev 11(2): 231–266
2. Cotter JD, Storfer A, Page RB, Beachy CK, Voss SR (2008) Transcriptional response of Mexican axolotls to Ambystoma tigrinum virus (ATV) infection. BMC Genomics 9: 493. https://doi.org/10.1186/1471-2164-9-493
3. Koniski A, Cohen N (1994) Mitogen-activated axolotl (Ambystoma mexicanum) splenocytes produce a cytokine that promotes growth of homologous lymphoblasts. Dev Comp Immunol 18(3):239–250. https://doi.org/10.1016/0145-305x(94)90016-7
4. McDonald CA, Longo AV, Lips KR, Zamudio KR (2020) Incapacitating effects of fungal coinfection in a novel pathogen system. Mol Ecol 29(17):3173–3186. https://doi.org/10.1111/mec.15452

5. Longo AV, Fleischer RC, Lips KR (2019) Double trouble: co-infections of chytrid fungi will severely impact widely distributed newts. Biol Invasions 21(6):2233–2245
6. Council NR (2010) Guide for the care and use of laboratory animals. National Academies Press, Washington, DC
7. Mähler M, Berard M, Feinstein R, Gallagher A, Illgen-Wilcke B, Pritchett-Corning K, Raspa M (2014) FELASA recommendations for the health monitoring of mouse, rat, hamster, guinea pig and rabbit colonies in breeding and experimental units. Lab Anim 48(3):178–192
8. Nicklas W (2008) International harmonization of health monitoring. ILAR J 49(3): 338–346
9. Ujszegi J, Molnár K, Hettyey A (2020) How to disinfect anuran eggs? Sensitivity of anuran embryos to chemicals widely used for the disinfection of larval and post-metamorphic amphibians. J Appl Toxicol 41:387–398
10. Reavill DR (2001) Amphibian skin diseases. Vet Clin North Am Exot Anim Pract 4(2): 413–440
11. Mitchell MA (2007) Parasites of amphibians. In: Flynn's parasites of laboratory animals, 2nd edn. Black-well Publishers Ltd., pp 117–175
12. Mutschmann F (2015) Parasite infestation in the axolotl (Ambystoma mexicanum)-recognition and therapy. Kleintierpraxis 60(9):461–472
13. Hallinger MJ, Taubert A, Hermosilla C (2020) Endoparasites infecting exotic captive amphibian pet and zoo animals (Anura, Caudata) in Germany. Parasitol Res 119(11): 3659–3673. https://doi.org/10.1007/s00436-020-06876-0
14. Klinger RE, Floyd RF (1998) Introduction to freshwater fish parasites. University of Florida Cooperative Extension Service, Institute of Food and Agriculture Sciences, EDIS
15. Kent ML, Fournie JW (2007) Parasites of fishes. In: Flynn's parasites of laboratory animals, 2nd edn. Black-well Publishers Ltd., pp 69–116
16. Pouder DB, Curtis EW, Yanong RP (2005) Common freshwater fish parasites pictorial guide: sessile ciliates. EDIS 2005 (9)
17. Pouder DB, Curtis EW, Yanong RP (2005) Common freshwater fish parasites pictorial guide: motile ciliates. EDIS 2005 (9)
18. Pouder DB, Curtis EW, Yanong RP (2005) Common freshwater fish parasites pictorial guide: nematodes. EDIS 2005 (9)
19. Pouder DB, Curtis EW, Yanong RP (2005) Common freshwater fish parasites pictorial guide: acanthocephalans, cestodes, leeches, & pentastomes. EDIS 2005 (9)
20. Pouder DB, Curtis EW, Yanong RP (2005) Common freshwater fish parasites pictorial guide: flagellates
21. Pouder DB, Curtis EW, Yanong RP (2005) Common freshwater fish parasites pictorial guide: digenean trematodes. EDIS 2005 (9)
22. Roberts HE, Palmeiro B, Weber ES III (2009) Bacterial and parasitic diseases of pet fish. Vet Clin North Am Exot Anim Pract 12(3):609–638
23. Grogan LF, Robert J, Berger L, Skerratt LF, Scheele BC, Castley JG, Newell DA, McCallum HI (2018) Review of the amphibian immune response to chytridiomycosis, and future directions. Front Immunol 9:2536
24. Poynton S, Whitaker B (2001) Protozoa and metazoa infecting amphibians. In: Amphibian medicine and captive husbandry. Krieger Publishing Company, Malabar, pp 193–221
25. Tinsley R (1995) Parasitic disease in amphibians: control by the regulation of worm burdens. Parasitology 111(S1):S153–S178
26. Shek WR, Smith AL, Pritchett-Corning KR (2015) Microbiological quality control for laboratory rodents and lagomorphs. In: Laboratory animal medicine. Elsevier, Amsterdam, pp 463–510
27. Philips BH, Crim MJ, Hankenson FC, Steffen EK, Klein PS, Brice AK, Carty AJ (2015) Evaluation of presurgical skin preparation agents in African clawed frogs (Xenopus laevis). J Am Assoc Lab Anim Sci 54(6):788–798
28. Buller NB (2004) Bacteria from fish and other aquatic animals: a practical identification manual. Cabi Publishing, Wallingford
29. Biswas S, Rolain JM (2013) Use of MALDI-TOF mass spectrometry for identification of bacteria that are difficult to culture. J Microbiol Methods 92(1):14–24. https://doi.org/10.1016/j.mimet.2012.10.014
30. Assis GBN, Pereira FL, Zegarra AU, Tavares GC, Leal CA, Figueiredo HCP (2017) Use of MALDI-TOF mass spectrometry for the fast identification of gram-positive fish pathogens. Front Microbiol 8:1492. https://doi.org/10.3389/fmicb.2017.01492
31. Bohme K, Fernandez-No IC, Barros-Velazquez J, Gallardo JM, Calo-Mata P, Canas B (2010) Species differentiation of seafood spoilage and pathogenic gram-negative bacteria by MALDI-TOF mass fingerprinting. J Proteome Res 9(6):3169–3183. https://doi.org/10.1021/pr100047q

32. Bohme K, Fernandez-No IC, Barros-Velazquez J, Gallardo JM, Canas B, Calo-Mata P (2011) Rapid species identification of seafood spoilage and pathogenic Gram-positive bacteria by MALDI-TOF mass fingerprinting. Electrophoresis 32(21): 2951–2965. https://doi.org/10.1002/elps.201100217
33. Bohme K, Fernandez-No IC, Pazos M, Gallardo JM, Barros-Velazquez J, Canas B, Calo-Mata P (2013) Identification and classification of seafood-borne pathogenic and spoilage bacteria: 16S rRNA sequencing versus MALDI-TOF MS fingerprinting. Electrophoresis 34(6):877–887. https://doi.org/10.1002/elps.201200532
34. Kralik P, Ricchi M (2017) A basic guide to real time PCR in microbial diagnostics: definitions, parameters, and everything. Front Microbiol 8:108
35. Compton SR (2020) PCR and RT-PCR in the diagnosis of laboratory animal infections and in health monitoring. J Am Assoc Lab Anim Sci 59(5):458–468
36. Li H, Bai R, Zhao Z, Tao L, Ma M, Ji Z, Jian M, Ding Z, Dai X, Bao F (2018) Application of droplet digital PCR to detect the pathogens of infectious diseases. Biosci Rep 38(6):1–8
37. Wong YP, Othman S, Lau YL, Radu S, Chee HY (2018) Loop-mediated isothermal amplification (LAMP): a versatile technique for detection of micro-organisms. J Appl Microbiol 124(3):626–643
38. Chertow DS (2018) Next-generation diagnostics with CRISPR. Science 360(6387): 381–382
39. Chiu C (2018) Cutting-edge infectious disease diagnostics with CRISPR. Cell Host Microbe 23(6):702–704
40. Crim MJ, Lawrence C, Livingston RS, Rakitin A, Hurley SJ, Riley LK (2017) Comparison of antemortem and environmental samples for zebrafish health monitoring and quarantine. J Am Assoc Lab Anim Sci 56(4): 412–424
41. Bauer BA, Besch-Williford C, Livingston RS, Crim MJ, Riley LK, Myles MH (2016) Influence of rack design and disease prevalence on detection of rodent pathogens in exhaust debris samples from individually ventilated caging systems. J Am Assoc Lab Anim Sci 55(6):782–788
42. Feldman SH, Ramirez MP (2014) Molecular phylogeny of Pseudocapillaroides xenopi (Moravec et Cosgrov 1982) and development of a quantitative PCR assay for its detection in aquarium sediment. J Am Assoc Lab Anim Sci 53(6):668–674
43. Burke RL, Whitehouse CA, Taylor JK, Selby EB (2009) Epidemiology of invasive Klebsiella pneumoniae with hypermucoviscosity phenotype in a research colony of nonhuman primates. Comp Med 59(6):589–597
44. Miller M, Sabrautzki S, Beyerlein A, Brielmeier M (2019) Combining fish and environmental PCR for diagnostics of diseased laboratory zebrafish in recirculating systems. PLoS One 14(9):e0222360. https://doi.org/10.1371/journal.pone.0222360
45. Miller M, Brielmeier M (2018) Environmental samples make soiled bedding sentinels dispensable for hygienic monitoring of IVC-reared mouse colonies. Lab Anim 52(3):233–239. https://doi.org/10.1177/0023677217739329
46. Heil N (2009) National wild fish health survey—laboratory procedures manual. US Fish and Wildlife Service, Warm Springs
47. Phosphate-buffered saline (PBS) (2006) Cold Spring Harb Protoc. https://doi.org/10.1101/pdb.rec8247
48. Underwood W, Anthony R (2020) AVMA guidelines for the euthanasia of animals: 2020 edition. American Veterinary Medical Association
49. Sheehan DC, Hrapchak BB (1980) Theory and practice of histotechnology. Battelle, Columbus, pp 190–192
50. Heyderman E (1992) Histotechnology. A self-instructional text. Histopathology 20(1): 91–91
51. Worthylake KM, Hovingh P (1989) Mass mortality of salamanders (Ambystoma tigrinum) by bacteria (Acinetobacter) in an oligotrophic seepage mountain lake. Great Basin Nat 49:364–372
52. Boyer CI Jr, Blackler K, Delanney LE (1971) Aeromonas hydrophila infection in the Mexican axoloti, Siredon mexicanum. Lab Anim Sci 21(3):372–375
53. Carey C, Bryant CJ (1995) Possible interrelations among environmental toxicants, amphibian development, and decline of amphibian populations. Environ Health Perspect 103(suppl 4):13–17
54. Taylor S, Green D, Wright K, Whitaker B (2001) Bacterial diseases. In: Amphibian medicine and captive husbandry. Krieger Publishing Company, Malabar
55. Raffel TR, Bommarito T, Barry DS, Witiak SM, Shackelton LA (2008) Widespread infection of the Eastern red-spotted newt (Notophthalmus viridescens) by a new species

of Amphibiocystidium, a genus of fungus-like mesomycetozoan parasites not previously reported in North America. Parasitology 135(2):203–215. https://doi.org/10.1017/S0031182007003708

56. Del Valle JM, Eisthen HL (2019) Treatment of chytridiomycosis in laboratory axolotls (Ambystoma mexicanum) and rough-skinned newts (Taricha granulosa). Comp Med 69(3):204–211. https://doi.org/10.30802/AALAS-CM-18-000090
57. Michaels CJ, Rendle M, Gibault C, Lopez J, Garcia G, Perkins MW, Cameron S, Tapley B (2018) Batrachochytrium dendrobatidis infection and treatment in the salamanders Ambystoma andersoni, A. dumerilii and A. mexicanum. Herpetol J 28(2):87–92
58. Duffus AL, Cunningham AA (2010) Major disease threats to European amphibians. Herpetol J 20(3):117–127
59. Frias-Alvarez P, Vredenburg VT, Familiar-Lopez M, Longcore JE, Gonzalez-Bernal E, Santos-Barrera G, Zambrano L, Parra-Olea G (2008) Chytridiomycosis survey in wild and captive mexican amphibians. EcoHealth 5(1):18–26. https://doi.org/10.1007/s10393-008-0155-3
60. Davidson EW, Parris M, Collins JP, Longcore JE, Pessier AP, Brunner J (2003) Pathogenicity and transmission of chytridiomycosis in tiger salamanders (Ambystoma tigrinum). Copeia 3:601–607
61. Hidalgo-Vila J, Díaz-Paniagua C, Marchand MA, Cunningham AA (2012) Batrachochytrium dendrobatidis infection of amphibians in the Doñana National Park, Spain. Dis Aquat Org 98(2):113–119
62. Kumar R, Malagon DA, Carter ED, Miller DL, Bohanon ML, Cusaac JPW, Peterson AC, Gray MJ (2020) Experimental methodologies can affect pathogenicity of Batrachochytrium salamandrivorans infections. PLoS One 15(9):e0235370. https://doi.org/10.1371/journal.pone.0235370
63. Martel A, Spitzen-van der Sluijs A, Blooi M, Bert W, Ducatelle R, Fisher MC, Woeltjes A, Bosman W, Chiers K, Bossuyt F, Pasmans F (2013) Batrachochytrium salamandrivorans sp.nov. causes lethal chytridiomycosis in amphibians. Proc Natl Acad Sci U S A 110(38):15325–15329. https://doi.org/10.1073/pnas.1307356110
64. Martel A, Blooi M, Adriaensen C, Van Rooij P, Beukema W, Fisher MC, Farrer RA, Schmidt BR, Tobler U, Goka K, Lips KR, Muletz C, Zamudio KR, Bosch J, Lotters S, Wombwell E, Garner TW, Cunningham AA, Spitzen-van der Sluijs A, Salvidio S, Ducatelle R, Nishikawa K, Nguyen TT, Kolby JE, Van Bocxlaer I, Bossuyt F, Pasmans F (2014) Wildlife disease. Recent introduction of a chytrid fungus endangers Western Palearctic salamanders. Science 346(6209):630–631. https://doi.org/10.1126/science.1258268
65. Sabino-Pinto J, Veith M, Vences M, Steinfartz S (2018) Asymptomatic infection of the fungal pathogen Batrachochytrium salamandrivorans in captivity. Sci Rep 8(1):11767. https://doi.org/10.1038/s41598-018-30240-z
66. Baker BB, Meyer DN, Llaniguez JT, Rafique SE, Cotroneo TM, Hish GA, Baker TR (2019) Management of multiple protozoan ectoparasites in a research colony of axolotls (Ambystoma mexicanum). J Am Assoc Lab Anim Sci 58(4):479–484. https://doi.org/10.30802/AALAS-JAALAS-18-000111
67. Borland S (2000) Practical axolotl. Axolotl Newsl 28:17–22
68. Farkas JE, Monaghan JR (2015) Housing and maintenance of Ambystoma mexicanum, the Mexican axolotl. Methods Mol Biol 1290:27–46. https://doi.org/10.1007/978-1-4939-2495-0_3
69. Schaefer DA, Eisthen HL (1990) Treatment of columnaris disease in aquatic salamanders. Axolotl Newsl 19:31–32
70. Björklund N, Duhon S (1999) The Mexican axolotl as a pet and a laboratory animal. Biology, husbandry and health care of reptiles and amphibians. Tropical Fish Hobbyist, Jersey City
71. Rankin JS (1937) An ecological study of parasites of some North Carolina salamanders. Ecol Monogr 7(2):169–269
72. Duhon ST (1989) Diseases of axolotls. In: Developmental biology of the axolotl. Oxford University Press, New York, pp 264–269
73. Herman RL (1984) Ichthyophonus-like infection in newts (Notophthalmus viridescens Rafinesque). J Wildl Dis 20(1):55–56. https://doi.org/10.7589/0090-3558-20.1.55
74. Mikaelian I, Ouellet M, Pauli B, Rodrigue J, Harshbarger JC, Green DM (2000) Ichthyophonus-like infection in wild amphibians from Quebec, Canada. Dis Aquat Org 40(3):195–201. https://doi.org/10.3354/dao040195
75. Raffel TR, Dillard JR, Hudson PJ (2006) Field evidence for leech-borne transmission of amphibian Ichthyophonus sp. J Parasitol 92(6):1256–1264. https://doi.org/10.1645/GE-808R1.1

76. Ware JL, Viverette C, Kleopfer JD, Pletcher L, Massey D, Wright A (2008) Infection of spotted salamanders (Ambystoma maculatum) with Ichthyophonus-like organisms in Virginia. J Wildl Dis 44(1):174–176. https://doi.org/10.7589/0090-3558-44.1.174
77. Black H, Rush-Munro FM, Woods G (1971) Mycobacterium marinum infections acquired from tropical fish tanks. Australas J Dermatol 12(3):155–164. https://doi.org/10.1111/j.1440-0960.1971.tb00004.x
78. Chemlal K, De Ridder K, Fonteyne P-A, Meyers WM, Swings J, Portaels F (2001) The use of IS2404 restriction fragment length polymorphisms suggests the diversity of Mycobacterium ulcerans from different geographical areas. Am J Trop Med Hyg 64(5): 270–273
79. Clark HF, Shepard CC (1963) Effect of environmental temperatures on infection with Mycobacterium marinum (Balnei) of mice and a number of poikilothermic species. J Bacteriol 86:1057–1069. https://doi.org/10.1128/JB.86.5.1057-1069.1963
80. Reavill DR, Schmidt RE (2012) Mycobacterial lesions in fish, amphibians, reptiles, rodents, lagomorphs, and ferrets with reference to animal models. Vet Clin North Am Exot Anim Pract 15(1):25–40, v. https://doi.org/10.1016/j.cvex.2011.10.001
81. Wright KM, Whitaker BR (2001) Amphibian medicine and captive husbandry. Krieger Publishing Company, Malabar
82. DeLanney LE, Collins NH, Cohen N, Reid R (1975) Transplantation immunogenetics and MLC reactivities of partially inbred strains of salamanders (A. mexicanum): preliminary studies. In: Immunologic phylogeny. Springer, Berlin, pp 315–324
83. Reavill DR (2001) Amphibian skin diseases. Vet Clin North Am Exot Anim Pract 4(2): 413–440, vi. https://doi.org/10.1016/s1094-9194(17)30038-5
84. Reichenbach-Klinke H, Elkan E (2013) The principal diseases of lower vertebrates. Elsevier
85. Bollinger TK, Mao J, Schock D, Brigham RM, Chinchar VG (1999) Pathology, isolation, and preliminary molecular characterization of a novel iridovirus from tiger salamanders in Saskatchewan. J Wildl Dis 35(3):413–429. https://doi.org/10.7589/0090-3558-35.3.413
86. Collins JP, Brunner JL, Jancovich JK, Schock DM (2004) A model host-pathogen system for studying infectious disease dynamics in amphibians: tiger salamanders (Ambystoma tigrinum) and Ambystoma tigrinum virus. Herpetol J 14:195–200
87. Donnelly TM, Davidson EW, Jancovich JK, Borland S, Newberry M, Gresens J (2003) What's your diagnosis? Ranavirus infection. Lab Anim 32(3):23–25
88. Flechoso MF, Alarcos G, Jara R (2019) Primer caso documentado de ranavirosis en Castilla y León en una población de gallipato, Pleurodeles waltl Michahelles, 1830. Bol Asoc Herpetol Esp 30(1):48–52
89. Green DE, Converse KA, Schrader AK (2002) Epizootiology of sixty-four amphibian morbidity and mortality events in the USA, 1996-2001. Ann N Y Acad Sci 969(1): 323–339
90. Chambers DL, Hulse AC (2006) Salmonella serovars in the herpetofauna of Indiana County, Pennsylvania. Appl Environ Microbiol 72(5):3771–3773
91. Loh R (2015) Common disease conditions in axolotls. In: 40th World Small Animal Veterinary Association Congress, Bangkok, Thailand, 15–18 May, 2015. Proceedings book. World Small Animal Veterinary Association, pp 640–642
92. Recuero E, Cruzado-Cortes J, Parra-Olea G, Zamudio KR (2010) Urban aquatic habitats and conservation of highly endangered species: the case of Ambystoma mexicanum (Caudata, Ambystomatidae). Ann Zool Fenn, JSTOR 47:223–238
93. Wardrip CL, Seps SL, Skrocki L, Nguyen L, Waterstrat PR (1999) Diagnostic exercise: fluffy, white, cotton candylike growth on the gills, fins, and skin of salamanders (Ambystoma tigrinum). J Am Assoc Lab Anim Sci 38(1):81–83
94. Pritchett KR, Sanders GE (2007) Epistylididae ectoparasites in a colony of African clawed frogs (Xenopus laevis). J Am Assoc Lab Anim Sci 46(2):86–91
95. Del-Pozo J, Girling S, Pizzi R, Mancinelli E, Else R (2011) Severe necrotizing myocarditis caused by Serratia marcescens infection in an axolotl (Ambystoma mexicanum). J Comp Pathol 144(4):334–338
96. Bryan LK, Baldwin CA, Gray MJ, Miller DL (2009) Efficacy of select disinfectants at inactivating Ranavirus. Dis Aquat Org 84(2): 89–94
97. Webb R, Mendez D, Berger L, Speare R (2007) Additional disinfectants effective against the amphibian chytrid fungus Batrachochytrium dendrobatidis. Dis Aquat Org 74(1):13–16

98. Johnson ML, Berger L, Philips L, Speare R (2003) Fungicidal effects of chemical disinfectants, UV light, desiccation and heat on the amphibian chytrid Batrachochytrium dendrobatidis. Dis Aquat Org 57(3):255–260
99. Gold KK, Reed PD, Bemis DA, Miller DL, Gray MJ, Souza MJ (2013) Efficacy of common disinfectants and terbinafine in inactivating the growth of Batrachochytrium dendrobatidis in culture. Dis Aquat Org 107(1):77–81
100. Van Rooij P, Pasmans F, Coen Y, Martel A (2017) Efficacy of chemical disinfectants for the containment of the salamander chytrid fungus Batrachochytrium salamandrivorans. PLoS One 12(10):e0186269
101. Miguel E, Grosbois V, Caron A, Pople D, Roche B, Donnelly CA (2020) A systemic approach to assess the potential and risks of wildlife culling for infectious disease control. Commun Biol 3(1):1–14
102. Farkas JE, Monaghan JR (2015) Housing and maintenance of Ambystoma mexicanum, the Mexican axolotl. In: Salamanders in regeneration research. Springer, New York, pp 27–46
103. Nugas CA (1996) Axolotl Larvae housing methods. Axolotl Newsl 25:6–9
104. Keller LR, Evans JH, Keller TC (1999) Experimental developmental biology: a laboratory manual. Academic Press, San Diego
105. Nichols DK (2000) Amphibian respiratory diseases. Vet Clin North Am Exot Anim Pract 3(2):551–554
106. Gresens J (2004) An introduction to the Mexican axolotl (Ambystoma mexicanum). Lab Anim 33(9):41–47
107. Blooi M, Pasmans F, Rouffaer L, Haesebrouck F, Vercammen F, Martel A (2015) Successful treatment of Batrachochytrium salamandrivorans infections in salamanders requires synergy between voriconazole, polymyxin E and temperature. Sci Rep 5: 11788. https://doi.org/10.1038/srep11788
108. Blooi M, Martel A, Haesebrouck F, Vercammen F, Bonte D, Pasmans F (2015) Treatment of urodelans based on temperature dependent infection dynamics of Batrachochytrium salamandrivorans. Sci Rep 5(1):1–4
109. Duhon S (1994) Short guide to axolotl husbandry. Indiana University Axolotl Colony Press. USA, 1994
110. Ginsburg MF, Twersky LH, Cohen WD (1987) Ambystoma embryo development after cold storage. Axolotl Newsl 16:3
111. Becker MH, Brucker RM, Schwantes CR, Harris RN, Minbiole KP (2009) The bacterially produced metabolite violacein is associated with survival of amphibians infected with a lethal fungus. Appl Environ Microbiol 75(21):6635–6638. https://doi.org/10.1128/AEM.01294-09
112. Harris RN, Lauer A, Simon MA, Banning JL, Alford RA (2009) Addition of antifungal skin bacteria to salamanders ameliorates the effects of chytridiomycosis. Dis Aquat Org 83(1): 11–16
113. Muletz-Wolz CR, Almario JG, Barnett SE, DiRenzo GV, Martel A, Pasmans F, Zamudio KR, Toledo LF, Lips KR (2017) Inhibition of fungal pathogens across genotypes and temperatures by amphibian skin bacteria. Front Microbiol 8:1551
114. Muletz Wolz CR, Yarwood SA, Campbell Grant EH, Fleischer RC, Lips KR (2018) Effects of host species and environment on the skin microbiome of Plethodontid salamanders. J Anim Ecol 87(2):341–353
115. Spitzen-van der Sluijs A, Stegen G, Bogaerts S, Canessa S, Steinfartz S, Janssen N, Bosman W, Pasmans F, Martel A (2018) Post-epizootic salamander persistence in a disease-free refugium suggests poor dispersal ability of Batrachochytrium salamandrivorans. Sci Rep 8(1):1–8
116. Smith SA (2007) Appendix: compendium of drugs and compounds used in amphibians. ILAR J 48(3):297–300. https://doi.org/10.1093/ilar.48.3.297
117. Inoue S, Singer M (1970) Lymphosarcomatous disease of the newt, Triturus pyrrhogaster. In: Comparative leukemia research 1969, vol 36. Karger Publishers, Basel, pp 640–641
118. Li WT, Chang HW, Pang VF, Wang FI, Liu CH, Chen TY, Guo JC, Wada T, Jeng CR (2017) Mycolactone-producing Mycobacterium marinum infection in captive Hong Kong warty newts and pathological evidence of impaired host immune function. Dis Aquat Org 123(3):239–249. https://doi.org/10.3354/dao03092
119. Fukano H, Yoshida M, Shimizu A, Iwao H, Katayama Y, Omatsu T, Mizutani T, Kurata O, Wada S, Hoshino Y (2018) Draft genome sequence of Mycobacterium montefiorense isolated from Japanese black salamander (Hynobius nigrescens). Genome Announc 6(21):e00448-18. https://doi.org/10.1128/genomeA.00448-18
120. Collymore C, Crim MJ, Lieggi C (2016) Recommendations for health monitoring and reporting for zebrafish research facilities. Zebrafish 13(Suppl 1):S138–S148. https://doi.org/10.1089/zeb.2015.1210

121. Cheatsazan H, de Almedia APLG, Russell AF, Bonneaud C (2013) Experimental evidence for a cost of resistance to the fungal pathogen, Batrachochytrium dendrobatidis, for the palmate newt, Lissotriton helveticus. BMC Ecol 13(1):1–11
122. Kent ML, Harper C, Wolf JC (2012) Documented and potential research impacts of subclinical diseases in zebrafish. ILAR J 53(2): 126–134. https://doi.org/10.1093/ilar.53.2.126
123. Crim MJ, Riley LK (2012) Viral diseases in zebrafish: what is known and unknown. ILAR J 53(2):135–143. https://doi.org/10.1093/ilar.53.2.135
124. Lyte M, Varcoe JJ, Bailey MT (1998) Anxiogenic effect of subclinical bacterial infection in mice in the absence of overt immune activation. Physiol Behav 65(1):63–68
125. Kent ML, Bishop-Stewart JK, Matthews JL, Spitsbergen JM (2002) Pseudocapillaria tomentosa, a nematode pathogen, and associated neoplasms of zebrafish (Danio rerio) kept in research colonies. Comp Med 52(4): 354–358
126. Tomás A, Fernandes LT, Sánchez A, Segalés J (2010) Time course differential gene expression in response to porcine circovirus type 2 subclinical infection. Vet Res 41(1):1–16
127. Yu F, Bruce L, Calder AG, Milne E, Coop R, Jackson F, Horgan G, MacRae J (2000) Subclinical infection with the nematode Trichostrongylus colubriformis increases gastrointestinal tract leucine metabolism and reduces availability of leucine for other tissues. J Anim Sci 78(2):380–390
128. Fosgate GT (2009) Practical sample size calculations for surveillance and diagnostic investigations. J Vet Diagn Investig 21(1): 3–14
129. Simon R, Schtll W (1984) Tables of sample size requirements for detection of fish infected by pathogens: three confidence levels for different infection prevalence and various population sizes. J Fish Dis 7(6):515–520
130. Dell RB, Holleran S, Ramakrishnan R (2002) Sample size determination. ILAR J 43(4): 207–213
131. Crim MJ (2020) Viral diseases. In: Cartner S, Eisen JS, Farmer S, Guillemin K, Kent ML, Sanders GE (eds) The zebrafish in biomedical research: biology, husbandry, diseases, and research applications, 1st edn. Academic Press, pp 509–526. https://doi.org/10.1016/B978-0-12-812431-4.00042-7
132. Armstrong JB, Malacinski GM (1989) Developmental biology of the axolotl. Oxford University Press, Oxford
133. Brothers A (1977) Instructions for the care and feeding of axolotls. Axolotl Newsl 3:9–16
134. Becker MH, Harris RN (2010) Cutaneous bacteria of the redback salamander prevent morbidity associated with a lethal disease. PLoS One 5(6):e10957. https://doi.org/10.1371/journal.pone.0010957

Chapter 4

Husbandry, Captive Breeding, and Field Survey of Chinese Giant Salamander (*Andrias davidianus*)

Wansheng Jiang, Haifeng Tian, and Lu Zhang

Abstract

The Chinese giant salamander (*Andrias davidianus*) is the largest extant amphibian species in the world with adults capable of reaching 2 m in length. Wild populations of *A. davidianus* have declined dramatically during the last century, making it also one of the top threatened species globally. Fortunately, aquaculture for this species developed in China during the 1970s has been extremely successful. Many relevant commercial products of *A. davidianus* have been produced in recent years on account of its nutritional and medicinal values. Balancing conservation and utilization will be key to the future destiny of *A. davidianus*. In this chapter, we describe detailed protocols for husbandry in indoor and outdoor facilities, captive breeding under natural-imitative conditions and using artificial insemination, and surveying and monitoring *A. davidianus* in the field. The protocols presented here aim to make the practices of *A. davidianus* operative and increase public awareness of this mystical and precious species.

Key words Cryptobranchidae, Conservation, Utilization, Reproduction behavior, Field monitoring, China

1 Introduction

Andrias davidianus (Fig. 1a), known colloquially as the Chinese giant salamander, is the largest extant amphibian species in the world, reaching a maximum length of 2 m and a weight of 50 kg [1]. *Andrias* belongs to the order Caudata (also called Urodela) and is a member of the Cryptobranchidae. Crowned as living fossils [2], the Cryptobranchidae, are widely recognized by three surviving species: two Asian species in the genus *Andrias*, *A. davidianus* and *A. japonica* (Japanese giant salamander), and one species in the genus *Cryptobranchus* in North America: *C. alleganiensis* (hellbender). Wild populations of all three giant salamanders have declined dramatically in recent decades due to human activities, making them global conservation icons for large, threatened amphibians and for the sustainable management of watersheds [3].

Ashley W. Seifert and Joshua D. Currie (eds.), *Salamanders: Methods and Protocols*,
Methods in Molecular Biology, vol. 2562, https://doi.org/10.1007/978-1-0716-2659-7_4,

Fig. 1 *A. davidianus* and its habitats. Pictures showing a living specimen (**a**) and two typical habitats: surrounding the exits of underground springs (**b**) and hiding under large boulders (**c**, **d**)

The *A. davidianus* usually lives in mountain creeks, streams, and rivers, especially water systems consisting of underground rivers and surface aquatic networks in limestone and karst areas (Fig. 1b–d). They have been recorded at moderate elevations between 300 m and 1500 m above sea level [1], with one enigmatic population also reported at 4200 m elevation in Qinghai Province [4]. The *A. davidianus* is nearly fully aquatic and generally nocturnal, hiding under large boulders to ambush its prey using smell and touch. It plays an important role in freshwater ecosystems as a top-level predator, where it preys upon crabs, fish, insects, frogs, water birds, and small mammals such as shrews [5]. It is also cannibalistic, consuming oocytes, eggs, and small-sized conspecifics [5].

The *A. davidianus* normally matures after 5 or 6 years when it measures 40–50 cm in length [6]. During the breeding season, females lay eggs in dens, and males guard the eggs until hatching [7]. Larval *A. davidianus* born in caves will flow out in late winter or early spring, where they will take up residence in the surrounding and proper aquatic refuges. Larval *A. davidianus* have short external gills until they reach 10–20 cm in length when their external gills drop off and the gill clefts close. Adult *A. davidianus* have small eyes without eyelids, four short and stout limbs with four fingers and five toes, paired small tubercles arranged in rows on heads and necks, and thick skin folds on their lateral sides with 12–15 costal grooves. The epidermal folds on their flanks are extensive, which is one of the diagnostic characteristics and thought to serve as a body length "gill" for oxygen exchange [3].

The *A. davidianus* once occurred widely in the three major river systems in China, including the Yellow, Yangtze, and Pearl rivers from north to south. However, their historical range contracted greatly during the last century with the population declining dramatically [8], mainly due to infectious disease, habitat loss, environmental pollution, and overharvesting [1]. The species has been classified as a class II state protected species in China since 1988, listed as Critically Endangered on the International Union for the Conservation of Nature and Natural Resources (IUCN) Red List, and recorded in Appendix I of the Convention on International Trade of Endangered Species (CITES). To protect the wild populations of *A. davidianus*, China has already set up 53 nature reserves to promote in-situ conservation [9]. Meanwhile, reinforcement and reintroduction have also been applied widely in order to augment the wild populations [10]. Recent genetic studies, however, have challenged the current conservation strategy that treated the possible species complex as a single species with the identification of several lineages originating from historically geographic-isolated populations [11], representing cryptic species [12], or proposed as newly recognized species [13, 14].

Historically, the *A. davidianus* was used as a traditional food in China for thousands of years [15], which may be attributed to its large body size with relatively more edible elements, as well as its unique nutritional and medicinal values as recorded in many ancient books. The meat of *A. davidianus* has traditionally been prepared as a delicacy and with a high-nutritional value known as "live ginseng in the water" [16]. There are a handful of studies about the nutritional composition of *A. davidianus*, which indicate frequently that *A. davidianus* is rich in high-quality proteins [16], an ideal proportion of amino acids and fatty acids [17], and desired mineral elements and vitamins [18]. Recent studies have shown that the mucus, skin, meat, and bone of *A. davidianus* contain many kinds of bioactive substances which have been suggested to possess various medicinal activities including antiaging, anti-fatigue, antitumor, antibacterial, and therapeutics for burns (Fig. 2, cited from [15]).

Because of the high conservation, nutritional, and medicinal values, captive breeding of *A. davidianus* was developed in the 1970s and has been successful. It has provided an essential way to augment wild populations through reinforcement and reintroduction, and brought a new field of product engineering through farmed stock development. According to Chinese law, although wild *A. davidianus* is prohibited from being sold, second and beyond generations of captive-bred *A. davidianus* can be utilized

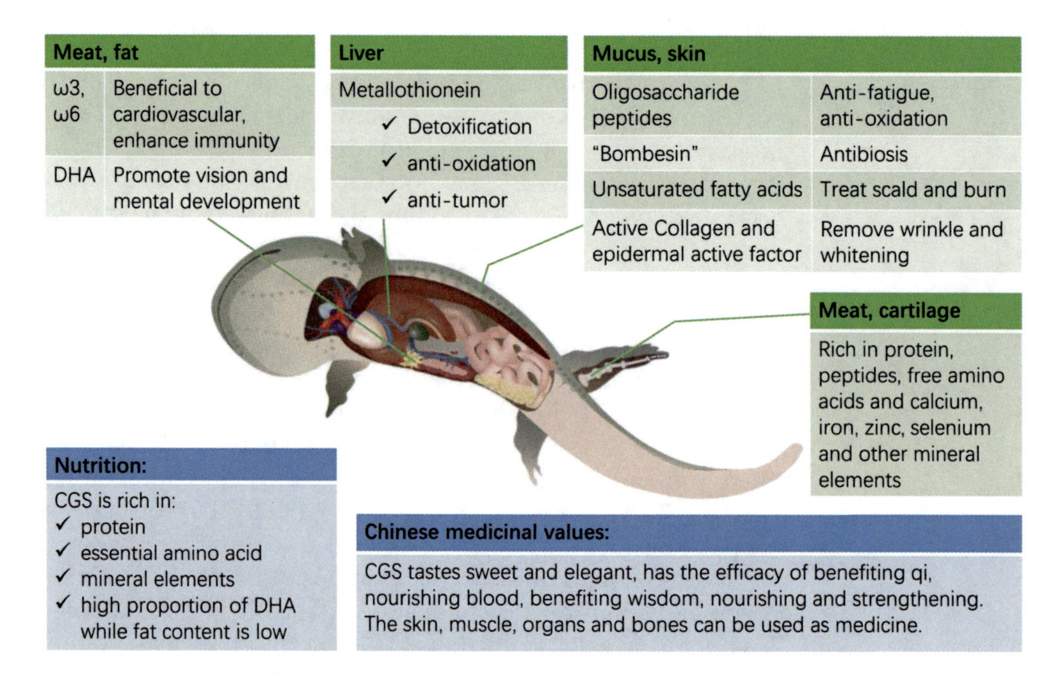

Fig. 2 Bioactive substances of captive-bred *A. davidianus* (cited from [15])

as resources for the food and pharmaceutical industries [19]. To date, the *A. davidianus* has been cultivated for more than 40 years and become an important aquaculture species in many provinces of China. However, the recovery of wild populations is still moving slowly even with the help of natural reserves and artificial reintroduction. Additionally, intensive *A. davidianus* farming has increased the risk of prevalence of many infectious diseases especially viral infections. Maintaining a balance between conservation and utilization will be key to the future of *A. davidianus* [20].

The present chapter describes the protocols for husbandry of *A. davidianus* in indoor and outdoor facilities, captive breeding under natural-imitative conditions and using artificial insemination, and surveying and monitoring *A. davidianus* in its natural habitats. We aim to facilitate the husbandry and breeding of *A. davidianus*, and increase public awareness of this mystical and precious species.

2 Materials

2.1 Setup and Equipment for Husbandry

The environmental conditions for the husbandry of *A. davidianus* are diverse; however, they can be classified into two basic types: indoor and outdoor. Indoor husbandry can better control water intake, temperature, dissolved oxygen, stocking density, etc., and is usually adopted by large-scale farms. Using natural caves or artificial tunnels can also be considered indoor husbandry. Outdoor husbandry is specifically constructed in the open area and usually

Fig. 3 Indoor and outdoor husbandry facilities. Pictures showing indoor three-layer tanks in an artificial tunnel (**a**) and a regular farm (**b**), juvenile *A. davidianus* reared in an indoor tank (**c**), an aeroview of a farm with both indoor and outdoor facilities (**d**), an outdoor husbandry system (**e**), artificial channel with dens at both sides (**f**), and an adult *A. davidianus* in a den entrance (**g**)

adopted by small-scale farms. It can facilitate captive breeding under natural-imitative conditions. However, some farms build both indoor and outdoor husbandry systems to maximize their advantages (Fig. 3). The following are key requirements for successful husbandry:

1. Water source: natural spring/stream (preferential) or tap water (not recommended, *see* **Note 1**).
2. Impounding reservoir: good for adjusting the water storage capacity (optional but recommended).
3. Filter pool: good for filter and sedimentation (optional but recommended).
4. Disinfection room and tank: for disinfecting *A. davidianus* and tools (usually at the entrance of the farm).
5. Water pipeline/channel: iron or plastic pipe (indoor); artificial channel or ditch (outdoor).

6. Container/tank: different sized cement tanks (indoor) with inner-wall tiles (*see* **Note 2**); irregular stone/gravel cavity (outdoor) with cover plate for shading.
7. Air conditioner: only for extreme seasons at indoor facilities (optional).
8. Natural or artificial plants: for shading and decoration at outdoor facilities (optional).

2.2 Maintenance and Food Supply

1. Prefilters: for poor water conditions in indoor facilities (optional).
2. PVC pipe and faucet: for water intake and flow adjustment.
3. Nets, brushes, plastic basket, and bucket.
4. Waterproof boot and low-light torch.
5. Live or frozen larval Chironomidae (red worm) and Tubificidae (water earthworm): the regular food for larval *A. davidianus.*
6. Live fish, usually small cyprinids: the regular food for juvenile and adult *A. davidianus.*
7. Frozen fish, usually small marine fish through cold-chain transportation: the artificially purchased food for juvenile and adult *A. davidianus.*
8. Live or frozen shrimp, crab, or other kinds of animals: supplementary food for adult *A. davidianus.*
9. Bubble chamber, packing materials, electric digital scale, disposable ice packs: for shipping *A. davidianus.*
10. Refrigerators: for storing food, reagents, etc.

2.3 Supplies, Reagents, and Solutions for Disease Treatment and Artificial Insemination

1. NaCl (sodium chloride, salt solution): for disinfection of food and tank.
2. ClO_2 (chlorine dioxide): for disinfection of tank (especially for the new-built tanks).
3. 75% alcohol: for disinfection of injured or infected *A. davidianus.*
4. Tincture of iodine (with available iodine content ~2%, usually purchased): for disinfection of injured or infected *A. davidianus.*
5. KMnO solution (dissolve KMnO (potassium permanganate) into water under room temperature): for disinfection of infected *A. davidianus* (~0.1% solution for dipping bath and ~1% solution for scrubbing the wound).
6. Norfloxacin, Gentamicin (or other antibiotics): for treatment of bacterial or fungal diseases.

7. Ribavirin (or other antiviral agents): for treatment of viral diseases.
8. Multivitamins: as food additives.
9. hCG (human chorionic gonadotropin): for artificial inducement of spawning.
10. LRH-A (luteinizing-releasing hormone analog): for artificial inducement of spawning.

2.4 Tools and Equipment for Field Survey and Monitoring

1. Wader, head torch, and hand torch: for surveying *A. davidianus* in the field (high light is optimal).
2. Crab traps: for trapping *A. davidianus* with minimal physical injury (collapsible traps are optimal).
3. Baits: live fish, pork, or chicken liver (bait with heavy smell is optimal).
4. Container: usually a plastic box for holding the *A. davidianus* in the field.
5. Portable water quality analyzer (e.g., HACH, Colorado, USA).
6. Sample measurement tools: tapeline, tailor's rule, and electric digital scale.
7. Sample collection tools: latex gloves, face mask, gauze, swab, scissor, tweezer, and tube (all sterilized before use).
8. Tools and supplies for surgery and transmitter implantation: tricaine methanesulfonate (MS-222) buffered with sodium bicarbonate (at a ratio of 1:1 by mass in grams, for anesthetization), surgical suit (gloves, blades, scissors, tweezers, suture materials, enrofloxacin, etc.), radio transmitters (e.g., F1035, Advanced Telemetry Systems, Inc., Isanti, USA).
9. Monitoring equipment: radio receiver with Yagi antenna (e.g., R410, Advanced Telemetry Systems, Inc., Isanti, Minnesota, USA), hand-held global positioning system (GPS) units (e.g., GPSMAP 60CSx, Garmin, Ltd., New Taipei City, Taiwan), and underwater inspection camera (e.g., M12, Milwaukee Electric Tool, WI, USA).

3 Methods

3.1 Daily Rearing, Feeding, and Cleaning

The *A. davidianus* is nearly fully aquatic and remains in water all the year, so maintaining a stable water supply system is the top task of proper husbandry. The daily rearing, feeding, and cleaning can be managed as follows:

1. Patrol every day in the morning to check the status of *A. davidianus*, taking note of possible water supply fluctuations, temperature conditions, physical injury, and disease symptoms.

2. Ensure indoor space dimly lit (<300 lx) but well ventilated with windows or exhaust fans.
3. Ensure proper temperature (10–25 °C) conditions. Keep the indoor temperature below 25 °C in summer (using an air conditioner or water cooling in extremely hot weather). Avoid freezing outdoor containers in extremely cold winter (cover them with dry grass to keep warm).
4. Select appropriate containers to rear different kinds of *A. davidianus*, according to the number and size of animals to be housed (*see* **Note 3**), and prevent *A. davidianus* from escaping.
5. Feed *A. davidianus* with appropriate food at specific intervals, according to its developmental stages (*see* **Note 4**).
6. Disinfect food before feeding by rinsing in 3–5% salt solution for 30–60 min or with 20 mg/kg ClO_2 solution for 10–20 min, then flushing continuously with clean water for 3–5 min.
7. Clean the debris of food and dead individuals timely.
8. Brush tanks once a month using a 3% salt solution or 10 ppm ClO_2. Rinse new-built or reused tanks with the same solutions for more than a week, then brush and flush.
9. Ensure a healthy, controllable, manageable husbandry system from time to time, and record routine events every day.

3.2 Health Inspection and Disease Treatment

As stocking density increases, disease is the top threat to farmed *A. davidianus*. Bacterial- and viral-borne diseases have increased along with the development of *A. davidianus* cultures. Specifically, virus-mediated disease has caused catastrophic losses to *A. davidianus* aquaculture in recent years [21]. Health inspection is a basic but critical step to detect disease at an early and potentially treatable stage. If diseases occur, proper treatments could save many individuals in most cases.

1. Keep *A. davidianus* well fed and in clean water for their entire life.
2. Disinfect transported animals with 10% salt solution for 10–15 min, or 0.01% KMnO solution for 30 min before settling them into new tanks.
3. Observe the status of *A. davidianus* and catch signs of physical injury or poor health early before the disease spreads (*see* **Note 5**). If a sick animal is detected (bleeding, ulceration of the skin, fungus, abnormal posture, lethargy, loss of righting reflex, etc.), isolate it from others.
4. Use separate gloves and sticks to handle sick animals. If possible, place and treat the sick animals in quarantine.

5. Treat physically injured animals with 75% alcohol or tincture of iodine and scrub the infected part 3–5 times a day.
6. Treat fungus-infected animals with 0.1% KMnO solution or tincture of iodine and scrub the infected part 3–5 times a day.
7. Treat parasite-infected animals with a high salt bath of 35 g/L for 5 min for three consecutive days.
8. To treat bacterial-infected animals (*see* **Note 6**): (1) use an antibiotics solution (e.g., 2 mg/L Norfloxacin) bath for 15–30 min daily for a week, (2) mix 30–100 mg Norfloxacin per kilogram animal weight with food daily for a week, or (3) administer intraperitoneal injections of 1–4 million international units Gentamicin solution per kilogram animal weight daily for a week. (1)–(3) are suitable for light, moderate, and heavy symptoms, respectively.
9. At present, there are no effective treatments for virus-infected animals (Fig. 4) (*see* **Note 7**). Ribavirin (or other antiviral agents) can be helpful but rarely presents a cure.
10. Follow the status of sick animals daily and check other animals in the tank in which the sick one was found. Stop treatment 2 days after the symptoms disappear.

3.3 Natural-Imitative Breeding and the Following Maintenance

Natural-imitative breeding requires outdoor husbandry facilities. Sexually matured *A. davidianus* kept in a natural-imitative environment will eventually produce spawns if both sexes are reared together. A series of breeding behaviors can be observed [7].

1. Construct and maintain a well-designed outdoor husbandry system that contains an artificial stream. In brief, (1) keep the stream water running slowly; (2) build several pairs of dens on both sides of the stream; (3) cover the dens with a plate and leave one small entrance toward the stream; (4) install a digital monitoring system in each den to observe breeding behaviors if possible.
2. Put sexually matured males and females in the channel streams at least 3 months (better for 1 year) before breeding season to acclimate.
3. The optimal number of animals should be ~70% of the number of dens and set the sex ratio to 1:1.
4. Feed these animals with good-quality food like live fish, and be sure to provide them adequate food to decrease their aggressive behaviors.
5. Observe the pre- and post-reproduction behaviors of *A. davidianus* (*see* **Note 8**) and look for eggs in the dens.

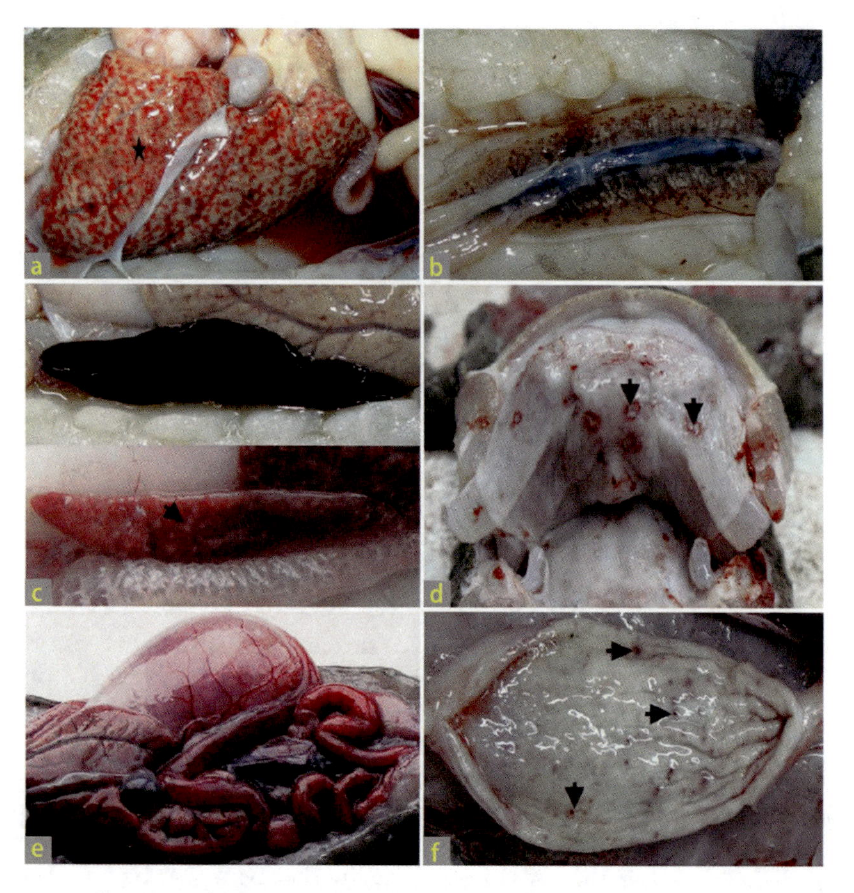

Fig. 4 Gross anatomic lesions of virus-infected *A. davidianus* (cited from [21]). Notes: (**a**) the liver was pale and swollen with multifocal hemorrhages (star), and bloody fluid (arrow) in the abdominal cavity; (**b**) the kidney was pale and swollen with petechial hemorrhages; (**c**) the spleen was swollen with congestion and focal necrosis (arrow); (**d**) the oral mucosa showing hemorrhage, erosion, and ulceration (arrows); (**e**) the gastrointestinal tract was congested and empty, and the gall bladder was enlarged; (**f**) the gastric mucosa showing hemorrhage, erosion, and ulceration (arrows)

6. Let the males care for the eggs until they hatch and free-swimming larvae flow out. Then collect all the larvae into a plastic container for rearing indoors (recommended, *see* Subheading 3.1 for other operations).
7. Or, collect fertilized eggs from dens. Transfer these eggs into an indoor hatching system for hatching.

3.4 Artificial Insemination and the Following Maintenance

Artificial insemination is usually conducted at an indoor husbandry facility. In general, it is easier to control the process compared to breeding under natural-imitative conditions, and more fertilized eggs can be generated (Fig. 5). However, the quality of the eggs and offspring might not be as good as those generated from natural-imitative breeding.

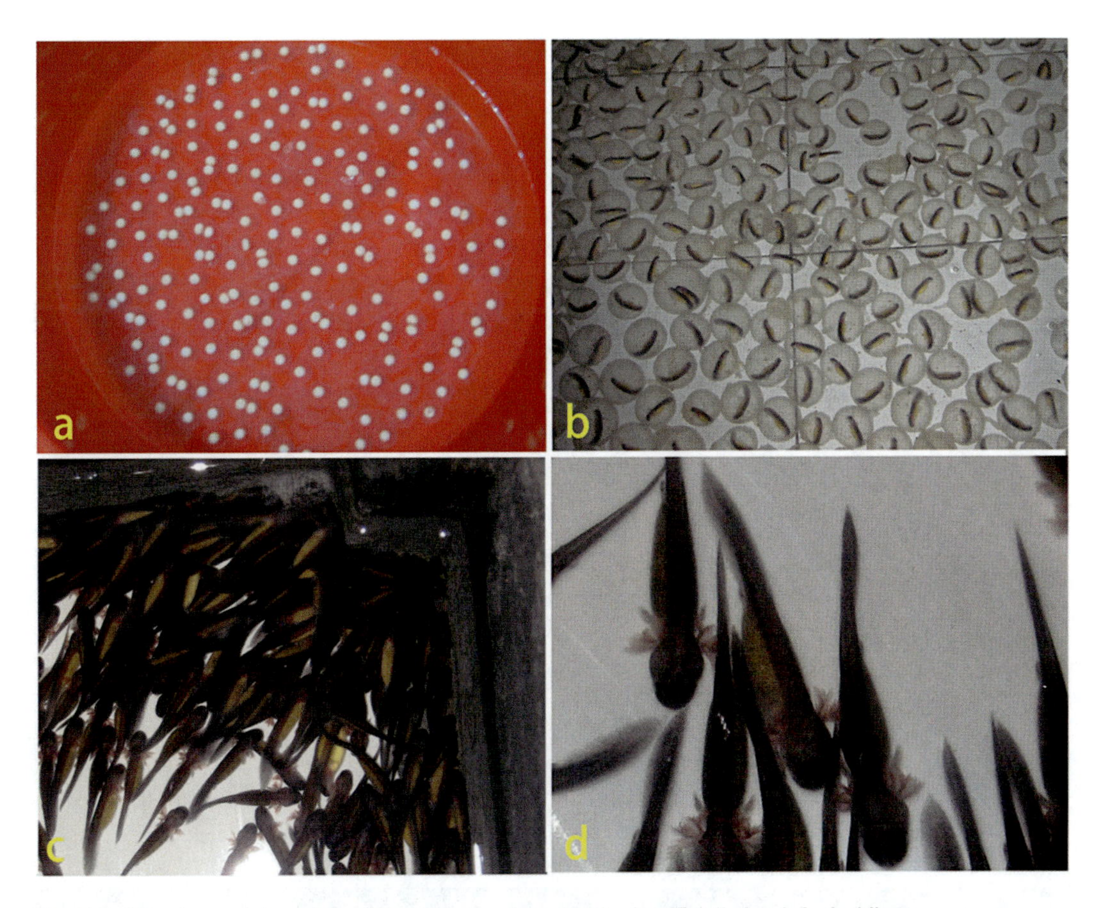

Fig. 5 Fertilized eggs (**a**), embryo (**b**) and larvae stage (**c**, **d**) of artificially bred *A. davidianus*

1. Raise progenitive animals with good-quality food like live fish for at least 2 months before the breeding season.
2. Select sexually mature *A. davidianus* that display breeding morphology and keep the males and females separated (*see* **Note 9**).
3. Observe the animals' pre-reproduction behaviors and check their breeding morphology (females have swollen abdomens; sperm can be squeezed out from males).
4. Give both the males and females a dose (<4 mL) of HCG (500–800 IU/kg) + LRH-A (5–10 mg/kg) by injecting it into the abdominal cavity from a point between two lateral costal grooves. Males could be injected 1–3 days earlier than females.
5. After ~80 h, sperm can be squeezed from males. After 96–120 h, females start to lay eggs.
6. Collect eggs and transfer them into a clean and dry plastic basin. Immediately, press gently on male's abdomen to collect sperm into a 2-mL tube containing normal saline (*see* **Note 10**).

7. Add the sperm solution over the eggs and mix them very gently with a small amount of water, and then hold for 2–3 min. Add water to fill half of the basin and stir slowly to let the eggs come in contact with the sperm and get fertilized. Then take the turbid water out and wash 3–4 times using clean water.
8. The eggs of *A. davidianus* will hydrate to ~15–20 mm diameter from the original spawned size of ~5–8 mm. Transfer the eggs into an indoor hatching system to hatch for 30–40 days.
9. After hatching, when larvae start swimming, monitor the development of the embryos daily. Feed them with live food (red worm or water earthworm) at appropriate timing.
10. Transfer the larvae to the containers/tanks (*see* Subheading 3.1 for other operations).

3.5 Field Survey, Capture, and Monitoring

According to Chinese law, the capture of wild *A. davidianus* is prohibited except for the purpose of scientific research, which still requires a local permit. The *A. davidianus* is generally nocturnal so collections are usually conducted at night. Here we present two simple and operable procedures for field survey based on our experiences, that is, using nocturnal encountering and trapping methods (Fig. 6). A simplified procedure of field monitoring through surgically implanted radio transmitters was also introduced here (Fig. 7). During field surveys, every participant should ensure safety and keep the risk to a minimum degree.

Fig. 6 Field survey and capture of *A. davidianus*. Pictures showing nocturnal encountering (**a**) and trapping practice (**b**), a wild *A. davidianus* captured by crab trap (**c**) and then released after measurements (**d**)

Fig. 7 Surgical implantation of radio transmitters (**a**, **b**) and field tracking using a three-element Yagi antenna and a handheld receiver after the animal is released (**c**, **d**)

1. Nocturnal encountering method: (1) wear a wader and a head torch and hold a high-light hand torch and a hand brail; (2) use a hand torch to search for animals along the river bank, focusing on suitable refuges; (3) once an animal is visualized, collect it immediately with fish-landing net if possible, and handle it carefully to avoid being bitten.
2. Trapping method: (1) around sunset, put crab traps into deep pools beside large rocks or other crevices that *A. davidianus* usually hide in; (2) use a combination of live fish and animal liver (heavily smelly) as bait; (3) check the traps before sunrise and retrieve them during daytime to avoid being accessed or stolen by locals; and (4) repeat (1)–(3) for at least 2 days or more depending on the purpose of survey.
3. Use a portable water quality analyzer to measure water temperature (WT), DO, and pH and record other habitat elements at each site.
4. Once an animal is collected, place it into a plastic box with some onsite water. Handle the animal with gloves to avoid transmission of any diseases.

5. Use tapeline or tailor's tape to measure the body length; use an electric digital scale to record the body weight; assign a discernable number to the specimen; and take digital photos of each individual.
6. Collect regular tissue samples (*see* **Note 11**): (1) swab the mucosal surface inside the animal's mouth to get buccal swab samples; (2) swab in the cloaca to get pathogen swab samples; (3) swab the limbs and tails to get skin swab samples; and (4) get other samples if necessary, which should be non-invasive or minimally invasive to minimize harm to the animal.
7. Release the *A. davidianus* onsite where it has been captured.
8. Surgical implantation of radio transmitters (*see* **Note 12**): (1) prepare a bath using 0.6 g/L MS-222 buffered with sodium bicarbonate to anesthetize the animal; (2) place the animal in dorsal recumbency of water containing 0.1 g/L MS-222 for the duration of the following surgical procedures; (3) make a paramedian incision about 2–3 cm in length using fine blades and scissors; (4) place the radio transmitter into the coelomic cavity carefully; (5) suture the body wall and skin with poliglecaprone or nylon; (6) treat the surgical animal with an intramuscular dose of enrofloxacin at 10 mg/kg postoperatively; and (7) allow up to 16 weeks for recovery before release the animal into the wild.
9. Release and track the animal using a three-element Yagi antenna and a handheld receiver.
10. Use standard datasheets to record information during each survey and keep the original datasheets and a digital copy of them.

4 Notes

1. In the natural habitats, *A. davidianus* relies on freshwater streams, so basically, they can be held in tap water conditions in the laboratory. However, tap water is usually chlorinated in cities, which could be harmful to the *A. davidianus*. If chlorinated tap water has to be used, it is suggested to be dechlorinated by aerating using fishing air pump (for several hours) or standing exposed to the air (for several days) to let the chlorine evaporate sufficiently.
2. Construct indoor tanks with three layers at most to maximize the use of space. For instance, use length × width × depth to 2 m × 1.2 m × 0.55 m, 1.2 m × 0.7 m × 0.4 m, and 0.6 m × 0.4 m × 0.3 m for the bottom-, medium-, and top-layer tanks, respectively.

3. The recommended containers for raising *A. davidianus* are small-sized tanks usually less than 2 m^2, because small tanks are easy to manage, and are beneficial to saving water, improving feed utilization rate, and speeding the product cycle. In general, the top-layer tanks are used for small-sized *A. davidianus* and the bottom-layer tanks are for large-sized individuals. The inner wall of the tank needs to be paved with tiles to decrease friction, and therefore to reduce the risk of disease.
4. Feed *A. davidianus* at dusk daily or once in several days according to temperature and efficiency of food intake. Feed larval *A. davidianus* with larval Chironomidae (red worm) or Tubificidae (water earthworm); feed juvenile *A. davidianus* with fish or other animal food with less hard elements like spines and bones; and feed adult *A. davidianus* with fish or other animal food of an appropriate size. Add multivitamins to the food twice a month. For any development stage, live food is generally better than frozen food.
5. *A. davidianus* is somewhat aggressive. They will bite each other especially when they are fighting for food, which will cause physical injuries (mostly on limbs). Signs indicating poor health include abnormalities of the skin (pigmentation abnormalities), behavior (slow or fast moving, floating), and body constitution (either bloating or extremely skinny).
6. Previous studies show that *Aeromonas hydrophila*, *Aeromonas sobria*, *Citrobacter freundii*, *Pseudomonas* spp., *Edwardsiella tarda*, *Cetobacterium somerae*, and *Hafnia alvei* are the main pathogens found on *A. davidianus* [22].
7. Viruses are the most serious diseases for farmed *A. davidianus*. To date, two nominally lethal pathogens have been reported, i.e., Chinese giant salamander ranavirus (CGSRV) and Chinese giant salamander iridovirus (CGSIV). Clinical signs of these virus infections are similar, including skin ulceration, anorexia, lethargy, occasionally edema, petechiae, erythema, toe necrosis, friable and gray–black liver, and friable lesions of the kidney and spleen. Morbidity can be higher than 95% in some affected areas [23].
8. Luo and colleagues studied reproductive behaviors of *A. davidianus* using a digital monitoring system [7]. They described a series of pre- and post-reproduction behaviors: (1) sand-pushing behavior is mainly carried out with the limbs, tail, head, and body of den-dominant males; (2) showering behavior includes rinsing the trunk, head, and tail; (3) courtship comprises a series of behaviors, including standing side-by-side, belly colliding, mounting, mouth-to-mouth posturing, chasing, inviting, cohabitating, and rolling over; and

(4) parental care includes tail fanning, agitation, shaking, and eating dead and unfertilized eggs.

9. In general, it is difficult to distinguish the genders of *A. davidianus*. For animals of the same age during non-breeding season, males can roughly be distinguished from females according to their relatively larger body size, wider mouth, and more humped head. However, during breeding season, the males can be distinguished from the females more distinctly based on their cloacal opening swelling with an oval-shaped cycle of white papillae (vs. cloacal opening sinking and smooth without any papillae).
10. Hold a matured male *A. davidianus* on a table with cotton gauze spread out. Turn over the animal to let the belly up, and touch the body gently to relax it. Squeeze the abdomen lightly to let the semen flow out, collect it into a 2-mL tube containing normal saline using a dropper. Usually, the sperm needs to be prepared in real time to avoid a decrease in sperm motility while stored too long.
11. Tapley and colleagues constructed a survey manual aiming at standardizing *A. davidianus* surveys in China [24]. In general, the methods they described are minimally invasive, logistically feasible, and robust. They wish these methods could be used by all researchers who are involved in *A. davidianus* studies, to facilitate comparable surveys. Please refer to this study if you are interested.
12. Marcec and colleagues provided a detailed procedure of surgical implantation of radio transmitters [25] and successfully tracked most of the animals in the wild for more than 1 year post-release [26]. They tested the concentration and dosage of the anesthetic procedure and discussed the materials used for suture, and the proper period of recovery. Please refer to these studies if you are interested.

Acknowledgments

We would like to thank Dr. Qinghua Luo and other members in our respective laboratories who helped develop the methods described in this chapter. Many appreciations to Dr. Xinhui Xing (Tsinghua University) and Dr. Yi Geng (Sichuan Agricultural University) for the permission of using the Figs. 2 and 4, respectively. This work was funded by the National Natural Science Foundation of China (32060128), the Talent Project of Hunan Provincial Science and Technology Department (2020RC3057), and Zhilan Foundation (2020040371B) to W.J.

References

1. Wang XM, Zhang KJ, Wang ZH, Ding YZ, Wu W, Huang S (2004) The decline of the Chinese giant salamander *Andrias davidianus* and implications for its conservation. Oryx 38: 197–202
2. Gao KQ, Shubin NH (2003) Earliest known crown-group salamanders. Nature 422: 424–428
3. Browne RK, Li H, Wang ZH, Hime P, McMillan A, Wu MY et al (2014) The giant salamanders (Cryptobranchidae): part A. palaeontology, phylogeny, genetics, and morphology. Amphib Reptile Conserv 5: 17–29
4. Pierson TW, Yan F, Wang YY, Papenfuss T (2014) A survey for the Chinese giant salamander (*Andrias davidianus*; Blanchard, 1871) in the Qinghai Province. Amphib Reptile Conserv 8:e74
5. Song M (1994) Food habit of giant salamander of China. Chin J Zool 29:38–41. (in Chinese)
6. Browne RK, Li H, Wang Z, Okada S, Hime P, McMillan A et al (2014) The giant salamanders (Cryptobranchidae): part B. Biogeography, ecology and reproduction. Amphib Reptile Conserv 5:30–50
7. Luo QH, Tong F, Song YJ, Wang H, Du ML, Ji HB (2018) Observation of the breeding behavior of the Chinese giant salamander (*Andrias davidianus*) using a digital monitoring system. Animals 8:1–13
8. Murphy RW, Fu JZ, Upton DE, De Lema T, Zhao EM (2000) Genetic variability among endangered Chinese giant salamanders, *Andrias davidianus*. Mol Ecol 9:1539–1547
9. Liang ZQ, Zhang SH, Wang CR, Wei QW, Wu YA (2013) Present situation of natural resources and protection recommendations of *Andrias davidianus*. Freshw Fish 43:13–17
10. Jiang WS, Lan XY, Wang JX, Xiang HM, Tian H, Luo QH (2022) Recent progress in the germplasm resources conservation and utilization of the Chinese giant salamander (*Andrias davidianus*). J Fish China 46: 683–705. (in Chinese)
11. Liang ZQ, Chen WT, Wang DQ, Zhang SH, Wang CR, He SP et al (2019) Phylogeographic patterns and conservation implications of the endangered Chinese giant salamander. Ecol Evol 9:3879–3890
12. Yan F, Lu JC, Zhang BL, Yuan ZY, Zhao HP, Huang S et al (2018) The Chinese giant salamander exemplifies the hidden extinction of cryptic species. Curr Biol 28:R590–R592
13. Turvey ST, Marr MM, Barnes I, Brace S, Cunningham AA (2019) Historical museum collections clarify the evolutionary history of cryptic species radiation in the world's largest amphibians. Ecol Evol 9:10070–10084
14. Chai J, Lu CQ, Yi MR, Dai NH, Weng XD, Di MX et al (2022) Discovery of a wild, genetically pure Chinese giant salamander creates new conservation opportunities. Zool Res 43: 469–480
15. He D, Zhu WM, Wen Z, Jun L, Yang J, Wang Y et al (2018) Nutritional and medicinal characteristics of Chinese giant salamander (*Andrias davidianus*) for applications in healthcare industry by artificial cultivation: a review. Food Sci Human Wellness 7:1–10
16. Luo QH (2010) Research advances in nutritional composition and exploitation of Chinese giant salamander (*Andrias davidianus*). Food Sci 31:390–393. (in Chinese)
17. Huang SY, Guo WT, Yang ZW, Ao MZ, Wang JW, Yu LJ (2009) Analysis of nutritional components of artificial breeding Chinese giant salamander (*Andrias davidianus*). Lishizhen Med Mater Med Res 20:1–2. (in Chinese)
18. Liu S, Liu HL, Zhou YH, Sun L, Yang AS, Cang DP (2010) Analysis of the nutritional composition of *Andrias davidianus*. Acta Nutr Sin 32:198–200. (in Chinese)
19. Zhu WM, Ji Y, Wang Y, He D, Yan YS, Su N et al (2018) Structural characterization and in vitro antioxidant activities of chondroitin sulfate purified from *Andrias davidianus* cartilage. Carbohydr Polym 196:398–404
20. Lu C, Chai J, Murphy RW, Che J (2020) Giant salamanders: farmed yet endangered. Science 367:989
21. Geng Y, Gray MJ, Wang KY, Chen DF, Ouyang P, Huang XL et al (2016) Pathological changes in *Andrias davidianus* infected with Chinese giant salamander ranavirus. Asian Herpetol Res 7:258–264
22. Yang CH, Gao HH, Gao Y, Yang CM, Dong WZ (2018) Bacteria in abnormal eggs of Chinese giant salamanders (*Andrias davidianus*) may derive from gut. Zool Sci 35:314–320
23. Li W, Zhang X, Weng SP, Zhao GX, He LG, Dong CF (2014) Virion-associated viral proteins of a Chinese giant salamander (*Andrias davidianus*) iridovirus (genus *Ranavirus*) and functional study of the major capsid protein (MCP). Vet Microbiol 172:129–139
24. Tapley B, Chen S, Turvey S, Redbond J, Okada S, Cunningham A (2017) A sustainable future for Chinese giant salamanders. Chinese

giant salamander field survey manual. IUCN SSC Amphibian Specialist Group, p 33

25. Marcec R, Kouba A, Zhang L, Zhang HX, Wang QJ, Zhao H et al (2016) Surgical implantation of coelomic radiotransmitters and postoperative survival of Chinese giant salamanders (*Andrias Davidianus*) following reintroduction. J Zoo Wildl Med 47:187–195
26. Zhang L, Jiang W, Wang Q, Zhao H, Zhang H, Marcec R et al (2016) Reintroduction and post-release survival of a living fossil: the Chinese giant salamander. PLoS One 11: e0156715

Part II

Traditional Molecular Techniques

Chapter 5

Whole-Mount In Situ Hybridization (WISH) for Salamander Embryos and Larvae

Sruthi Purushothaman and Ashley W. Seifert

Abstract

Whole-mount in situ hybridization (WISH) is widely used to visualize transcribed gene sequences (mRNA) in developing embryos, larvae, and other nucleotide probe permeable tissue samples. This methodology involves the hybridization of an antisense nucleotide probe to the target mRNA, followed by chromogen or fluorescence-based detection. Here we describe a protocol for the spatiotemporal analysis of mRNA transcripts in axolotl embryos/larvae using digoxigenin-labeled riboprobes, anti-digoxigenin alkaline phosphatase, Fab fragments antibody, and NBT/BCIP chromogen detection.

Key words In situ hybridization (ISH), Limb regeneration, Limb development, Salamander, In vitro transcription, Digoxigenin, Riboprobe

1 Introduction

Whole-mount in situ hybridization (WISH) is a foundational method used to detect spatiotemporal gene expression in developing embryos, larvae, and regenerating tissues. Whole-mount visualization of RNA transcripts is less labor intensive compared to performing hybridization in sectioned tissue and has the added benefit of providing three-dimensional information for gene expression domains. Various steps include synthesis of a labeled nucleotide probe complementary to an mRNA of interest, fixation and permeabilization of whole embryos, larvae or tissue pieces, hybridization of the probe within the cells, and probe detection. This methodology was initially developed for *Drosophila* embryos [1] using digoxigenin-labeled DNA probes, a phosphatase-coupled antibody against digoxygenin, and an NBT/X-phosphate color solution. The methodology was later adopted and modified for use in other embryos such as chicks [2], frogs [3], worms [4], mice [5], and fish [6, 7]. Subsequently, different types of labeled probes and visualization methods were developed.

Ashley W. Seifert and Joshua D. Currie (eds.), *Salamanders: Methods and Protocols*,
Methods in Molecular Biology, vol. 2562, https://doi.org/10.1007/978-1-0716-2659-7_5,

Axolotls (*Ambystoma mexicanum*) have been used extensively to study gene expression dynamics during limb development and regeneration [8–18]. However, clear visualization of gene expression in whole-mount embryos has been restricted to a small group of highly expressed genes and has been challenging for many labs to perform, especially during regeneration. Here we describe an optimized protocol to detect mRNA transcripts in fixed whole-mount axolotl embryos and larvae. This protocol elaborates on existing methods for embryo harvest and processing, riboprobe design from existing open-source online repositories for the axolotl genome, riboprobe synthesis, and hybridization and visualization of hybridized probes in axolotl tissue [5, 12]. Our protocol uses digoxigenin-labeled RNA probes, anti-digoxigenin-AP, Fab fragments antibody, and p-nitroblue tetrazolium chloride (NBT)/5-bromo-4-chloro-3-indolyl phosphate (BCIP) chromogen solution for visualization. Some of the major modifications to the existing standard WISH protocols include increased riboprobe length (1000–1200 bp), increased amount of plasmid used in the linearization step (20 μg), increased amount of linearized plasmid used for in vitro transcription (2.5 μg), increased riboprobe incubation time (2 days), gene and stage-specific proteinase-K treatments, and usage of higher riboprobe concentrations (0.05–0.5 μg/ml).

2 Materials

2.1 Reagents

1. Benzocaine: Millipore Sigma, *catalog number*: E1501-500G.
2. DIG RNA-labeling mix: Millipore Sigma, *catalog number*: 11277073910.
3. Protector RNase inhibitor: Millipore Sigma, *catalog number*: 3335399001.
4. T7 RNA polymerase: Millipore Sigma, *catalog number*: 10881775001.
5. SP6 RNA polymerase: Millipore Sigma, *catalog number*: 11487671001.
6. RQ1 RNase-free DNase: Promega, *catalog number*: M6101.
7. DNase stop solution (comes with the RNase-free DNase), *catalog number*: M199A.
8. RNeasy mini kit: Qiagen, *catalog number*: 74104.
9. Proteinase K recombinant, PCR grade: Millipore Sigma, *catalog number*: 3115836001.
10. CHAPS hydrate: Millipore Sigma, *catalog number*: C3023-5G.
11. Goat serum: Millipore Sigma, *catalog number*: G9023-10ML.
12. Anti-digoxigenin-AP, Fab fragments: Millipore Sigma, *catalog number*: 11093274910.

13. Formamide: Ambion, *catalog number*: AM9342-500G.
14. Blocking powder: Millipore Sigma, *catalog number*: 11096176001-50G.
15. tRNA, from brewer's yeast: Millipore Sigma, *catalog number*: 10109525001-500MG.
16. EDTA: Millipore Sigma, *catalog number*: 798681-100G.
17. Heparin sodium salt from porcine intestinal mucosa: Millipore Sigma, *catalog number*: H4784-250MG.
18. BCIP: Millipore Sigma, *catalog number*: 11383221001.
19. NBT: Millipore Sigma, *catalog number*: 11383213001.
20. DMF: Millipore Sigma, *catalog number*: D4551-250ML.

2.2 Reagent Preparation

1. 10× benzocaine: Dissolve 1 g benzocaine in 30 ml 95% ethanol and make up to 1 L with Diethyl pyrocarbonate (DEPC) treated water.
2. 4% paraformaldehyde.
   (a) Add 900 ml of DEPC-PBS and stir bar into a 1-L beaker and heat on a stirring plate to reach 62–64 °C.
   (b) Add 100 μl 10 N NaOH (made in DEPC water) and 40 g paraformaldehyde. Cover the beaker with aluminum foil. When the temperature reaches 62–64 °C, the fixative will gradually clear.
   (c) Once dissolved, remove the fixative from the hot plate, cool down to room temperature, and filter it through a 0.2-μm filter.
   (d) pH of the fixative to be 7.4 (if necessary), with 10 N NaOH and top off with DEPC-PBS.
   (e) Aliquot 40 ml fixative in 50-ml conical tubes and store at −80 °C.
3. DEPC-PBS: Fill 1-L glass bottles with 1-L double-distilled water, add 1 ml DEPC and 10 PBS tablets (without calcium and magnesium) and mix thoroughly. Place the bottles inside a fume hood overnight with the cap slightly open to vent. Autoclave the bottles the following morning.
4. DEPC-PBT: Add 1 ml of Triton X-100 into a 1-L bottle of sterile DEPC-PBS and mix thoroughly.
5. DEPC water: Fill 1-L glass bottles with 1-L double-distilled water, add 1 ml of DEPC and mix thoroughly. Place the bottles inside the hood overnight with the cap slightly open to vent. Autoclave the bottles the following morning.
6. 3 M sodium acetate, pH 5 (100 ml): Dissolve 24.6 g sodium acetate in 75 ml DEPC water. Adjust to pH 5 with 12 M HCl, top off with DEPC water and autoclave.

7. 20× SSC: 3 M NaCl, 0.3 M sodium citrate dihydrate (1 L). Dissolve 175.3 g of NaCl and 77.4 g of sodium citrate dihydrate in 900 ml of DEPC water. Adjust pH to 7 with 12 M HCl, top off with DEPC water and autoclave.
8. 0.5 M EDTA, pH 8 (100 ml): Dissolve 14.6 g EDTA in 75 ml of DEPC water. Adjust pH to 8 with 10 N NaOH, top off with DEPC water and autoclave.
9. Hybridization buffer (500 ml): 50% formamide, 5× SSC, 2% blocking powder, 0.1% Triton X-100, 0.1% CHAPS, 1 mg/ml tRNA, 5% dextran sulfate, 5 mM EDTA at pH 8, 50 μg/ml heparin sodium salt. To 500-ml bottle add 250 ml formamide, 125 ml 20× SSC, 10 g blocking powder, 500 μl Triton X-100, 0.5 g CHAPS, 500 mg tRNA, 25 g dextran sulfate, 5 ml 0.5 M EDTA at pH 8, 25 mg heparin sodium salt and stir to dissolve at 55 °C. Top off with DEPC water. Aliquot 40 ml in 50-ml conical tubes and store at −80 °C.
10. 2× SSC, 0.1% CHAPS (50 ml): In 50-ml conical tube, add 0.05 g CHAPS to 5 ml 20× SSC. Top off with DEPC water and vortex to mix.
11. 0.2× SSC, 0.1% CHAPS (50 ml): In 50-ml conical tube, add 0.05 g CHAPS to 0.5 ml 20× SSC. Top off with DEPC water and vortex to mix.
12. 1 M Tris–HCl pH 7.5, 9, and 9.5 (100 ml): Dissolve 12.1 g Tris–HCl in 75 ml of DEPC water. Adjust pH to 7.5 or 9 or 9.5 with 10 N NaOH, top off with DEPC water and autoclave.
13. 5 M NaCl (100 ml): Dissolve 29.2 g NaCl in 75 ml DEPC water. Top off with DEPC water and autoclave.
14. 1 M KCl (100 ml): Dissolve 7.4 g in 75 ml DEPC water. Top off with DEPC water and autoclave.
15. 1 M $MgCl_2$ (100 ml): Dissolve 9.5 g $MgCl_2$ in 75 ml DEPC water. Top off with DEPC water and autoclave.
16. KTBT buffer: 15 mM Tris–HCl pH 7.5, 150 mM NaCl, 10 mM KCl, 1% Tween-20 (50 ml). To a 50-ml conical tube, add 750 μl 1 M Tris–HCl, 1.5 ml 5 M NaCl, 500 μl 1 M KCl, 500 μl Tween-20, and 25 ml DEPC water. Vortex vigorously and top off with DEPC water.
17. NTMT buffer: 100 mM Tris–HCl pH 9.5, 50 mM $MgCl_2$, 100 mM NaCl, 1% Tween-20 (50 ml). To a 50-ml conical tube, add 5 ml 1 M Tris–HCl pH 9.5, 2.5 ml 1 M $MgCl_2$, 1 ml 5 M NaCl, 500 μl Tween-20, and 25 ml DEPC water. Vortex vigorously and top off with DEPC water.
18. Color solution (prepared in NTMT): 0.175 mg/ml BCIP, 0.3 mg/ml NBT, 10% DMF, NTMT (50 ml). To a 50-ml conical tube, add 175 μl 50 mg/ml BCIP, 168.8 μl 100 mg/ml NBT, 5 ml DMF, and bring to 50 ml with NTMT.

3 Methods

3.1 Tissue Harvest and Processing (See Notes 1–3)

1. For RNA extraction (*see* **Note 4**): Anesthetize each axolotl embryo/larva/juvenile at desired stages in 1× benzocaine to acquire target tissue. Rinse well in 1× DEPC-PBS. Put the sample into a 1.5-ml tube, snap freeze in liquid nitrogen, and store at −80 °C until RNA extraction.
2. For the whole-mount in situ:

   Anesthetize axolotl embryos/larvae at desired stages in 1× benzocaine. Rinse well in DEPC-PBS and fix in 4% PFA overnight at 4 °C (with gentle rocking if possible).

   (a) Post fixation, wash the fixed samples with DEPC-PBT three times, 10 min each.

   (b) Dehydrate samples in graded methanol/PBT series: 25%, 50%, 75%, and 100% methanol, 20 min each. Store samples in 100% methanol at −20 °C until further use.

3.2 Primer Design

1. For the gene of interest, pull out the human coding sequence from NCBI and locate the last exon.
2. Pull out the corresponding axolotl coding sequence from ambystoma.uky.edu, www.axolotl-omics.org, or Bryant et al. (2017) and align this sequence with the human sequence to locate the last exon.
3. Design primers against this region using Primer3Plus webtool with the following parameters while maintaining other default values:

   Product size: 1000–1200 bp
   Primer size: min = 18, opt = 25, max = 28
   Primer Tm: min = 57, opt = 60, max = 63
   Primer GC%: min = 40, max = 70
   Max Tm difference = 1 (should not exceed 5).
4. Blast the primer sequences in ambystoma.uky.edu or www.axolotl-omics.org to confirm target specificity.

3.3 RNA Extraction and cDNA Synthesis

1. Thaw previously collected tissue samples on ice and add 750 μl of TRIzol (Invitrogen) into the tubes and manually homogenize the sample using a plastic pestle.
2. Add 150 μl of chloroform into the tubes, shake vigorously for 15 s, and incubate for 2–3 min at room temperature.
3. Spin at 12,000 g for 15 min at 4 °C and transfer the aqueous phase into fresh 1.5 ml tubes.
4. Add 100 μl 3 M of sodium acetate (pH 5) and 800 μl isopropyl alcohol and incubate at room temperature for 10 min.

5. Spin at 14,000 g for 20 min at 4 °C and carefully discard the supernatant.
6. Wash the pellet with 1 ml 75% ethanol and vortex briefly.
7. Spin at 7500 g for 5 min at 4 °C, carefully discard the supernatant and air-dry the pellet in the hood for 15–20 min till ethanol evaporates.
8. Resuspend the pellet in 20–50 μl of DNase-/RNase-free water according to the size of the pellet (*see* **Note 5**).
9. Place the sample on the heating block at 55–60 °C for 15 min and quantify the sample.
10. Synthesize cDNA from 1 μg RNA using a standard kit.

3.4 Standardizing PCR for Target Amplification

Amplify the target gene using a high-fidelity polymerase kit at different annealing temperatures (50, 55, 60, 65, 70, and 75 °C) using different sources and concentrations of RNA and the number of cycles in the amplification step. Run the PCR product on 1% agarose gel and check for single bands at desired base pair size. If single bands are achieved, then run replicates of the corresponding PCR program and purify the required amount for the ligation step.

3.5 PCR Product Purification

1. Pool similar PCR products and purify using a standard PCR purification kit.
2. Elute in 20 μl DNase-/RNase-free water (preheated to 65–70 °C) and quantify.

3.6 Gel Excisions and Extraction of PCR Product

1. If you get multiple bands (including the desired base pair size) from all the standardized PCR reactions, run the products on a thick 1% agarose gel after replicating PCRs and perform gel excision followed by purification using a standard gel extraction kit.
2. Elute in 10 μl DNase-/RNase-free water (preheated to 65–70 °C) and quantify.

3.7 Ligation of PCR Product into Vector and Bacterial Transformation

1. Use a T-A cloning system for easy, rapid, and efficient ligation of the purified PCR product into a vector.
2. Add a deoxyadenosine to the 5′ end of the purified PCR product using a high-fidelity DNA polymerase.
3. Ligate this to a high copy number cloning vector with 3′T-overhangs on both ends, promoter sequences on both ends (e.g., T7 and SP6), blue/white colony screening capability and multiple unique restriction enzyme sites.
4. Amplify the ligated product via transformation using high-efficiency competent cells.

3.8 Colony PCR

1. Label and subculture the blue/white colonies obtained after transformation on a fresh plate.
2. Perform colony PCR on each colony using a high-fidelity polymerase kit and primer pair complementary to the promoter sequences on the vector backbone.
3. Verify insert size on a 1% agarose gel.

3.9 Plasmid Extraction (Miniprep) and Sequencing

1. Once the size of the insert is verified, make a miniprep of the specific colony and sequence verify the gene of interest.
2. Align the result sequence with the reference coding sequence (make sure it is in $5' \rightarrow 3'$orientation) (*see* **Note 6**).

3.10 Plasmid Extraction (Maxiprep) and Glycerol Stock

Once the sequence and orientation of the gene of interest have been verified, make a larger amount of plasmid using a standard maxiprep kit. Additionally, for long-term storage, make glycerol stocks of the positive bacterial colony (*see* **Note 7**).

3.11 Plasmid Linearization

For each gene of interest, find a unique restriction enzyme site (single copy) flanking the insertion site on the vector backbone that does not cut the gene itself. If using a vector with SP6/T7 promoter sequences, to make an SP6 probe you need to linearize the plasmid with a restriction enzyme that has a unique site near the T7 promoter site and vice versa. Use high-fidelity restriction enzymes that have proofreading capacities.

1. Reaction mix is as follows:

Reaction mix	
Plasmid (20 μg)	= *x* μl
Restriction enzyme (~40 U)	= *x* μl
10× buffer (in the restriction enzyme kit)	= 10 μl
DNase/RNase-free molecular grade water	= *x* μl
	100 μl (incubate overnight at 37 °C).

2. Run the reaction mix on 1% agarose gel and verify the single band.
3. Purify the reaction using a standard PCR purification kit, elute in 20 μl DNase/RNase-free water (preheated to 65–70 °C) and quantify.

3.12 In Vitro Transcription

1. Make the following reaction mix:

Reaction mix	
10× buffer (comes with T7/SP6 enzymes)	= 5 μl
Digoxigenin labeling mix	= 5 μl
1M DTT	= 0.5 μl
RNase inhibitor	= 1 μl
T7 or SP6 polymerase	= 2 μl
Linearized plasmid (2.5 μg)	= *x* μl
DNase/RNase-free molecular grade water	= *x* μl
	50 μl (incubate overnight at 37 °C)

2. Spin tubes.
3. Add 1 μl RNase-free DNase and incubate at 37 °C for 10 min.
4. Add 2 μl DNase stop solution (comes with the RNase-free DNase) and incubate at room temperature for 5 min.

3.13 Purification of In Vitro Transcription Product

1. Purify the reaction using RNeasy Minikit (Qiagen) according to the manufacturer's protocol (*see* **Note 8**).
2. Place column onto a 1.5-ml tube supplied, add 50 μl of RNase-free water (from the kit) and let it stand for 10 min.
3. Spin at 8000 g for 1 min, quantify the riboprobe, and store at −20 °C for short-term storage and −80 °C for long-term storage.

3.14 Whole-Mount In Situ Hybridization (WISH)

Unless specifically mentioned, perform every step at room temperature.

1. Day 1: Rehydration and Blocking
   (a) Transfer two dehydrated embryo/larvae (from Subheading 3.1) into a 2 ml of Eppendorf tube.
   (b) Graded methanol treatment (*ON ICE*):
   Rehydrate tissue by washing 10 min each in graded methanol series (75%, 50%, 25%) in DEPC-PBT. Wash 1× in DEPC-PBT for 5 min.
   (c) Bleach samples in 6% hydrogen peroxide in DEPC-PBS for 1 h (*ON ICE*).
   (d) Wash 1× in DEPC-PBT for 5 min (*ON ICE*).
   (e) Treat samples with 20–40 μg/ml proteinase K in DEPC-PBS for 7–10 min (*see* **Note 9**).

(f) Rinse 1× in DEPC-PBT for 5 min.

(g) Refix tissue with 0.2% glutaraldehyde/4% paraformaldehyde for 20 min.

(h) Rinse 1× in DEPC-PBT for 5 min.

(i) Remove DEPC-PBT, add hybridization buffer (for blocking), and incubate overnight at 60 °C in a shaking hybridization oven.

2. Day 2: Probe Hybridization

   (a) Remove overnight hybridization buffer, add fresh 2 ml of hybridization buffer (pre-warmed to 60 °C), and antisense probe or sense probe (negative control) (*see* **Notes 10** and **11**).

   (b) Incubate at 60 °C for at least 2 days (*see* **Note 12**).

3. Day 4: Washing and Addition of Antibody

   (a) Prepare wash solutions and warm to 60 °C.

   (b) Wash 3× with 2× SSC, 0.1% CHAPS for 20 min at 60 °C (low stringency wash).

   (c) Wash 4× with 0.2× SSC, 0.1% CHAPS for 25 min at 60 °C (high stringency wash).

   (d) Rinse 1× in KTBT buffer for 10 min at room temperature.

   (e) Pre-block the samples with 20% goat serum in KTBT for 3 h at 4 °C.

   (f) Remove pre-block and add fresh 20% goat serum in KTBT plus anti-digoxigenin-AP, Fab fragments antibody (1/3000 dilution). Incubate overnight at 4 °C.

4. Day 5: Washing

   Wash 5× with KTBT 1 h each at 4 °C. Incubate in KTBT overnight at 4 °C.

5. Day 6: Color Reaction and Imaging

   (a) Move the samples into 6- or 12-well plates and remove KTBT.

   (b) Wash 2× with NTMT buffer 10 min each.

   (c) Make fresh color solution in NTMT.

   (d) Add 1–2 ml color solution per well (avoid air bubbles) and check for signals every 15 min.

   (e) Once desired signal is achieved, image the samples in NTMT solution on a 1% agarose plate (*see* **Note 13**).

   (f) If signals do not show up on *Day 6*, leave the samples in NTMT overnight at 4 °C and check the next day.

   (g) After imaging, wash samples 2× with TE buffer, fix and store in 4% paraformaldehyde at 4 °C.

4 Notes

1. Clean the surfaces using RNAZap or equivalent before starting the protocol.
2. Autoclave all reaction tubes prior to use. This can be done beforehand and tubes stored in the beaker they were autoclaved in.
3. Use DNase-/RNase-free filter tips and make all solutions in DEPC-treated water/1× PBS/1× PBT.
4. For RNA extraction: The stage of harvest will depend on the gene of interest. For early developmental genes like *Shh* and *Fgfs*, use embryos at stages 30–40 [19] or early to late blastema of a regenerating limb. For genes involved in later events like skeletal differentiation (e.g., *Sox9*), use embryos at stages 35–40 [19]. To obtain close to 5 μg RNA, use 1 stage 30–40 embryo or pool 2–3 early to late blastema.
5. Use commercial DNase/RNase-free molecular grade water for all PCR reactions and elution steps.
6. Note the alignment directions. If you get a +/+ alignment with the SP6 primer, then SP6 polymerase will give you the sense probe after the in vitro transcription step. If you get a +/− alignment with the SP6 primer, then SP6 polymerase will give you the antisense probe after the in vitro transcription step.
7. Do not let the glycerol stock thaw.
8. Traditional phenol-chloroform precipitation method can also be used to purify the in vitro transcription product.
9. Proteinase-K concentration and duration of treatment should be standardized for each probe and stage of the embryo/larvae. For the *Shh* antisense probe at stage 48, 20 μg/μl proteinase-K treatment for 7 min is the ideal condition while 30-min treatment results in tissue degradation (Fig. 1). At a similar stage, for skeletal/precursor genes like *Sox9* and *Ihh*, 7 min treatment with 40 μg/μl proteinase-K gives a cleaner signal with minimal background (Fig. 2).
10. If you are running a WISH experiment for a novel gene, always run the sense and antisense probes side by side (Fig. 3) to differentiate between true signal and background.
11. Probe concentration must be standardized for each gene. Use between 0.05 and 0.5 μg/ml to begin with.
12. Duration of probe treatment should also be standardized according to the gene of interest. Start with 2 days.
13. The duration of the color reaction is dependent on the probe type and stage of embryo/larvae. You will have to standardize this for each probe to avoid background staining. For instance,

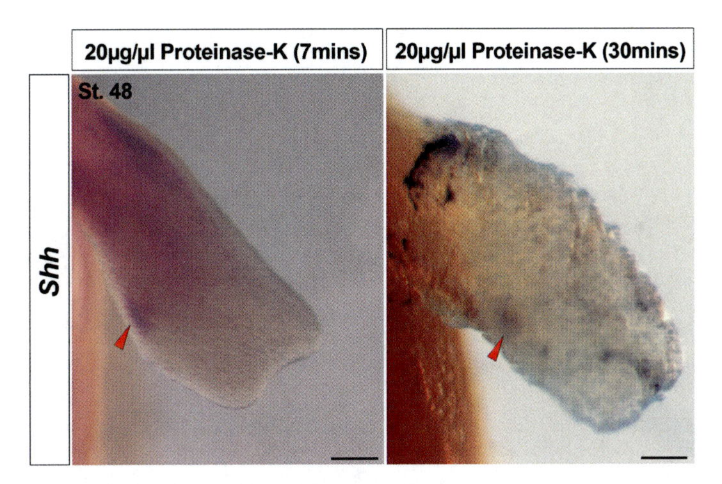

Fig. 1. Whole-mount in situ hybridization of *Shh* in stage 48 axolotl limbs. The image on left depicts a limb that was treated with 20 μg/μl proteinase-K treatment for 7 min and the image on right depicts limb that was treated with 20 μg/μl proteinase-K treatment for 30 min. Note the high degree of tissue degradation that occurred in the right panel. Red arrows: gene expression domains. Scale bar = 100 μm

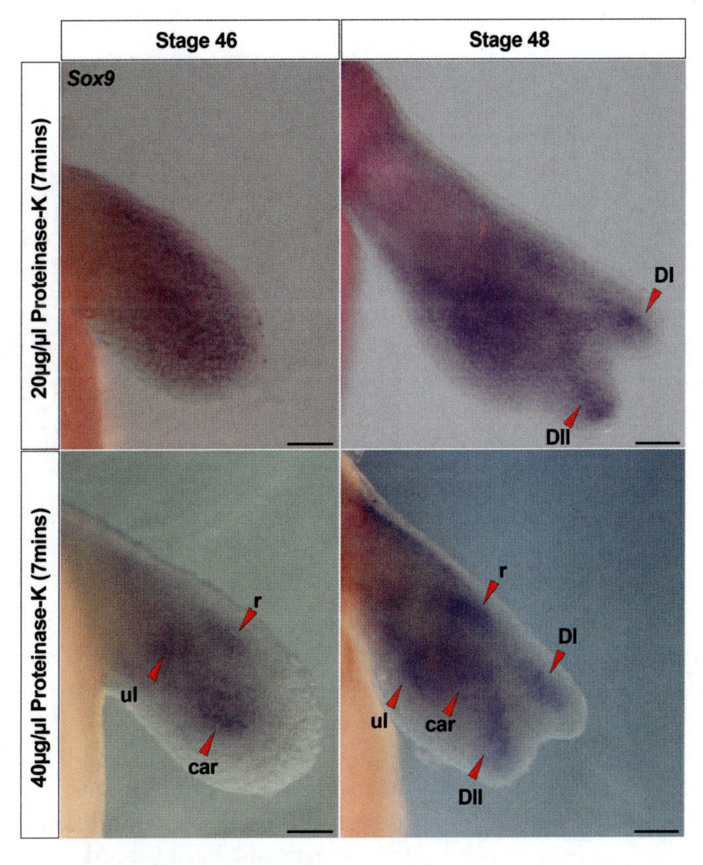

Fig. 2. Whole-mount in situ hybridization of *Sox9* in stages 46 and 48 axolotl limbs. This specific antisense probe requires a higher concentration (40 μg/μl) of proteinase-K treatment for clearer staining of the expression domains. Red arrows: gene expression domains. Scale bar = 100 μm. ul: ulna, r: radius, car: carpal, DI: digit 1, DII: digit 2

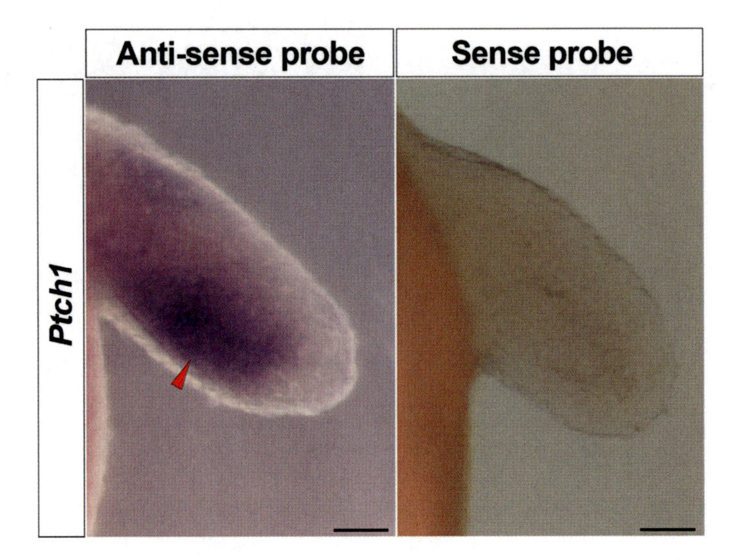

Fig. 3. Whole-mount in situ hybridization of *Ptch1* gene in stage 46 axolotl limbs. The image on the left depicts positive staining with antisense probe and the image on the right represents the absence of staining with a sense probe (negative control). Red arrows: gene expression domains. Scale bar = 100 μm

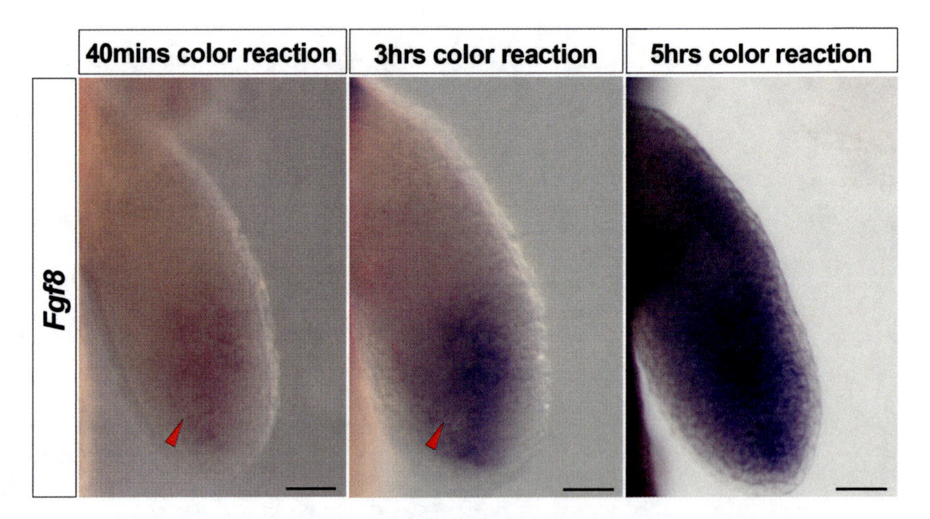

Fig. 4. Whole-mount in situ hybridization of *Fgf8* in stage 46 axolotl limbs. The images (from left to right) represent increased duration of color reactions (40 min, 3 h, and 5 h). Red arrows: gene expression domains. Scale bar = 100 μm

Fig. 4 shows the color development for an *Fgf8* antisense probe in a stage 46 axolotl limb bud. For this probe, the color within the expression domain gradually develops in 3 h and is followed by non-specific background staining at later time points.

Acknowledgments

This work was partially supported by NIH R01AR070313 to Ashley W. Seifert.

References

1. Tautz D, Pfeifle C (1989) A non-radioactive in situ hybridization method for the localization of specific RNAs in Drosophila embryos reveals translational control of the segmentation gene hunchback. Chromosoma 98(2):81–85
2. Nieto MA, Patel K, Wilkinson DG (1996) Chapter 11 In situ hybridization analysis of chick embryos in whole mount and tissue sections. In: Methods in cell biology, vol 51. Elsevier, pp 219–235. https://doi.org/10.1016/s0091-679x(08)60630-5
3. Hemmati-Brivanlou A, Frank D, Bolce M, Brown B, Sive H, Harland R (1990) Localization of specific mRNAs in Xenopus embryos by whole-mount in situ hybridization. Development 110(2):325–330
4. Umesono Y, Watanabe K, Agata K (1997) A planarian orthopedia homolog is specifically expressed in the branch region of both the mature and regenerating brain. Develop Growth Differ 39(6):723–727
5. Rosen B, Beddington RS (1993) Whole-mount in situ hybridization in the mouse embryo: gene expression in three dimensions. Trends Genet 9(5):162–167
6. Hauptmann G, Gerster T (1994) Two-color whole-mount in situ hybridization to vertebrate and Drosophila embryos. Trends Genet 10(8):266–266
7. Schulte-Merker S, Ho R, Herrmann B, Nusslein-Volhard C (1992) The protein product of the zebrafish homologue of the mouse T gene is expressed in nuclei of the germ ring and the notochord of the early embryo. Development 116(4):1021–1032
8. Torok MA, Gardiner DM, Izpisúa-Belmonte JC, Bryant SV (1999) Sonic hedgehog (shh) expression in developing and regenerating axolotl limbs. J Exp Zool 284(2):197–206
9. Torok MA, Gardiner DM, Shubin NH, Bryant SV (1998) Expression ofHoxDGenes in developing and regenerating axolotl limbs. Dev Biol 200(2):225–233
10. Shimokawa T, Yasutaka S, Kominami R, Shinohara H (2013) Lmx-1b and Wnt-7a expression in axolotl limb during development and regeneration. Okajimas Folia Anat Jpn 89(4):119–124
11. Han MJ, An JY, Kim WS (2001) Expression patterns of Fgf-8 during development and limb regeneration of the axolotl. Dev Dyn 220(1):40–48
12. Gardiner DM, Blumberg B, Komine Y, Bryant SV (1995) Regulation of HoxA expression in developing and regenerating axolotl limbs. Development 121(6):1731–1741
13. Koshiba K, Kuroiwa A, Yamamoto H, Tamura K, Ide H (1998) Expression of Msx genes in regenerating and developing limbs of axolotl. J Exp Zool 282(6):703–714
14. Carlson M, Komine Y, Bryant S, Gardiner D (2001) Expression of Hoxb13 and Hoxc10 in developing and regenerating Axolotl limbs and tails. Dev Biol 229(2):396–406
15. Hutchison C, Pilote M, Roy S (2007) The axolotl limb: a model for bone development, regeneration and fracture healing. Bone 40(1):45–56
16. Ghosh S, Roy S, Séguin C, Bryant SV, Gardiner DM (2008) Analysis of the expression and function of Wnt-5a and Wnt-5b in developing and regenerating axolotl (Ambystoma mexicanum) limbs. Develop Growth Differ 50(4):289–297
17. Johnson AD, Bachvarova RF, Drum M, Masi T (2001) Expression of axolotl DAZL RNA, a marker of germ plasm: widespread maternal RNA and onset of expression in germ cells approaching the gonad. Dev Biol 234(2):402–415
18. Metscher B, Northcutt RG, Gardiner DM, Bryant SV (1997) Homeobox genes in axolotl lateral line placodes and neuromasts. Dev Genes Evol 207(5):287–295
19. Schreckenberg G, Jacobson A (1975) Normal stages of development of the axolotl, Ambystoma mexicanum. Dev Biol 42(2):391–399
20. Bryant DM, Johnson K, DiTommaso T, Tickle T, Couger MB, Payzin-Dogru D, et al. (2017) Cell Rep 18(3):762–776 Accession Number: 28099853 PMCID: PMC5419050 https://doi.org/10.1016/j.celrep.2016.12.063

Chapter 6

Hybridization Chain Reaction Fluorescence In Situ Hybridization (HCR-FISH) in *Ambystoma mexicanum* Tissue

Alex M. Lovely, Timothy J. Duerr, David F. Stein, Evan T. Mun, and James R. Monaghan

Abstract

In situ hybridization is a standard procedure for visualizing mRNA transcripts in tissues. The recent adoption of fluorescent probes and new signal amplification methods have facilitated multiplexed RNA imaging in tissue sections and whole tissues. Here we present protocols for multiplexed hybridization chain reaction fluorescence in situ hybridization (HCR-FISH) staining, imaging, cell segmentation, and mRNA quantification in regenerating axolotl tissue sections. We also present a protocol for whole-mount staining and imaging of developing axolotl limbs.

Key words In situ hybridization, Hybridization chain reaction, Axolotl, Limb regeneration, Limb development

1 Introduction

In situ hybridization (ISH) has been the primary technique for imaging DNA and RNA in cells and tissues for over 50 years [1, 2]. Although the original approaches relied upon radiolabeled probes, the most widespread ISH approach in modern research utilizes digoxigenin-tagged probes, which were first demonstrated using antisense DNA probes [3, 4]. Eventually, antisense RNA probes became the dominant method for ISH in tissue sections [5, 6] and whole-mount tissues [7–9] because they showed high specificity and were relatively straightforward and inexpensive to generate [10]. Fluorescent probes have some advantages over digoxigenin-labeled probes, such as their multiplexing capabilities and high resolution but have traditionally been expensive to produce. The cost of oligonucleotide synthesis has decreased in recent years, which could facilitate the widespread adoption of fluorescently tagged oligonucleotide probes and emerge as a solid alternative to digoxigenin-based ISH. Yet, a large variety of FISH methods

Ashley W. Seifert and Joshua D. Currie (eds.), *Salamanders: Methods and Protocols*,
Methods in Molecular Biology, vol. 2562, https://doi.org/10.1007/978-1-0716-2659-7_6,

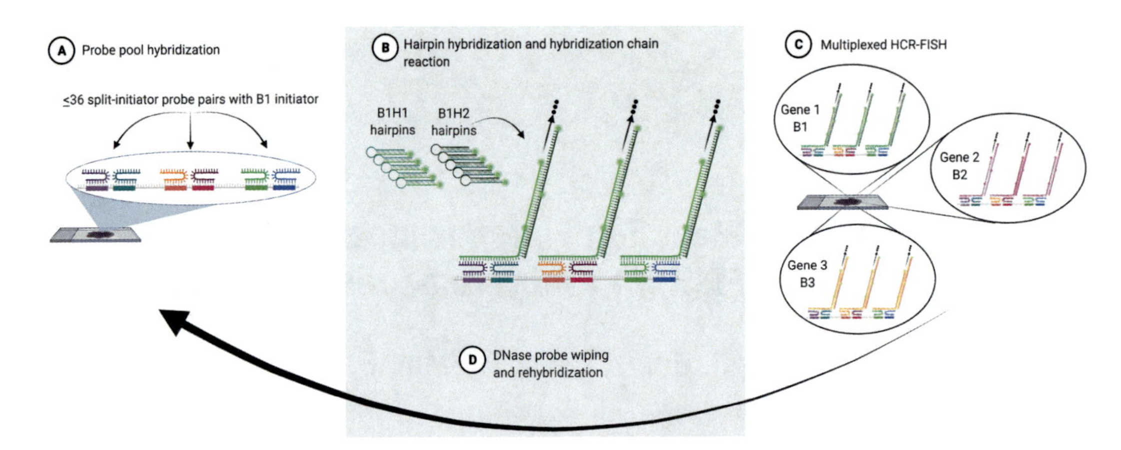

Fig. 1 HCR schematic diagram. Schematic of v3.HCR-FISH on tissue sections. Transcripts of interest are hybridized with pools of split initiator probes containing unique initiators, B1–B5 (**a**). Following hybridization, initiator-specific fluorescently labeled hairpins are applied to the samples and a hybridization chain reaction is initiated (**b**). Sections are imaged for each gene of interest (**c**). Probes with hairpins are removed from the sections with a DNase treatment and probe wash (**d**). Subsequent rounds of hybridization are performed using new probe pools. (Adapted from "Loop-Mediated Isothermal Amplification (LAMP)," by BioRender.com (2021). Retrieved from https:/app.biorender.com/biorender-templates)

make it difficult to evaluate the most user-friendly and robust option for any particular animal model [11–18]. Commercial resources generally market mice or human samples, making non-model probe sets prohibitively expensive to commercially design and optimize. Furthermore, colorimetric ISH traditionally does not use quantitative image analysis, so learning new computational image analysis approaches may be intimidating or challenging to implement.

One new method that has shown considerable promise to work in many animal types and large tissues is hybridization chain reaction (HCR). Dirks and Pierce initially developed HCR in 2004 [19], subsequently to be used for multiplexed FISH [20]. The third iteration of HCR (v3.HCR-FISH) has a considerably higher signal to background ratio over the previous methods due to the implementation of split-initiator oligonucleotide probes [18] (Fig. 1a). Two split-initiator probes become whole when the pair hybridizes to their target mRNA, significantly increasing specificity over unsplit initiator probes. Probe pair hybridization then triggers the hybridization chain reaction (Fig. 1b). Five different initiators can be used that pair with five types of fluorescently labeled hairpins, enabling the staining of up to five gene targets in one hybridization, depending upon the microscope's capabilities (Fig. 1c). The number of genes analyzed with HCR-FISH has also been increased through subsequent rounds of staining [21] (Fig. 1d). Lastly, v3.HCR-FISH has been successful in many different species, corroborating the robustness of the protocol.

ISH has been the primary tool of choice for studying the development and regeneration of amphibians. For example, the axolotl is a powerful model for studying vertebrate development, physiology, and regeneration [22]. Recent advances in transgenics and genomic sequencing are driving new investigation into the molecular basis of axolotl regeneration [23–27]. Although ISH has been utilized for decades in axolotls [28, 29], it has been challenging to implement consistently across tissues and in whole mount. Newer FISH methods are beginning to be used in axolotls [27, 30, 31], including v3.HCR-FISH [32, 33], but broadly applicable probe sets and protocols are not established.

With these considerations in mind, we present here v3.HCR-FISH protocols that have worked well for our lab using axolotl tissues. We provide multiplexed v3.HCR-FISH protocols in axolotl tissue sections and whole-mount samples for visualizing gene expression.

2 Materials

2.1 Axolotls

The supplier for research axolotls in the United States is the Ambystoma Genetic Stock Center, University of Kentucky, USA.

2.2 Equipment and Consumables

1. Microscope slides.
2. Water bath sonicator.
3. Number 1.5 coverslips.
4. Disposable cryomolds.
5. Cryostat.
6. Slide staining boxes (EMS 62010-37).
7. 50-ml conical tubes.
8. Hybridization chambers (Grace Biolabs 621101).
9. Incubation oven at 37 °C.
10. Thermocycler.
11. Zeiss LSM 880 or equivalent confocal microscope.
12. Glass capillaries for lightsheet microscopy.
13. Zeiss Z.1 lightsheet microscope or equivalent.

2.3 Buffers and Solutions

1. DEPC-treated water: Add 1 ml of 97% pure diethyl pyrocarbonate (DEPC) to 1 L of filtered DI water, mix well, and incubate at room temperature overnight. Autoclave the solution the following day (*see* **Note 1**).
2. 1 M KOH: Weigh 2.8 g KOH and add to a 50-ml conical tube. Fill to 50 ml with DEPC water.

3. Coverslip functionalization solution: Add 1 ml of glacial acetic acid to 18.4 ml of 100% methanol. Then add 600 μl of 3-animopropyltriethoxysilane [34].
4. Tissue clearing solution: 4% SDS, 200 mM boric acid, pH 8.5. For 1 L, measure 12.37 g of boric acid. Add 500 ml of DEPC-treated water and pH to 8.5 with NaOH. Add 200 ml of 20% SDS and fill to 1 L with DEPC-treated water [35].
5. 5×SSCT: Combine 12.5 ml of 20×SSC and 500 μl of 10% Tween-20 solution in a 50-ml conical tube. Fill with DEPC water to 50 ml.
6. Purchase HCR-FISH reagents including hybridization buffer, wash buffer, and amplification buffer through Molecular Instruments (https://www.molecularinstruments.com/). Dispose of all waste following appropriate regulations.
7. Probe solution: 50 μl of probe solution at 5 nM. Resuspend lyophilized oPool oligonucleotide pools in 50 μl TE buffer to make a 1-μM stock probe solution. Dilute stock probe solution 1:200 in hybridization buffer to generate a 5-nM probe solution.
8. Hairpin solution: 50 μl of hairpins at 60 nM. Pipet 1 μl of each fluorescently labeled H1 hairpin (3 μM) into a 0.2-ml PCR tube. Hairpins with different fluorophores can be combined in the same tube (for example, H1B1-Alexa488 and H1B2-Alexa594). Use a separate tube for the H2 hairpins. Heat the hairpins to 95 °C for 90 s in a thermocycler. Remove from the thermocycler and cool in the dark for 30 min. Combine all hairpins together in 50 μl amplification solution. These amounts can be scaled up as long as the ratio of hairpin to amplification solution is 1:50 vol/vol.
9. DNase solution: 50 μl of 500 units/ml DNase. Combine 5 μl of 10× DNase buffer and 32.5 μl of DEPC water. Then add 12.5 μl of 2000 units/ml DNase (NEB M0303).
10. 1.5% low melt (LM) agarose: Add 0.15 g of LM agarose to 10 ml 1×PBS. Heat the solution in a heat block or microwave until agarose is completely dissolved. Aliquot into 1.5 ml tubes and store in 4 °C.

2.4 Oligo Probe Design and Ordering

The oPools scale that accommodates 50 pmol/oligo can include up to 36 split-initiator DNA oligonucleotide pairs per mRNA target. Probe pairs are designed to be two bases apart and have half of an HCR initiator sequence appended to the end (Fig. 1). We have developed a web application that includes the axolotl genome for designing custom probe pools (*see* **Note 2**). The web app, called Probegenerator, utilizes Oligominer [36] and Bowtie2 [37] to generate probe set pools (*see* **Note 3**). First, OligoMiner generates all possible 25 base sequences that fall within specified melting temperature and GC content ranges. Probe candidates are then filtered into pairs of sequences separated by two nucleotides and then aligned to the axolotl genome using Bowtie2 to screen for off-target hits. Qualifying probe pairs are written to separate .csv files for each user specified v3.HCR-FISH initiator sequence. The Probegenerator web application can be found here: pro begenerator.herokuapp.com. Sometimes more than 36 probe pairs are available. Probes are selected in the coding sequence first, then 3′UTR, and last the 5′UTR. Once 36 probe pairs are selected, they are ordered as an oPool from Integrated DNA Technologies. oPools contain 50 pmol of each oligonucleotide in a dry pellet. This ordering scheme provides enough probes for 10 ml of hybridization solution (*see* **Note 4**).

2.5 Ordering Hairpins

We order fluorescently labeled hairpins directly from Molecular Instruments (https://www.molecularinstruments.com/). Be sure to match the hairpins (B1–B5) with the initiator on the mRNA of interest. It is up to the investigator to best match initiators with fluorophores (for example, our common ordering scheme is 600 pmol B1 Alexa Fluor 647, 600 pmol B2 Alexa Fluor 594, and 600 pmol B3 Alexa Fluor 488).

3 Methods

3.1 Coverslip Functionalization

Multiround HCR-FISH is performed on a functionalized coverslip with a hybridization chamber adhered to the coverslip. Coverslips need to be functionalized to promote tissue adhesion. Coverslip functionalization protocol is based upon a previous protocol [34]. All steps are performed at room temperature.

1. Submerge coverslips into 1 M KOH solution in a 50-ml conical tube. Float the tube in a water bath sonicator and sonicate for 20 min.
2. Wash coverslips with DEPC-treated water 3 × 5 min.
3. Submerge coverslips in 100% MeOH for 10 min.
4. Pipet the coverslip functionalization solution onto the coverslips and incubate for 2 min. This step can be done in a petri dish or slide staining box.
5. Wash coverslips with DEPC-treated water 3 × 5 min.
6. Dry coverslips and store at room temperature in a slide box or conical tube with silica for up to 2 weeks.

3.2 Tissue Cryosectioning and Section Dehydration

1. Tissue collection: Anesthetize axolotl in 0.01% benzocaine. Surgically remove tissue from animal, mount immediately in cryomold filled with 100% OCT, and flash freeze in dry ice/isopentane bath.

Fig. 2 Hybridization chamber-well locations. Areas on the slide correlating to the hybridization chambers are marked with a sharpie. Sections are collected onto a functionalized coverslip placed on top of the marked slide. This ensures proper placement of the sections for the hybridization chamber

2. Tissue sectioning: For multiround HCR, sections are collected on a coverslip rather than a microscope slide (*see* **Note 5**). Draw a template grid on a microscope slide that matches with the hybridization chambers (Fig. 2). This ensures that cryosections line up with the hybridization chamber that is used in the protocol. Adhere the coverslip to the microscope slide by adding a small drop of water on the microscope slide and placing the coverslip on top of the templated slide at room temperature. Cryosection the tissue at 10 μm and adhere to the coverslip that is on top of the templated slide. Once the tissue section is on the coverslip, do not remove it from the cryostat. Let it dry in the cryostat for 30 min. Immediately perform tissue fixation after cryosectioning.
3. Fixation and tissue clearing: Without letting the slide thaw, apply 4% PFA to cover sections and incubate for 15 min. Wash sections with 2×SSC, 3 × 5 min by placing solutions directly on the coverslip. Apply 4% SDS, 200 mM boric acid solution to sections, 2 × 5 min to clear the tissue. Wash sections with 2×SSC, 3 × 5 min. All steps are performed at room temperature.
4. Dehydration: Incubate sections in 100% EtOH for 10 min at room temperature and then allow sections to dry completely. Once dry, peel the plastic cover off of the hybridization chamber and adhere the chamber onto the coverslip on top of the tissue sections. Wash sections with 50 μl 2×SCC 2 × 5 min.

3.3 v3.HCR-FISH on Tissue Sections

1. Prehybridization: Pipet 50 μl of hybridization buffer into a hybridization chamber and incubate at 37 °C for 15 min. Place the prepared probe solution at 37 °C to preheat during prehybridization.
2. Hybridization: Remove the hybridization buffer from the chamber and pipet 50 μl of preheated probe solution into the hybridization chamber. Hybridize overnight at 37 °C in an incubator using a coverslip to seal the chambers.
3. Probe washing: Preheat probe wash solution at 37 °C. Remove the probe solution from the chamber and pipet 50 μl of preheated probe wash solution to the hybridization chamber 3 × 15 min at 37 °C. Then wash sections with 5×SSCT for 15 min at 37 °C followed by a 5-min 5×SSCT wash at room temperature.
4. Amplification: Remove 5×SSCT and pipet 50 μl of amplification buffer to the hybridization chamber. Incubate for 30 min at room temperature. Remove preamplification solution and pipet 50 μl of hairpin solution into the chamber. Cover hybridization chambers with a coverslip and leave for 3 h at room temperature in a humidified slide staining box.
5. Amplification washing and mounting: Remove hairpin solution and wash sections with 50 μl 5×SSCT, 2 × 30 min at room temperature. Apply a DNA stain such as DAPI or Hoechst to the chambers for 5 min at room temperature and then wash with PBS for 5 min. Fill the hybridization chamber with antifade mounting media and image immediately or store at 4 °C for later imaging.
6. Imaging: Acquire images using a Zeiss LSM 880 confocal microscope with a 63×/1.4 oil Plan Apochromat objective or similar microscope. The laser lines and associated gating that are recommended for each fluorophore are as follows: DAPI/405 (378–497), 488 (493–520), 514 (580–624), 561 (612–661), and 647 (668–755).

3.4 Probe Wiping and Rehybridization of New Probes

The first probe set can be removed and new probes added if you plan to image more genes than can be imaged in a single-staining experiment (more than 4).

1. Remove mounting media and pipet 50 μl of 2×SSC, 3 × 15 min at room temperature.
2. Wipe DNA probes and hairpins with DNase solution for 1 h at room temperature. Then pipet 50 μl of probe wash solution 3 × 5 min at 37 °C followed by 3 × 5 min 2×SSC washes at room temperature.

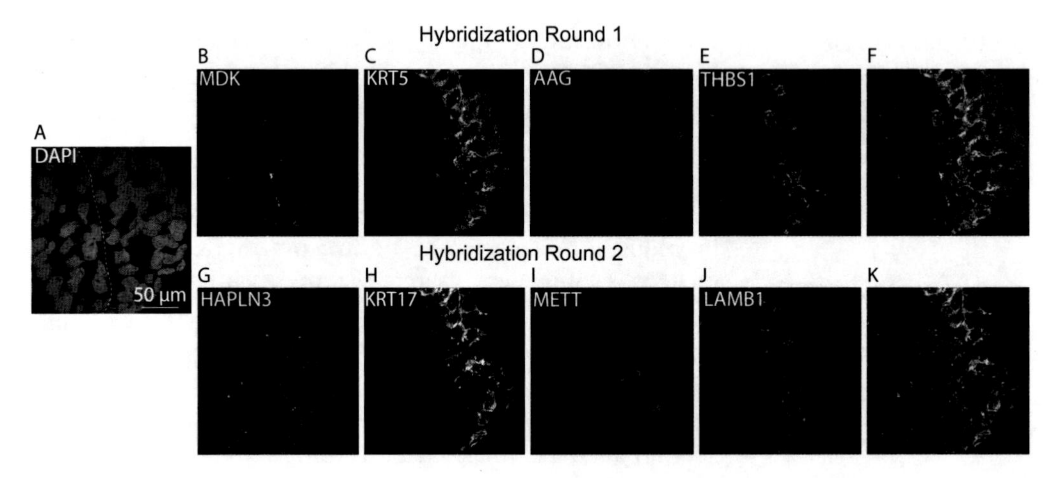

Fig. 3 Multiplexed FISH. Two sequential rounds of v3.HCR-FISH performed on a regenerating axolotl limb tissue section. The DAPI image is shown for reference (**a**) along with four genes stained in round 1: *Midkine* (*Mdk*) (**b**), *Keratin 5* (*Krt5*) (**c**), *Axolotl anterior gradient* (*Aag*) (**d**), *Thrombospondin 1* (*Thbs1*) (**e**) and an overlay of all four genes (**f**). Four genes stained in round 2: *Hyaluronan And Proteoglycan Link Protein 3* (**g**), *Keratin 17* (**h**), *Methyltransferase-like* (**i**), *Laminin beta 1* (**j**) and an overlay of all 4 genes (**k**). 50-µm scale bar is included on DAPI image and a dotted line represents the epithelium/mesenchyme boundary

3. Repeat steps in Subheading 3.3 with new probes. Skip the nuclear staining step in subsequent rounds of v3.HCR-FISH.
4. Repeat steps in Subheading 3.4 for each subsequent round of staining.

3.5 Multiround v3. HCR-FISH Image Analysis: Segmentation

Images of multiround experiments are aligned using Zen software. We provide an example data analysis pipeline for counting fluorescent foci for two rounds of HCR-FISH. This analysis consists broadly of two major steps: cell segmentation and dot counting. We perform cell segmentation using the deep-learning method, Cellpose [38] (Fig. 3). Cellpose can be utilized in a web app (www.cellpose.org) or through local installation (https://github.com/mouseland/cellpose).

1. To open the installed Cellpose on your local machine, first open Anaconda.
2. Open command prompt through base directory.
3. Type -run "conda activate cellpose."
4. Type -run "python -m cellpose."
5. Drag the image into Cellpose window.
6. Change cell diameter based on sample type and image acquisition.
7. Change parameters if needed for optimal segmentation.
8. Once desired mask is generated, save mask as a .png.

9. It may be necessary to convert this mask file to a .tif depending on what program you wish to use for your analysis.
10. It may be necessary to hand annotate images if settings do not catch every cell, this is easily accomplished in the cellpose GUI.

3.6 Multiround v3. HCR-FISH Image Analysis: Dot Counting

Perform dot detection and counting using CellProfiler [39]. This pipeline consists of identifying cell regions of interest from the Cellpose mask, isolating puncta from raw FISH images, counting dots, and assigning dots to a cell. The CellProfiler pipeline used is provided here (https://github.com/Lovelya-NEU/Hybridization-chain-reaction-fluorescence-in-situ-hybridization-in-Ambystoma-mexicanum-tissue). Raw image files were put through a gaussian filter of sigma radius 1 prior to uploading into the pipeline (Fig. 4).

3.7 Whole-Mount v3. HCR-FISH

1. Sample collection and preparation: For developmental limb buds, cut the embryo in half and collect only the trunk and limb bud. Immediately fix the tissue in at least 10× the volume of 4% PFA in a 1.5-ml tube overnight at 4 °C. Wash tissue 3 × 5 min in 1×PBST at room temperature.
2. Sample dehydration: Dehydrate samples on ice in an increasing methanol series for 5 min at each step: 25% MeOH (diluted in 1×PBST), 50% MeOH, 75% MeOH, and 100% MeOH. Replace the 100% MeOH and store at −20 °C overnight (*see* **Note 6**).
3. Sample rehydration: Rehydrate samples on ice in a decreasing methanol series for 5 min at each step: 75% MeOH (diluted in 1×PBST), 50% MeOH, 25% MeOH, followed by two 5-min washes in PBST.
4. Proteinase K treatment: Treat samples with 10 μg/ml of proteinase K for 15 min at room temperature (*see* **Note 7**). Terminate proteinase K activity by fixing samples in 4% PFA at room temperature for 3 × 20 min followed by 3 × 5 min washes with 1×PBST at room temperature.
5. Hybridization: Briefly incubate tissue in hybridization buffer at 37 °C for 5 min in a 1.5-ml tube. Prehybridize tissue in preheated hybridization buffer at 37 °C for 30 min. Apply enough hybridization probe solution to cover the tissues (usually 50 μl) in a 1.5-ml tube and incubate overnight at 37 °C.
6. Probe washing: Wash the sample in preheated probe wash solution for 4 × 15 min in a 37 °C incubator. Wash samples 2 × 5 min in 5×SSCT at room temperature. Pre-amplify tissue in amplification buffer for 30 min at room temperature.

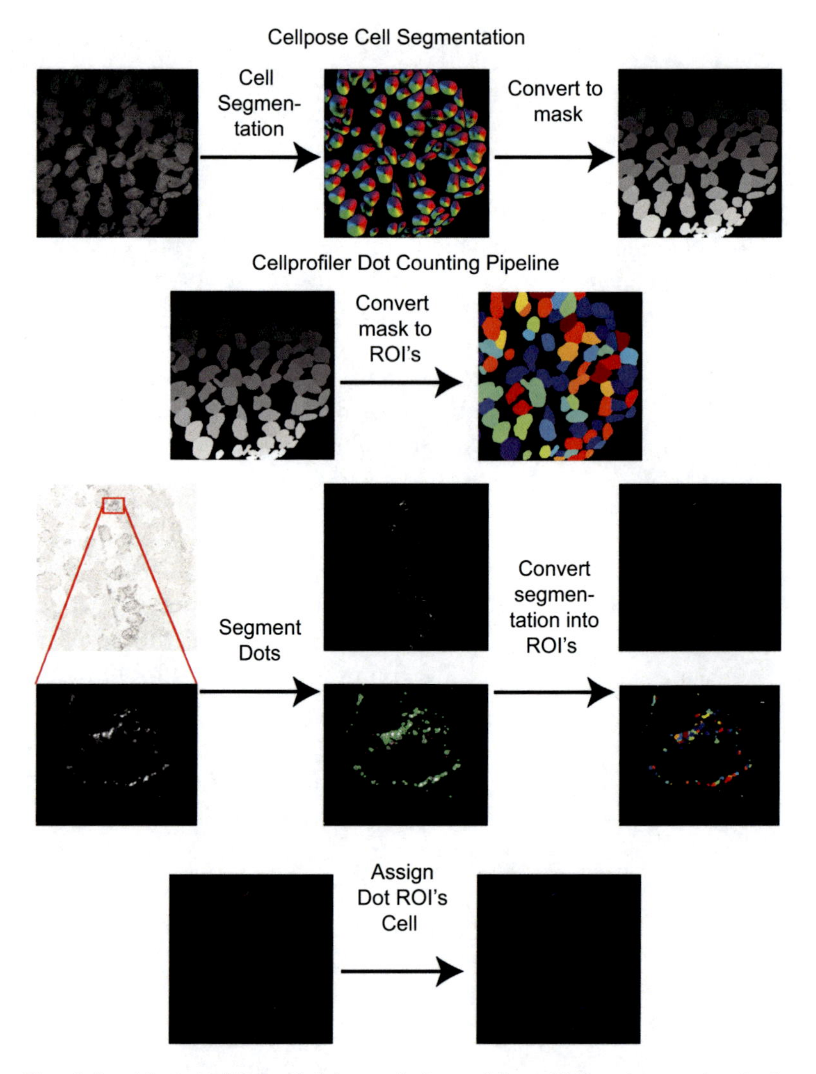

Fig. 4 Graphical depiction of data analysis workflow. Major steps and outputs from HCR analysis pipeline. First, nuclei/cells are segmented using Cellpose (**a**). This mask is saved and then imported into CellProfiler where it is converted into regions of interest (**b**). HCR-FISH puncta are then identified in CellProfiler and assigned to a parent cell (**c**). The data can be outputted into a spreadsheet for further analysis

7. Amplification: Prepare the hairpin amplification solution as described in Subheading 2.3. Make enough hairpin to submerge the tissue (usually 25–50 μl). Apply hairpin amplification solution to sample and incubate overnight at room temperature in the dark.

8. Amplification washing: Wash samples with 1 ml of 5×SSCT at room temperature for 5, 30, 30, and 5 min each. Sample can be stored at 4 °C prior to imaging.

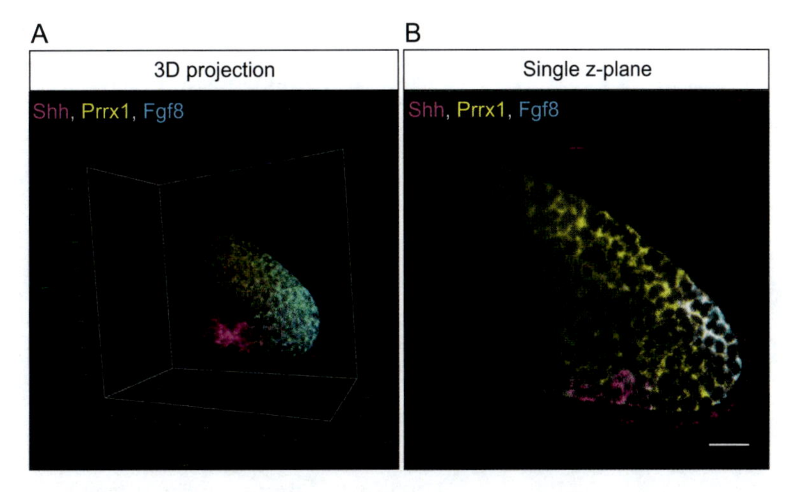

Fig. 5 Whole-mount FISH. Whole-mount, 3D visualization of *Shh* (magenta), *Prrx1* (yellow), and *Fgf8* (cyan) in a stage 46 axolotl limb bud (**a**). A single z-slice of the limb bud (**b**). Scale bar = 100 μm

3.8 Imaging with Lightsheet Fluorescence Microscopy (LSFM)

1. Mount sample in LSFM capillary with 1.5% low-boiling point agarose (*see* **Note 8**).
2. Once the sample solidifies, place agarose-embedded sample in 1×PBS for 5 min.
3. Whole-mount imaging is performed on a Zeiss Lightsheet Z.1 microscope using a 10× water objective with laser lines 488, 561, and 647 (Fig. 5) (*see* **Note 9**). The imaging chamber is filled with 1×PBS.

4 Notes

1. Perform the protocol using RNase-free techniques. Clean work area with RNase AWAY and use fresh gloves. Use disposable plasticware or RNase-free glassware for all the solutions. Prepare all solutions using Diethyl pyrocarbonate (DEPC)-treated water or water from a high-end water purification system that generates 18-megohm water. When handling DEPC, make sure you are wearing gloves and only open the bottle inside a fume hood.
2. Probes can also be designed through Molecular Instruments. They will design and synthesize the probe pools and ship hybridization-ready pools.
3. For further description of the website design, *see* https://github.com/davidfstein/probegenerator.
4. Alternatively, plates of individual oligos that contain a single oligo in each well can be purchased, which will provide a large amount of oligo (4 nM per oligo for example).

5. If one round of HCR is going to be performed, sections can be collected on Superfrost Plus microscope slides instead of coverslips and the hybridization chambers are not used.
6. Samples can be left at −20 °C for long-term storage.
7. The concentration and duration of proteinase K treatment should be optimized for each tissue type. Our lab has found that this treatment is sufficient for mild clearing of tissue without significant tissue damage.
8. It is important to orient the sample such that no tissue obscures the region of interest, such as the body cavity. Forceps can be inserted into the LSFM capillary as the agarose is solidifying to manipulate the tissue into position.
9. Refractive index matching can be performed by embedded tissue in EasyIndex overnight at 4 °C and imaging with a 5× or 20× clearing-enabled objective in EasyIndex.

Acknowledgements

All images were acquired in the Northeastern Chemical Imaging of Living Systems facility. We thank the Institute for Chemical Imaging of Living Systems at Northeastern University for consultation and imaging support. The authors would like to thank Harry Choi at Molecular Instruments for feedback during protocol development. Grant sponsors include NSF grants 1558017 and 1656429 to JRM. NIH grant HD099174 to JRM. Material and information obtained from the Ambystoma Genetic Stock Center funded through NIH grant: P40-OD019794.

Author Contributions AML, TJD, ETM, and JRM designed the experiments and conducted the experiments. DFS designed the probe design pipeline. AL, TJD, and JRM analyzed data and wrote the paper. All authors contributed to editing the manuscript and gave approval of the final version.

Competing Interests Authors declare no competing interests.

References

1. Gall JG, Pardue ML (1969) Formation and detection of RNA-DNA hybrid molecules in cytological preparations. Proc Natl Acad Sci U S A 63(2):378–383. https://doi.org/10.1073/pnas.63.2.378
2. John HA, Birnstiel ML, Jones KW (1969) RNA-DNA hybrids at the cytological level. Nature 223(5206):582–587. https://doi.org/10.1038/223582a0
3. Herrington CS, Burns J, Graham AK, Evans M, McGee JO (1989) Interphase cytogenetics using biotin and digoxigenin labelled probes I: relative sensitivity of both reporter molecules for detection of HPV16 in CaSki

cells. J Clin Pathol 42(6):592–600. https://doi.org/10.1136/jcp.42.6.592
4. Tautz D, Pfeifle C (1989) A non-radioactive in situ hybridization method for the localization of specific RNAs in Drosophila embryos reveals translational control of the segmentation gene hunchback. Chromosoma 98(2):81–85. https://doi.org/10.1007/bf00291041
5. Aigner S, Pette D (1990) In situ hybridization of slow myosin heavy chain mRNA in normal and transforming rabbit muscles with the use of a nonradioactively labeled cRNA. Histochemistry 95(1):11–18. https://doi.org/10.1007/bf00737222
6. Komminoth P, Merk FB, Leav I, Wolfe HJ, Roth J (1992) Comparison of 35S- and digoxigenin-labeled RNA and oligonucleotide probes for in situ hybridization. Expression of mRNA of the seminal vesicle secretion protein II and androgen receptor genes in the rat prostate. Histochemistry 98(4):217–228. https://doi.org/10.1007/bf00271035
7. Hemmati-Brivanlou A, Frank D, Bolce ME, Brown BD, Sive HL, Harland RM (1990) Localization of specific mRNAs in Xenopus embryos by whole-mount in situ hybridization. Development 110(2):325–330
8. Herrmann BG (1991) Expression pattern of the Brachyury gene in whole-mount TWis/TWis mutant embryos. Development 113(3):913–917
9. Schulte-Merker S, Ho RK, Herrmann BG, Nüsslein-Volhard C (1992) The protein product of the zebrafish homologue of the mouse T gene is expressed in nuclei of the germ ring and the notochord of the early embryo. Development 116(4):1021–1032
10. Melton DA, Krieg PA, Rebagliati MR, Maniatis T, Zinn K, Green MR (1984) Efficient in vitro synthesis of biologically active RNA and RNA hybridization probes from plasmids containing a bacteriophage SP6 promoter. Nucleic Acids Res 12(18):7035–7056. https://doi.org/10.1093/nar/12.18.7035
11. Raj A, van den Bogaard P, Rifkin SA, van Oudenaarden A, Tyagi S (2008) Imaging individual mRNA molecules using multiple singly labeled probes. Nat Methods 5(10):877–879. https://doi.org/10.1038/nmeth.1253
12. Kishi JY, Beliveau BJ, Lapan SW et al (2018) SABER enables highly multiplexed and amplified detection of DNA and RNA in cells and tissues. bioRxiv. https://doi.org/10.1101/401810
13. Rouhanifard SH, Mellis IA, Dunagin M et al (2018) ClampFISH detects individual nucleic acid molecules using click chemistry-based amplification. Nat Biotechnol. https://doi.org/10.1038/nbt.4286
14. Codeluppi S, Borm LE, Zeisel A et al (2018) Spatial organization of the somatosensory cortex revealed by osmFISH. Nat Methods 15(11):932–935. https://doi.org/10.1038/s41592-018-0175-z
15. Wang F, Flanagan J, Su N et al (2012) RNAscope: a novel in situ RNA analysis platform for formalin-fixed, paraffin-embedded tissues. J Mol Diagn 14(1):22–29. https://doi.org/10.1016/j.jmoldx.2011.08.002
16. Nagendran M, Riordan DP, Harbury PB, Desai TJ (2018) Automated cell-type classification in intact tissues by single-cell molecular profiling. Elife 7. https://doi.org/10.7554/eLife.30510
17. Gyllborg D, Langseth CM, Qian X et al (2020) Hybridization-based in situ sequencing (HybISS) for spatially resolved transcriptomics in human and mouse brain tissue. Nucleic Acids Res 48(19):e112. https://doi.org/10.1093/nar/gkaa792
18. Choi HMT, Schwarzkopf M, Fornace ME et al (2018) Third-generation in situ hybridization chain reaction: multiplexed, quantitative, sensitive, versatile, robust. Development 145(12). https://doi.org/10.1242/dev.165753
19. Dirks RM, Pierce NA (2004) Triggered amplification by hybridization chain reaction. Proc Natl Acad Sci U S A 101(43):15275–15278. https://doi.org/10.1073/pnas.0407024101
20. Choi HM, Chang JY, Trinh LA, Padilla JE, Fraser SE, Pierce NA (2010) Programmable in situ amplification for multiplexed imaging of mRNA expression. Nat Biotechnol 28(11):1208–1212. https://doi.org/10.1038/nbt.1692
21. Shah S, Lubeck E, Zhou W, Cai L (2016) In situ transcription profiling of single cells reveals spatial organization of cells in the mouse hippocampus. Neuron 92(2):342–357. https://doi.org/10.1016/j.neuron.2016.10.001
22. Voss SR, Epperlein HH, Tanaka EM (2009) Ambystoma mexicanum, the axolotl: a versatile amphibian model for regeneration, development, and evolution studies. Cold Spring Harb Protoc 2009(8). https://doi.org/10.1101/pdb.emo128
23. Khattak S, Tanaka EM (2015) Transgenesis in axolotl (Ambystoma mexicanum). Methods Mol Biol 1290:269–277. https://doi.org/10.1007/978-1-4939-2495-0_21
24. Flowers GP, Timberlake AT, McLean KC, Monaghan JR, Crews CM (2014) Highly efficient targeted mutagenesis in axolotl using Cas9 RNA-guided nuclease. Development

141(10):2165–2171. https://doi.org/10.1242/dev.105072

25. Fei JF, Schuez M, Knapp D, Taniguchi Y, Drechsel DN, Tanaka EM (2017) Efficient gene knockin in axolotl and its use to test the role of satellite cells in limb regeneration. Proc Natl Acad Sci U S A 114(47):12501–12506. https://doi.org/10.1073/pnas.1706855114
26. Nowoshilow S, Schloissnig S, Fei JF et al (2018) The axolotl genome and the evolution of key tissue formation regulators. Nature 554(7690):50–55. https://doi.org/10.1038/nature25458
27. Smith JJ, Timoshevskaya N, Timoshevskiy VA, Keinath MC, Hardy D, Voss SR (2019) A chromosome-scale assembly of the axolotl genome. Genome Res 29(2):317–324. https://doi.org/10.1101/gr.241901.118
28. Gardiner DM, Blumberg B, Komine Y, Bryant SV (Jun 1995) Regulation of HoxA expression in developing and regenerating axolotl limbs. Development 121(6):1731–1741
29. Sun HB, Neff AW, Mescher AL, Malacinski GM (1995) Expression of the axolotl homologue of mouse chaperonin t-complex protein-1 during early development. Biochim Biophys Acta 1260(2):157–166. https://doi.org/10.1016/0167-4781(94)00187-8
30. Freitas PD, Lovely AM, Monaghan JR (2019) Investigating Nrg1 signaling in the regenerating axolotl spinal cord using multiplexed FISH. Dev Neurobiol. https://doi.org/10.1002/dneu.22670
31. Woodcock MR, Vaughn-Wolfe J, Elias A et al (2017) Identification of mutant genes and introgressed tiger salamander DNA in the laboratory axolotl, Ambystoma mexicanum. Sci Rep 7(1):6. https://doi.org/10.1038/s41598-017-00059-1
32. Leigh ND, Sessa S, Dragalzew AC et al (2020) von Willebrand factor D and EGF domains is an evolutionarily conserved and required feature of blastemas capable of multitissue appendage regeneration. Evol Dev 22(4):297–311. https://doi.org/10.1111/ede.12332
33. Schloissnig S, Kawaguchi A, Nowoshilow S et al (2021) The giant axolotl genome uncovers the evolution, scaling, and transcriptional control of complex gene loci. Proc Natl Acad Sci U S A 118(15). https://doi.org/10.1073/pnas.2017176118
34. Goh JJL, Chou N, Seow WY et al (2020) Highly specific multiplexed RNA imaging in tissues with split-FISH. Nat Methods 17(7):689–693. https://doi.org/10.1038/s41592-020-0858-0
35. Chung K, Deisseroth K (2013) CLARITY for mapping the nervous system. Nat Methods 10(6):508–513. https://doi.org/10.1038/nmeth.2481
36. Beliveau BJ, Kishi JY, Nir G et al (2018) OligoMiner provides a rapid, flexible environment for the design of genome-scale oligonucleotide in situ hybridization probes. Proc Natl Acad Sci U S A 115(10):e2183–e2192. https://doi.org/10.1073/pnas.1714530115
37. Langmead B, Salzberg SL (2012) Fast gapped-read alignment with Bowtie 2. Nat Methods 9(4):357–359. https://doi.org/10.1038/nmeth.1923
38. Stringer C, Wang T, Michaelos M, Pachitariu M (2021) Cellpose: a generalist algorithm for cellular segmentation. Nat Methods 18(1):100–106. https://doi.org/10.1038/s41592-020-01018-x
39. McQuin C, Goodman A, Chernyshev V et al (2018) CellProfiler 3.0: next-generation image processing for biology. PLoS Biol 16(7):e2005970. https://doi.org/10.1371/journal.pbio.2005970

Chapter 7

Ethyl Cinnamate-Based Tissue Clearing Strategies

Wouter Masselink and Elly M. Tanaka

Abstract

Tissue clearing turns otherwise turbid and opaque tissue transparent, enabling imaging deep within tissues. The nontransparent nature of most tissues is due to the refractive index mismatch between its three major constituent components (lipids, proteins, and water). All tissue clearing methods rectify this mismatch by homogenizing the refractive index within the tissue and carefully matching it to the surrounding media. Here we describe a detailed protocol to clear a wide range of salamander tissues. We also include several optional steps such as depigmentation, antibody staining, and tissue mounting. These steps are optional, and do not change anything in the steps needed for tissue clearing. Depending on the fluorescent signal and optics employed, images up to several millimeters inside of the tissue can be acquired.

Key words Tissue clearing, Dehydration, Ethyl cinnamate, Antibody staining

1 Introduction

Tissue clearing is the process of manipulating the refractive index of a tissue to turn otherwise opaque tissues transparent enabling researchers to image deep within whole-mount tissue. The anatomist Werner Spalteholz was the first to show the principles of refractive index matching on biological tissues in the early twentieth century. Spalteholz used a combination of benzyl alcohol and methyl salicylate of human and animal tissues [1]. The power of tissue clearing in the context of axolotl regeneration was already appreciated over 40 years ago. Jonathan Slack used a slight modification of the original Spalteholz method, using methyl salicylate as a high-refractive index matching solution to refractive index match dehydrated axolotl samples [2, 3]. While these particular chemicals are incompatible with modern-day fluorescent imaging, these principles continue to be applied to this day.

The turbidity and opaqueness of tissues are caused by the refractive index mismatch by its three major constituents, water (RI: 1.3), lipids (RI: 1.4), and protein (RI: 1.5). Since all these components are present at slightly different ratios throughout a

Ashley W. Seifert and Joshua D. Currie (eds.), *Salamanders: Methods and Protocols*,
Methods in Molecular Biology, vol. 2562, https://doi.org/10.1007/978-1-0716-2659-7_7,

tissue, light is quickly scattered and dispersed resulting in the appearance of an opaque tissue. To combat this and turn a tissue transparent, the refractive index of the tissue needs to be homogenized and matched to the surrounding immersion media. Broadly speaking, two different types of tissue clearing approaches were developed to achieve this: aqueous and solvent-based methods. Aqueous-based methods attempt to increase the refractive index of the water by immersion of the tissue in a high-refractive index watery solution. This can be as simple as submerging the tissue in sucrose solution and allowing for passive diffusion to equilibrate the RI throughout the tissue [4]. Other methods attempt to force clearing solutions into the tissue using electrophoresis [5, 6]. Generally, these methods have the advantage that they preserve endogenous fluorescence and do not result in tissue shrinkage. However, they often require extensive optimization for a given tissue, use specialized chambers, and especially when applied to larger tissues are rather laborious. On the other hand, solvent-based methods such as the Spalteholz method are more rudimentary in their approach. In principle, solvent-based methods attempt to turn a tissue transparent in a two-step process of dehydrating and delipidating, followed by refractive index matching of the remaining tissue with a high-refractive index solution. These methods are fast, easy, and broadly applicable. However, they result in tissue shrinkage, often use toxic and/or odorous organic solvents, and fail to preserve endogenous fluorescence.

The recent optimization of solvent-based methods has resulted in the development of a nontoxic, solvent-based method called 2Eci (second-generation ethyl cinnamate) that preserves a broad range of fluorescent proteins [7, 8]. This has been applied successfully to a wide range of tissues and organisms including axolotl, Xenopus, Drosophila, and cerebral organoids. In addition, Adrados and colleagues reported the optimization of a similar ethyl cinnamate-based method [9]. Ethyl cinnamate is used extensively in our day-to-day research on axolotl, exploring a wide variety of biological tissues, including brain, spinal cord, tail, and limb.

It is worth mentioning that other non-dehydration-based strategies have also been used successfully on Axolotl. Duerr and colleagues have used a 2,2′-thiodiethanol (TDE)-based clearing strategy to clear axolotl tissue [10]. Furthermore, Pende and colleagues used a CUBIC-based clearing strategy to clear a wide range of organisms including axolotl [11]. Each method has its own peculiarities, and we strongly encourage users to explore the wide range of existing methods. Choosing a suitable clearing strategy depends on the experimental questions, necessary downstream analysis, and the available equipment. However, for the purpose of this method, we will discuss the use of the broadly applicable ethyl cinnamate-based clearing approach.

1.1 Planning the Tissue Clearing and Imaging Process

A successful clearing and imaging experiment requires a consideration of the peculiarities of a given sample and availability of microscopes. Prior to processing your samples, consideration should be given to the labeling approach, if there is a need to remove pigmentation, and finally what microscopes are available for image acquisition. Depending on those needs, the following basic outline is suggested: fixation, depigmentation (optional), antibody labeling (optional), tissue embedding (optional), dehydration, refractive index matching, and finally mounting and imaging (Fig. 1).

Dehydration is an essential aspect of this clearing method and is absolutely required. Any tissue processing steps which require aqueous solution should be done prior to these steps. After fixing samples (and optionally applying depigmentation, antibody labeling, and mounting), samples are transferred to a sequential dehydration series of 30%, 50%, 70%, and 2× 100% 1-Propanol (99%; Sigma Cat.W292818, 99.7% anhydrous: Sigma Cat. 27,944): PBS solution which was set to a pH of 9.0–9.5 with trimethylamine (Sigma Cat. T0886). Dehydration is performed at 4 °C on a gyratory rocker to maintain high levels of fluorescence and took at least 4 h per step for small samples but can take up to 24 h for large samples. The use of pH-adjusted 1-propanol solutions is essential to maintain most fluorescent proteins (GFP and its derivatives as well as mCherry are validated to be compatible with this clearing method, while tdTomato is not).

After dehydration, the samples are transferred to ECi (≥98%: Sigma Cat. W243000) in VOA glass vials and incubated on a gyratory rocker at room temperature until transparent. This can require 1 h up to 24 h depending on the size of the sample and whether samples are mounted in phytagel. For large samples or samples mounted in phytagel, we found it useful to refresh the ethyl cinnamate after several hours of clearing. Samples should be stored in light-protected and air-sealed containers, and recordings should be performed within the next days after clearing. Ethyl cinnamate is an FDA-approved food additive and generally safe with a pleasant smell. It has a high-refractive index and efficiently allows for refractive index matching of a dehydrated tissue.

Depending on the tissue of interest, and the available microscopes, different mounting and imaging strategies may be used. We will briefly discuss three common examples, including free-floating and phytagel-embedded samples to be imaged on microscopes with conventional stages, as well as lightsheet microscopes such as the Zeiss lightsheet 7.

This protocol should be treated as a starting off point which should provide adequate clearing for most conditions. However, based on the sample type, size, the signal that should be recorded, and microscope availability, users would need to optimize several aspects. Important aspects to optimize for each sample include fixation and dehydration time as well as sample mounting methods.

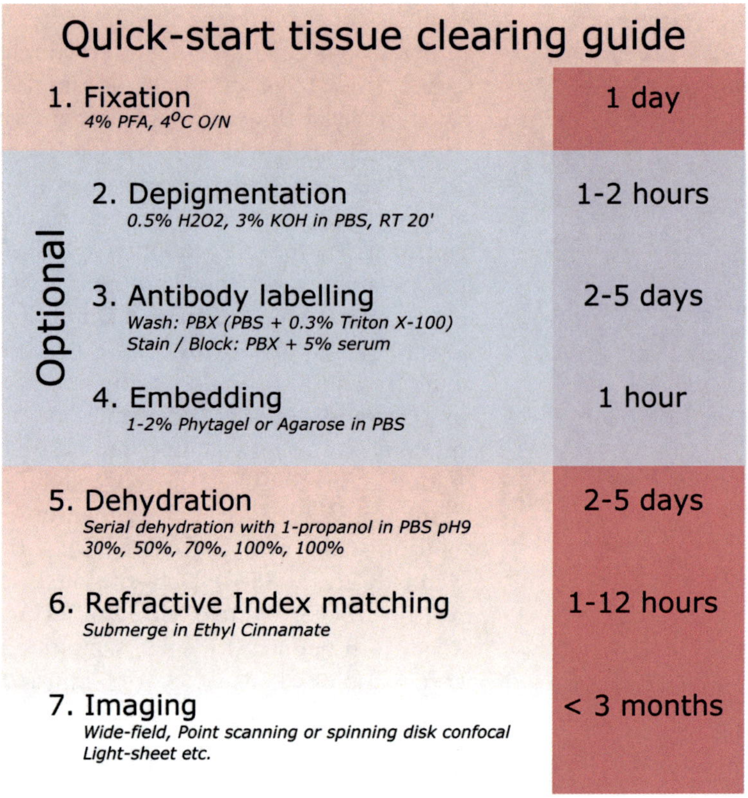

Quick-start tissue clearing guide

Step	Duration
1. Fixation *4% PFA, 4°C O/N*	1 day
Optional: 2. Depigmentation *0.5% H2O2, 3% KOH in PBS, RT 20'*	1-2 hours
Optional: 3. Antibody labelling *Wash: PBX (PBS + 0.3% Triton X-100)* *Stain / Block: PBX + 5% serum*	2-5 days
Optional: 4. Embedding *1-2% Phytagel or Agarose in PBS*	1 hour
5. Dehydration *Serial dehydration with 1-propanol in PBS pH9* *30%, 50%, 70%, 100%, 100%*	2-5 days
6. Refractive Index matching *Submerge in Ethyl Cinnamate*	1-12 hours
7. Imaging *Wide-field, Point scanning or spinning disk confocal* *Light-sheet etc.*	< 3 months

Fig. 1 Step-wise overview of ethyl cinnamate-based tissue clearing approaches. Approximate duration for each step is indicated but can be varied depending on the tissue type and size

2 Materials

1. Gyratory rocker.
2. pH paper, Whatman indicator papers (pH 4.5–10.0) (Whatman Cat. WHA2614991).
3. Glass VOA vials (Thermo Scientific: C126-0020).
4. 37 °C incubator.
5. Fixation and staining vessels (Eppendorf tubes, conical tubes, etc.).
6. 1-mL syringes.
7. Scalpel blades.
8. High-grit sandpaper.
9. Forceps (Dumont #5).
10. Pattex super glue gel.
11. Disposable base mold (Fisherbrand Cat. 22-363-555).

12. Lint-free tissue paper (KimTECH Kimwipes).
13. Fine gage needles (g30).
14. Imaging chambers: Ibidi μ-Dish 35 mm high, glass bottom (81158), as well as Ibidi μ-plates 96-well black uncoated (89621) for high throughput imaging.
15. Air displacement pipettes.
16. Pasteur pipettes.
17. Beaker glasses.
18. Graduated cylinders.
19. High-precision laboratory scale.
20. 0.45-μm screw top rapid flow sterile filter (Nalgene Cat. 564-0020).
21. Schott bottles.
22. Fluorescence microscope (confocal, spinning disk, and lightsheet).

Prepare all solutions fresh unless otherwise stated.

2.1 Fixation

1. *Sodium phosphate buffer, 240 mM, pH 7.4.* To make 1 L of sodium phosphate buffer, add 27.52 g of disodium phosphate (Na_2HPO_4) and 5.52 g of monosodium phosphate (NaH_2PO_4) to 800 mL of deionized water. Adjust the pH to 7.4 using 5 N sodium hydroxide (NaOH) to fill to 1 L with deionized water.
2. *PFA solution, 4%, pH 7.4.* To make 2 L of PFA solution, weigh 80 g of paraformaldehyde into a 2-L bottle. Add 1 L of deionized water and several drops of 5 N sodium hydroxide (NaOH). Heat to 60 C in a water bath with occasional shaking until dissolved. Cool on ice. Add 1 L of sodium phosphate buffer. Using pH paper, confirm that the PFA solution is still at pH 7.4. Adjust pH with 5 N sodium hydroxide (NaOH) if necessary. Sterile filter and aliquot into 15 and 50 mL batches and store at −20 °C.

2.2 Depigmentation (Optional)

1. *Phosphate-buffered saline (10×).* To make 1 L of PBS (10×) combine the following: 80 g of sodium chloride (NaCl), 2 g of potassium chloride (KCl), 14.4 g of disodium phosphate (Na_2HPO_4), 2.4 g of monopotassium phosphate (KH_2PO_4), and 800 mL of deionized water. Adjust pH to 7.4 with 5 N sodium hydroxide (NaOH). Add deionized water to 1 L. Sterile filter or autoclave and store at room temperature.
2. *Depigmentation solution.* To make 100 mL of depigmentation solution, combine the following: 10 mL of PBS (10×), 0.5 mL of hydrogen peroxide (H_2O_2), and 3 g of potassium hydroxide (KOH). Add deionized water to 100 mL.

2.3 Antibody Staining (Optional)

1. *Phosphate-buffered saline* (*1*×). To make 1 L of PBS (1×), combine 100 mL of PBS (10×) with 900 mL of deionized water. Sterile filter or autoclave and store at room temperature.
2. *PBX* (*1*×). To make 1 L of PBX (PBS + 0.3% triton X-100), combine 30 mL of Triton X-100, with 100 mL of PBS 10×. Add deionized water to 1 L. Store at room temperature.
3. *Blocking solution* (*100 mL*). To make 100 mL of blocking solution, combine 5 mL goat serum with 95 mL PBX.

2.4 Embedding (Optional)

1. *Embedding solution*. To make 100 mL embedding solution add 1 g phytagel (Sigma P8169) to 100 mL of PBS (1×). Heat until dissolved under continuous stirring. Use while warm.

2.5 Dehydration

1. *30% 1- propanol* (*pH 9*). To make 100 mL of 30% 1-propanol (pH 9), combine 30 mL of 1-propanol (Sigma W292818) with 70 mL PBS (1×). Adjust the pH using triethylamine (Sigma T0886) to pH 9 (*see* **Note 1**).
2. *50% 1- propanol* (*pH 9*). To make 100 mL of 50% 1-propanol (pH 9), combine 50 mL of 1-propanol (Sigma W292818) with 50 mL PBS (1×). Adjust the pH using triethylamine (Sigma T0886) to pH 9 (*see* **Note 1**).
3. *70% 1- propanol* (*pH 9*). To make 100 mL of 70% 1-propanol (pH 9), combine 70 mL of 1-propanol (Sigma W292818) with 30 mL PBS (1×). Adjust the pH using triethylamine (Sigma T0886) to pH 9 (*see* **Note 1**).
4. *100% 1- propanol* (*pH 9*). To make 100 mL of 100% 1-propanol anhydrous (pH 9), adjust the pH of 100 mL 1-propanol anhydrous (Sigma 279544) using triethylamine (Sigma T0886) to pH 9 (*see* **Note 1**).

2.6 Refractive Index Matching

1. Refractive index matching solution is a 98% solution of ethyl cinnamate (Sigma W243000) and is bought directly from the manufacturer (*see* **Note 2**).

3 Methods

3.1 Fixation

1. Thaw an aliquot of 4% PFA pH 7.4. Fix fresh tissue in PFA solution at 4 °C overnight on a rocker (*see* **Note 3**).

3.2 Depigmentation (Optional, See Note 4)

1. Wash out fixative in PBS (1×) at room temperature twice for 20 min each.
2. Transfer samples into depigmentation solution. Exact time required will depend on the sample size and type. To depigment axolotl embryos, we commonly treat for 20 min.
3. Stop depigmentation by transferring samples back into PBS (1×).

3.3 Antibody Labeling (Optional, See Note 5)

1. Wash sample extensively in PBS (2 × 20 min) at room temperature in appropriately sized vessels (Eppendorf tubes, VOA glass vials, or conical tubes) on a gyratory rocker.
2. Permeabilize the sample in PBX (PBS + 0.3% Tween X-100) for 3 h at room temperature on a gyratory rocker.
3. Incubate in blocking solution overnight at 37 °C on a gyratory rocker.
4. Incubate sample in primary antibody in blocking solution for 2 days at 37 °C on a gyratory rocker.
5. Warm PBX to 37 °C.
6. Rinse sample briefly twice in preheated PBX at 37 °C.
7. Wash sample in PBX 4× 2 h at 37 °C on a gyratory rocker.
8. Incubate sample in blocking solution overnight at 37 °C on a gyratory rocker.
9. Incubate sample in secondary antibody in blocking solution for 2 days at 37 °C on a gyratory rocker.
10. Rinse sample twice briefly in PBX at 37 °C.
11. Wash sample in PBX 2× 1 h at 37 °C on a gyratory rocker.
12. Wash sample in PBS 2× 1 h at room temperature on a gyratory rocker.

3.4 Embedding (Optional, See Note 6)

Mounting on conventional stages.

1. Heat 1% phytagel solution in a microwave until completely liquid.
2. Place sample in disposable base mold, orientated correctly for imaging.
3. Confirm phytagel is still liquid and warm to the touch (approx. 35–45 °C)
4. Flood base mold with 1% phytagel solution and leave to set.
5. Trim excess phytagel, leaving a small lip adjacent to the sample which can be used to glue the sample in place during image acquisition (*see* **Note 7**).

Mounting in cylinders for use on systems such as the Z1 light-sheet microscope.

1. Cut the end from a 1-mL syringe using a scalpel blade, and smooth out the edge using high-grit sandpaper.
2. Submerge sample in warm 1% phytagel solution and using the plunger suck into the syringe. Carefully orientate the sample if needed using fine forceps. Manually rotate the syringe for a couple of minutes or until the phytagel has solidified.

3. Leave phytagel to set and push out of the syringe. You should now have a sample mounted in a cylinder of phytagel.
4. Trim excess phytagel using a scalpel blade, taking care to leave enough to glue the sample in place during mounting.

3.5 Dehydration

1. Dehydrate sample in up to 24 h using 30% 1-propanol pH 9 using triethylamine at 4C in glass VOA vials (*see* **Note 8**).
2. Replace 30% 1-propanol with 50% 1-propanol pH 9 using triethylamine at 4 °C in glass VOA vials. Dehydrate up to 24 h (*see* **Notes 9** and **10**).
3. Replace 50% 1-propanol with 70% 1-propanol pH 9 using triethylamine at 4 °C in glass VOA vials. Dehydrate up to 24 h (*see* **Notes 9** and **10**).
4. Replace 70% 1-propanol with 100% 1-propanol pH 9 using triethylamine at 4 °C in glass VOA vials. Dehydrate up to 24 h (*see* **Notes 9** and **10**).
5. Dehydrate sample up to 24 h using 100% 1-propanol pH 9 using triethylamine at 4 °C in glass VOA vials (*see* **Note 10**).
6. Dehydrate sample another 24 h using 100% 1-propanol pH 9 using triethylamine at 4 °C in glass VOA vials to completely remove residual water molecules (*see* **Note 11**).

3.6 Refractive Index Matching

1. Replace 100% 1-propanol pH 9 with ethyl cinnamate (Sigma Cat. W243000). Incubate for 1–12 h at room temperature in light-protected and air-sealed containers on a gyratory rocker (*see* **Note 12**).
2. Replace ethyl cinnamate with fresh ethyl cinnamate to remove any residual 1-propanol at room temperature (*see* **Note 12**).
3. Store at room temperature protected from ambient light (*see* **Note 13**).

3.7 Mounting and Imaging

Free-floating samples

1. Place sample in an imaging dish and flood dish with ethyl cinnamate.
2. Remove any air bubbles using a fine gauge needle.
3. Image samples on inverted confocal or spinning disk microscope using low NA, low magnification air objectives (*see* **Note 14**).
4. Correct for the fish tank effect during image processing by multiplying the recorded Z-depth with the refractive index ratio of the immersion solution (ethyl cinnamate; RI 1.559) and the light path (air; RI 1).

Samples in a phytagel matrix on confocal or spinning disk systems

1. Place sample containing phytagel on a tissue paper to remove excess ethyl cinnamate.
2. Using Pattex super glue gel, glue the lip of the phytagel (*see* Subheading 3.4, **step 5**) to the imaging dish. Ensure no glue end up in the light path.
3. Flood imaging dish with ethyl cinnamate.
4. Mount fine gauge needles on a 1 mL syringe and bend needle at a 90° angle. Remove any air bubbles using a fine gauge needle.
5. Image samples on inverted confocal or spinning disk microscope using low NA, low magnification air objectives (*see* **Note 14**).
6. Correct for the fish tank effect during image processing by multiplying the recorded Z-depth with the refractive index ratio of the immersion solution (ethyl cinnamate; RI 1.559) and the light path (air; RI 1).

Lightsheet imaging (for example, Zeiss lightsheet 7):

7. Place sample containing phytagel on a tissue paper to remove excess ethyl cinnamate.
8. Glue the end of the phytagel cylinder to the sample holder using super glue (Pattex super glue gel).
9. Image samples according to manufacturer's recommendations (*see* **Note 15**).

4 Notes

1. Triethylamine is toxin and should be handled in accordance with your relevant local safety guidelines; all handling should be done in a fume hood. Carefully add small amounts of triethylamine, and it is easy to overshoot the recommended pH.
2. Higher purity ethyl cinnamate (for example, Sigma Cat. 112372) can be used but are significantly more expensive but do not provide any improvement.
3. Optimal fixation time is ultimately dependent on tissue size. We find that 4 °C overnight is a good catch all for most tissues. Consider fixing particularly large samples for longer periods, for example, adult heads at 4 °C for 24 h.
4. Depigmentation is optional and should only be done if samples are pigmented which are expected to interfere with image acquisition. Depigmentation will permanently destroy fluorescent proteins and would therefore require subsequent antibody labeling.

5. Antibody labeling is another optional step and could be combined with the presence of endogenous fluorescent proteins such as GFP and mCherry, Alexa-fluorophores, DAPI, SYTOX Green, Edu-Click chemistry as well as Abberior fluorophores. Alternatively, it can also be used to label fluorescent proteins after depigmentation. Common antibody staining procedures are used. Generally, we find that existing staining protocols for sections can be easily adapted to whole-mount antibody staining by increasing the concentrations of primary and secondary antibody two- to fivefold while increasing antibody incubation temperature and time. Exact methods will differ depending on the sample size, tissue type, and antibodies used.
6. Tissue embedding is an optional step and will depend largely on the tissue and imaging setup used. Samples that are easily orientated and are imaged on a microscope with conventional stages do not require embedding and can simply float in an imaging chamber or dish compatible with ethyl cinnamate. In all other instances, we suggest embedding the samples.
7. Do not trim any phytagel surfaces that will be part of the light path during imaging. The rough surface left by the trimming would otherwise disrupt the light path and negatively affect the final image quality.
8. Large samples will require an appropriate volume of dehydrating agent, and hence we recommend using VOA glass vials throughout the dehydration and clearing procedure as it does not react with either alcohols or ethyl cinnamate (Thermo Scientific: C126-0020).
9. 30% and 50% 1-Propanol can be prepared in advance, as solutions are stable at room temperature. For long-time storage, we recommend adjusting the pH at least to 9.2, and confirming the pH prior to use as the pH slightly drops over time.
10. Use 99% 1-propanol (Sigma Cat.W292818) to prepare all dehydrating solutions from 30% to 70% 1-propanol dehydration steps. Anhydrous 1-propanol is more costly and should be exclusively used during the final dehydration step.
11. Use anhydrous 1-propanol (Sigma 279544).
12. 1-propanol and ethyl cinnamate have different refractive indexes, as samples equilibrate streaks of 1-propanol will appear in the ethyl cinnamate. Continue to wash until all streaks are gone and the sample is transparent.
13. Do not store samples at 4 °C. Ethyl cinnamate is a solid at 4 °C and this could result in degradation of tissue morphology.
14. While not designed for large volume image acquisition, we have had good results using conventional confocal and spinning disk microscopes equipped with low NA and low

magnification (i.e., 10× 0.3NA) air objectives. Overall image quality can be dramatically improved by increasing pixel dwell times (in the case of point-scanning confocal microscopes). The low NA of such an objective makes it more robust to refractive index mismatching.

15. Not all lightsheet microscopes are compatible with organic solvents. Notable examples of commercially available organic solvent compatible lightsheet microscopes include the LaVision UltraMicroscope II, the Luxendo LCS SPIM, and the Zeiss lightsheet 7.

Acknowledgments

This work was supported by an Austrian Science Fund Lise Meitner Fellowship (M2444) to W.M. and a European Research Council Advanced Grant (742046) and a Deutsche Forschungsgemeinschaft grant (TA 274/13-1) to E.M.T.

References

1. Spalteholz W (1914) Uber das Durchsichtig machen von menschlichen und tierischen Präparaten und seine theoretischen Bedingungen, Nebst Anhang, über Knochenfärbung. S. Hirzel, Leipzig, p 93
2. Slack MJ (1980) Morphogenetic properties of the skin in axolotl limb regeneration. J Embryol Exp Morphol 58:265–288
3. Slack JM (1980) Regulation and potency in the forelimb rudiment of the axolotl embryo. J Embryol Exp Morphol 57:203–217
4. Tsai PS et al (2009) Plasma-mediated ablation: an optical tool for submicrometer surgery on neuronal and vascular systems. Curr Opin Biotechnol 20(1):90–99
5. Chung K et al (2013) Structural and molecular interrogation of intact biological systems. Nature 497(7449):332–337
6. Yang B et al (2014) Single-cell phenotyping within transparent intact tissue through whole-body clearing. Cell 158(4):945–958
7. Masselink W et al (2019) Broad applicability of a streamlined Ethyl cinnamate-based clearing procedure. Development 146(3):dev166884
8. Gerber T et al (2018) Single-cell analysis uncovers convergence of cell identities during axolotl limb regeneration. Science (New York, N.Y.) 362(6413):eaaq0681-13
9. Subiran Adrados C et al (2020) Salamander-Eci: an optical clearing protocol for the three-dimensional exploration of regeneration. Dev Dyn. https://doi.org/10.1002/dvdy.264
10. Duerr TJ et al (2020) 3D visualization of macromolecule synthesis. elife 9:e60354
11. Pende M et al (2020) A versatile depigmentation, clearing, and labeling method for exploring nervous system diversity. Sci Adv 6(22): eaba0365

Chapter 8

Induction and Characterization of Cellular Senescence in Salamanders

Qinghao Yu, Hannah E. Walters, and Maximina H. Yun

Abstract

Cellular senescence is a permanent proliferation arrest mechanism induced following the detection of genotoxic stress. Mounting evidence has causally linked the accumulation of senescent cells to a growing number of age-related pathologies in mammals. However, recent data have also highlighted senescent cells as important mediators of tissue remodeling during organismal development, tissue repair, and regeneration. As powerful model organisms for studying such processes, salamanders constitute a system in which to probe the characteristics, physiological functions, and evolutionary facets of cellular senescence. In this chapter, we outline methods for the generation, identification, and characterization of salamander senescent cells in vitro and in vivo.

Key words Cellular senescence, Regeneration, Salamander

1 Introduction

Upon recognition of genotoxic stress such as exhaustive telomere attrition or oncogene activation, cells can undergo a highly conserved proliferative arrest mechanism, termed cellular senescence [1]. While many stressors can trigger senescence, proliferation arrest is coordinated through common molecular mediators such as p53, and enforced by upregulation of cell cycle inhibitors including $p16^{INK4A}$ and $p21^{CDKN1A}$ [2, 3]. By preventing the propagation of potentially damaged cells, senescence constitutes an important defense against tumorigenesis [4–6]. However, senescent cells remain metabolically active and through a highly variable secretory phenotype (the Senescence-Associated Secretory Phenotype, SASP) can exert non-cell-autonomous effects on the local microenvironment [6]. Indeed, recent evidence indicates that in mammals, senescent cells accumulate in various tissues where they

Qinghao Yu and Hannah E. Walters contributed equally with all other contributors.

Ashley W. Seifert and Joshua D. Currie (eds.), *Salamanders: Methods and Protocols*,
Methods in Molecular Biology, vol. 2562, https://doi.org/10.1007/978-1-0716-2659-7_8,

are implicated in the progression of age-related deterioration and pathologies. Accordingly, removal of these senescent cells by genetic or targeted pharmacological means results in increased health and lifespan in mammals [7–19].

Conversely, recent reports have also highlighted important physiological roles for cellular senescence. In salamander limb regeneration, senescence is dynamically induced in the blastema, before cells are promptly cleared by macrophages [20]. Similarly, local induction of senescence occurs during zebrafish fin regeneration [21]. Further, senescent cells have also been observed in transient organ forms during embryonic development in salamanders as well as mammals, where they contribute to structural degeneration [22–24]. Thus, senescent cells are hypothesized to be important coordinators of these physiological processes through their non-cell-autonomous effects, including regulation of immune recruitment and activity, the proliferation (or arrest) of nearby cells, cellular plasticity, and possibly also tissue patterning [25].

These findings highlight cellular senescence as an important, highly conserved tissue remodeling mechanism with important consequences for both regeneration and organismal development. Salamanders, therefore, offer a valuable system to uncover the potential physiological functions of senescence and robust mechanisms of senescent cell clearance in contexts including regeneration and organismal development. In this chapter, we provide different methods for the induction and detection of cellular senescence in salamanders, both in vitro and in vivo. First, we describe the induction of senescence in salamander cell culture, by eliciting sublethal doses of DNA damage, using either the topoisomerase II inhibitor etoposide or through UV irradiation. Secondly, we present different methods for evaluating senescence-associated phenotypes in vitro and expand these to in vivo analysis where applicable. Namely, we provide protocols for the analysis of DNA damage foci through immunocytochemistry to detect the common DNA damage marker γH2AX and describe quantification of proliferation using EdU staining. We also describe protocols for the analysis of mitochondrial networks and ROS in senescent cells; constitutive activation of growth pathways including mTOR signaling, a failure of mitophagy and altered fission/fusion dynamics of the mitochondrial network result in both enlargement and dysfunction of mitochondria during cellular senescence. Further, we provide the methods for analyzing lysosomal changes which occur in senescence, including lysosomal expansion and senescence-associated β-galactosidase activity, a classic biomarker of senescence which relies on the activity of lysosomal β-galactosidase at suboptimal pH 6.0 [26].

2 Materials

2.1 Induction of Cellular Senescence in Salamander Cell Culture

1. Humidified incubator at 2.5% CO_2 and 25 °C.
2. Tissue culture-grade plasticware.
3. UV Stratalinker 2400 or equivalent.
4. 500 ml bottle top vacuum filter, 0.45 μm pore.
5. Porcine gelatin.
   (a) To prepare porcine gelatin, dissolve 0.75 g per 100 ml diH_2O in a shaking water bath at 65 °C. Filter sterilize using a 0.45 μm filter and store in 50 ml aliquots at 4 °C. Heat to 37 °C prior to use.
6. Amphibian-PBS (A-PBS).
   (a) Add 25 ml of sterile dH_2O to 100 ml of PBS.
7. Amphibian (A)-trypsin.
   (a) Add 1 ml of 1× trypsin and 10 ml of A-PBS.
8. Amphibian growth media (A-MEM):

Component	Volume (ml)
Sterile dH_2O	48
MEM	126
Heat-inactivated FCS	20
Insulin-transferrin-selenium, 100×	2
L-glutamine, 200 mM	2
Penicillin/streptomycin, 10,000 U/ml	2
Total volume	*200 ml*

9. Cell culture-grade DMSO.
10. 100 mM etoposide (*see* **Note 1**).
    (a) To prepare a 100 mM stock, dissolve 25 mg of etoposide in 425 μl of DMSO. Aliquot and store at −20 °C.
11. 10 mM nutlin-3a (*see* **Note 1**).
    (a) To prepare a 10 mM stock, dissolve 1 mg of nutlin-3a in 172 μl DMSO. Aliquot and store at −20 °C.

2.2 Evaluating Senescence-Associated Phenotypes in Tissue Culture

1. A-MEM (as in Subheading 2.1).
2. A-PBS (as in Subheading 2.1).
3. MitoTracker Red FM (Invitrogen # M22425).
   (a) To prepare a 1 mM stock, dissolve 50 μg of lyophilized MitoTracker in 69 μl of DMSO.

4. Dihydrorhodamine 123, 5 mM (Invitrogen # D23806).
5. LysoTracker Red DND-99, 1 mM (Invitrogen # L7528).
6. Fluorescence microscope.
7. 4% formaldehyde.
   (a) Add 25 ml of A-PBS for every 1 g of PFA. Dissolve at 55 °C in a water bath for 3–4 h. Shake vigorously every hour until no precipitates are visible and the solution is clear. Cool to 4 °C and use on the same day.
8. Blocking buffer.
   (a) Add 5 ml of goat serum and 150 μl of Triton X100 to 45 ml of A-PBS.
9. Antibody dilution buffer.
   (a) Add 2.5 ml of goat serum and 150 μl of Triton X100 to 47.5 ml of A-PBS.
10. Primary antibody: α-γH2AX-Ser139 antibody, mouse monoclonal (Merck-Millipore 05-636).
11. Secondary antibody: goat α-mouse IgG Alexa Fluor 488.
12. 10 mM EdU (Invitrogen #C10637, component A).
    (a) Add 2 ml of DMSO to 5 mg of EdU for a 10 mM stock. Mix well, aliquot, and store at −20 °C.
13. EdU Click-iT imaging kit (Invitrogen #C10637).
    (a) Prepare reagents and store according to the manufacturer's instructions.
14. Hoechst (Invitrogen # C10637, Component G).
    (a) Dilute Hoechst stock 1:10,000 in PBS.
15. Coverslips.
16. Clear nail varnish.

2.3 In Vitro SA-β-Gal Staining

1. 4% formaldehyde (as in Subheading 2.2).
2. N,N-dimethylformamide.
3. X-gal, 20 mg/ml (Cell Signaling # 9860, item # 11678) (*see* **Note 2**).
   (a) To prepare X-gal solution, dissolve X-gal in an appropriate volume of DMF for a working concentration of 20 mg/ml in a microcentrifuge tube. Vortex well to mix and use immediately.
4. SA-β-gal staining kit (Cell Signaling # 9860).
5. 37 °C atmospheric incubator.
6. pH meter.
7. 37% HCl.
8. 70% glycerol.

2.4 *In Vivo* Senescence Analysis

2.4.1 Tissue Collection, Processing, and Fixation

1. 0.03% benzocaine solution.
   (a) To prepare 500 ml of 10% (wt/vol) benzocaine stock, add 50 g of benzocaine to 500 ml 100% ethanol and mix. This solution can be stored at room temperature for up to 1 year.
   (b) To prepare 1 L of 10× TBS, add 24.2 g Tris and 90 g NaCl to 990-ml distilled water. Mix using a stirrer bar. Adjust the pH to 8.0 using 37% HCl. This solution can be stored at room temperature for up to 6 months.
   (c) For 1 L of 400% (wt/vol) Holtfreter's solution, add 0.2875 g KCl, 0.536 g $CaCl_2 \cdot 2H_2O$, 1.1125 g $MgSO_4 \cdot 7H_2O$, and 15.84 g NaCl. Bring volume to 1 L with distilled water. This solution can be stored at room temperature for up to 6 months.
   (d) For 2 L of anesthetic 0.03% benzocaine, mix 100 ml of 10× TBS, 100 ml of 400% Holtfreter's solution, and 6 ml of 10% (wt/vol) benzocaine, and bring to 2 L with distilled water. This solution can be stored at room temperature for up to 6 months.
2. Fast Green FCF.
3. EdU injection solution, 1 mg/ml.
   (a) To prepare EdU injection solution, dissolve 5 mg of EdU and 1 mg of Fast Green FCF in 5 ml of A-PBS. Mix well, aliquot, and store at −20 °C.
4. 30 G needle.
5. 1 ml syringe.
6. Clean, sterile surgical tools.
7. 4% formaldehyde (*see* Subheading 2.3).
8. OCT compound or equivalent embedding media.
9. SuperFrost Plus adhesive slides.
10. Cryostat.

2.4.2 SA-β-Gal Staining

1. Paraffin wax.
2. Super-PAP pen.
3. X-gal, 20 mg/ml (*see* Subheading 2.3; *see* **Note 2**).
4. SA-β-gal staining kit (Cell Signalling # 9860).
5. Parafilm.
6. 37 °C atmospheric incubator.
7. pH meter.
8. 37% HCl.
9. 70% glycerol.

2.4.3 EdU and Antibody Staining

1. EdU Click-iT imaging kit (Invitrogen #C10637) (*see* Subheading 2.2).
2. IHC permeabilization buffer.
   (a) Mix 3 ml of Triton X100 in 1 L PBS.
3. IHC blocking buffer.
   (a) Mix 10 ml of goat serum with 90 ml of 0.3% Triton X100: PBS.
4. Primary antibody.
5. Secondary antibody.
6. Hoechst (Invitrogen # C10637, Component G) (*see* Subheading 2.2).
7. Coverslips.
8. Clear nail varnish.

3 Methods

Carry out all the procedures at room temperature unless otherwise specified.

3.1 Induction of Cellular Senescence in Salamander Cell Culture

3.1.1 Cellular Expansion

1. Using routine cell culture methods, grow a flask of healthy, proliferating amphibian cells, e.g., axolotl AL1 [27] or newt A1 [28], to confluence in culture. Aspirate the growth media and wash the cells briefly in A-PBS (e.g., 4 ml for T75 flask). Aspirate the A-PBS wash.
2. To lift the cells, add an appropriate volume of A-trypsin (e.g., 1.5 ml for T75) and incubate for 3–5 min. During this time, gently rock the flask back and forth to ensure the entire growth area is covered by the A-trypsin and to lift the cells. Cell detachment can be followed under a microscope.
3. When the cells have detached, add an appropriate volume of complete media to the flask to quench the A-trypsin (e.g., 4.5 ml for T75). Using a 10 ml pipette, pass the cell suspension over the growth area of the flask around 10 times to collect all cells in suspension.
4. Then, transfer the cell suspension to a 15 ml Falcon tube, and centrifuge the cells at 200 RCF, for 3 min at room temperature.
5. Gently but thoroughly resuspend the cell pellet in a small volume A-MEM (e.g., 2 ml).
6. Seed amphibian cells onto gelatin-coated tissue culture dishes or multi-well plates in an appropriate volume of A-MEM (e.g., 10 ml for a 10 cm dish) at 50% cell confluence, i.e., a 1: 2 dilution of a confluent harvested monolayer. Seed sufficient vessels to include proliferating controls in each experiment. For

UV induction of senescence, control cells must be seeded in a separate plate or dish. Proliferating controls may require seeding at a lower cell density or subculturing during the experiment to avoid confluence (*see* **Notes 3–6**).

7. Leave cells to adhere to tissue culture surface overnight. Verify cell health and density by microscopy before proceeding with the induction of senescence.

3.1.2 Etoposide-Induced DNA damage (Day 1)

1. Without removing the tissue culture medium, treat cells with etoposide and 1 μM nutlin-3a. Incubate for 24 h (*see* **Notes 1** and **5**).
   (a) For A1 cells, we recommend a dose of 20 μM etoposide.
   (b) For AL1 cells, we recommend lowering the dose of etoposide to 10 μM, as higher concentrations result in an increased induction of apoptosis.
2. After 24 h treatment, aspirate the drug-supplemented culture medium and replace with fresh A-MEM, containing nutlin-3a at 1 μM (*see* Fig. 1 and **Notes 1** and **5**).

3.1.3 UV-Induced DNA damage (Day 1)

1. Using a tissue soaked in 70% ethanol, thoroughly disinfect the insides of the UV Stratalinker or equivalent.
2. Following overnight cell adherence, remove the tissue culture medium and place cells in an appropriate volume of A-PBS (e.g., 2 ml of A-PBS for a 3.5 cm dish or 100 μl per well of a 96-well plate).
3. Place the culture dish inside the UV Stratalinker and remove the lid. Expose the cells to UV irradiation (*see* **Note 6**).
   (a) For A1 cells, we recommend a dose of 10 J/m^2.
   (b) For AL1 cells, we recommend lowering the dose to 5 J/m^2 to circumvent extensive induction of apoptosis.
4. Remove the A-PBS and replace with fresh A-MEM, supplemented with 1 μM nutlin-3a (*see* Fig. 1).
5. Proliferating control cells should be mock-treated, i.e., media aspirated, incubated in A-PBS for the duration of UV irradiation, then A-PBS aspirated and supplied with fresh A-MEM containing equivalent volumes of DMSO as used for nutlin-3a treatment.

3.1.4 p53 Stabilization and Senescence Induction (Days 2–12)

1. Incubate UV or etoposide-treated cells for 11 further days, sequentially adding 1 μM nutlin-3a to the culture medium every 2 days without changing the media (*see* **Note 1**).
2. For proliferating control populations, cells should be mock-treated during senescence induction with identical volumes of cell culture-grade DMSO only. Note that proliferating controls

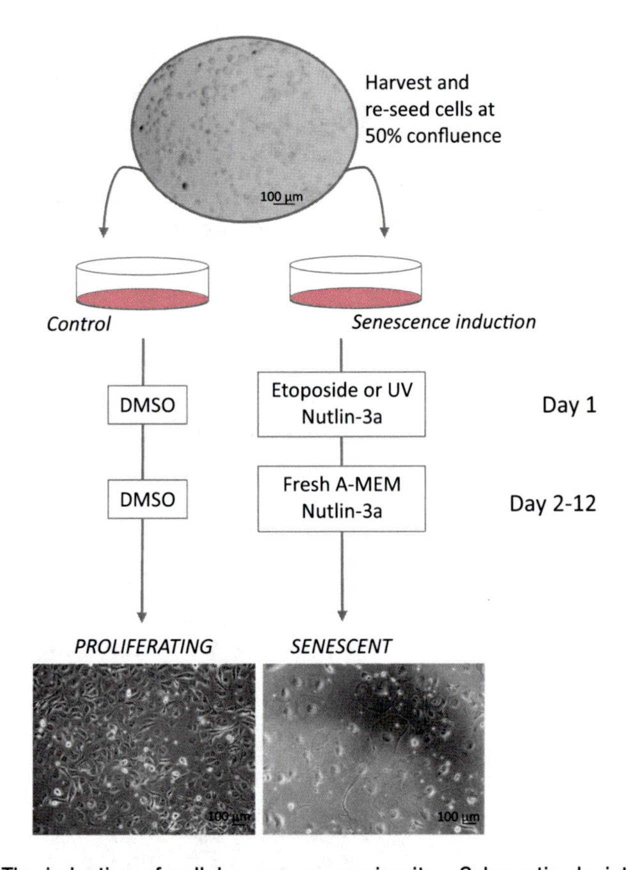

Fig. 1 The induction of cellular senescence in vitro. Schematic depicting the protocol for etoposide and UV-induced senescence. Proliferating cells are harvested and seeded into fresh culture dishes, before treatment with etoposide and nutlin-3a for 24 h, or exposure to UV as indicated. Note that etoposide/UV-treated cells exhibit classic enlarged, flattened morphology after 12 days, whereas control cells maintain a smaller, rounded morphology

may require seeding at a lower cell density or subculturing during the 12-day senescence induction period to avoid confluence or myotube formation in the case of A1 cells.

3. For applications such as western blotting or RNA analysis, harvest the senescent and proliferating control cells as in Subheading 3.1, **step 1** and proceed with downstream analysis as appropriate (*see* **Notes 7–9**).

3.2 Detection of Senescence-Associated Phenotypes In Vitro

3.2.1 Analysis of Mitochondrial Networks and ROS Production

1. MitoTracker and DHR123 staining enable visualization of mitochondrial networks and reactive oxygen species, respectively (increases in which suggest mitochondrial dysfunction). To examine mitochondrial networks:
   (a) Dilute 1 mM of MitoTracker stock solution 1:5000 in A-MEM to give a working concentration of 200 nM.

   (b) Incubate senescent and control proliferating cells for 1 h at 25 °C.

   (c) Exchange the staining solution for fresh A-PBS for live imaging (*see* Fig. 2a and **Note 10**).

2. For detection of ROS production:

   (a) Dilute 5 mM of DHR123 stock solution 1:500 in A-MEM to give a working concentration of 10 μM.

   (b) Incubate senescent and control proliferating cells for 2 h at 25 °C with working solution. Cells can be imaged to live in A-PBS or fixed with 4% PFA and mounted for imaging (*see* Fig. 2b).

3.2.2 Analysis of Lysosomal Networks

1. In addition to measuring SA-β-gal using the protocol discussed below (*see* Subheading 3.3), it is possible to verify the increase in lysosomal networks in amphibian cells accompanying senescence induction using LysoTracker Red DND99.

   (a) Dilute 1 mM of LysoTracker stock solution 1:2000 in A-MEM to give a working concentration of 500 nM. Incubate senescent and proliferating cells for 1 h at 25 °C.

   (b) Exchange for fresh A-PBS and image live (*see* Fig. 2c).

3.2.3 Analysis of DNA Damage by γH2AX Immunostaining

1. Aspirate growth media and wash the cells briefly in A-PBS.
2. Fix cells in 4% PFA solution for 15 min at 4 °C (e.g., 100 μl per well in a 96-well plate).
3. Wash the cells in A-PBS.
4. Aspirate the A-PBS wash, and incubate with blocking buffer for 1 h at room temperature (e.g., 100 μl per well in a 96-well plate).
5. Incubate in 1:500 primary antibody (α-γH2AX-Ser139) diluted in antibody-dilution buffer overnight at 4 °C (e.g., 50 μl per well in a 96-well plate).
6. Wash twice in A-PBS and then incubate in 1:1000 secondary antibody (goat α-mouse IgG Alexa Fluor 488) in antibody dilution buffer for 1–4 h at room temperature (e.g., 50 μl per well in a 96-well plate).
7. Wash twice in A-PBS.
8. Counterstain using Hoechst (1:10,000 in PBS, e.g., 50 μl per well in a 96-well plate).
9. Aspirate staining solution and replace with fresh A-PBS. Cells can now be imaged using fluorescence microscopy (*see* Fig. 2d). Alternatively, if cells have been grown on coverslips, these can be mounted prior to imaging, by placing 1–2 drops of DAKO fluorescence mounting medium onto a glass slide, and carefully

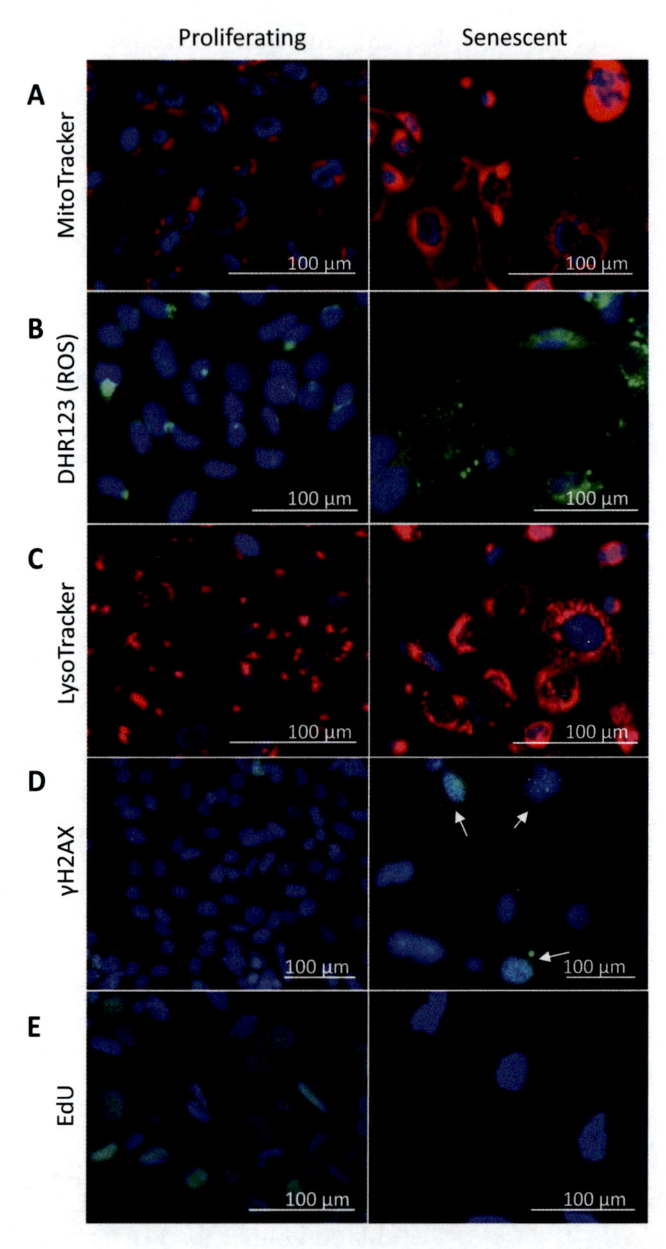

Fig. 2 Assessing senescence-associated phenotypes in vitro. Representative images of senescent AL1 cells after etoposide treatment, alongside vehicle-only treated (DMSO) proliferating controls. Cells were analyzed 12 days after senescence induction. (**a–c**) for analysis of mitochondrial content, ROS, and lysosomal networks, cells were stained and imaged live (MitoTracker Red, DHR123 and LysoTracker, respectively). (**d**) for DNA damage analysis, cells were fixed, blocked, and immunostained against γH2AX (nuclear foci indicated by white arrows). (**e**) for proliferation analysis, cells were incubated with EdU for 24 h before fixation and staining

placing the coverslip on top (side with cells facing down) using fine tweezers. The edges can be sealed using clear nail varnish to secure coverslip for imaging.

3.2.4 Proliferation Analysis by EdU Staining

1. Prepare a 2× EdU solution by diluting 10 mM EdU stock 1: 1000 in A-MEM for 10 μM.
2. Without aspirating the growth media, add 1 volume of 2× EdU solution (e.g., to 100 μl media per well of a 96-well plate, add a further 100 μl 2× EdU solution).
3. Incubate cells for the desired length of time, e.g., 24 h (*see* **Note 11**).
4. Aspirate media and rinse the cells briefly with A-PBS.
5. Aspirate the A-PBS wash and fix cells in 4% PFA for 15 min at room temperature (e.g., 100 μl per well of 96-well plate).
6. Rinse in A-PBS.
7. Permeabilize cells in 0.5% Triton X100 in A-PBS for 20 min at room temperature.
8. During this incubation, prepare a sufficient volume of fresh Click-iT reaction cocktail for staining, in the following order:

Reaction component	Volume (μl)
1× Click-iT reaction buffer	860
CuSO4	40
Alexa Fluor azide	2.5
1× reaction buffer additive	100
Total volume	*1 ml*

9. Following permeabilization, rinse cells in A-PBS.
10. Aspirate the A-PBS wash and incubate cells in the Click-iT reaction cocktail for 30 min at room temperature, in the dark (e.g., 50 μl per well of a 96-well plate). From this point onwards, samples must be protected from light.
11. Rinse in A-PBS.
12. Counterstain using Hoechst (1:10,000 in PBS).
13. Aspirate staining solution and replace with fresh A-PBS. Cells can now be imaged using fluorescence microscopy (*see* Fig. 2e), or coverslips can be mounted as in analysis of DNA damage section.
14. To quantify the proportion of cells undergoing S-phase during the EdU incubation period, image several hundred nuclei per replicate, and count the proportion of these which stain positive for EdU incorporation.

3.3 SA-β-Gal Staining In Vitro

1. To the population of cells to be analyzed, remove tissue culture medium and wash briefly in A-PBS (*see* **Note 12**).
2. Fix cells in either 0.5% glutaraldehyde or 4% formaldehyde in A-PBS for 10–15 min at room temperature (*see* **Note 13**).
3. Rinse cells in A-PBS.
4. Warm SA-β-gal staining solution components to 37 °C before use. Prepare SA-β-gal staining solution fresh as follows:

Reaction component	Volume (μl)
10× staining solution	93
dH_2O	837
X-gal, 20 mg/ml	50
Solution A, 100×	10
Solution B, 100×	10
Total volume	*1 ml*

5. Adjust the SA-β-gal staining solution to pH 6.0 with concentrated HCl or NaOH before use (*see* **Note 14**).
6. Aspirate the A-PBS wash and add staining solution to cells, e.g., 1 ml per well in a 24-well plate, and seal the plate with parafilm to prevent evaporation. Incubate cells for 16 h at 37 °C, in the dark, with atmospheric CO_2 in SA-β-gal staining solution (*see* **Note 15**).
7. *Optional.* It is possible to combine SA-β-gal and EdU staining. To do so, incubate cells with EdU as described in Subheading 3.2. Following EdU treatment and fixation, stain for SA-β-gal, and subsequently proceed from **step 4** of EdU staining as described in Subheading 3.2, **step 4**.
8. Overlay with 70% glycerol and image (*see* Fig. 3). SA-β-gal signal is stable for at least 12 months when stored at 4 °C.

3.4 SA-β-Gal Staining, EdU, and Immunostaining: In Vivo

3.4.1 EdU Treatment, Tissue Collection, Fixation, and Processing

1. To calculate the appropriate dose of EdU for i.p. treatment, first weigh the animals. Place a tank of water on a scale, zero the scale and then using a net, transfer the animal into the tank. Record the animal weight and calculate the appropriate volume of EdU: 5 μl of 1 mg/ml stock per gram of animal weight.
2. Place animal into an appropriate dilution of amphibian anesthetic.
   (a) To prepare anesthetic, dilute 0.03% benzocaine with an appropriate volume of dechlorinated tap water. For >1 year-old animals, we recommend diluting 1:3 with tap water, and for juvenile animals, we recommend diluting 1:6 with tap water.

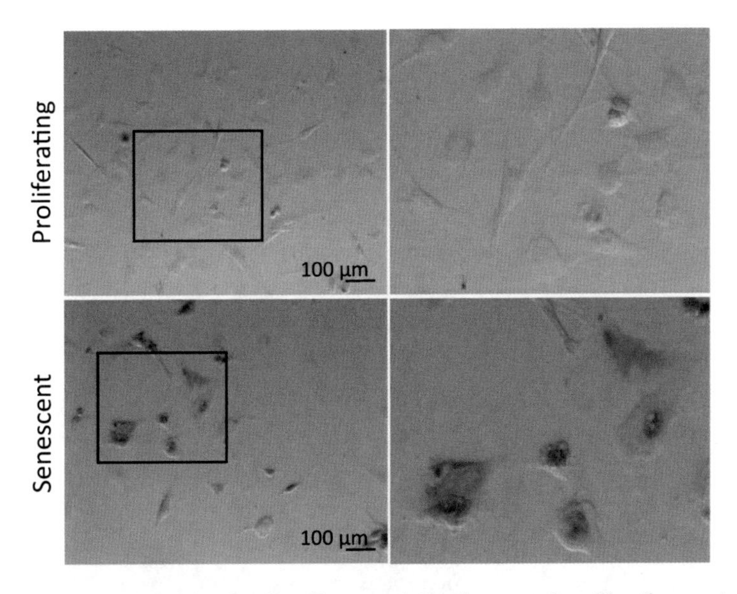

Fig. 3 SA-β-gal staining in vitro. Representative images of proliferating control (DMSO) and etoposide-induced senescent A1 cells, fixed and stained for SA-β-gal activity 12 days after senescence induction. Right column shows a magnification of the area indicated by the black square

3. Inject EdU solution intraperitoneally with a 30-gauge needle and syringe. Return animal to fresh tap water to recover from anesthesia and to allow EdU to distribute systemically.
4. Following a desired length of EdU pulse, return the animal to anesthetic solution (*see* **Note 11**). Using sterilized surgical tools, collect tissue samples to be analyzed into A-PBS. Following sample collection, sacrifice animal by locally approved procedures.
5. Transfer the tissue sample into freshly prepared 4% PFA solution. Incubate for 20 h at 4 °C with rocking (*see* **Note 16**).
6. Following fixation, wash the tissue sample twice with A-PBS for 5 min at room temperature with rocking.
7. Carefully embed the sample in OCT media and flash-freeze with dry ice and store at −80 °C until further processing.
8. Section the OCT-embedded sample using a cryostat at 8 µm for one cell-thick sections. Collect sections onto SuperFrost glass slides.
9. Store slides at −20 °C. Analysis of SA-β-gal activity should be performed as soon as possible after sample collection. Activity is retained for at least 1 month when slides are stored under these conditions.

3.4.2 SA-β-Gal Staining

1. Prior to staining, air-dry the slides for 15 min.
2. Wash slides twice in A-PBS to remove embedding media and rehydrate the sections.
3. Encircle the area on the slide containing tissue sections using paraffin or super PAP pen to ensure staining solution will not drain off during incubation.
4. Prepare the β-gal staining solution (as above) and adjust pH to 6.0 with 37% HCl. Measure pH using pH strips or a pH electrode (*see* **Note 14**).

Reaction component	Volume (μl)
10× staining solution	93
dH_2O	837
X-gal, 20 mg/ml	50
Solution A, 100×	10
Solution B, 100×	10
Total volume	*1 ml*

5. Prepare a humidified staining chamber using wet tissue paper to prevent evaporation of staining solution. Place slides flat in the chamber and distribute 1 ml of staining solution per slide.
6. Overlay slides with parafilm to ensure an even distribution of staining solution.
7. Incubate slides in an atmospheric 37 °C incubator in the dark for 16 h (*see* **Note 15**).
8. Wash slides twice in A-PBS for 15 min.
9. *Optional.* If only SA-β-gal staining is required, it is possible to end the protocol here and overlay with 70% glycerol and mount for imaging (*see* Fig. 4).

3.4.3 EdU and Antibody Staining

1. Wash slides 3× in PBS for 10 min to remove the SA-β-gal staining solution.
2. Reinforce the paraffin protection with the super PAP pen.
3. Permeabilize the sections in 0.5 ml IHC permeabilization solution for 15 min.
4. Block with 1 ml IHC blocking buffer for 10 min.
5. Warm Click-iT EdU reagents to room temperature 30 min prior to use. Dilute 10× reaction buffer additive 1:10 with dH_2O immediately before use. Prepare staining solution in the following order:

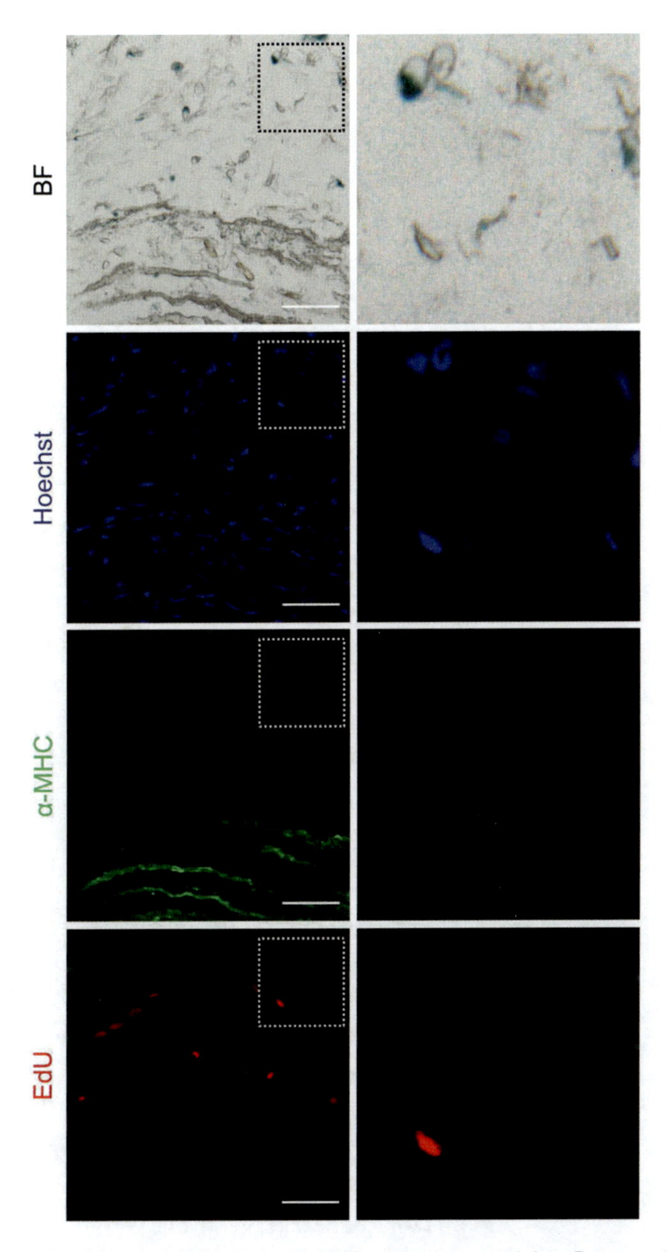

Fig. 4 In vivo SA-β-gal, EdU, and anti-MHC antibody co-staining. Representative images of midbud blastema sections following SA-β-gal, EdU, and MHC (myosin heavy chain) co-staining, as detailed in the Methods section. Axolotls received an EdU pulse 4 h prior to sample collection. Right column shows a magnification of the area indicated by the dashed square

Reaction component	Volume (μl)
1× Click-iT reaction buffer	860
CuSO4	40
Alexa Fluor azide	2.5
1× reaction buffer additive	100
Total volume	*1 ml*

6. Incubate for 2 h, protected from light.
7. *Optional.* If no antibody staining is required, then it is possible to wash slides with PBS, counterstain with Hoechst, and mount for imaging.
8. Wash twice in PBS for 5 min.
9. Permeabilize in IHC permeabilization buffer for 15 min.
10. Block in IHC blocking buffer for 1 h.
11. Incubate in 200 μl of primary antibody appropriately diluted in IHC blocking buffer at 4 °C overnight. Overlay slides with parafilm to aid distribution of the staining solution and to prevent evaporation.
12. Wash in PBS for 10 min.
13. Wash in IHC permeabilization buffer for 5 min.
14. Incubate in 200 μl of secondary antibody appropriately diluted in IHC blocking buffer for 2 h.
15. Wash twice in PBS for 5 min.
16. Counterstain with Hoechst 1:10,000 for 20 min.
17. Wash with PBS for 5 min.
18. To mount, overlay with 3–4 drops of DAKO fluorescence mounting medium, add coverslip, and secure for imaging using clear nail varnish.
19. Image slides (*see* Fig. 4).

4 Notes

1. Care should be taken not to exceed 0.1% of the total culture volume as DMSO to avoid toxicity. Drug stocks should thus be made up at appropriate concentrations (e.g., etoposide at 10–100 mM, nutlin at 10 mM).
2. We recommend making up fresh solutions of X-gal in DMF at 20 mg/ml on the day of staining (only in polypropylene plastic

or glass tubes), to avoid X-gal crystal formation during staining.

3. For in situ analysis, such as compound toxicity testing or imaging analysis, we recommend performing the outlined protocol for generating senescent cells in the same culture dishes that would be appropriate for analysis. Harvesting and reseeding following senescence induction into dishes for analysis is not advised as senescent cells do not adhere well following trypsinization, resulting in loss of a substantial proportion of cells.
4. Cell density at the point of DNA damage induction is crucial. High cell density can compromise the induction of senescence as quiescent cells are less sensitive to DNA-damaging agents. Low cell density will limit the amount of biological material available for downstream applications, as cells will not proliferate following the induction of senescence. Therefore, we recommend a range of 50–70% confluence at the start of senescence induction.
5. To avoid disturbing cells at the onset of senescence induction, we recommend adding the appropriate volume of etoposide and nutlin-3a stock solutions directly to the culture media and gently swirling the culture dish. Alternatively, when using small well sizes such as 96-well plates, we recommend seeding cells in ½ volume (i.e., 100 μl/well) for overnight bedding, then subsequently adding 100 μl of 2× etoposide/nutlin-3a in A-MEM per well for the 24 h incubation, before proceeding with 11 day incubation and repetitive nutlin-3a treatment in fresh media.
6. For UV induction of senescence, doses should be optimized according to the specific equipment used. As cells must be placed inside the UV irradiator for treatment, proliferating cell controls should be seeded in separate plates or dishes, so that these cells can be mock-treated only. For mock treatment, cells should be removed from the incubator in parallel to those undergoing UV treatment, and then treated with equal volumes of DMSO to the volume of nutlin-3a added to the UV-treated cells.
7. Conditioned media (CM) can be generated by incubating senescent and control proliferating cell cultures with fresh, drug-free A-MEM for 24 h (or an alternative time period) following senescence induction. Depending on the downstream application, it may be necessary to generate CM with a lower concentration of FCS. For example, testing the secretory profile of senescent cells by cytokine array requires serum-free or low-serum media (e.g., 0.2% FCS), as serum itself often contains cytokines or growth factors.

8. Following the induction of senescence, senescent cells can be maintained in vitro in drug-free A-MEM without loss of viability or resuming proliferation. Fresh media should be supplemented weekly.
9. As the senescence program is highly dynamic, it is important that a consistent time point is used for senescent cell analysis.
10. MitoTracker-derived fluorescence is lost upon fixation so imaging should be performed on live cells.
11. In vitro and in vivo EdU pulse lengths should be determined according to the application. In vitro, a 24 h EdU pulse labels >50% of AL1 and A1 proliferating control cells, but <5% of etoposide-induced senescent AL1 or A1 cells following our protocol above. In vivo, a 4 h EdU pulse labels ~40% and ~15% of cells in the mesenchyme of a midbud blastema in a 2-month and a 1-year-old axolotl, respectively.
12. Negative controls should be included in every SA-β-gal assay to verify that positive staining is senescence-specific rather than a result of extended incubation times or incorrect staining solution pH. For this, we recommend using fixed proliferating control cultures at <80% confluence as negative controls for in vitro stainings. Mature limb tissue can be used as a negative control for in vivo stainings. It is normal to observe some non-specific staining in the epidermis.
13. We routinely fix cells in 4% formaldehyde prior to SA-β-gal staining, but we have also tested fixation of 0.5% glutaraldehyde, which does not affect the assay. The fixative should therefore be chosen on the basis of compatibility with any further staining required, e.g., for optimized protocols for particular primary antibodies.
14. Care should be taken to adjust the pH of the staining solution to precisely pH 6.0. A lower pH will result in false positives, and a higher pH will result in false negatives.
15. Incubation time in SA-β-gal staining solution should be optimized according to the specific tissue. Staining intensity should be monitored at regular intervals to optimize the incubation times.
16. Under-fixation and loss of SA-β-gal enzyme from tissue sections represents a bigger problem than over-fixation. We recommend fixing large samples, such as from blastemas collected from 1-year-old animals (e.g., with stump tissue diameter of >0.5 cm but <1 cm), for a minimum of 20 h.

Acknowledgments

MHY is supported by the DFG Research Center and Cluster of Excellence—Center for Regenerative Therapies Dresden (DFG FZ 111, DFG EXC 168). HEW is an Alexander von Humboldt post-doctoral fellow.

References

1. Hayflick L, Moorhead PS (1961) The serial cultivation of human diploid cell strains. Exp Cell Res 25:585–621. https://doi.org/10.1016/0014-4827(61)90192-6
2. Beauséjour CM, Krtolica A, Galimi F et al (2003) Reversal of human cellular senescence: roles of the p53 and p16 pathways. EMBO J 22:4212–4222. https://doi.org/10.1093/emboj/cdg417
3. Muñoz-Espín D, Serrano M (2014) Cellular senescence: from physiology to pathology. Nat Rev Mol Cell Biol 15:482–496. https://doi.org/10.1038/nrm3823
4. Michaloglou C, Vredeveld LCW, Soengas MS et al (2005) BRAFE600-associated senescence-like cell cycle arrest of human naevi. Nature 436:720–724. https://doi.org/10.1038/nature03890
5. Collado M, Gil J, Efeyan A et al (2005) Senescence in premalignant tumours. Nature 436:642–642. https://doi.org/10.1038/436642a
6. Chen Z, Trotman LC, Shaffer D et al (2005) Crucial role of p53-dependent cellular senescence in suppression of Pten-deficient tumorigenesis. Nature 436:725–730. https://doi.org/10.1038/nature03918
7. Baker DJ, Wijshake T, Tchkonia T et al (2011) Clearance of p16 Ink4a-positive senescent cells delays ageing-associated disorders. Nature 479:232–236. https://doi.org/10.1038/nature10600
8. Baker DJ, Childs BG, Durik M et al (2016) Naturally occurring p16 Ink4a-positive cells shorten healthy lifespan. Nature 530:184–189. https://doi.org/10.1038/nature16932
9. Childs BG, Baker DJ, Wijshake T et al (2016) Senescent intimal foam cells are deleterious at all stages of atherosclerosis. Science (80-) 354:472–477. https://doi.org/10.1126/science.aaf6659
10. Ogrodnik M, Miwa S, Tchkonia T et al (2017) Cellular senescence drives age-dependent hepatic steatosis. Nat Commun 8:15691. https://doi.org/10.1038/ncomms15691
11. Jeon OH, Kim C, Laberge RM et al (2017) Local clearance of senescent cells attenuates the development of post-traumatic osteoarthritis and creates a pro-regenerative environment. Nat Med 23:775–781. https://doi.org/10.1038/nm.4324
12. Xu M, Pirtskhalava T, Farr JN et al (2018) Senolytics improve physical function and increase lifespan in old age. Nat Med 24:1246–1256. https://doi.org/10.1038/s41591-018-0092-9
13. Aguayo-Mazzucato C, Andle J, Lee TB et al (2019) Acceleration of β cell aging determines diabetes and senolysis improves disease outcomes. Cell Metab 30:129–142.e4. https://doi.org/10.1016/j.cmet.2019.05.006
14. Xu M, Palmer AK, Ding H et al (2015) Targeting senescent cells enhances adipogenesis and metabolic function in old age. Elife 4:e12997. https://doi.org/10.7554/eLife.12997
15. Schafer MJ, White TA, Evans G et al (2016) Exercise prevents diet-induced cellular senescence in adipose tissue. Diabetes 65:1606–1615. https://doi.org/10.2337/db15-0291
16. Farr JN, Xu M, Weivoda MM et al (2017) Targeting cellular senescence prevents age-related bone loss in mice. Nat Med 23:1072–1079. https://doi.org/10.1038/nm.4385
17. Yosef R, Pilpel N, Tokarsky-Amiel R et al (2016) Directed elimination of senescent cells by inhibition of BCL-W and BCL-XL. Nat Commun 7:11190. https://doi.org/10.1038/ncomms11190
18. Gevaert AB, Shakeri H, Leloup AJ et al (2017) Endothelial senescence contributes to heart failure with preserved ejection fraction in an aging mouse model. Circ Heart Fail 10:e003806. https://doi.org/10.1161/CIRCHEARTFAILURE.116.003806
19. Bussian TJ, Aziz A, Meyer CF et al (2018) Clearance of senescent glial cells prevents tau-dependent pathology and cognitive decline. Nature 562:578–582. https://doi.org/10.1038/s41586-018-0543-y

20. Yun MH, Davaapil H, Brockes JP (2015) Recurrent turnover of senescent cells during regeneration of a complex structure. Elife 4: 1–16. https://doi.org/10.7554/eLife.05505
21. Da Silva-Álvarez S, Guerra-Varela J, Sobrido-Cameán D et al (2019) Cell senescence contributes to tissue regeneration in zebrafish. Aging Cell 19(1):e13052. https://doi.org/10.1111/acel.13052
22. Davaapil H, Brockes JP, Yun MH (2017) Conserved and novel functions of programmed cellular senescence during vertebrate development. Development 144:106–114. https://doi.org/10.1242/dev.138222
23. Muñoz-Espín D, Cañamero M, Maraver A et al (2013) Programmed cell senescence during mammalian embryonic development. Cell 155:1104–1118. https://doi.org/10.1016/j.cell.2013.10.019
24. Storer M, Mas A, Robert-Moreno A et al (2013) Senescence is a developmental mechanism that contributes to embryonic growth and patterning. Cell 155:1119–1130. https://doi.org/10.1016/j.cell.2013.10.041
25. Walters HE, Yun MH (2020) Rising from the ashes: cellular senescence in regeneration. Curr Opin Genet Dev 64:94–100. https://doi.org/10.1016/j.gde.2020.06.002
26. Dimri GP, Leet X, Basile G et al (1995) A biomarker that identifies senescent human cells in culture and in aging skin in vivo. PNAS 92:9363–9367. https://doi.org/10.1073/pnas.92.20.9363
27. Roy S, Gardiner DM, Bryant SV (2000) Vaccinia as a tool for functional analysis in regenerating limbs: ectopic expression of Shh. Dev Biol 218:199–205. https://doi.org/10.1006/dbio.1999.9556
28. Ferretti P, Brockes JP (1988) Culture of newt cells from different tissues and their expression of a regeneration-associated antigen. J Exp Zool 247:77–91. https://doi.org/10.1002/jez.1402470111

Chapter 9

Methods for Studying Appendicular Skeletal Biology in Axolotls

Camilo Riquelme-Guzmán and Tatiana Sandoval-Guzmán

Abstract

The axolotl is a great model for studying cartilage, bone and joint regeneration, fracture healing, and evolution. Stainings such as Alcian Blue/Alizarin Red have become workhorses in skeletal analyses, but additional methods complement the detection of different skeletal matrices. Here we describe protocols for studying skeletal biology in axolotls, particularly Alcian Blue/Alizarin Red staining, microcomputed tomography (μCT) scan and live staining of calcified tissue. In addition, we describe a method for decalcification of skeletal elements to ease sectioning.

Key words Calcification, Live staining, Skeleton, Cartilage/bone

1 Introduction

The skeletal tissue is a critical structural component of limbs in tetrapods, creating support for the movement and other species-specific adaptations. In the axolotl (*Ambystoma mexicanum*), the appendicular skeleton occupies over 50% of the surface on a cross-section [1]. Although skeletal cells do not participate in regeneration [2–4], the efficiency of regeneration and/or possible defects can be easily assessed by analyzing skeletal morphology [5–8].

Advances in understanding the mechanisms of regeneration in axolotls have opened new possibilities for using this model to address bone biology. The axolotl limb has been a model for fracture healing [1] regeneration of critical size defect injuries [9], and joint regeneration [10, 11]. Moreover, understanding the process of ossification in different species has become a powerful tool for unraveling the evolutionary dynamics within different taxonomic ranks. In the axolotl, the early larval stages are characterized by a cartilaginous skeleton, contrasting with the skull skeleton, that begins ossification early in development [12]. As the animals reach sexual maturity, the strongly calcified cartilage in

Ashley W. Seifert and Joshua D. Currie (eds.), *Salamanders: Methods and Protocols*,
Methods in Molecular Biology, vol. 2562, https://doi.org/10.1007/978-1-0716-2659-7_9,

Fig. 1 Alcian Blue/Alizarin Red staining of animals from a 5-cm total length (TL) axolotl, a 10-cm TL and a 16-cm TL axolotl. Staining by Yuka Taniguchi-Sugiura

the limbs is replaced by ossified bone (Fig. 1). The changing cellular and structural landscape of the skeleton, as the animal grows and ages, should be studied in more detail to better understand regeneration.

In this chapter, we report methods for studying skeletal biology in axolotls. First, we describe a protocol to perform Alcian Blue/ Alizarin Red staining in axolotl limbs. This staining allows the rapid identification of morphological features or changes in the skeleton, relying on the labeling of calcified tissue (mainly found in skeleton) with Alizarin Red and polysaccharides in cartilage with Alcian Blue. Secondly, we provide a 48-h protocol to decalcify samples for paraffin or cryosections. Skeletal elements are usually difficult to section due to the intrinsic hardness of the tissue, but also because of the differences in the mechanical properties with the surrounding softer tissue. Thus, performing a correct decalcification is critical for the sectioning of skeletal elements and later processing. Moreover, skeletal architecture and bone volume can be assessed using a μCT scan. Here, we present a brief method for sample preparation and the scanning methodology used to analyze axolotl bone volume. Finally, we introduce a rapid and non-invasive method for staining of calcified matrices in vivo in the axolotl, which allows the evaluation of skeletal dynamics during development and regeneration.

To help the researcher decide on a suitable protocol, the following should be considered: (1) Alizarin Red is a calcium-binding dye that detects calcification patterns, it is usually combined with Alcian Blue, which stains polysaccharides in cartilage. While an Alcian Blue/Alizarin Red nicely differentiates strongly calcified tissue from cartilage, Alizarin Red does not discriminate between calcified cartilage and an ossified skeleton. (2) To distinguish calcification/mineralization from ossification, we recommend μCT scans. (3) In addition, tissue processing during the combined Alcian Blue/Alizarin Red staining, quenches a weak Alizarin Red staining. Thus, if the researcher needs to identify early stages of calcification (or a mild calcification), we recommend using Alizarin Red or calcein alone (*see* Subheading 3.4).

2 Materials

2.1 General Tools and Reagents

1. Microsurgery scissors and forceps.
2. 15- and 50-mL conical tubes.
3. 60-mm petri dish.
4. Phosphate-buffered saline (PBS).
5. Formaldehyde 10% in water.
6. Plastic container with lid (preferably opaque for light protection).
7. Plastic film.

2.2 Alcian Blue/Alizarin Red Staining

1. Glycerol.
2. Acetic acid.
3. Potassium hydroxide (KOH) 1% (w/v).
4. 25%, 50%, 70%, 80%, 90%, and 100% ethanol.
5. Glycerol/ethanol solution (1:3, 1:1, 3:1). Mixing produces several bubbles which can remain in the sample. It is advisable to let the solution sit for a day after mixing to reduce bubbles to a minimum.
6. Alcian Blue solution: Alcian Blue (Sigma A3157) 0.0001% (w/v) in ethanol 60% and acetic acid 40%. For 50 mL: 0.005 g of Alcian Blue, 30 mL ethanol, and 20 mL acetic acid.
7. Alizarin Red solution: Alizarin Red (Sigma A5533) 0.0001% (w/v) in KOH 1%. For 50 mL: 0.005 g Alizarin Red, 50 mL KOH 1%.
8. Borax saturated solution: 5 g Borax anhydrous (Sigma 71996) in 50 mL of water. The solution is warmed up at 65 °C and stirred until dissolved. Cool down at RT and remove any sediments before use.

9. Trypsin solution: 1% Trypsin (Sigma 85450C, w/v) in 30% borax. For 10 mL: 0.1 g Trypsin, 3 mL Borax, 7 mL water.
10. KOH 1% (w/v) / glycerol 20% solution. For 200 mL: 2 g KOH, 40 mL glycerol, 160 mL water.

2.3 Sample Preparation for μCT Scan

1. Ethanol at 25%, 50%, 70%, and 100%.
2. Micro-CT scan, vivaCT40 (SCANCO Medical, Switzerland).

2.4 Bone Decalcification

1. MEMFA 1×: MOPS 0.1 M pH 7.4, EGTA 2 mM, $MgSO_4 7H_2O$ 1 mM, 3.7% formaldehyde.
2. EDTA 0.5 M pH 7.5: Autoclave and keep at 4 °C. For 500 mL: 93 g EDTA (Supelco 108418) and 10 g NaOH pellets in water.

2.5 In Vivo Skeletal Staining

1. Calcein stock solution: calcein (Sigma C0875) 0.4% in animal holding water (*see* **Note 1**).
2. Alizarin Red stock solution: Alizarin Red 1% in animal holding water.

3 Methods

3.1 Alcian Blue/ Alizarin Red Staining

Staining can be done in any container that allows a volume at least three times the size of the sample. Although limbs can be pooled in the same container, this may hinder staining. It is advisable to use individual containers and perform all incubating steps in a rocking platform.

1. Fix samples in formaldehyde 10% overnight at 4 °C in a rocking platform.
2. Move samples into a new container and wash them in PBS three times for 30 min each at RT in a rocking platform.
3. When using the whole animal, carefully remove viscera using fine forceps.
4. Dehydrate samples by successively moving them into solutions of 25%, 50%, and 70% ethanol for 20 min each on a rocking platform (*see* **Note 2**).
5. Move samples into a new container and stain them with Alcian Blue solution for 1–3 days at RT (*see* **Notes 3** and **4**). Samples can be left in a rack during the whole incubation. The solution is discarded afterward.
6. Using a new container, rehydrate samples by successively moving them into serial washes of 80%, 70%, 50%, 25% ethanol, and finally water for 20 min each. Ethanol solutions might turn blue due to the Alcian Blue washing.

7. Using a new container, treat samples with trypsin solution for 30 min (*see* **Note 5**).
8. Wash samples with KOH 1% for 30 min.
9. Stain samples with Alizarin Red solution for 1–3 days (*see* **Note 4**). Samples can be left in a rack during the whole incubation.
10. Transfer samples to a new container and wash them with KOH 1% twice for 30 min each.
11. Clear samples with KOH 1%/glycerol 20% overnight or until samples look cleared (*see* **Note 6**) using a rocking platform.
12. Dehydrate samples by successively moving them into solutions of 25%, 50%, 70%, 90%, and 100% ethanol for 20 min each.
13. Transfer samples into serial washes of glycerol/ethanol (1:3, 1:1, 3:1) for 30 min each.
14. Finally, transfer samples into a new container with 100% glycerol and store them at room temperature (Figs. 1 and 2a).

3.2 Sample Preparation for μCT Scan

1. Fix samples in formaldehyde 10% overnight at 4 °C in a rocking platform.
2. Wash samples in PBS three times for 30 min each.
3. Dehydrate samples by successively moving them into solutions of 25%, 50%, and 70% EtOH for 20 min each. Samples can be stored at −20 °C for a long term before imaging.
4. Prior to imaging, our samples are positioned vertically in a 15-mL falcon tube without ethanol. Samples are fixed to the inner surface of the tube by filling the container with plastic film. It is critical that samples are completely fixed during the imaging. For μCT scan, we use vivaCT40 equipment from SCANCO Medical. Samples are transferred to the instrument's sample holder. For measuring the total volume of bones and determining microarchitecture (Fig. 2b), the instrument's predefined "script 6" can be used. The isotropic voxel size was 10.5 μm (70 kVp, 114 μA, 200 ms integration time). Scanning time is around 1 h per skeletal element for a limb from a 14-cm total length animal.

3.3 Bone Decalcification

Bone decalcification is performed after sample fixation with MEMFA 1×. The protocol was adapted from [13]. Samples are immersed in a container with a generous amount of solution (*see* **Note 7**)

1. Wash samples in PBS three times for 30 min each.
2. Incubate samples with EDTA at 4 °C in rotation with washes of fresh EDTA solution as needed (for periods of incubation, *see* **Note 8**).
3. Wash samples with H_2O three times for 30 min each.

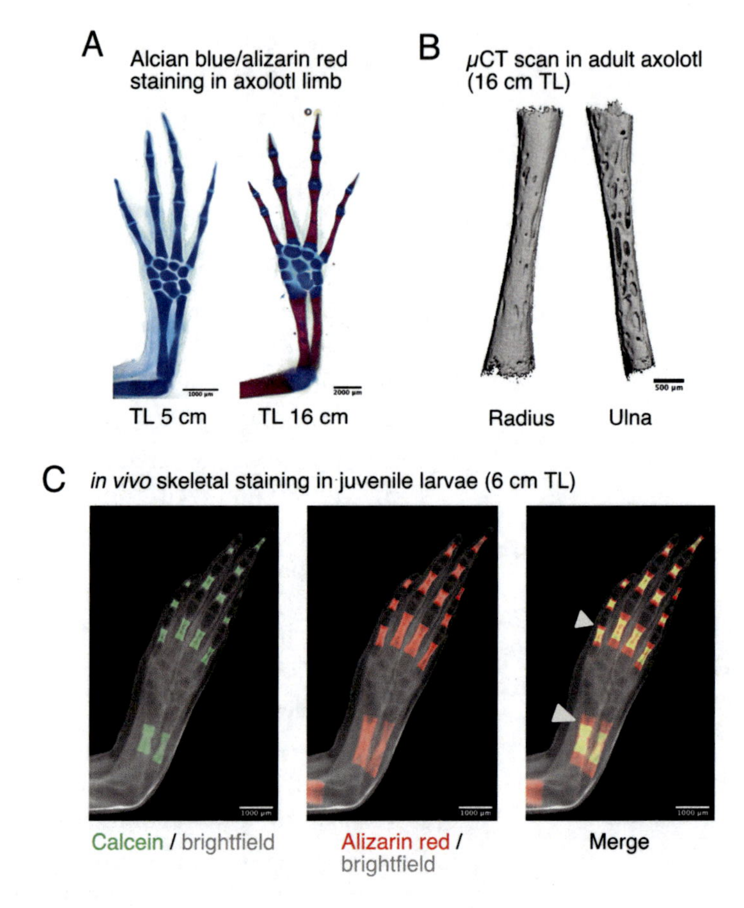

Fig. 2 (**a**) Alcian Blue/Alizarin Red staining of limbs from a 5-cm total length (TL) axolotl (left panel) and a 16-cm TL axolotl (right panel). Absence of calcified tissue (Alizarin Red staining) can be seen in the 5-cm animal, although calcein or Alizarin Red alone stain appendicular skeleton of animals from 1.5–1.8 cm TL on. (**b**) μCT scan from zeugopodial bones (radius and ulna) from a 16-cm total length axolotl. 3D reconstruction from μCT scan is shown. (**c**) in vivo skeletal staining in juvenile larvae of 6-cm TL. Animals were stained with calcein (left) and 45 days later with Alizarin Red (middle). Both dyes stain the calcified cartilage in the limb. New calcification is observed in the merge image (arrow heads)

4. Samples can be embedded or used further for other applications (*see* **Note 9**).

3.4 In Vivo Skeletal Staining

Calcium-binding dyes are non-toxic, and thus can be used to stain live animals. In axolotls, this has been very useful for longitudinal studies and to complement live imaging of transgenic animals.

1. Working calcein solution is prepared by diluting the stock solution in holding water to a 0.1% final concentration. If Alizarin Red solution is preferred, the stock solution is also diluted to a 0.1% final concentration (*see* **Note 10**).

2. Animals with a total length up to 6 cm are transferred into a light-protected container with calcein 0.1% and kept inside for 5–10 min (for timing, *see* **Note 11**). It is recommended to keep animals in the dark to protect the solution during the incubation time.
3. Animals are washed to remove the excess calcein. Washing can be done by letting animals swim in their holding container with the holding water. Change the water when it is heavily stained. Change enough times until the water is no longer stained (*see* **Note 12**).
4. For imaging calcein, a standard GFP filter or preset can be used (excitation: 470 nm; emission: 509 nm). Alizarin Red can be imaged with a mCherry filter or preset (excitation: 530–560 nm; emission: 580 nm).

4 Notes

1. We recommend to prepare calcein solution using the holding water in which animals are kept (e.g., Holtfreter's solution or dechlorinated tap water).
2. Samples can be stored at −20 °C in 70% ethanol. No solution decay was observed in samples up to 1 year.
3. For big samples, such as limbs from animals over 10-cm total length, is it advisable to remove the epidermis. The epidermis can be easily removed by carefully pulling with forceps.
4. Time of incubation depends on the size of samples. For limbs from 4–6 cm total length animals, 1 day is sufficient. For limb from animals over 15-cm total length, 3 days might produce better results. Timing should be optimized for each sample, as overstaining could produce noise in the muscles and dermis.
5. Trypsin treatment is not necessary for limbs of animals 4–6 cm total length. Time should be adjusted according to the sample size. Importantly, the sample becomes fragile after treatment and can be easily destroyed when handled with forceps. When staining the whole animal (3–7 cm total length), the incubation time should be between 1 and 2 h.
6. If samples are overstained with Alizarin Red or if there is too much noise in the soft tissues, a clearing step is critical. This step should be done by changing the clearing solution as many times as necessary until the solution is no longer stained with Alizarin Red.
7. For isolated bones of animals up to 20-cm total length, samples were decalcified in 10 mL of solution (ca. 10 times more volume). For complete limbs of similar size animals, samples were decalcified in a 50 mL of solution.

8. In general, limbs from animals smaller than 7-cm total length do not need to be decalcified for optimal sectioning. Limbs from animals up to 20-cm total length can be decalcified for 48 h, changing once the solution. Limbs from animals older than 2 year old can be decalcified for 1–2 weeks by changing the solution every 2–3 days.
9. This decalcification protocol has been used for OCT and paraffin embedding and for whole-mount immunofluorescence of limbs and isolated bones.
10. Calcein and Alizarin Red have the same staining pattern. Both stainings can be used sequentially for assessing differences in skeletal growth (Fig. 2c), or for complementary use in reporter transgenic animals.
11. Incubation length should be optimized for specific needs. In general, animals up to 6-cm total length are incubated with calcein or Alizarin Red for 5–10 min. Animals with a total length of 14 cm are incubated for 30 min with calcein. Additionally, we have observed that calcein working solution, but not Alizarin Red, can be reused a maximum of 10 times. Incubation periods can also be shortened when calcein is used. Nevertheless, we recommend preparing new working solutions often to avoid contamination and significant decreases in the quality of staining.
12. For better imaging after staining with calcein, animals should be kept in holding water for about 10 min to properly remove the excess calcein from the tissue. For the best result, it is advisable to stain a day before starting with imaging. We have observed that skeletal elements remained stained for more than 60 days.

Acknowledgments

We are thankful to Beate Gruhl, Anja Wagner and Dr. Judith Konantz for their dedication to the axolotls. We would like to thank all members of the Sandoval-Guzmán Lab for their unconditional support. We thank Yuka Taniguchi-Sugiura for her advice on Alcian Blue/Alizarin Red stainings. This work was supported by the Center for Regenerative Therapies Dresden (CRTD) and the German Research Council (DFG). CRG was supported by the DIGS-BB program.

References

1. Hutchison C, Pilote M, Roy S (2007) The axolotl limb: a model for bone development, regeneration and fracture healing. Bone 40: 45–56. https://doi.org/10.1016/j.bone.2006.07.005
2. Currie JD, Kawaguchi A, Traspas R, Schuez M, Chara O, Tanaka EM (2016) Live imaging of axolotl digit regeneration reveals spatiotemporal choreography of diverse connective tissue progenitor pools. Dev Cell 39:411–423. https://doi.org/10.1016/j.devcel.2016.10.013
3. McCusker CD, Diaz-Castillo C, Sosnik J, Phan AQ, Gardiner DM (2016) Cartilage and bone cells do not participate in skeletal regeneration in Ambystoma mexicanum limbs. Dev Biol 416:26–33. https://doi.org/10.1016/j.ydbio.2016.05.032
4. Muneoka K, Fox WF, Bryant SV (1986) Cellular contribution from dermis and cartilage to the regenerating limb blastema in axolotls. Dev Biol 116:256–260. https://doi.org/10.1016/0012-1606(86)90062-X
5. Bothe V, Mahlow K, Fröbisch NB (2020) A histological study of normal and pathological limb regeneration in the Mexican axolotl Ambystoma mexicanum. J Exp Zool B Mol Dev Evol. https://doi.org/10.1002/jez.b.22950
6. Vieira WA, Wells KM, Milgrom R, McCusker CD (2018) Exogenous vitamin D signaling alters skeletal patterning, differentiation, and tissue integration during limb regeneration in the axolotl. Mech Dev. https://doi.org/10.1016/j.mod.2018.08.004
7. Koriyama K, Sakagami R, Myouga A, Hayashi T, Takeuchi T (2018) Newts can normalize duplicated proximal–distal disorder during limb regeneration. Dev Dyn 247: 1276–1285. https://doi.org/10.1002/dvdy.24685
8. Stock GB, Bryant SV (1981) Studies of digit regeneration and their implications for theories of development and evolution of vertebrate limbs. J Exp Zool 216:423–433. https://doi.org/10.1002/jez.1402160311
9. Chen X, Song F, Jhamb D, Li J, Bottino MC, Palakal MJ, Stocum DL (2015) The axolotl fibula as a model for the induction of regeneration across large segment defects in long bones of the extremities. PLoS One 10:e0130819. https://doi.org/10.1371/journal.pone.0130819
10. Lee J, Gardiner DM (2012) Regeneration of limb joints in the Axolotl (Ambystoma mexicanum). PLoS One 7:e50615. https://doi.org/10.1371/journal.pone.0050615
11. Cosden RS, Lattermann C, Romine S, Gao J, Voss SR, MacLeod JN (2011) Intrinsic repair of full-thickness articular cartilage defects in the axolotl salamander. Osteoarthr Cartil 19: 200–205. https://doi.org/10.1016/j.joca.2010.11.005
12. Riquelme-Guzmán C, Schuez M, Böhm A, Knapp D, Edwards-Jorquera S, Ceccarelli AS, Chara O, Rauner M, Sandoval-Guzmán T (2021) Postembryonic development and aging of the appendicular skeleton in Ambystoma mexicanum. Dev Dyn. https://doi.org/10.1002/dvdy.407
13. Kusumbe AP, Ramasamy SK, Starsichova A, Adams RH (2015) Sample preparation for high-resolution 3D confocal imaging of mouse skeletal tissue. Nat Protoc 10: 1904–1914. https://doi.org/10.1038/nprot.2015.125

Chapter 10

Fluorescence In Situ Hybridization of DNA Probes on Mitotic Chromosomes of the Mexican Axolotl

Melissa Keinath and Vladimir Timoshevskiy

Abstract

Fluorescence in situ hybridization (FISH) is used extensively for visual localization of specific DNA fragments (and RNA fragments) in broad applications on chromosomes or nuclei at any stage of the cell cycle: metaphase, anaphase, or interphase. The cytogenetic slides that serve as a target for the labeled DNA probe might be prepared using any approach suitable for obtaining cells with appropriate morphology for imaging and analysis. In this chapter, we focus on the application of molecular cytogenetic methods such as DNA labeling, slide preparation, and in situ hybridization related to cells from Mexican axolotl.

Key words Fluorescence in situ hybridization, Chromosomal in situ suppression (CISS) hybridization, Axolotl mitotic chromosomes, BAC, Nick translation

1 Introduction

Since the development of the technique by Joseph Gall's group [1], in situ hybridization of nucleic acids has been practiced broadly in basic research and used as a clinical diagnostic tool. In situ hybridization (ISH) is a cytogenetic technique that effectively permits the detection, localization, and quantification of target nucleic acid sequence on chromosomes, in cells, or in tissues. The technique requires the hybridization of antisense DNA or RNA probes that complement the sequence of the nucleic acid targets. These probes can be designed to target DNA in metaphase and interphase chromosomes and/or RNA to evaluate gene expression in cells or tissue. Direct detection of the target can be visualized with a radiographic or fluorescent probe [2] or indirectly detected through a reaction between the probe and a chromogenic molecule or label-bound antigen [3]. Those ISH methods that include a fluorescent probe are referred to as fluorescence in situ hybridization (FISH). Clinical applications for FISH primarily include prenatal diagnostics and oncology. Probes can detect aneuploidy,

Ashley W. Seifert and Joshua D. Currie (eds.), *Salamanders: Methods and Protocols*,
Methods in Molecular Biology, vol. 2562, https://doi.org/10.1007/978-1-0716-2659-7_10,

translocations, insertions, or deletions associated with hereditary disease or somatic changes in blood and solid tumors. In contrast to applications in human cytogenetics, where specifically designed probes are widely accessible and techniques for slide preparation are well established, the application of FISH technology and cytological slide preparation methods in other species must be carefully developed based on the project and cell type/properties. A major consideration for the Mexican axolotl is the genome size and number of chromosomes. The axolotl genome is comprised of 14 haploid chromosomes with a genome size about 10 times the size of the human genome, ~32Gb [4, 5]. Thus, metaphase chromosomes of axolotl have a remarkable size. Chromosomes 1–7 are large metacentrics, 8–12 are medium-sized metacentrics and submetacentrics, 13 and 14 are the smallest submetacentrics in complement [6, 7].

Bacterial artificial clones (BACs) have been used on these giant mitotic chromosomes obtained from embryos to verify chromosome and assign chromosome-length scaffolds to physical chromosomes, respectively [4, 7, 8]. A selection of these BAC probes was used on lampbrush chromosomes derived from axolotl oocytes in addition to centromere probes developed by the author to correlate metaphase chromosomes with lampbrush chromosomes [9]. The methods used for chromosomes will be outlined in this chapter.

In this chapter, we present methods for FISH on mitotic chromosomes. Spreads of mitotic chromosomes could be obtained using any tissue with highly proliferative activity (brain, intestinal epithelium, AL-1 cell cultures) or whole early-stage embryos [4]. A successful slide preparation depends on the disaggregating cells, the hypotonic treatment, and the spreading technique. Here we provide the protocol for chromosome preparations using whole embryos and subsequent FISH using DNA probes. FISH using labeled BAC-DNA probes are also described with an additional step of background fluorescence reduction through repressing repeats which may be present in the probe with unlabeled repetitive fraction of genomic DNA (Cot-*x* DNA).

2 Materials

2.1 Spring Water (10% Holtfreter's Solution)

For 100 L of spring water, dilute the following components in DI water [10]:

NaCl	–35.0 g
$NaHCO_3$	–2.0 g

(continued)

KCl	–0.5 g
$MgSO_4$	–4.0 g
CaCl2	–2.0 g
AmQuel	–3.20
Novaqua	–6.67 mL

2.2 Hypotonic Solution

0.4% KCl, 10 mM HEPES

Prepare 10x HEPES stock solution (100 mM) in advance, pH = 6.8. For 100 ml of HEPES stock solution, dissolve 23.8 g HEPES, free acid in 80 ml of ddH2O, adjust pH to 6.8 with NaOH pellets or using 10 M NaOH. Fill to a final volume of 100 ml with ddH2O and sterilize with filtration. 10× HEPES can be stored at 4 °C for several months.

The hypotonic solution should be freshly made before use: weigh 4 g KCl, add 10 ml of HEPES stock solution, adjust volume to 100 ml with ddH2O (*see* **Note 1**).

2.3 Fixative Solution

Combine three parts of methanol with one part of glacial acetic acid. Use fresh (*see* **Note 2**).

2.4 DNase I Working Solution

Dilute 2 μL of DNase I (1 U/μL) in 98 μL of cold H2O, mix by inverting (do not vortex) and spin on centrifuge. Keep on ice and use when freshly prepared.

2.5 BSA Working Solution 0.5 mg/ml

Make stock solution: weigh 10 mg of BSA and dissolve in 1 mL of H_2O. Use 1:20 dilution for working BSA solution. Working solution may be aliquoted and kept at −20 °C.

2.6 dNTPs Mixture

0.5 mM each of unlabeled dATP, dCTP, and dGTP and 0.15 mM of dTTP. In order to make the mixture, use 100 mM dNTP kit. Mix in a 1.5-mL plastic tube 984 μL of H2O, 5 μL of each dATP, dGTP, dCTP, and 1.5 μL of dTTP. Aliquot and store at −20 °C.

2.7 RNase Solution

Dissolve 10 mg RNase in 1-mL stock solution (10 μL 1 M Tris–HCl pH = 7.5, 3 μL of 5 M NaCl, 987 μL H2O). Warm at 100 °C for 15 minutes then chill slowly to ambient temperature. Store at 4 °C. In order to make a working solution, dilute 10 μL of stock RNase in 990 μL of 2× SSC.

2.8 Pepsin Solution

Stock solution: weigh 0.1 g pepsin in 1.5-mL plastic tube and dilute in ddH2O to final a volume of 1 mL. Divide into 50 μL aliquots and store at −20 °C. For working solution, use one aliquot per 50 mL of ddH_2O combined with 50 μL of concentrated HCl.

2.9 Hybridization Mix

10% dextran sulfate, 60% formamide, and 1.2 × SSC.

For 2 mL of hybridization mix, combine 575 μL ddH_2O and 125 μL 20× SSC, and place in heating block at 60 °C. Weigh 0.2 g dextran sulfate and dissolve in diluted, warm SSC solution from the previous step. After dextran sulfate dissolves, add 1.2 mL of formamide from a frozen aliquot (thaw before use) to the solution and mix thoroughly on shaker or nutator until homogenized. Store at −20 °C.

2.10 Denaturing Solution

70% formamide, 2 × SSC.

Aliquot purchased or manually deionized 100% formamide by 25 mL in 50 mL falcon tubes and store at −20 °C. Formamide aliquots should be frozen prior to use. An excessive liquid fraction indicates poor quality of the formamide, which can have adverse effects on the denaturation of DNA. Thaw a single 25-mL formamide aliquot, add 4 mL of 20 × SSC and 6 ml of water. Store at 4 °C.

2.11 Washing Solutions

WS1: 0.4 × SSC, 0.3% NP-40. WS2: 2 × SSC, 0.1% NP-40.

Make 500 mL of each washing solution. For WS1, combine 10 mL of 20 × SSC, 1.5 mL of IGEPAL® CA-630, and water to 500 mL. For WS2, combine 50 mL 20 × SSC, 0.5 mL of IGEPAL® CA-630, and water to 500 mL.

3 Methods

3.1 Slide Preparation

1. Axolotl eggs can be obtained from stock facilities (in our case, *Ambystoma* Genetic Stock Center (AGSC), University of Kentucky) and maintained in an 18 °C incubator until they reach neurula stage (stage 17) [11].
2. Coat the bottom of a petri dish with agarose gel. Using fine-tip dissecting forceps, release embryos from chorion into the agarose gel-coated petri dish containing 0.1% colchicine in 10% Holtfreter's solution. Keep the embryos in this solution for 48 hours at 18 °C for metaphase accumulation.
3. After incubation, wash embryos once in 10% Holtfreter's solution and transfer with a cut-tip pipette to a Dounce homogenizer. Remove as much Holtfreter's solution as possible and replace with 25 mL of 0.4% buffered KCl hypotonic solution from **Step 2.1** and disaggregate them using loose pestle for about 5 passes. For 25 mL solution, use 25–30 embryos. Transfer suspension to a 50-ml tube and allow the cells to swell by incubating for 45 minutes at 18 °C.
4. After incubation, gradually add 10-mL cold fixative solution and prefix for 10 min. Spin in centrifuge at 500 × g and discard supernatant, leaving 1–2 ml above the pellet. Gently

disaggregate the pellet and fix cells with 40-mL fixative solution. At this step, the cells can be stored at −20 °C or proceed to next step after a 10-minute incubation.

5. Make several changes of fixing solutions. Spin at 1000 × g for 10 min, remove the supernatant, leaving 1–2 mL above the pellet, then resuspend and add 40 mL of fresh fixative. Repeat this step 2–3 times in order to completely replace the water from the cell suspension with fixative. At this point, it is suitable to transfer the cell suspension into a 2.0- or 1.5-mL tube for freezer storage. Store the suspension at −20 °C.
6. Preheat the water bath to 60 °C, submerge a heating block and cover the bath. Spread the cells on a glass slide by applying a 25–30 μL drop, and immediately placing in a steam chamber at 60 °C for 1 minute (*see* **Note 3**).
7. For better chromosome morphology and specific hybridization, chromosome preparations can be aged 24–48 hours on a slide warmer at 37 °C or at room temperature and then stored in −20 °C freezer.

3.2 Probe Labeling

For the large molecules such as BACs, nick translation labeling using directly labeled fluorescent nucleotides provides good results for subsequent imaging. In this two-enzyme system, DNase is used for making single-strand brakes (nicks) and DNA polymerase I for replacing "nicked"-strand using dNTPs (including fluorescent ones) from the reaction mix, forming a tagged DNA sequence.

1. Isolate BAC-DNA using an appropriate technique, for example: QIAGEN Large-Construct Kit.
2. Prepare reaction mix for nick translation on ice with final volume of 50 μl (*see* **Note 4**): 1–1.5 μg isolated BAC-DNA, 5 μL of dNTP mixture, 5 μL of 0.5 mg/mL BSA, 5 μL of 10× buffer for DNA-polymerase I, 1 μL of Cy3-dUTP (or another fluorochrome-conjugated dUTP), 5 μL of working DNase solution, and 1 μL (10 U) of DNA-polymerase .
3. Mix by tapping, briefly spin, and incubate in thermocycler at 15 °C for 3 hours.
4. Stop reaction by adding 1 μL of 0.5 M EDTA. Store probe at −20 °C and protect from light while handling.

3.3 Cot-2–3 DNA Isolation

Normally, the BAC, consistent of 100,000–300,000 base pairs of cloned DNA, inevitably includes repetitive sequences, which when labeled, contributes to unspecific binding with random chromosome targets and increasing background fluorescence. To increase signal/background ratio, the unlabeled fraction of highly repetitive DNA (Cot-x) should be used to suppress off-target probe hybridization. Desired pool of repetitive DNA can be obtained by

Table 1
DNA concentration and reannealing times for isolation of Cot-2 and Cot-3 fractions [13]

	DNA concentration μg/μl	Time reannealing, minutes
Cot-2	0.1	100
	0.3	33
	0.5	20
	0.7	14
	0.9	11
	1	10
Cot-3	0.1	150
	0.3	50
	0.5	30
	0.7	21
	0.9	17
	1	15

reassociating denatured salamander genomic DNA and digesting the low copy sequences with S1-nuclease.

1. Isolate 400–500 μg genomic DNA from salamander tissue using any available method. A basic phenol/chloroform extraction using blood or tail tip works well. At the final step, dissolve DNA pellet in H2O and determine concentration (see Table 1).
2. Add 1/9 volume of 12× SSC.
3. Denature and shear the extracted DNA by placing safe-lock or screw cup tube with genomic DNA dissolved in 1.2× SSC in the heat block at 120 °C for 2 minutes.
4. Reassociate DNA at 60 °C 1 using the times listed in Table 1 or time can be calculated according to the formula:

$$t = \text{Cot} - x \times 4.98/\text{Co}$$

where t = time of incubation, Cot-x = Cot fraction (Cot-1 = 1, Cot-2 = 2, etc.), and Co = initial DNA concentration in μg/μ L [12].

5. Stop DNA reassociation by placing the tube with DNA on ice.
6. Add S1 nuclease buffer and S1 nuclease to a final concentration of 100 units per 1 mg DNA calculated by the amount used from initial isolation. Incubate at 42 °C for 1 hour.
7. Precipitate DNA by adding 0.1 volumes of 3 M sodium acetate and 1 volume isopropanol.
8. Centrifuge at max speed for 20 minutes at 18–20 °C. Pour off the supernatant and wash the DNA pellet with 70% ethanol. Centrifuge at max speed for 10 minutes.

9. Air-dry pellet and dissolve in TE buffer. Measure the DNA concentration by spectrophotometer and visualize by gel electrophoresis (the final amount of Cot-2–3 fraction is typically 35–50% of the original DNA content).

3.4 Fluorescence In Situ Hybridization

Prehybridization

1. Incubate slides in 2×SSC for 30 minutes at 37 °C.
2. Wash slides in a series of ethanol: 70%, 80%, and 100% at RT for 5 minutes each, and then air-dry.
3. Apply 100 μL of working RNase solution (from **Step 2.3**) to the slides, cover with a coverslip, and incubate in a 37 °C humidity chamber for 30 minutes.
4. Wash slides in 2×SSC at 37 °C for 5 minutes and rinse in H_2O.
5. Incubate slides in prewarmed pepsin solution (from **Step 2.4**) at 37 °C for 5 minutes and then wash in 1×PBS at RT for 5 minutes.
6. Fix slides in 4% formaldehyde at RT for 2 minutes and then wash twice for 10 minutes in 1×PBS at RT.
7. Dehydrate slides in a series of ethanol: 70%, 80%, and 100% at RT for 5 minutes each and then dry on a slide warmer at 37 °C.

BAC-DNA Probe Preparation

1. For each slide, mix 200-ng-labeled DNA probe (from **Step 3.3**.), 2–4 μg unlabeled repetitive fraction (Cot-2-3, from **Step 3.2**.), and 1 μg of sheared salmon sperm DNA (Thermo Fisher # 15632011) in a tube.
2. Precipitate DNA with 0.1 volumes 3 M sodium acetate, pH = 5.2 and 2 volumes 95% ethanol. Dissolve pellet in 10-μL hybridization mix (Subheading 2.8).

Hybridization

1. Denature the chromosome DNA by placing in prewarmed 70% formamide at 72 °C for 2 minutes (*see* **Note 5**).
2. Wash slides in a series of cold (−20 °C) ethanol: 70% and 80% for 5 minutes each then in 100% ethanol at RT for 5 minutes and dry at 37 °C.
3. Denature prepared hybridization mix for 5 minutes at 95 °C, cool on ice, and then prehybridize at 37 °C for 30 min before applying to slide.
4. Place 10 μL (for 20 × 20 mm coverslip) of hybridization mix to the slide and apply a coverslip avoiding bubble formation. If necessary, remove the bubbles applying slight pressure to coverslip with a pipette tip.
5. Seal the coverslip edges with rubber cement and incubate slides overnight in a humidity chamber at 37 °C.

Washing

1. Using forceps, carefully remove rubber cement from the slides, rinse in prewarmed (72 °C) WS1, agitate the slide slightly, and the coverslip will fall off. Once the coverslip is removed, wash for 2 minutes in the prewarmed (72 °C) WS1.
2. Transfer the slide into the WS2 and incubate for 5 minutes at RT.
3. Remove excess liquid using a Kimwipe without touching the preparation and mount with antifade media and DAPI (Thermo Fisher Sci #P36935, Vector Laboratories #H-2000) under coverslip.

Imaging

Chromosomes can be visualized with fluorescent microscope using a filter set for DAPI and 10× or 20× objectives. The bigger magnification (immersion 40× or 60× objectives) might be needed for signal search and image capture using filter sets corresponding to fluorophore used for probe labeling (for example, FITC, Cy3, and Cy5). BAC-specific signals on salamander spreads are represented by comparatively small paired bright dots clearly distinguishable from background fluorescence (see Fig. 1).

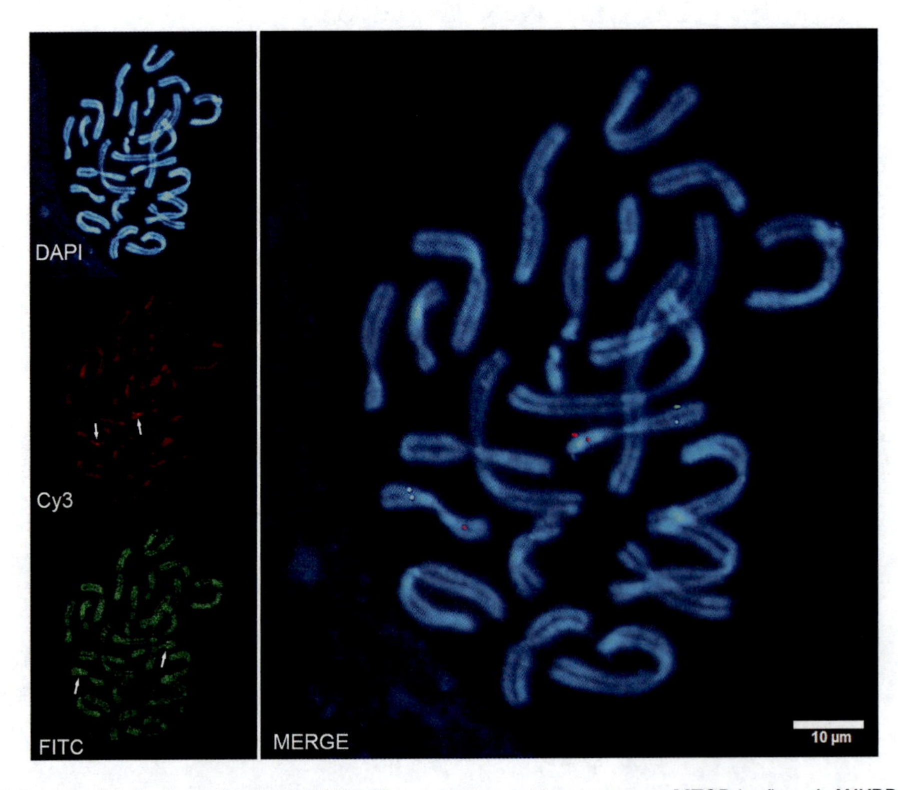

Fig. 1 Two-color in situ hybridization of BAC clones corresponding to genes *MTOR* (red) and *ANKRD1* (green) localized on chromosome 8. Left panels show separate images for different fluorescence channels – DAPI, Cy3, and FITC; these channels are filtered from background and merge in one image present in the right panel

4 Notes

1. The concentration of hypotonic solution can be empirically optimized for obtaining well-spread chromosomes and may vary in the range of 0.4–0.2% KCl.
2. To increase the degree of spreading, 3:2 methanol: acetic acid fixative can be used.
3. The humidity inside of the water bath can be increased by placing a wet paper towel on the hot surface of the aluminum block.
4. The volume of the reaction mix can be scaled up or down.
5. Longer denaturing times might be adversarial to chromosome integrity but acceptable if the interphase nuclei are the object of main interest.

References

1. Pardue ML, Gall JG (1969) Molecular hybridization of radioactive DNA to the DNA of cytological preparations. Proc Natl Acad Sci U S A 64(2):600–604
2. Raj A, van den Bogaard P, Rifkin SA, van Oudenaarden A, Tyagi S (2008) Imaging individual mRNA molecules using multiple singly labeled probes. Nat Methods 5(10):877–879
3. Jin L, Lloyd RV (1997) In situ hybridization: methods and applications. J Clin Lab Anal 11(1):2–9
4. Keinath MC, Timoshevskiy VA, Timoshevskaya NY, Tsonis PA, Voss SR, Smith JJ (2015) Initial characterization of the large genome of the salamander Ambystoma mexicanum using shotgun and laser capture chromosome sequencing. Sci Rep 5:16413
5. Schloissnig S, Kawaguchi A, Nowoshilow S, Falcon F, Otsuki L, Tardivo P et al (2021) The giant axolotl genome uncovers the evolution, scaling, and transcriptional control of complex gene loci. Proc Natl Acad Sci U S A 118(15):e2017176118
6. Callan HG (1966) Chromosomes and nucleoli of the axolotl, Ambystoma mexicanum. J Cell Sci 1(1):85–108
7. Smith JJ, Timoshevskaya N, Timoshevskiy VA, Keinath MC, Hardy D, Voss SR (2019) A chromosome-scale assembly of the axolotl genome. Genome Res 29(2):317–324
8. Woodcock MR, Vaughn-Wolfe J, Elias A, Kump DK, Kendall KD, Timoshevskaya N et al (2017) Identification of mutant genes and introgressed tiger Salamander DNA in the laboratory axolotl, Ambystoma mexicanum. Sci Rep 7(1):6
9. Keinath MC, Davidian A, Timoshevskiy V, Timoshevskaya N, Gall JG (2021) Characterization of axolotl lampbrush chromosomes by fluorescence in situ hybridization and immunostaining. Exp Cell Res 401(2):112523
10. Armstrong JB, Duhon ST, Malacinski GM (1989) Raising the axolotl in captivity. In: Armstrong JB, Malacinski GM (eds) Developmental biology of the axolotl. Oxford University Press, New York, pp 220–227
11. Schreckenberg GM, Jacobson AG (1975) Normal stages of development of the axolotl. Ambystoma mexicanum. Dev Biol 42(2): 391–400
12. Trifonov V, Vorobieva N, Rens W (2009) FISH with and without COT1 DNA. In: Liehr T (ed) Fluorescence In Situ Hybridization (FISH) – application guide. Springer, Jena
13. Timoshevskiy VA, Sharma A, Sharakhov IV, Sharakhova MV (2012) Fluorescent in situ hybridization on mitotic chromosomes of mosquitoes. J Vis Exp 67:e4215

Chapter 11

The Use of Small Molecules to Dissect Developmental and Regenerative Processes in Axolotls

Stéphane Roy

Summary

The axolotl provides an interesting model organism to study different biological processes that are of interest to basic biological sciences and biomedical research. Although axolotls have been in labs for close to 160 years, genetic manipulations still represent a major challenge for most labs. The use of small molecules to target specific signaling pathways allows studies to proceed in animals that are difficult to manipulate genetically. This chapter provides a description of how we administer these chemicals to axolotls.

Key words Axolotl, Regeneration, Signaling pathways, Small molecules, Chemicals

1 Introduction

The discovery of DNA and the advent of modern molecular and cellular biology have opened the door to understanding the actual mechanisms of action of how cells respond to cues they receive during development, in response to injury or directly from the environment. For many model organisms, the ability to manipulate specific genes provides the means to test what function specific genes have in regard to the questions being addressed. If we take the mouse for example, it is now possible to create animals in which we can induce the deletion of a specific gene in a specific tissue at a precise time. This provides an incredible ability to assess gene function in a spatiotemporal manner. Unfortunately, the axolotl is not quite there yet (it is moving towards this level of sophistication and should be there in a few years) and since developing such techniques requires much time and effort, one has to approach questions in a different manner. Fortunately, the development of combinatorial chemistry for drug discovery has led to the creation of millions of small molecules which can be screened for specific functions. These small molecules are grouped in chemical libraries which are tested for various effects [1]. The screening of these

Ashley W. Seifert and Joshua D. Currie (eds.), *Salamanders: Methods and Protocols*,
Methods in Molecular Biology, vol. 2562, https://doi.org/10.1007/978-1-0716-2659-7_11,

libraries has been very successful and has led to the discovery of thousands of molecules targeting hundreds of specific proteins. A majority of these compounds are liposoluble and act in an inhibitory manner; however, some can be activators. The first small molecule inhibitors were found in nature (e.g., cyclopamine and paclitaxel) and the pharmaceutical industry has been mining plants for such molecules for decades [2–5].

Many labs, including my own, have used small molecule inhibitors to interrogate the role of different signaling pathways during regeneration in salamanders [6–17]. My lab first used a small molecule inhibitor to assess the role of the sonic hedgehog (Shh) signaling pathway during the process of limb regeneration in axolotls [12]. This molecule was cyclopamine, isolated from the plant *Veratrum californicum* [4], purportedly one of the most specific hedgehog pathway inhibitors available [18]. We have also taken advantage of this approach to dissect the signaling of TGF-beta, BMP, p53, p38, JNK, and Tak-1 signaling during limb regeneration [7–9, 13, 15, 16]. The advantage of using small molecules is that there is no genetic manipulation required, and any stage/age axolotls can be used directly without prior treatment of any kind. Another advantage of using small molecules is that they can be administered at specific time points for the desired amount of time during an experiment. In addition, different doses can be used to test partial inhibition of a given pathway [16]. The Voss group has done a chemical screen to identify molecules that could affect tissue regeneration using axolotl embryos [10].

2 Materials

All solutions are prepared using ultrapure water and molecular-grade reagents.

1. Stocks of 10 mM are prepared for small molecules (*see* **Note 1**), aliquoted (to prevent repeated freeze-thaw), and stored at −20 °C. Unless specified by the vendor (*see* **Note 2**), the drugs are diluted in 100% DMSO (*see* **Note 3**). DMSO is a versatile solvent for drugs since it can be used to dissolve liposoluble chemicals as well as many hydrophilic compounds. Another advantage of DMSO is that it is very well tolerated by animals (DMSO can make up to about 1% of the volume of water the animals are kept in) [19].
2. For limb regeneration experiments, wild-type and/or white axolotls are used (*see* **Note 4**). We use larvae that are between 4 and 5 cm in total length kept at 19–22 °C, with a 12-h lights-on, 12-h lights-off cycle.

3. For developmental studies, wild-type and/or white axolotl eggs are used.
4. 20% Holtfreter's solution (0.29 mM $MgSO_4$, 0.19 mM $CaCl_2$, 0.17 mM KCl, 12.5 mM NaCl, and 0.45 mM Trizma fish grade).
5. 0.1% MS222 (diluted in 20% Holtfreter's solution, pH 7.0).

3 Methods

3.1 Larvae

1. Dilute the drug stock at desired concentration (*see* **Note 5**) in Holtfreter's solution and dispense it in containers that will house the animals.
2. Prior to surgeries, animals are anesthetized by transferring them to 0.1% MS222 until they no longer react to hind leg pinch stimulus.
3. Animals are measured from tip of the nose to end of the tail before amputation and again once the experiment is completed (*see* **Notes 6** and **7**).
4. Animals are amputated either through the upper (humerus) or lower (radius/ulna) arms.
5. The animals are housed individually in 5 or 10 mL of Holtfreter's solution in disposable medicine cups or small disposable plastic containers (see Fig. 1a, b).
6. Change the drug solution every day (around the same time of day +/− 1 h) for the duration of the treatment (7 days a week if required) (*see* **Note 8**).
7. Animals are monitored during daily feeding just before changing the Holtfreter's solution (*see* **Note 9**).
8. Control animals are treated with Holtfreter's solution containing the same volume of DMSO that is used to administer the drugs to experimental animals (*see* **Note 10**).
9. Control animals are changed daily just like the drug-treated animals.

3.2 Eggs

1. Eggs are used for developmental studies (*see* **Note 11**).
2. The eggs are manually decapsulated using sharp forceps before any treatment to prevent the layer of jelly and capsule from interfering with the embryo's drug exposure.
3. The eggs are housed individually in a 24-well tissue culture plate (at room temperature) in either 500 or 1000 μL of Holtfreter's solution (see Fig. 1c, d).
4. The drugs are changed daily around the same time of day for the duration of the experiment (*see* **Note 12**).

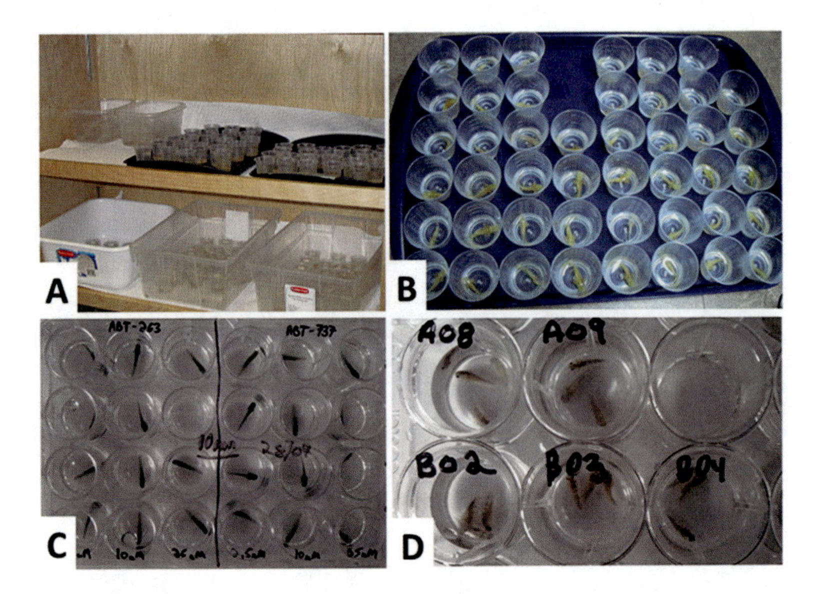

Fig. 1 Pictures showing the ease of scaling up the number of animals that can be treated using either (**a**, **b**) medicine cups (1 oz. size). Typically, 5 mL of 20% Holtfreter's solution is used per cup (works well for animals 4–5 cm in total length). If using animals that are larger, then you have to increase the volume and if need be larger containers can be used. We use disposable plastic containers that can be discarded after each use. This reduces the risk of contamination. Animals are switched to another cup using a plastic sieve. Embryos can also be treated easily using 24-well plates by putting one embryo per well (**c**). If using small embryos and the drug is in limited quantity, they can be pooled 2–3 per well (**d**). The solution is changed daily when using 24-well plates and if the experiment runs longer than 5 days the plate if changed for a new one. One reason for changing once every 5 days is to reduce the risk of injuring the embryos in the transfer as they are fragile at that size

5. The 24-well plates are used for 5 days and then the eggs are transferred to a new one (*see* **Note 13**).
6. All the solutions containing the drugs are disposed following the institutional health and safety recommendations.
7. Control animals are treated with Holtfreter's solution containing the same volume of DMSO that is used to administer the drugs to experimental animals.
8. Control animals are changed daily just like the drug-treated animals.

4 Notes

1. We found that small inhibitors purchased from Sigma-Aldrich (now Millipore Sigma) yielded consistent results from lot to lot. Another vendor that provided small molecules of good

Table 1
Validated small molecule inhibitors used for salamander and newt regeneration experiments

Molecule	Optimal concentration	Function	Reference
SB431542	25 μM	TGF-β inhibitor	[7, 9, 13]
SiS3	2 and 3 μM	Smad3 inhibitor	[7]
Naringenin	35 μM	Smad3 inhibitor	[7]
Cyclopamine (in ethanol)	1 and 2 μg/mL	Shh inhibitor	[8, 12]
SD208	15 μM	TGF-β inhibitor	Unpublished data (inhibited limb regeneration like SB431542)
Pifithrin-α	5 μM	P53 inhibitor	[15, 17]
SB203580	50 μM	P38 inhibitor	[13]
SP600125	5 μM	JNK inhibitor	[13]
5Z-7-oxozeanol	1 μM	Tak-1 inhibitor	[13]
LDN193189	0.2, 0.5, and 1 μM	BMP inhibitor	[16]

quality was Selleckchem. We encountered a lot of reproducibility problems with drugs purchased from other vendors and hence we strongly recommend the two aforementioned vendors. Table 1 contains a list of the small molecules that we have used in the lab.

2. One exception that we encountered was cyclopamine which has to be diluted in 100% anhydrous ethanol.
3. When possible, we aim to have a stock solution that needs to be diluted 1:1000 so that the total amount of DMSO is around 0.1%. If the compound used is diluted in 100% ethanol, then it is very important to keep the dilution at least to 1:1000 as animals do not tolerate ethanol as well as DMSO (personal observation).
4. Axolotls are obtained from the Ambystoma Genetic Stock Center in Lexington KY.
5. In order to determine the optimal drug concentration to use, it is useful to have some types of marker that can be assessed to confirm that the small molecule is indeed working. For example, when we used the LDN 193189, we looked at the levels of p-Smad1/5 in regenerating tissues [16]. This was possible because LDN 193189 is an ATP competitor for the BMP type II receptor and inhibits its ability to phosphorylate Smad1/5. In addition, there was an antibody available that cross-reacted with the axolotl p-Smad1/5. When we used pifithrin-α to inhibit p53, we looked at the toxicity of the drug on

the axolotls and the expression of known p53 target genes (using RT-PCR) in AL-1 cells [15]. We also consider the EC50 or ID50 to start testing for optimal concentration ranges for the drugs when available. However, since most of the small molecules being used are new, many do not have in vivo data regarding toxicity. In addition, the EC50 is often based on cell culture experiments done in a single cell type. This makes extrapolation to in vivo organisms somewhat difficult and a certain level of trial and error is required to find the optimal concentrations. In these situations, we will first test three concentrations with a ten-fold increase between each concentration (e.g., 0.1 μM, 1 μM, and 10 μM). Finally, for some drugs, we use expected phenotypes to assess whether or not they are working. It happens sometimes that there are no clear targets to measure or antibodies that cross-react with axolotl proteins are not available (e.g., cyclopamine causes a reduction in digits because it blocks Shh signaling which is important for digit development [20, 21]).

6. Measuring animals at the beginning and end of each experiment allows us to determine whether the small molecule treatment had an adverse effect on the growth of the animal.
7. In order to avoid being misguided by toxic levels of drugs (which could affect regeneration if the animals become sick) being used, we have decided to use concentrations that did not prevent the growth of the animals.
8. We found that although the cost of the drugs is higher when changed every day, the results are more consistent.
9. To monitor the health of animals, we assess feeding, color, and texture of the skin, movement when given food and when their water is changed. Animals that look sick and do not behave like DMSO control animals are euthanized.
10. DMSO is used as a control to ascertain that any effect observed in animals treated with drugs is not caused by DMSO itself.
11. Axolotl eggs can be obtained from the axolotl colony in Lexington KY at the required stages.
12. A p-1000 is used to aspirate slowly the Holtfreter's solution from each plate taking care not to aspirate the embryos.
13. We use a 7-mL plastic transfer pipet cut midway up the stem in order to create a larger opening to make it easier to aspirate the embryos without causing any injuries.

Acknowledgement

This work was supported by CIHR grant 111013 and NSERC grant rgpin/03912-2017 to S.R.

References

1. Liu R, Li X, Lam KS (2017) Combinatorial chemistry in drug discovery. Curr Opin Chem Biol 38:117–126
2. Goodman J, Walsh V (2001) The story of taxol: nature and politics in the pursuit of an anti-cancer drug. Cambridge University Press, Cambridge; New York, 282 p
3. Incardona JP et al (2000) Cyclopamine inhibition of sonic hedgehog signal transduction is not mediated through effects on cholesterol transport. Dev Biol 224(2):440–452
4. Keeler RF (1973) Teratogenic compounds of Veratrum californicum (Durand). XIV. Limb deformities produced by cyclopamine. Proc Soc Exp Biol Med 142(4):1287–1291
5. Sieber SM, Mead JA, Adamson RH (1976) Pharmacology of antitumor agents from higher plants. Cancer Treat Rep 60(8):1127–1139
6. Currie JD et al (2016) Live imaging of axolotl digit regeneration reveals spatiotemporal choreography of diverse connective tissue progenitor pools. Dev Cell 39(4):411–423
7. Denis JF et al (2016) Activation of Smad2 but not Smad3 is required for mediating TGF-beta signaling during limb regeneration in axolotls. Development 143:3481–3490
8. Guimond JC et al (2010) BMP-2 functions independently of SHH signaling and triggers cell condensation and apoptosis in regenerating axolotl limbs. BMC Dev Biol 10(1):15
9. Levesque M et al (2007) Transforming growth factor: Beta signaling is essential for limb regeneration in axolotls. PLoS One 2(11): e1227
10. Ponomareva LV et al (2015) Using Ambystoma mexicanum (Mexican axolotl) embryos, chemical genetics, and microarray analysis to identify signaling pathways associated with tissue regeneration. Comp Biochem Physiol C Toxicol Pharmacol
11. Purushothaman S, Elewa A, Seifert AW (2019) Fgf-signaling is compartmentalized within the mesenchyme and controls proliferation during salamander limb development. elife 8
12. Roy S, Gardiner DM (2002) Cyclopamine induces digit loss in regenerating axolotl limbs. J Exp Zool 293(2):186–190
13. Sader F et al (2019) Epithelial to mesenchymal transition is mediated by both TGF-β canonical and non-canonical signaling during axolotl limb regeneration. Sci Rep 9(1):1144
14. Vieira WA et al (2019) FGF, BMP, and RA signaling are sufficient for the induction of complete limb regeneration from non-regenerating wounds on Ambystoma mexicanum limbs. Dev Biol 451(2):146–157
15. Villiard E et al (2007) Urodele p53 tolerates amino acid changes found in p53 variants linked to human cancer. BMC Evol Biol 7:180
16. Vincent E et al (2020) BMP signaling is essential for sustaining proximo-distal progression in regenerating axolotl limbs. Development 147(14)
17. Yun MH, Gates PB, Brockes JP (2013) Regulation of p53 is critical for vertebrate limb regeneration. Proc Natl Acad Sci U S A 110(43):17392–17397
18. Incardona JP et al (1998) The teratogenic Veratrum alkaloid cyclopamine inhibits sonic hedgehog signal transduction. Development 125(18):3553–3562
19. Wexler, P. and B.D. Anderson, *Encyclopedia of toxicology.* Third edition. ed. 2014, Amsterdam ; Boston: Elsevier/AP, Academic Press is an imprint of Elsevier. 4 volumes
20. Chiang C et al (1996) Cyclopia and defective axial patterning in mice lacking sonic hedgehog gene function. Nature 383(6599):407–413
21. Riddle RD et al (1993) Sonic hedgehog mediates the polarizing activity of the ZPA. Cell 75(7):1401–1416

Chapter 12

COMET Assay for Detection of DNA Damage During Axolotl Tail Regeneration

Belfran Carbonell, Jennifer Álvarez, Gloria A. Santa-González, and Jean Paul Delgado

Abstract

The purpose of this chapter is to evaluate DNA damage during axolotl tail regeneration using an alkaline comet assay. Our method details the isolation of cells from regenerating and non-regenerating tissues and the isolation of peripheral blood for single-cell gel electrophoresis. Also, we detail each of the steps for the development of the comet assay technique which includes mounting the isolated cells on an agarose matrix, alkaline electrophoresis, and DNA damage detection.

Key words DNA damage, Alkaline comet assay, Tail regeneration, single-cell gel electrophoresis

1 Introduction

Throughout life, organisms are exposed to various insults of an endogenous or exogenous nature that threaten the integrity of DNA. Maintaining DNA integrity is critical to tissue homeostasis, life span, and health span [1]. One of the major events affecting genomic integrity is the generation of DNA damage. DNA damage has been studied for several decades in different contexts including aging, cancer, oxidative stress, degenerative diseases, and exposure to different physical and chemical genotoxic agents that affect DNA integrity [2]. These types of DNA damage include a wide group of injuries: oxidative damage, depurination and depirimidation, deamination of cysteines, and single- and double-stranded DNA breaks [3]. Despite the high frequency of these lesions, cells are equipped with machinery that can repair DNA damage in response to injury [4]. Nevertheless, the repair processes are not always perfect and the damage can persist. Accumulated DNA damage can affect the processes of replication and transcription of genes involved during embryonic development programs, cell cycle regulation, adult tissue homeostasis, and even genes involved in the

Ashley W. Seifert and Joshua D. Currie (eds.), *Salamanders: Methods and Protocols*,
Methods in Molecular Biology, vol. 2562, https://doi.org/10.1007/978-1-0716-2659-7_12,

same reparative response, thus affecting decisions of cell fate and the development of pathological conditions such as cancer and other diseases [5, 6]. According to the above, the identification of DNA lesions, particularly DNA chain breaks, has become a focus of interest due to their unquestionable implications in genomic integrity.

The comet assay (single-cell gel electrophoresis) is a technique described in 1984 by Ostling and Johanson to detect DNA breaks. Subsequently, new versions have been described, including the alkaline version proposed by Singh et al. in 1989 [7, 8]. This technique allows for the detection of single- and double-strand DNA breaks as well as alkali-sensitive sites in DNA. For this purpose, cells are embedded in an agarose matrix, exposed to lysis solution (alkaline or neutral solution) and subsequent electrophoresis [9]. The comet assay represents a very economical, practical, and sensitive technique. Also, the use of specific repair enzymes in this technique allows to discriminate the type of DNA damage, thus showing the great potential that this technique has in the study of DNA damage and repair response [9, 10]. This technique, although few, has drawbacks such as technical variability and lack of standardized automation for DNA damage detection and quantification. Other methods for quantification of DNA damage include phosphorylation status of γH2AX, Halo assay, DNA breakage detection (DBD) FISH, micronuclei, and extra-long quantitative polymerase chain reaction (XL-QPCR) [11–13]. However, the use of the comet assay has been extended both in vitro and in vivo tests in different animal models both terrestrial and aquatic in the field of ecotoxicology, genotoxicity, and other physiological contexts of interest such as regeneration [14–18].

Within the field of regeneration, the study of how cells and tissues respond to DNA damage has become an area of interest in regenerative species including planaria and *Ambystoma mexicanum* [16, 18]. The axolotl has an exceptional capacity to regenerate amputated structures morphologically and functionally equivalent to the lost structures [19, 20]. Regeneration involves high DNA replication/cell proliferation events and cell turnover which may correlate with the presence of DNA damage. [16, 18, 21, 22]. Additionally, increased production of ROS (Reactive Oxygen Species) occurs during the regeneration of amputated structures representing another potential inducer of DNA damage during regeneration [23]. Despite this, the regeneration of lost structures do not present morphological alterations, suggesting that regenerative species can regulate different threats to DNA integrity and efficiently repair DNA damage that may be generated in response to injury [18].

Here we present the identification of DNA damage during tail regeneration of *Ambystoma mexicanum* during the first 7 days post injury (dpa) using an alkaline comet assay including respective basal DNA damage controls such as peripheral blood and non-regenerating tissue. The protocol was carried out following the guidelines of the International Workshop on Genotoxicity Test Procedures [24]. The following protocol is used in our laboratory to detect DNA damage in vitro cell models exposed to hydrogen peroxide and their adaptive response of these cells [13]. Finally, we consider that the development of the comet assay in this animal model not only promises the identification of DNA damage during regeneration process, but it can also be used to evaluate the eco-toxicological and genotoxic effects of some environmental pollutants, considering that this species has also been postulated as an excellent model for environmental biomonitoring.

2 Materials

2.1 Maintenance and Surgical Procedures

1. The procedure described here can be performed on axolotls of different sizes according to the researcher's interest.
2. Stock 40% Holtfreter's solution (wt/vol): Mix 0.2875 g of KCl, 0.536 g of $CaCl_2$ -$2H_2O$, 1.1125 g of $MgSO_4$ -$7H_2$ O, and 15.84 g of NaCl and fill up to 10 L with distilled water. Prepare 20% Holtfreter's solution from stock solution with filtered and dechlorinated water.
3. Stock solution of 1% tricaine methanesulfonate (MS-222).

   Mix 5 g of tricaine methanesulfonate, 50 μl of phenol red, 100 μl of EDTA 1 M pH 7.5, and adjust to 100 ml with Holtfreter's 40% solution adjusting pH to 7.5 with NaOH. Using the stock MS-222 solution, prepare a 0.1% working solution from with 40% Holtfreter's solution.
4. Scalpel blade # 21, ophthalmic scissors, and fine forceps (Dumont #5).
5. Stereomicroscope.

2.2 Tissue Dissociation

1. Amphibian dissection medium (A-PBS): Prepare A-PBS by first mixing 25 mL sterile H_2O with 100 mL PBS (phosphate-buffered saline) to obtain A-PBS, a PBS equivalent to amphibian osmolarity (225 $\pm$ 5 mOsm/L). Subsequently, to obtain amphibian dissection medium, take 10 mL A-PBS and add 10 μL of 50 mg/mL gentamicin. This medium is required to maintain the tissues before the process of cell dissociation.
2. Complete medium L-15: Prepare Leibovitz's L-15 medium (L-15) by obtaining a 64.8% L-15 solution with sterile water. The pH must be adjusted to 6.4 with 0.1 M HCl and sterilized through a 0.22-μm filter. Add 94 mL of prepared L-15, 5 mL of FBS, 1× L-glutamine (1 mL of 100× stock), 0.2× ITS (200 μl of 100× stock), and 1× P/S, according to Denis et al. 2015 [25]. This L-15 complete medium is stable for

5–6 weeks at 4 °C. Because L-15 lacks proteins and growth factors, it is necessary to supplement with commonly used fetal bovine serum (FBS) and other components such as glutamine, insulin-transferrin-selenium (ITS), and penicillin/streptomycin (P/S).

3. Micropipettes and tips (100 μL, 10 μL).
4. Refrigerated centrifuge (2000 × g).

2.3 Peripheral Blood Collection

1. Insulin 1 ml syringes 27 G × 1/2″ - 12× 0.4 mm.
2. 10× Hank's balanced salt solution (HBSS):
3. HBSS-EDTA solution (0.7×): To prepare a 50 mL solution, add 3.5 mL 10× HBSS to 46 mL sterile water. Add 0.5 mL of 5 M EDTA to this solution.

2.4 Pretreatment of Microscope Slides for Electrophoresis and Sample Assembly

1. 1" × 3" (25.4 × 76.2 mm) microscope slides.
2. 100% Methanol.
3. Alcohol burner.
4. 1% Agarose normal melting point: Dissolve 0.1 g in 10 mL distilled water.
5. Heating plate.
6. 0.6% Low-melting point agarose (LMPA) (Invitrogen, 15,517–022): Weigh 0.06 g LMPA and dissolve in 10 mL of PBS 1X. This volume is sufficient to prepare 50 samples.
7. Glass coverslips 24 × 50 mm, 1.5 thickness.

2.5 Lysis and Single-Cell Electrophoresis

1. Stock lysis solution: Mix 146 g of NaCl, 37.2 g of EDTA, and 12 g of Tris–HCl and with 700 ml of distilled water. Add 12 g of NaOH and mix until the components are completely dissolved. Add 10 g of sodium lauryl sarcosinate in small portions to favor its dissolution. Add distilled water until 950 ml, adjust pH to 10 and top up to 1000 ml with distilled water. Keep at room temperature.
2. Working lysis solution: For 100 ml, mix 89 ml of stock lysis solution, 10 ml of DMSO, and 1 ml of Triton X-100 (100%). Working solution must be prepared fresh and refrigerated at 4 °C for at least 1 h.
3. Neutralizing buffer (Tris 0.4 M pH 7.5): For 1 L, solubilize 48.5 g of Tris–HCl in 1 L of distilled water. Adjust pH to 7.5.
4. EDTA 200 mM: Dissolve 3.72 g of EDTA in distilled water and complete up to 50 ml. Adjust pH to 8.0 with NaOH.
5. Electrophoresis buffer: For 1 L, dissolve 14.4 g of NaOH in 950 ml of distilled water and add 5 ml of EDTA 200 mM. Adjust pH to 13.0 and top up to 1000 ml with distilled water (*see* **Note 1**).

3 Methods

For a general understanding of this protocol, a summary of the main steps is illustrated in Fig. 1.

3.1 Axolotl Tail Amputation to Induce Regeneration

1. Anaesthetize the axolotl in 0.1% MS-222 for 12–15 min. Check the depth of anesthesia by puncturing the most distal portion of the tail with a fine needle.
2. Place the anesthetized animal on a flat surgical surface under the stereomicroscope and amputate 8 mm of the tail with scalpel blade # 21.
3. Apply 0.5% antibiotic sulfamerazine topically with a Pasteur pipette over the amputation site and leave the animal in 20% Holtfreter's solution until collection time point for the regenerating tissue.

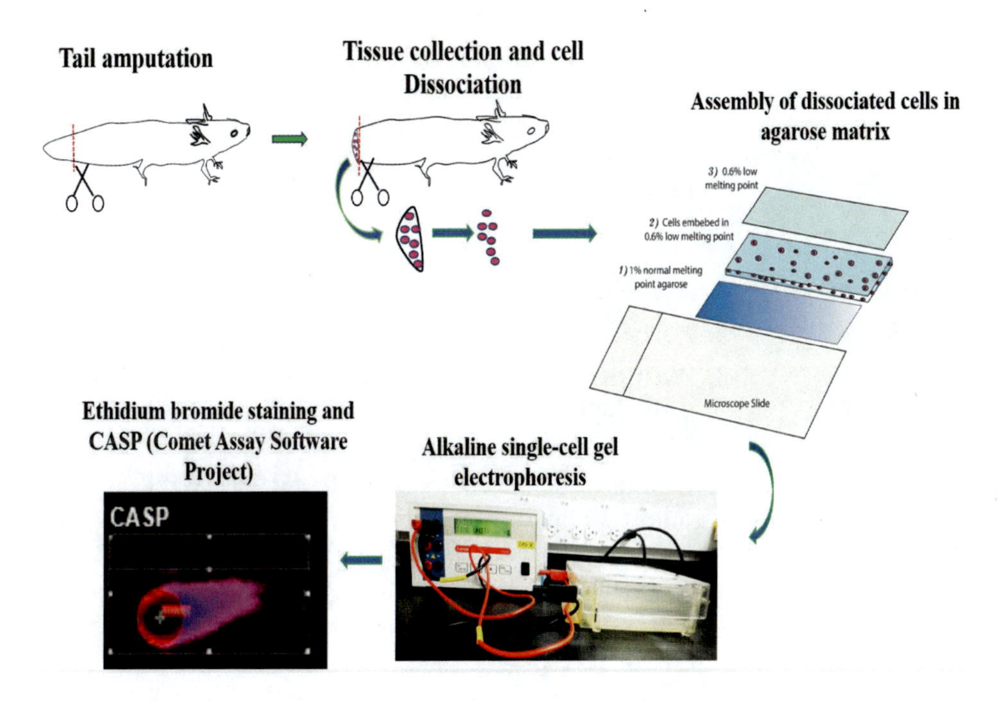

Fig. 1 Comet assay (single-cell gel electrophoresis) of regenerating tail cells from axolotls. Summary of the main steps for the development of the comet assay. A more detailed description of each step will be provided in the next sessions

3.2 Recovery of Regenerating and Non-regenerative Tissue and Dissociation into Single-Cell Suspension

1. Select the stage of regeneration for tissue collection. (For this assay, tissues were collected at 0 dpa (control), 1 dpa, 3 dpa, 5 dpa, and 7 dpa from the previously amputated animals described in **step 3.1**).
2. Anaesthetize the animal in 0.1% MS-222 for 12–15 min. With scalpel blade # 21 perform a new amputation plane approximately 0.5 mm distal to the previously performed amputation plane and collect the tissue of interest (*see* **Note 2**).
3. Transfer tissue with Dumont forceps #5 to a 1.5 ml Eppendorf tube containing 200 μl of L-15 complete medium solution (*see* **Note 3**).
4. Perform tissue dissociation by constant pipetting for up to 5 minutes until maximum cell dissociation is achieved (start with 200 μl tips and finish with 10 μl tips) (*see* **Note 4**).
5. Centrifuge dissociated cells at 900 × g × 5 min, discard the supernatant and resuspend in 20 μl of fresh L-15 complete medium (preferably cold at 4 °C). In this step, the cells are ready to be transferred onto the electrophoresis plates.
6. For the collection of non-regenerative control tissue, anesthetize as in **step 2** and amputate 7 mm of the tail. Transfer the amputated tissue to a sterile Petri dish containing L-15 complete medium and dissect a portion of the epithelial tissue. Follow **steps 3, 4**, and **5** to obtain dissociated cells ready to be mounted on the electrophoresis plates.

3.3 Peripheral Blood Cell Isolation

1. Anaesthetize the animal by immersion in 0.1% for 12–15 min.
2. Place the animal with its back on a flat surface.
3. Wash the insulin syringe (1 ml 27 G × 1/2″ – 12× 0.4 mm) with Hanks' balanced salt solution (HBSS-EDTA 0.7×) twice prior to blood sample collection.
4. Carefully channel the main efferent gill vessel, aspirate, and collect 100 μl of blood with the insulin syringe that was previously washed with HBSS-EDTA.
5. Gently withdraw the needle and deposit the blood sample in a 1.5 ml Eppendorf tube containing 300 μl of 0.7× HBSS-EDTA.
6. Centrifuge the collected sample at 900 × g × 5 min and discard the supernatant.
7. Resuspend cells in 20 μL of L-15 complete medium. Cells are ready to be loaded onto the electrophoresis plates (see Subheading 3.4) (*see* **Note 5**).

3.4 Preparation of Individual Cells from Regenerating Tissues and Controls (Peripheral Blood and Non-regenerative Tissue) on Microscope Slides "Electrophoresis Plates"

3.4.1 Pretreatment of Microscope Slides for Single-Cell Electrophoresis

1. Select microscope slides of 25.4 × 76.2 mm (1" × 3") and score stripes on one of their faces (two to three lines parallel spaced 5 mm to each other with a pen diamond point).
2. Wash the slide with hot water (90 °C) to remove grease residues and other impurities. Allow the slide to dry and immerse it in 100% methanol for 1 minute. Flame the slide using the alcohol burner with a lighter until it is completely dried.
3. Immerse the slide in a beaker containing 1% normal melting point agarose preheated to 37 °C in order to form a layer of agarose on the surface where the stripes were created.
4. Heat the slides on a heating plate at 60 ° C until the agarose film is completely dry (approximately 10 minutes).
5. Store the slides at room temperature until further use (*see* **Note 6**).

3.4.2 Assembly of Dissociated Cells onto Pretreated Microscope Slides

1. Add 80 μl of 0.6% low-melting point agarose to the dissociated cells and resuspended in 20 μl of L-15 complete medium (obtained in Subheadings 3.2 and 3.3). Transfer to an Eppendorf tube and mix gently until a homogeneous sample is obtained (*see* **Note** 7).
2. Distribute 80 μl of the mixture homogeneously on the electrophoresis slide and cover with a 24 × 50 mm, 1.5 thickness glass coverslip.
3. Incubate the slide at 4 °C for 5 min. Then remove the coverslip by moving it gently with a slight displacement. Be careful not to break the agarose layer containing the cells.
4. Add 70 μl of 0.6% low-melting point agarose to the previous layer and cover it again with a coverslip. Incubate for 5 min at 4 °C. Remove the coverslip with a slight displacement. At this point, the glass slide is covered by three layers of agarose: 1% normal melting point agarose, 0.6% low-melting point agarose (containing the cells), and 0.6% low-melting point agarose (does not contain the cells). The assembly is ready for electrophoresis (*see* Fig. 2).

3.5 Lysis and Electrophoresis of Individual Axolotl Cells

1. Incubate the slides in a staining dish coplin containing cold lysis buffer overnight. Note: Incubation in this buffer should be done in a cold room or incubated in a refrigerator at 4 °C.
2. Remove the slides from the lysis solution and wash three times for 5 min with PBS pH 7.4.
3. Place the slides horizontally with frosted part of slide toward the cathode in the electrophoresis chamber and cover them with sufficient cold electrophoresis buffer to cover the slides. Incubate the slides in this buffer for at least 20 minutes.

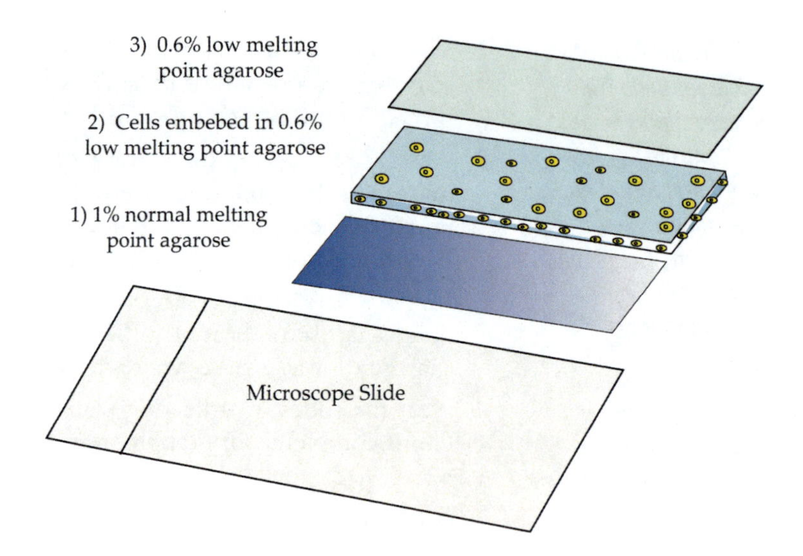

Fig. 2 Preparing dissociated cells on slides for the comet assay Dissociated cells are mixed with 0.6% low-melting point agarose preheated to 37 °C. This mixture is placed on a microscope slide pretreated with 1% low-melting point agarose. A final layer of 0.6% low-melting point agarose is then added after cooling of the layer with cells. Assembled in this way, the result is a sandwich of agarose layers where the intermediate layer contains the cell population to be evaluated

4. Run electrophoresis at 25 Volts and 300 mAMP for 30 minutes (*see* **Note 8**).
5. Immerse the slides in a staining dish coplin containing pH 7.5 neutralizing buffer and leave here for 5 minutes.
6. Perform two additional washes in neutralizing buffer for 5 minutes each.
7. Slides are ready to be stained with ethidium bromide (*see* **Note 9**). Alternatively, slides could be stored for up to 48 hours at 4 °C until staining procedure.

3.6 DNA Damage Detection and Analysis

1. Wash the slides with distilled water if the slides were stored at 4 °C.
2. Add between 40 and 50 μl of 0.2 mg/ ml ethidium bromide on the face of the laminate containing the agarose ensuring that the whole area is covered. Be careful to perform this procedure in an area intended to work with ethidium bromide and avoid leaving residues of this reagent on work surfaces (*see* **Note 10**).
3. Observe the slides in a fluorescence microscope with a Rhodamine filter or with a filter that covers the range of emission of the ethidium bromide and take images of the number of cells that you consider sufficient for the respective analysis of the DNA damage. At least 50–100 cells should be evaluated.

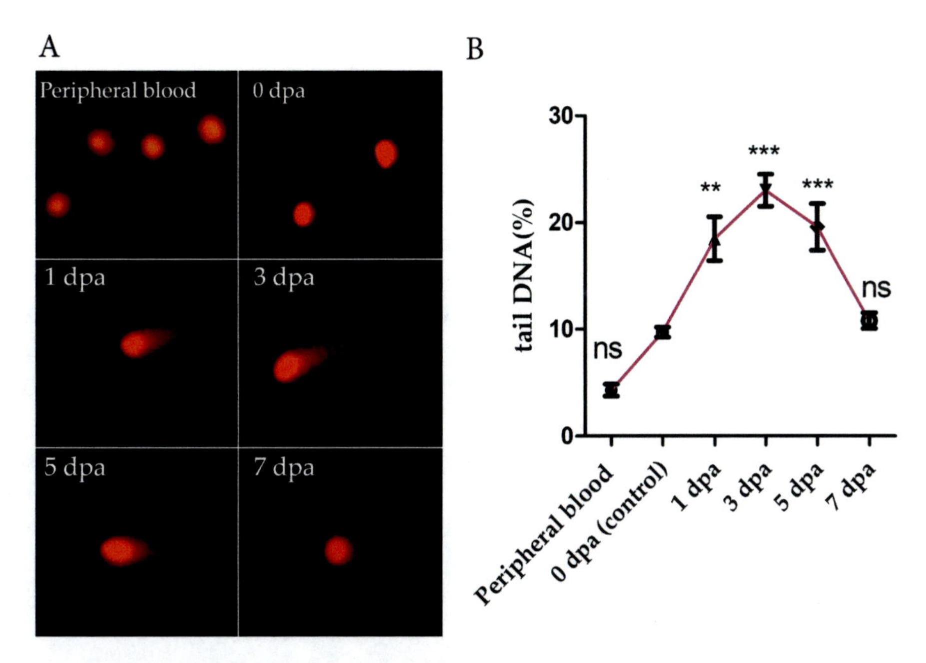

Fig. 3 Detecting DNA damage in axolotl tail cells using the comet assay (**a**, **b**) Dynamics of DNA damage from 0 to 7 dpa (days post amputation) when blastema development takes place. (**a**) DNA damage observed at 1, 3, and 5 dpa evidenced by the formation of a comet-like structure. (**b**) Quantification of DNA damage assessed as % tail DNA. When comparing all samples with each other by ANOVA and subsequent application of Tukey post hoc, the results show a statistically significant difference in tail percentage at 1, 3 and 5 dpa respect to 0dpa and to peripheral blood control. (*** $P < 0{,}001$; ** $P < 0{,}01$; * $P < 0.05$)

4. Analyze the DNA damage by free CASP software to determine the DNA damage evidenced by the % tail DNA (relative size of the tail to the head or cell nucleus) that will give the appearance of the comet (*see* Fig. 3; **Note 11**).

4 Notes

1. To maintain optimal conditions for electrophoresis, it is recommended to keep the electrophoresis buffer at 4 °C.
2. To dissociate and separate mesenchymal cells from early or late blasts from the epithelial tissue, collect the blasts according to **step 2** and transfer them onto a circular glass cover glass containing 100 ul L-15 complete medium that completely covers the blasts. Under the stereomicroscope, make an incision on the axial plane of the blastema with Vannas scissors. Next, suspend the blastema from the distal end with a Dumont #5 forceps and gently allow the mesenchymal cells of the blastema to be released into the L-15 complete medium. Proceed with dissociation of the remaining tissue from the blastema as in **steps 4** and **5**. In this step, the cells are ready to be served on the electrophoresis plates.

3. During dissection and tissue dissociation, it is recommended that the media be preferably cold (4 °C).
4. We recommend performing mechanical dissociation of the tissues. To achieve a single-cell suspension, first, start pipetting the tissue with 200 ul tips for approximately 2 minutes until no tissue remnants of considerable size are observed. Next, pipette with 10 ul tips for approximately 3 minutes until the sample looks as homogeneous as possible (no tissue clumps).
5. The time between isolating blood cells and preparing them with low-melting point agarose for mounting on the plates should be as short as possible in order to avoid conditions that alter genomic integrity.
6. In order to ensure that the agarose layer is completely dry, it is recommended that the preparation of microscope slides for single-cell electrophoresis be carried out 1 day before the assembly of the cells.
7. The low-melting point agarose must be preheated to 37 °C before mixing with the dissociated cells.
8. Take care to keep these running conditions stable. It is recommended to be kept at a current of 300 mAmp and a temperature not exceeding 14 ° C. These conditions can be influenced by a low volume of electrophoresis buffer covering the slides and sudden changes in current can affect the appearance of the comet.
9. For better visualization of the comets in axolotls, it is recommended not to overload the gel with more than $1x10^4$ cells.
10. In addition to ethidium bromide, there are other molecules that can be used to visualize DNA in the comet assay such as 4,6-diamino-2-phenylindole, acridine range, Vista Green DNA, and SybrGreen among others. However, we have obtained excellent results with ethidium bromide.
11. Several parameters are used to calculate the DNA damage between these tail length, percentage of DNA in the tail, and tail momentum. In our case, we use percentage of DNA in the tail. The other parameters can also be used according to the available software for such quantification. To work with CASP software, we recommend downloading the version available at https://casplab.com/. In addition, information on installation and image processing can be accessed at this address. We also invite you to consult the article by K. Ko'nca et al. [26]. Additionally in ImageJ software available in free version at https://imagej.nih.gov/ij/download.html, it is possible to run the OpenComet plugins to load and analyze images of the comet assay. OpenComet is a tool designed with a user-friendly interface that allows image processing in different formats. The

most remarkable thing is that the operator can load and analyze many images simultaneously and the results are obtained quickly. The OpenComet plug-in can be downloaded from http://cometbio.org/download.html and uploaded to ImageJ. For a more detailed explanation of the handling of the OpenComet plugin, we recommend consulting the paper by Benjamin M. Gyori et al., where you will find the detailed steps for analyzing the comet assay images [27].

Acknowledgments

Belfran Carbonell was funded by Minciencias grant 936-2019. This work was funded by the University of Antioquia funding.

References

1. Franzke B, Neubauer O, Wagner KH (2015) Super DNAging-new insights into DNA integrity, genome stability and telomeres in the oldest old. Mutat Res Rev Mutat Res 766:48–57
2. Turgeon MO, Perry NJS, Poulogiannis G (2018) DNA damage, repair, and cancer metabolism. Front Oncol 8:15
3. Chatterjee N, Walker GC (2017) Mechanisms of DNA damage, repair, and mutagenesis. Environ Mol Mutagen 58:235–263
4. Friedberg EC (2003) DNA damage and repair. Nature 421:436–440
5. Lin X, Kapoor A, Gu Y et al (2020) Contributions of DNA damage to Alzheimer's disease. Int J Mol Sci 21
6. Kermi C, Aze A, Maiorano D (2019) Preserving genome integrity during the early embryonic DNA replication cycles. Genes (Basel) 10
7. Ostling O, Johanson KJ (1984) Microelectrophoretic study of radiation-induced DNA damages in individual mammalian cells. Biochem Biophys Res Commun 123:291–298. https://doi.org/10.1016/0006-291X(84)90411-X
8. Singh NP, McCoy MT, Tice RR, Schneider EL (1988) A simple technique for quantitation of low levels of DNA damage in individual cells. Exp Cell Res 175:184–191. https://doi.org/10.1016/0014-4827(88)90265-0
9. Collins AR (2004) The comet assay for DNA damage and repair: principles, applications, and limitations. Appl Biochem Biotechnol Part B Mol Biotechnol 26:249–261
10. Langie SAS, Azqueta A, Collins AR (2015) The comet assay: past, present, and future. Front Genet 6:266
11. Firsanov D V., Solovjeva L V., Mikhailov VM, Svetlova MP (2016) Methods for the detection of DNA damage. In: genome stability: from virus to human application. Elsevier Inc., pp 635–649
12. Figueroa-González G, Pérez-Plasencia C (2017) Strategies for the evaluation of DNA damage and repair mechanisms in cancer. Oncol Lett 13:3982–3988
13. Santa-Gonzalez GA, Gomez-Molina A, Arcos-Burgos M et al (2016) Distinctive adaptive response to repeated exposure to hydrogen peroxide associated with upregulation of DNA repair genes and cell cycle arrest. Redox Biol 9:124–133. https://doi.org/10.1016/j.redox.2016.07.004
14. de Lapuente J, Lourenço J, Mendo SA et al (2015) The comet assay and its applications in the field of ecotoxicology: a mature tool that continues to expand its perspectives. Front Genet 6:180
15. Burlibaşa L, Gavrilacaron L (2011) Amphibians as model organisms for study environmental genotoxicity. Appl Ecol Environ Res 9: 1–15. https://doi.org/10.15666/aeer/0901_001015
16. Wouters A, Ploem JP, Langie SAS et al (2020) Regenerative responses following DNA damage - β-catenin mediates head regrowth in the planarian Schmidtea mediterranea. J Cell Sci 133. https://doi.org/10.1242/jcs.237545
17. Barghouth PG, Thiruvalluvan M, LeGro M, Oviedo NJ (2019) DNA damage and tissue repair: what we can learn from planaria. Semin Cell Dev Biol 87:145–159
18. Sousounis K, Bryant DM, Fernandez JM et al (2020) Eya2 promotes cell cycle progression by

regulating DNA damage response during vertebrate limb regeneration. elife 9. https://doi.org/10.7554/eLife.51217

19. Ponomareva LV, Athippozhy A, Thorson JS, Voss SR (2015) Using Ambystoma mexicanum (Mexican axolotl) embryos, chemical genetics, and microarray analysis to identify signaling pathways associated with tissue regeneration. Comp Biochem Physiol Part C Toxicol Pharmacol 178:128–135. https://doi.org/10.1016/j.cbpc.2015.06.004
20. Echeverri K, Tanaka EM (2002) Ectoderm to mesoderm lineage switching during axolotl tail regeneration. Science (80) 298:1993–1996. https://doi.org/10.1126/science.1077804
21. Gauron C, Rampon C, Bouzaffour M et al (2013) Sustained production of ROS triggers compensatory proliferation and is required for regeneration to proceed. Sci Rep 3. https://doi.org/10.1038/srep02084
22. McCusker CD, Athippozhy A, Diaz-Castillo C et al (2015) Positional plasticity in regenerating Amybstoma mexicanum limbs is associated with cell proliferation and pathways of cellular differentiation regeneration and repair. BMC Dev Biol 15:1–17. https://doi.org/10.1186/s12861-015-0095-4
23. Al Haj Baddar NW, Chithrala A, Voss SR (2019) Amputation-induced reactive oxygen species signaling is required for axolotl tail regeneration. Dev Dyn 248:189–196. https://doi.org/10.1002/dvdy.5
24. Tice RR, Agurell E, Anderson D et al (2000) Single cell gel/comet assay: guidelines for in vitro and in vivo genetic toxicology testing. Environmental and Molecular Mutagenesis. Environ Mol Mutagen, In, pp 206–221
25. Denis JF, Sader F, Ferretti P, Roy S (2015) Culture and transfection of axolotl cells. Methods Mol Biol 1290:187–196. https://doi.org/10.1007/978-1-4939-2495-0_15
26. Końca K, Lankoff A, Banasik A et al (2003) A cross-platform public domain PC image-analysis program for the comet assay. Mutat Res Genet Toxicol Environ Mutagen 534:15–20. https://doi.org/10.1016/S1383-5718(02)00251-6
27. Gyori BM, Venkatachalam G, Thiagarajan PS et al (2014) OpenComet: an automated tool for comet assay image analysis. Redox Biol 2:457–465. https://doi.org/10.1016/j.redox.2013.12.020

Part III

Experimental Manipulations and Surgeries

Chapter 13

In Vivo and Ex Vivo View of Newt Lens Regeneration

Georgios Tsissios, Anthony Sallese, Weihao Chen, Alyssa Miller, Hui Wang, and Katia Del Rio-Tsonis

Abstract

Lens regeneration in the adult newt illustrates a unique example of naturally occurring cell transdifferentiation. During this process, iris pigmented epithelial cells (iPECs) reprogram into a lens, a tissue that is derived from a different embryonic source. Several methodologies both in vivo and in culture have been utilized over the years to observe this phenomenon. Most recently, Optical Coherence Tomography (OCT) has been identified as an effective tool to study the lens regeneration process in continuity through noninvasive, real-time imaging of the same animal. Described in this chapter are three different methodologies that can be used to observe the newt lens regeneration process both in vivo and ex vivo.

Key words Reprogramming, Transdifferentiation, Explant, Optical Coherence Tomography

1 Introduction

One of the most impressive paradigms of naturally occurring adult cell transdifferentiation comes from the newt eye. Following the removal of the lens, iPECs lose their identity (dedifferentiate) and then differentiate into lens cells [1, 2]. This is astonishing because iPECs, which are packed with melanosomes and normally function to block light, undergo cell-type conversion to become transparent lens cells that focus light on the retina. After an injury or complete removal of the lens from the ocular cavity, the iPECs re-enter the cell cycle and start shedding pigment. By day 8 (in Eastern newts), following lens removal, iPECs from the dorsal apical region depigment and undergo further proliferation to form a new lens vesicle. The remaining dorsal iPECs from the central and basal regions, along with all ventral iPECs, withdraw from the cell cycle, resynthesize their melanosomes, and do not contribute to the regenerating lens (Fig. 1A, B). For more extensive reviews on this process, refer to [3–5].

Ashley W. Seifert and Joshua D. Currie (eds.), *Salamanders: Methods and Protocols*,
Methods in Molecular Biology, vol. 2562, https://doi.org/10.1007/978-1-0716-2659-7_13,

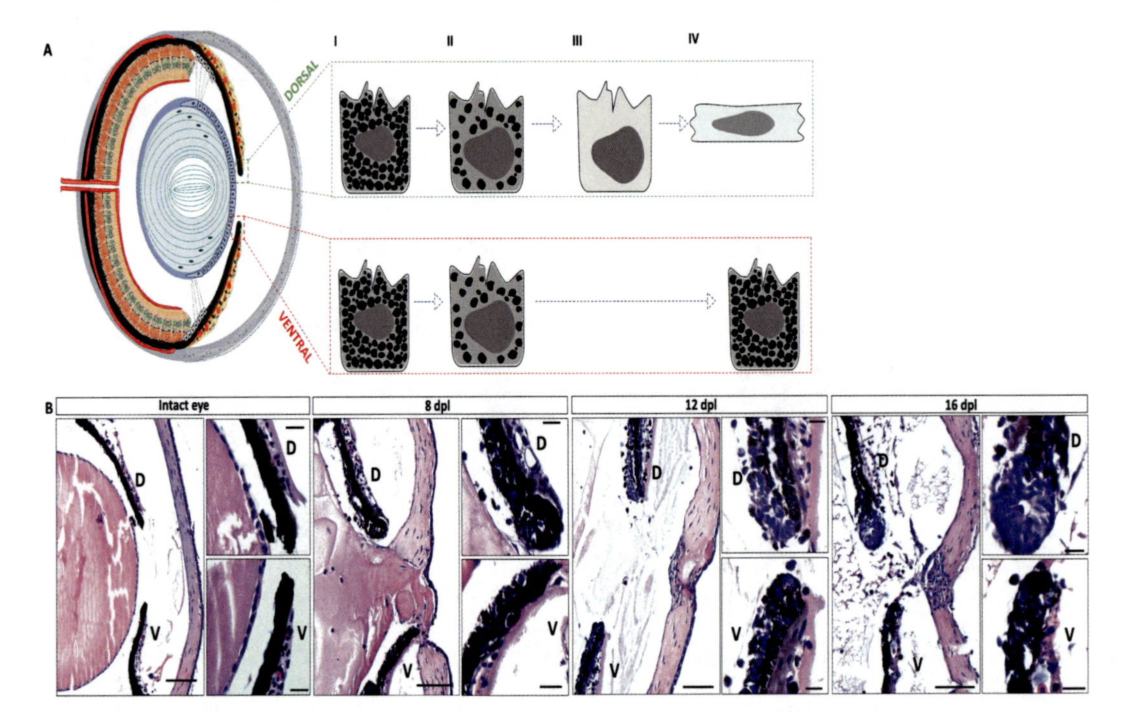

Fig. 1 Iris pigmented epithelial cell (iPEC) transdifferentiation into lens cells. (**A**) Cartoon schematic depicting the cellular transformations of dorsal and ventral iPECs during lens regeneration. At rest, intact iPECs are filled with melanosomes, and the nuclei appear condensed (I). After lens removal, both dorsal and ventral iPECs re-enter the cell cycle and start shedding their pigment (dedifferentiation phase) (II). Enlargement of the nuclei and nucleoli is also observed at that time. Dorsal iPECs that become completely depigmented, proliferate and differentiate into lens cells that elongate to form lens fibers (differentiation phase), whereas the ventral cells withdraw from the cell cycle and replenish their melanosomes (III & IV). (**B**) The early stages of newt lens regeneration (depicting the Eastern newt) are shown by Hematoxylin and Eosin staining. D: dorsal iris, V: ventral iris. Scale bars for large images are 50 μm and for magnified images are 10 μm

The intrinsic ability of the dorsal iPECs to transdifferentiate into a lens, even when placed in environments outside of the eye, has been well demonstrated by a number of transplantation and culturing experiments such as whole organ cultures, whole iris tissue transplantation, and cell culture [6–13]. Furthermore, the replication efficiency and duration of the cell cycle during transdifferentiation of iPECs in culture were similar to the observations made *in vivo* [14]. Many researchers have taken advantage of the fact that cultured iPECs mimic *in vivo* conditions to interrogate the molecular mechanisms of lens regeneration in a controlled environment that is easy to manipulate [15–17]. Here we describe a simple and fast process to culture whole iris tissue in Matrigel to study lens regeneration.

Despite the tremendous advances in modern microscopy, detailed *in vivo* imaging of the newt lens regeneration process has proven a challenging task. In contrast to the newt limb and tail, the iris is an internal tissue not accessible by conventional microscopy.

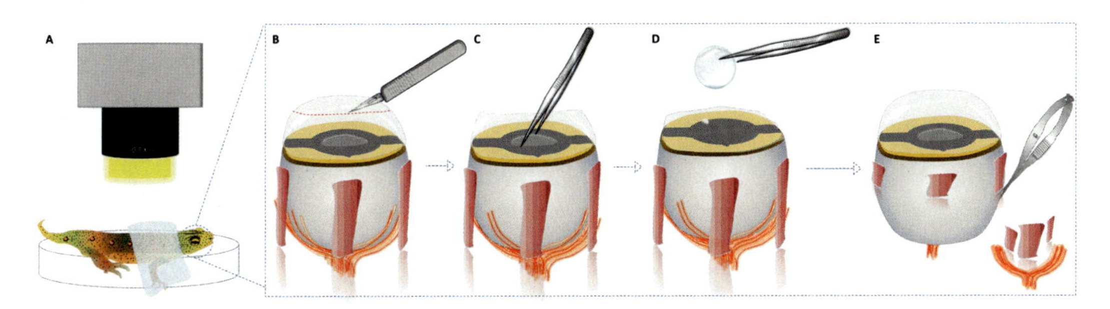

Fig. 2 Lentectomy schematic. (**A**) A moist paper towel helps hold the animal stationary. (**B**) Incision to the cornea with a scalpel. (**C**, **D**) Removal of the lens with tweezers. (**E**) Enucleation of the eye

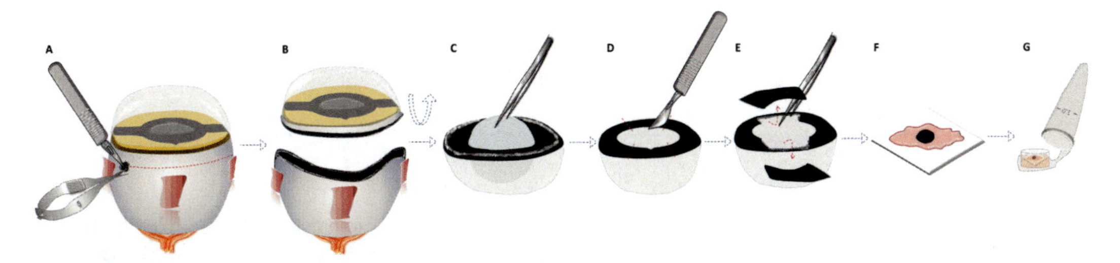

Fig. 3 Explant culture schematic. (**A**) Incision through the sclera below the irido-corneal ring. (**B**) Anterior/posterior separation. (**C**) Lens and other tissue (ciliary body and retina) removal. (**D, E**) Separation of the dorsal/ventral iris segments. (**F**, **G**) Embedment of the iris explant in Matrigel and transfer into the culture medium

Recently, we adapted optical coherence tomography (OCT), a noninvasive imaging modality, to bypass these limitations and observe the entire process of lens regeneration in continuity [18]. This not only minimizes the use of live animals but also allows for efficient screening of regeneration-inducing or inhibiting compounds *in vivo*.

In this chapter, we provide detailed protocols for lentectomy and iPEC explant cultures (Figs. 2 and 3). Establishing an explant culture system that faithfully imitates *in vivo* behaviors of the iris allows researchers to easily manipulate the cells and screen for regeneration-inducing factors such as FGF2 (Fig. 4). Furthermore, the process of OCT *in vivo* imaging is described (Fig. 5). Using this new and exciting imaging technique, we were able to observe the entire lens regeneration process in continuity for the first time since Colucci and Wolff first described the process in the 1890s (Fig. 6A). Finally, by processing OCT data, we can reconstruct three-dimensional images which offer a multi-angle view of morphological changes during lens regeneration and allows for volumetric quantification of different tissues (Fig. 6B). The experimental procedures outlined here can allow researchers to shed light on one of the most remarkable naturally occurring biological processes, the transdifferentiation of newt iPECs to lens cells.

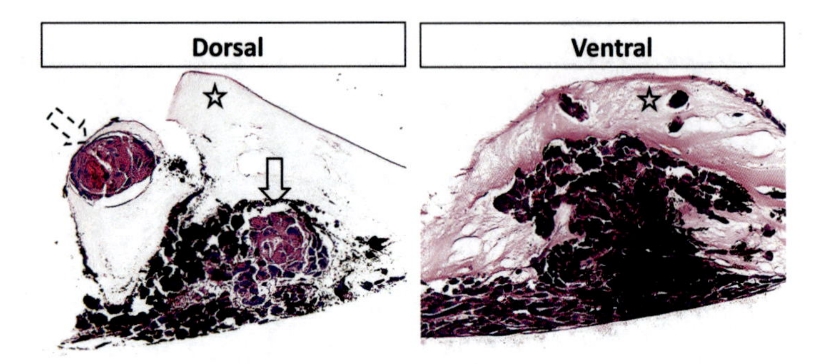

Fig. 4 Hematoxylin and Eosin staining of dorsal and ventral explants cultured in L-15 supplemented with 100 ng/ml of FGF2 for 30 days. Arrows point to the regenerated lens. The dashed arrow points to a mature lens with cuboidal lens epithelial cells in the anterior layer and the elongated lens fibers in the central area. Asterisk shows the area covered by Matrigel

2 Materials

2.1 Reagents

1. 20x Modified Amphibian Phosphate-Buffered Solution (APBS): Weigh 120 g of NaCl, 4.5 g of KCl, 0.525 g of Na_2HPO_4, 0.3 g of KH_2PO_4. Add 900 ml of water, stir, and adjust pH to 7.4 using HCl. Add Water to 1 L and autoclave to sterilize.
2. 0.1% MS222: Weigh 0.5 g of Ethyl 3-Aminobenzoate-Methane Sulfonic Acid Salt. Add 500 ml of 1x modified APBS solution.
3. 5% Lugol's: Add 5 ml of Lugol's iodine solution and 95 ml of 70% ethanol.
4. Modified calcium magnesium-free (CMF) Hank's solution: Weigh 6.4 g of NaCl, 0.32 g of KCl, 0.08 g of $MgSO_4$•7H2O, 0.05 g of KH_2PO_4, 0.15 g of Na_2HPO_4•$7H_2O$, 1.19 g of HEPES and Add 900 ml of water. Stir and adjust pH to 7.4 using HCl. Add water to 1 L and autoclave to sterilize.
5. 100x Modified Hank's solution: Weigh 1.1 g of $CaCl_2$ and 1.6 g of $MgCl_2$•$6H_2O$ and add 100 ml of modified CMF Hank's solution.
6. Modified L-15 media: Weigh 9 g of Leibovitz's L-15 media (Gibco # 41300–039). Add 10 ml of 1 M HEPES, 100 ml of fetal bovine serum, 10 ml of antibiotic-antimycotic and amphotericin B at 2.5 μg/ml concentration. Add 700 ml of water and adjust pH to 7.4 with HCl. Add water to 1 L and sterilize by filtration.

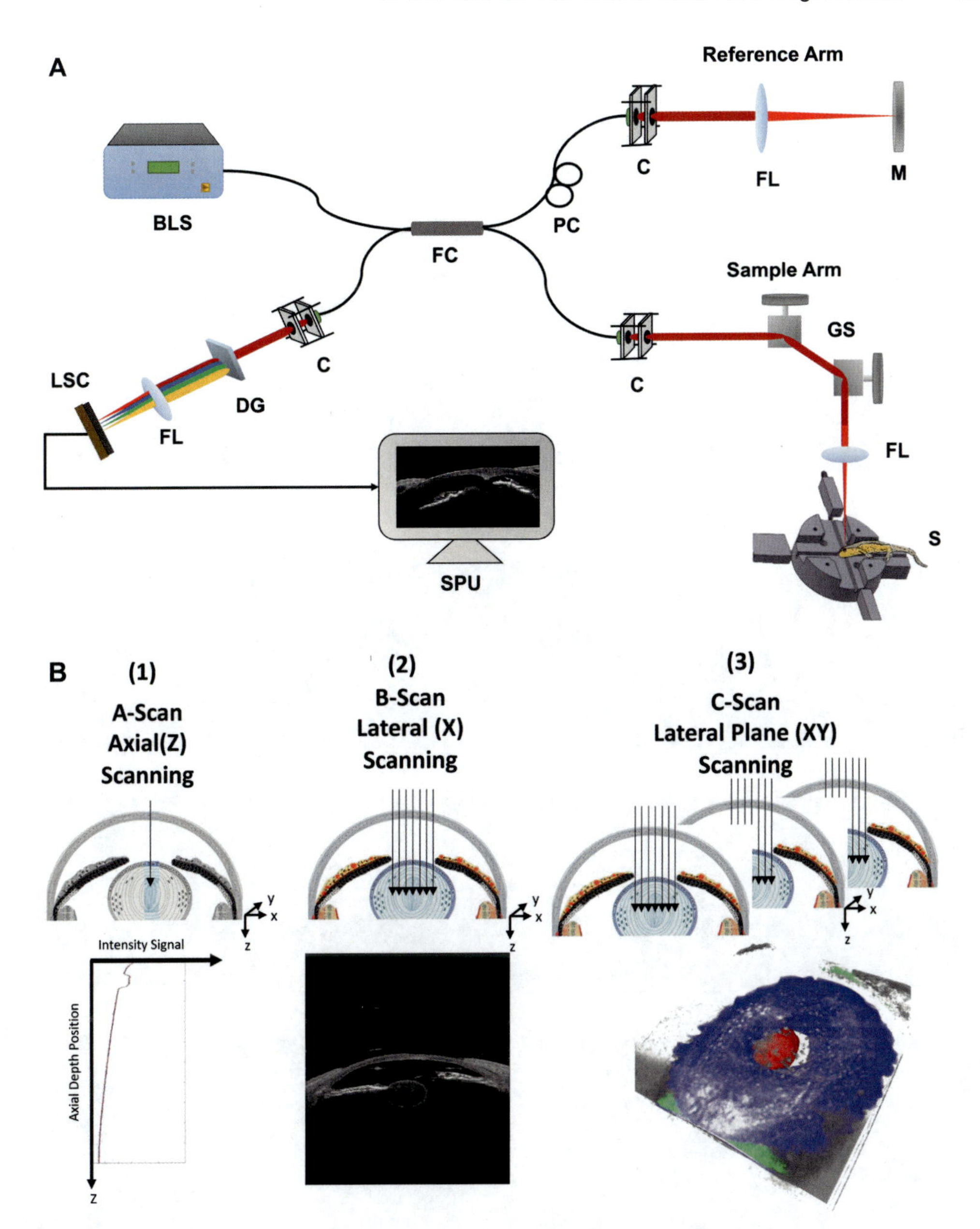

Fig. 5 OCT imaging. (**A**) Optical coherence tomography platform. BLS: band light source, FC: fiber coupler, LSC: line scanning camera, SPU: signal processing unit, PC: polarization controller, FL: focusing lens, C: collimator, GS: galvanometer scanner, M: mirror, DG: diffractive grating, S: Stage. (From Chen et al., 2021[18]). (**B**) Three scanning modalities: (1) A scan, (2) B scan, and (3) C scan

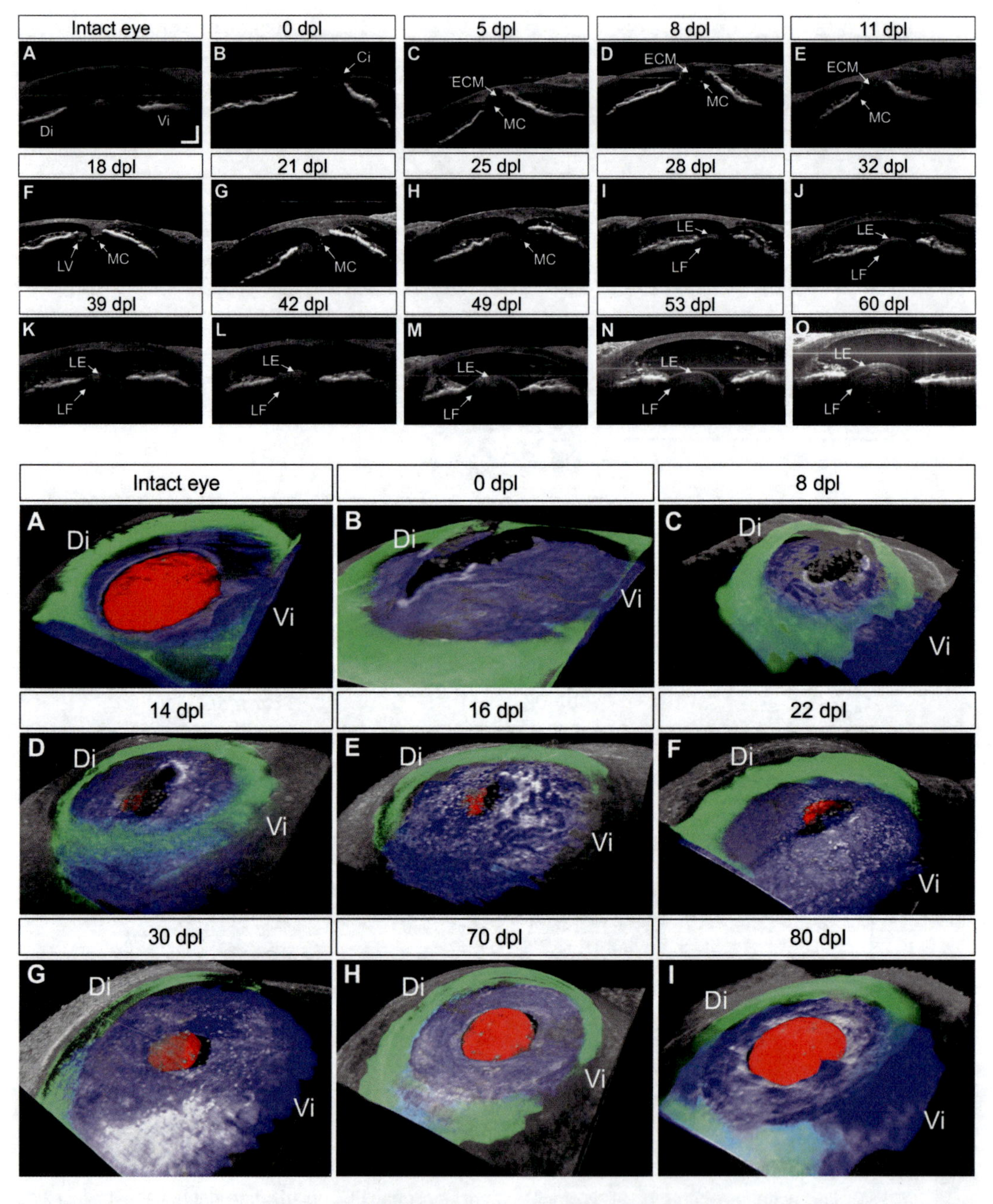

Fig. 6 In vivo imaging of newt lens regeneration using optical coherence tomography. (**Top A-O**) Kinetic observation of the lens regeneration process by following the same eye from a single newt for a period of 60 days. Di: dorsal iris, Vi: ventral iris, ECM: extracellular matrix, MC: migrating cells, Ci: corneal incision, LV: lens vesicle, LE: lens epithelial, LF: lens fiber. (**Bottom A-I**) Three-dimensional reconstruction of C-scans offer a different view for observing the lens regeneration process. (From Chen et al., 2021 [18])

7. 10x Dispase: Add 0.3 g of dispase (Thermo Fisher # 17105041) into 9 ml of modified L-15 media. Sterilize by filtration. Use right away or store at −20 °C.
8. Matrigel matrix (Corning # 354277).

2.2 Equipment

1. Micro dissecting tweezers #55 (Roboz # RS-4984).
2. Micro dissecting spring scissors vannas straight (Roboz # RS-5620).
3. Micro dissecting spring scissors vannas curved (Roboz # RS-5621).
4. Surgical blade #15(Electron Microscopy Sciences # 72044–15).
5. Surgical blade #11(Electron Microscopy Sciences # 72044–11).
6. Membrane filter paper (Millipore # GSWP 04700).
7. Stereomicroscope.
8. Tissue culture hood.
9. Incubator set at 27 °C.
10. Eppendorf tubes 1.5 ml.

2.3 Optical Coherence Tomography

1. Broad band light source (BLS) with a total of 20 mW of power output and ~ 150 Nm full width at half maximum.
2. Fiber coupler (FC).
3. Line scanning camera (LSC).
4. Signal processing unit (SPU) includes an image acquisition card, a data acquisition card, and a computer.
5. Polarization controller (PC).
6. Focusing lens (FL).
7. Collimator (C).
8. 2D galvanometer scanner (GS).
9. Mirror (M).
10. Diffractive grating (DG).
11. Imaging processing software (MATLAB R2019a, ICY, ImageJ, Imaris).
12. Sample holder (S).

3 Methods

3.1 Housing

1. We used adult Eastern newts, *Notophthalmus viridescens*, that were collected from the wild and housed at Miami University animal care facility according to the guidelines approved by the Institutional Animal Care and Use Committee (IACUC).
2. House animals in plastic containers filled with dechlorinated water 2 cm deep. Avoid overcrowding since it could lead to cannibalistic behavior. For a plastic container with an area of 613cm^2, house a maximum of five adult animals (*see* **Note 1**).
3. Feed newts with frozen blood worms (Live aquaria #CD-12957) three times a week (for example Monday/Wednesday/Friday) at night and change them into a clean container with water the following morning.
4. Keep room temperature at 22 °C and 12-hr light/dark cycles.
5. After the operation (lentectomy), keep animals for several days in a moist chamber (using a wet paper towel) and monitor their health before returning them to water.

3.2 Lentectomy

1. Anesthetize the animal by immersing the whole body in 0.1% MS222 (*see* **Note 2**).
2. Put a wet paper towel around the newt body to help stabilize the animal (Fig. 2A).
3. Under a stereomicroscope carefully make an incision to the cornea using a scalpel blade #11 without touching the iris or the lens (Fig. 2B) (*see* **Note 3**).
4. Insert tweezers through the open cornea to grab and remove the entire lens in one piece (Fig. 2C, D). Let the lens regenerate for a period of time (up to 80 days) and either visualize using OCT or collect for further analysis.
5. Return the animal into the housing container and monitor until recovered from anesthesia (~15–20 min).
6. To enucleate the eye, gently tear the thin conjunctiva and Tenon's capsule using tweezers, and using straight edge scissors, cut around the remaining ocular muscles. Using curved edge scissors, cut the optic nerve and then grab the eye from the sclera using tweezers to detach it from the eye socket (Fig. 2E). Incubate eyes with the appropriate fixative and proceed for histology.
7. Optional step: For easy identification of the dorsal/ventral iris, leave the dorsal eyelid intact when enucleating the eye.

3.3 Explant Tissue Cultures

1. Autoclave Eppendorf tubes and surgical tools to sterilize. Place filter paper under UV light for 30 min to sterilize.
2. Anesthetize the animal by immersing the whole body in 0.1% MS222.
3. Enucleate eyes following **step 3.2.6.**
4. Change gloves and transfer the eyes to a tissue culture hood.
5. Sterilize enucleated eyes by washing in Lugol's solution for 5 seconds.
6. Wash eyes in CMF Hank's solution 4 times for 30 seconds.
7. Transfer eyes (one at a time) to a clean petri dish with modified Hank's solution and begin dissection.
8. Under a stereomicroscope, poke the eye through the sclera below the iridocorneal ring with a scalpel blade #11, and then using microsurgical scissors cut around the cornea area to separate the eye into anterior and posterior halves (Fig. 3A, B).
9. Discard the posterior cup.
10. Flip the anterior cavity so that the iPECS are facing up and the cornea, lens epithelium, and stroma face down.
11. Using tweezers remove the lens, ciliary body and any retina left behind (Fig. 3C).
12. Using a scalpel blade #15 separate the dorsal and ventral segments from the remaining iridocorneal complex (Fig. 3D, E) (*see* **Note 4**).
13. *Optional step*: Separate the iPECS from the stroma by incubating the iris segments in dispase (7.5 units/ml) for 2 hrs at 27 °C. After dispase incubation, separate the iPECs from the stroma by pulling the tissues apart with tweezers (*see* **Note 5**).
14. Place a drop of Matrigel on filter paper and using tweezers transfer the iris segment inside the gel (Fig. 3F).
15. Transfer the entire Matrigel/filter paper/iris complex into a 1.5 ml Eppendorf tube cap containing 100 μl of L-15 medium (*see* **Note 6**).
16. Incubate iris explants at 27 °C until collection.
17. Change media every 3 days.

3.4 Optical Coherence Tomography

1. Anesthetize newt by immersing the whole body in 0.1% MS222.
2. Measure the output power of the afferent light spot from the fiber coupler with a power meter above the newt eye and adjust the light source power percentage until the power is around 2.3 mW.
3. Place the newt on a customized imaging stage that has two moving blocks for stabilizing its head during imaging (Fig. 5A).

4. Focus the light spot on the iris in the eye and match the optical path difference between the sample arm (where the newt is) and the reference arm (where the reflecting mirror is) by adjusting the mirror position on the reference arm and the focal plane on the sample arm.
5. Optimize the power of the reference arm to avoid saturation with proper numerical aperture.
6. Adjust the polarization controller to optimize the imaging depth and contrast.
7. Acquire tomographical images according to the required speed of application (27 k A-scan/second in this case).

3.5 Image Data Processing

1. Import coherence raw data into MATLAB (version R2019a).
2. Remove the DC noise by subtracting the average of all spectral fringes of the A-scans in a B-scan (Fig. 5B).
3. Linearize the spectral fringes data from wavelength to wavenumber domain.
4. Apply Fast Fourier Transform (FFT) to each A-scan.
5. Calculate the modulus and then the logarithm of each A-scan.
6. Import all B-scans into Fiji ImageJ (NIH,1.52p).
7. Adjust the B-scan size to 500x500 pixels and stack these B-scans into a C-scan.
8. Segmentate and label the lens, the iris, and the cornea with *Segmentation editor* in Fuji.
9. Reconstruct the image into a four-color multi-channel stack using the three labeled channels from **step 3.5.8** and the grayscale image from **step 3.5.5**.
10. Visualize the multi-channel image in ImarisViewer (Oxford Instruments, Version 9.9.1).

4 Notes

1. If newts are housed in an aquarium with a large volume of water, floating surface objects should be provided as adult newts are air breathers.
2. Time required to complete the anesthesia effect varies based on body weight. Position the animal upside down and observe movement for 2 minutes to ensure that the animal is fully anesthetized. In addition, the toe of the newt can be pinched and observed the animal for a reflex. If a level plane of anesthesia is reached, no reflex will be observed.
3. The incision made with the scalpel should cover ~3/4 of the entire cornea and should be achieved in one move and one

direction. If the incision is not big enough after the first cut, carefully lift the cornea tissue up with tweezers and use scissors to extend the cut area.

4. To separate dorsal and ventral iris segments, position the scalpel blade at a ~ 70-degree angle to avoid contaminating the dorsal segment with ventral iPECs and vice versa.
5. Depending on the experimental question, the stroma can be removed from the iPECs. Newt iris stroma contains a heterogeneous population of cells (fibroblasts, endothelial cells, macrophages, melanocytes, keratinocytes, mast cells, etc.) that could interfere with the lens transdifferentiation process. A non-enzymatic, but more time-consuming alternative way to remove the stroma, is by grabbing the iPECs with tweezers and then very gently rolling a scalpel blade # 15 to remove the stroma matrix.
6. Individually housing the explants in an Eppendorf tube cap helps to avoid spreading contamination and is cost efficient due to the minimum volume of media required. A cell culture dish or tissue plate can be used instead.

Acknowledgments

This work was supported by grants from the National Eye Institute RO1 EY027801 to KDRT; R21 EY031865 to HW and KDRT, and by the John W. Steube Professor endowment to KDRT. The Authors thank the Laboratory Animal Facility at Miami University and Erika Grajales-Esquivel, Tracy Haynes and Jared Tangeman for the critical reading of the protocol.

References

1. Barbosa-Sabanero K, Hoffmann A, Judge C, Lightcap N, Tsonis P, Del Rio-Tsonis K (2012) Lens and retina regeneration: new perspectives from model organisms. Biochem J 447(3):321–334
2. Call MK, Grogg MW, Tsonis PA (2005) Eye on regeneration. Anat Rec B New Anat 287(1): 42–48
3. Vergara MN, Tsissios G, Del Rio-Tsonis K (2018) Lens regeneration: a historical perspective. Int J Dev Biol 62(6-7-8):351–361
4. Yamada T (1977) Control mechanisms in cell-type conversion in newt lens regeneration. Monogr Dev Biol 13:1–126
5. Yamada T, McDevitt DS (1984) Conversion of iris epithelial cells as a model of differentiation control. Differentiation 27(1):1–12
6. Yamada T, Reese DH, McDevitt DS (1973) Transformation of Iris into Lens in vitro and its Dependency on Neural Retina. Differentiation 1(1):65–82
7. Reyer RW, Woolfitt RA, Withersty LT (1973) Stimulation of lens regeneration from the newt dorsal iris when implanted into the blastema of the regenerating limb. Dev Biol 32(2): 258–281
8. Okamoto M, Ito M, Owaribe K (1998) Difference between dorsal and ventral iris in lens producing potency in normal lens regeneration is maintained after dissociation and reaggregation of cells from the adult newt, Cynops pyrrhogaster. Dev Growth Differ 40(1):11–18
9. Reyer RW (1954) Regeneration of the lens in the amphibian eye. Q Rev. Biol 29(1):1–46
10. Hayashi T, Mizuno N, Owaribe K, Kuroiwa A, Okamoto M (2002) Regulated lens

regeneration from isolated pigmented epithelial cells of newt iris in culture in response to FGF2/4. Differentiation 70(2–3):101–108

11. Eguchi G, Abe SI, Watanabe K (1974) Differentiation of lens-like structures from newt iris epithelial cells in vitro. Proc Natl Acad Sci U S A 71(12):5052–5056
12. Bhavsar RB, Nakamura K, Tsonis PA (2011) A system for culturing iris pigment epithelial cells to study lens regeneration in newt. J Vis Exp 52
13. Hoffmann A, Nakamura K, Tsonis PA (2014) Intrinsic lens forming potential of mouse lens epithelial versus newt iris pigment epithelial cells in three-dimensional culture. Tissue Eng Part C Methods 20(2):91–103
14. Yamada T, Beauchamp JJ (1978) The cell cycle of cultured iris epithelial cells: its possible role in cell-type conversion. Dev Biol 66(1): 275–278
15. Grogg M, Call M, Okamoto M, Vergara N, Del Rio-Tsonis K, Tsonis P (2005) BMP inhibition-driven regulation of six-3 underlies induction of newt lens regeneration. Nature 438(7069):858–862
16. Hayashi T, Mizuno N, Takada R, Takada S, Kondoh H (2006) Determinative role of Wnt signals in dorsal iris-derived lens regeneration in newt eye. Mech Dev 123(11):793–800
17. Bhavsar RB, Tsonis PA (2014) Exogenous Oct-4 inhibits lens transdifferentiation in the newt Notophthalmus viridescens. PLoS One 9(7):e102510
18. Chen W, Tsissios G, Sallese A, Smucker B, Nguyen A, Chen J, Wang H, Del Rio-Tsonis K (2021) In vivo imaging of newt lens regeneration: Novel insights into the regeneration process. Trans Vision Sci Technol 10(10):4

Chapter 14

Bead Implantation and Delivery of Exogenous Growth Factors

Rena Kashimoto, Saya Furukawa, Sakiya Yamamoto, and Akira Satoh

Abstract

Genetic methods in axolotls (*Ambystoma mexicanum*) remain in their infancy which has hampered the study of limb regeneration. There is much room for advancement, especially with respect to spatiotemporal regulation of gene expression. Secreted growth factors play a major role in each stage of regeneration. The use of slow-release beads is one of the most effective methods to control the spatiotemporal expression of secretory gene products. The topical administration of secreted factors by slow-release beads may also prove effective for future applications in non-regenerative animals and for medical applications in humans, in which genetic methods are not available. In this chapter, we describe a methodology for using and implanting slow-release beads to deliver exogenous growth factors to salamanders.

Key words Limb regeneration, Sustained release, Gelatin, Regeneration, FGF

1 Introduction

Organ regeneration in urodele amphibians is known to be controlled, at least partially, by secreted molecules. Among them, nerve secreting molecules are considered to be important since nerves govern the early stage of regeneration in some organs, such as the limbs [1, 2]. Nerve secreting factors that can trigger limb regeneration in the absence of nerves have been reported. For instance, we previously identified fibroblast growth factor (FGF) and bone morphogenic protein (BMP) as nerve factors [3, 4]. The identification of these molecules was achieved using a unique experimental model called the accessory limb model (methodology described in this book) [5] and the grafting of sustained, slow-releasing beads. Bead grafting to deliver proteins of interest has been used extensively in developmental biology research. A relatively short application of an

Supplementary Information The online version contains supplementary material available at [https://doi.org/10.1007/978-1-0716-2659-7_14].

Ashley W. Seifert and Joshua D. Currie (eds.), *Salamanders: Methods and Protocols*,
Methods in Molecular Biology, vol. 2562, https://doi.org/10.1007/978-1-0716-2659-7_14,

individual protein is sufficient to test how it affects the embryonic development of a particular tissue. Thus, the protein retention of beads is not important. However, the time frame of regeneration requires a much longer exposure time. Even the initial phase of limb regeneration lasts several days. A sustained, slow-releasing bead has been developed to achieve the long-term effects of proteins of interest [6]. For instance, gelatin is a biodegradable and biocompatible molecule that has been used to develop a sustained scaffold that can hold proteins for release in vivo. The usage of gelatin allows sustained release due to its negative/positive charge [7]. Growth factors usually take on a positive or negative charge. Thus, the selection of an opposing charge results in an electrical coupling. Biodegradability, biocompatibility, and electrical coupling result in sustained release in amphibian tissues. In this chapter, we describe the gelatin bead preparation for axolotl tissues. The method of bead preparation is based on the previous report [6] and is modified to adjust to amphibian applications. We also emphasize that the method of bead preparation described here can be used for other animals, especially aquatic animals. Bead preparation needs modification if the bead is grafted in homeothermic animals, however, because the biodegradation rate is different.

Although the technology indicates feasibility, the technical proficiency of scientists is another factor in a successful investigation. Modern science pursues technical generality for all operators. Amphibian regeneration demands technical proficiency in certain areas, such as surgery, however. For this reason, in addition to bead preparation, we describe some of the technical tricks for bead grafting in this chapter.

2 Materials

2.1 Obtaining Axolotls and Axolotl Husbandry

1. Animals: Axolotls can either be purchased from colonies such as the Ambystoma Genetic Stock Center at the University of Kentucky, Lexington, or spawned in the laboratory. Newts are purchased from the Amphibian Research Center at Hiroshima University.
2. Animals are maintained at 25–27 °C on a 12-h light-dark cycle and fed daily. Excellent animal condition is critical for an investigation. For example, when animals are infected with bacteria, their regeneration potency is reduced.
3. Axolotls of 5–12 cm (snout-to-tail tip) are preferred because they can regenerate a limb relatively quickly and are easy to handle (*see* **Note 1**). Newts of 5–8 cm (snout-to-tail) are preferred for the same reason.

2.2 Limb Amputation

1. Anesthetic: 0.1% tricaine methanesulfonate (MS222) solution.
2. Surgical supplies: Fine forceps (Dumont #5 or similar), angled ophthalmic scissors with 2.5-mm cutting edge, fine needles (*see* **Note 2**). Sterilize the surgical instruments with 70% ethanol or in a glass bead sterilizer.

2.3 Bead Preparation

1. Gelatin beads: Type A or B are stored at room temperature and in a dry air condition (*see* **Note 3**).
2. Olive oil (any grade, but higher purity is recommended).
3. Siliconized Petri dish: Evenly coat the dish with a few drops of silicon solution (e.g., Sigmacote). Air-dry the dish at room temperature. Wash the dish with deionized distilled water (DDW) before use.
4. Glutaraldehyde fixative: 2–10 mM in DDW containing 0.1% Tween-20 (*see* **Note 4**).
5. Glycine: 750 μg/mL in DDW.
6. Proteins (e.g., FGF2) (*see* **Note 5**).

3 Methods

3.1 Preparation of Beads (Movie 1)

1. Warm 50 mL of olive oil to 40–60 °C in a 200-mL glass beaker and stir with a stir bar. Bead condition is fine anywhere in this temperature range.
2. Dissolve 0.5 g of gelatin in 5 mL of DDW using a microwave (*see* **Note 6**).
3. Add the gelatin solution into the warm olive oil to generate an emulsion. Keep stirring the emulsion gently (Note: The stirring speed makes the bead diameter. For axolotl tissues, ~100 rpm forms good size beads Fig. 1).
4. Continue to stir the emulsion for 2 h at room temperature.

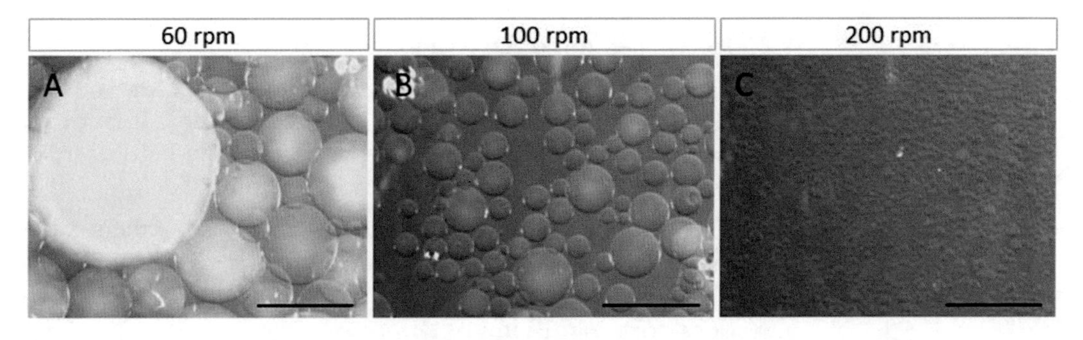

Fig. 1 The size of the bead depends on the stirring speed. Beads were prepared by stirring at 60 rpm (**a**), 100 rpm (**b**), or 200 rpm (**c**). The scale bar is 1 mm

5. Place the emulsion at 4 °C and continue stirring for another 2 h. A longer time (up to overnight) may be employed at this step. This cooling process solidifies the gelatin solution and forms the microspheres (*see* **Note 7**).
6. Add 50 mL of acetone into the beaker and continue stirring the mixture for 1 h at 4 °C. This step is intended to remove the oil from the beads.
7. Transfer the mixture into two 50-mL conical tubes and centrifuge at 250 × g for 5 min. Discard the supernatant.
8. Add 50 mL of acetone and invert the tubes several times to mix the solution and centrifuge at 250 × g for 5 min.
9. Discard the supernatant and add 50 mL of isopropanol to each tube, gently invert a few times, and centrifuge at 250 × g for 5 min.
10. Discard the supernatant and add 50 mL of DDW to each tube, gently invert a few times, and centrifuge at 250 × g for 5 min.
11. Transfer the beads of Step 10 into a glass beaker with 100 mL of 2–10 mM glutaraldehyde-Tween 20 solutions for fixation. Stir the solution overnight at 4 °C (*see* **Note 4**).
12. Centrifuge the solution in two 50 mL tubes at 250 × g for 5 min and collect the beads.
13. Add DDW and wash the beads by inverting them a few times. Centrifuge again at 250 × g to remove the supernatant.
14. Transfer the beads into a glass beaker with 100 mL of 750 μg/mL glycine/DDW solution to neutralize the remaining glutaraldehyde. Stir the solution at room temperature for 1 h.
15. Centrifuge the solution in a 50 ml tube at 250 × g for 5 min and discard the supernatant.
16. Add DDW and shake well to wash the beads. Centrifuge and discard the supernatant.
17. Wash the beads twice more with DDW as described in Step 16.
18. Store the beads in DDW at 4 °C (*see* **Note 8**).

3.2 Protein Absorption into the Beads

1. Pipette 3–5 mL of DDW into a 35-mm Petri dish.
2. Transfer beads into the Petri dish using a plastic transfer pipette. Do not transfer a large number of beads. If an excess number of beads is transferred, the water in the Petri dish will become cloudy, which makes bead selection difficult.
3. Select beads with the appropriate size for the experiment. Stick the bead with a fine needle and transfer it into a new 35-mm siliconized Petri dish. Select beads of 200–400 μm in diameter (ideal for axolotl's implantation).

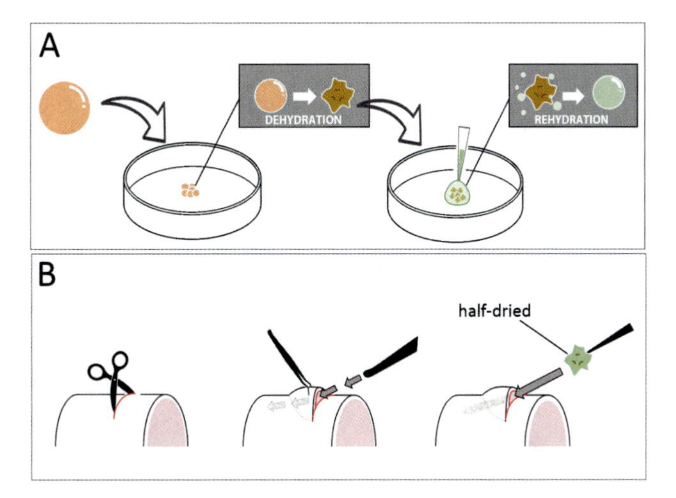

Fig. 2 (**a**) Schematic diagram of bead grafting. Beads are transferred from a stock tube onto a plastic dish (ideally, a silicone-coated dish). The transferred beads are dehydrated in air, and then re-swollen with protein solutions. The swollen beads are stored in a refrigerator until usage. (**b**) For bead grafting, the skin is cut using micro-scissors. Next, a tunnel is created with forceps to make a route for a bead. Then, the protein-soaked bead is picked up by a needle. The picked-up bead is exposed to the air so it can dry out slightly. Ensure that the bead is not completely dried out. The shrunk bead swells in tissues, which increases its stability

4. Allow the beads to dry out completely (Fig. 2a). Dried beads can be stored at room temperature for a couple of days.
5. Drop 1 uL of proteins onto the 5–8 dried beads. The beads will begin to swell as they rehydrate (Fig. 2a). Allow the beads to absorb the proteins for at least 2 h on ice or in a refrigerator. Place a wet tissue wipe along the inside rim of the dish to prevent the dehydration of beads.
6. The beads impregnated with proteins can be stored at 4 °C for 2–10 days, but it is optimal to use beads as close to the preparation time as possible.

3.3 Delivery of a Bead into a Blastema (Movie 2)

1. Anesthetize animals with MS222. It usually takes 5 min. If anesthetic recovery difficulties are encountered, decrease the concentration of MS222 from 0.1% to 0.01%.
2. Limbs were amputated by scissors, and the cutting plane should be flattened by trimming the protruding bone(s) (*see* **Note 9**).
3. To insert a bead, make a small incision proximal to the amputation plane (Fig. 2b).
4. Make a tunnel toward the target site (any place in a blastema) (Fig. 2b). Do not pin through the blastema epidermis. Do not

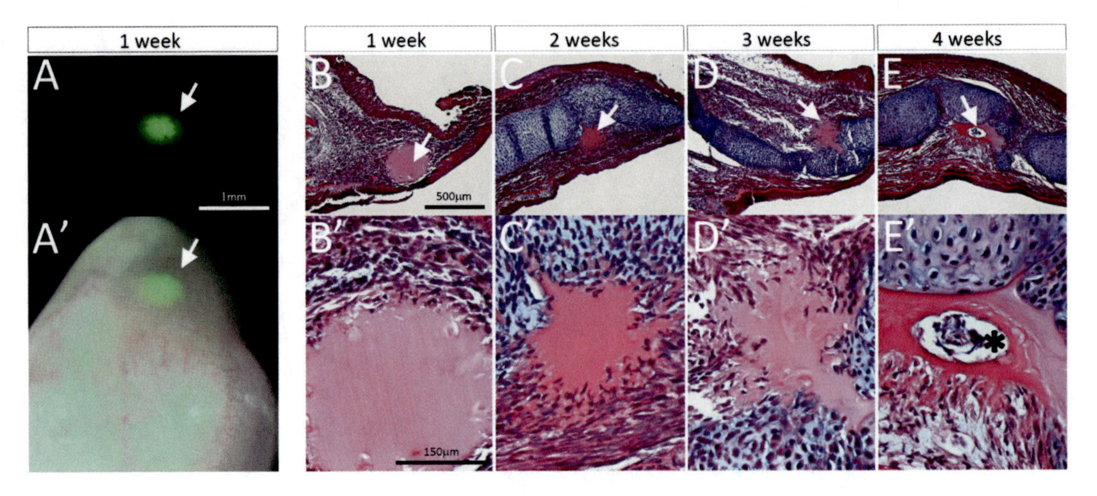

Fig. 3 (**a**) Fluorescence image of a bead-grafted blastema. (**a'**) Merged image of a bright-field image and (**a**). (**b–e**) Histology of blastemas 1–4 weeks after bead grafting stained with Alcian blue stain in combination with hematoxylin and eosin (HE) stain. White arrows indicate the grafted beads. The asterisk indicates migrated lymphocytes

make a big (wide) tunnel. A round bead will slip off from the tunnel.

5. Pick up a bead using a fine needle or fine forceps. Protein-soaked beads are usually slippery. Pin the bead from directly above (*see* **Note 10**).
6. Insert the bead into the incision and deliver it to the destination (Fig. 2b).
7. After pin removal, wipe the water around the incision so that it closes quickly.
8. The animal is then placed on ice for 2 h before being placed back in housing water.
9. The bead placement can be confirmed 24 h after grafting. Grafted gelatin beads can be readily seen under a fluorescent microscope due to their strong autofluorescence (Fig. 3) (*see* **Notes 11** and **12**).

4 Notes

1. We recommend not using larger sizes of animals for experiments. Leucistic and albino mutants are useful since it is easier to visualize tissues through their unpigmented skin.
2. We used fine needles made from tungsten wire [8]. If tungsten needles are not available, an injection needle (gauge #27 ~ 30) can be used instead.

3. Gelatin is a type of protein produced by the partial hydrolysis of native collagen. Depending on the process used, two types of gelatin, namely, type A (acid hydrolysis; isoelectric point 7–9) and type B (alkaline hydrolysis; acid hydrolysis; isoelectric point 4.7–5.4) are generally obtained. The basic and acidic FGF possess isoelectric points of 9.6 and 5 to 6, respectively. Therefore, in general, it seems appropriate to use Type B gelatin to obtain a sustained release of bFGF in an environment around pH 7 [6, 7].
4. The concentration of fixative affects the stiffness of the beads. Generally, 2 mM of glutaraldehyde is suitable for experiments in axolotls.
5. Protein activities vary according to companies. We strongly recommended proteins from R&D systems. Usually, stock solutions are prepared in higher concentrations from 0.5 ~ 1 ug/ul range. To induce a limb, 0.3 ~ 1 ug/ul concentration is sufficient.
6. When heating the gelatin solution in a microwave, the solution may boil over. To avoid this, the condition of the solution should be checked every 3 s. It is important that the gelatin must be completely dissolved.
7. This step is the most important. If insufficient time is taken, the beads will collapse in later steps. The time required for cooling depends on the condition. If you encounter collapsed beads, use a longer time here.
8. Beads are stable for 1 month. If long storage is needed, use 0.05% sodium azide. In this case, the beads should be washed with DDW very well before experiments.
9. Trimming of the tips of bones is important for successful regeneration. The protruding bone ends slows wound closure by surrounding the epidermis, resulting in delayed regeneration.
10. Use a semi-dried bead (Fig. 2b). Make sure the bead is not over-dried under a microscope. If the bead is slightly shrunk, there are no significant effects on the activity of proteins. This procedure gives an advantage in grafting. A small size is better for grafting and the semi-dried bead will swell and recover in size in a blastema. This bead size recovery enables the bead to stabilize inside a blastema.
11. The autofluorescence can indicate its position and successful anchoring in tissues in an experiment. Thus, it is not necessary to label beads by staining, such as a fluorescent dye. Proteins bind to gelatin beads electrostatically. Usage of additional chemicals may cause changes in protein release.

12. A limb blastema is observable in an axolotl (10 cm) 1 week after implantation of beads soaked in FGF2 or FGF8. The formation of a well-patterned limb induced by a skin graft is completed in approximately 2 months.

Acknowledgments

This work was supported by Grant-in-Aid for Scientific Research (B) (KAKENHI #20H03264 to AS).

References

1. Goss R.J (1969) Chapter 9 – The amphibian limb. In: Goss RJ, ed Principles of Regeneration. Academic Press, pp 140–190. https://doi.org/10.1016/B978-1-4832-3250-8.50013-Xpp
2. Todd TJ (1823) On the process of reproduction of the members of the aquatic salamander. Quart J Sci Liter Arts 16:84–96
3. Makanae A, Mitogawa K, Satoh A (2014) Co-operative bmp- and Fgf-signaling inputs convert skin wound healing to limb formation in urodele amphibians. Dev Biol 396:57–66. https://doi.org/10.1016/j.ydbio.2014.09.021
4. Satoh A, Mitogawa K, Makanae A (2015) Regeneration inducers in limb regeneration. Develop Growth Differ 57:421–429. https://doi.org/10.1111/dgd.12230
5. Endo T, Gardiner DM, Makanae A, Satoh A (2015) The accessory limb model: an alternative experimental system of limb regeneration. Meth Mol Biol (Clifton, N.J.) 1290:101–113. https://doi.org/10.1007/978-1-4939-2495-0_8
6. Tabata Y, Hijikata S, Muniruzzaman M, Ikada Y (2012) Neovascularization effect of biodegradable gelatin microspheres incorporating basic fibroblast growth factor. J Biomater Sci Polymer Ed 10:79–94
7. Tabata Y, Ikada Y (1999) Vascularization effect of basic fibroblast growth factor released from gelatin hydrogels with different biodegradabilities. Biomaterials 20:2169–2175. https://doi.org/10.1016/s0142-9612(99)00121-0
8. Brady J (1965) A simple technique for making very fine, durable dissecting needles by sharpening tungsten wire electrolytically. Bull World Health Organ 32:143–144

Chapter 15

The Accessory Limb Model Regenerative Assay and Its Derivatives

Michael Raymond and Catherine D. Mccusker

Abstract

When the Accessory Limb Model (ALM) regenerative assay was first published by Endo, Bryant, and Gardiner in 2004, it provided a robust system for testing the cellular and molecular contributions during each of the basic steps of regeneration: the formation of the wound epithelium, neural induction of the apical epithelial cap, and the formation of a positional disparity between blastema cells. The basic ALM procedure was developed in the axolotl and involves deviating a limb nerve into a lateral wound and grafting skin from the opposing side of the limb axis into the site of injury. In this chapter, we will review the studies that lead to the conception of the ALM, as well as the studies that have followed the development of this assay. We will additionally describe in detail the standard ALM surgery and how to perform this surgery on different limb positions.

Key words Axolotl, Regeneration, Blastema, Accessory Limb Model, Positional Plasticity, Nerve, Wound healing

1 Introduction

1.1 History of the Study of Nerve-Induced Ectopic Limb Structures

The Accessory Limb Model (ALM) was developed from experimental observations over the last century involving the formation of ectopic limb structures from nerve deviations in amphibians. The term "accessory limb" was first defined by Margaret Egar in 1988 as a way to distinguish between supernumerary limbs that formed as a result of amputation, from ectopic limb structures that develop after a nerve is diverted into a wound [1]. The earliest published example of an accessory limb was in 1925 when Locatelli reported that nerve deviation was necessary to induce ectopic limb patterns in newts [2]. Over 30 years later, Charles Bodemer performed a similar manipulation in *Tritus viridescens* (Fig. 1a) and discovered that most of the animals grew accessory limbs when the trauma was induced in the pectoral muscle beneath the wound site (Fig. 1b) [3]. Phenomenological characterization showed that these induced

Ashley W. Seifert and Joshua D. Currie (eds.), *Salamanders: Methods and Protocols*,
Methods in Molecular Biology, vol. 2562, https://doi.org/10.1007/978-1-0716-2659-7_15,

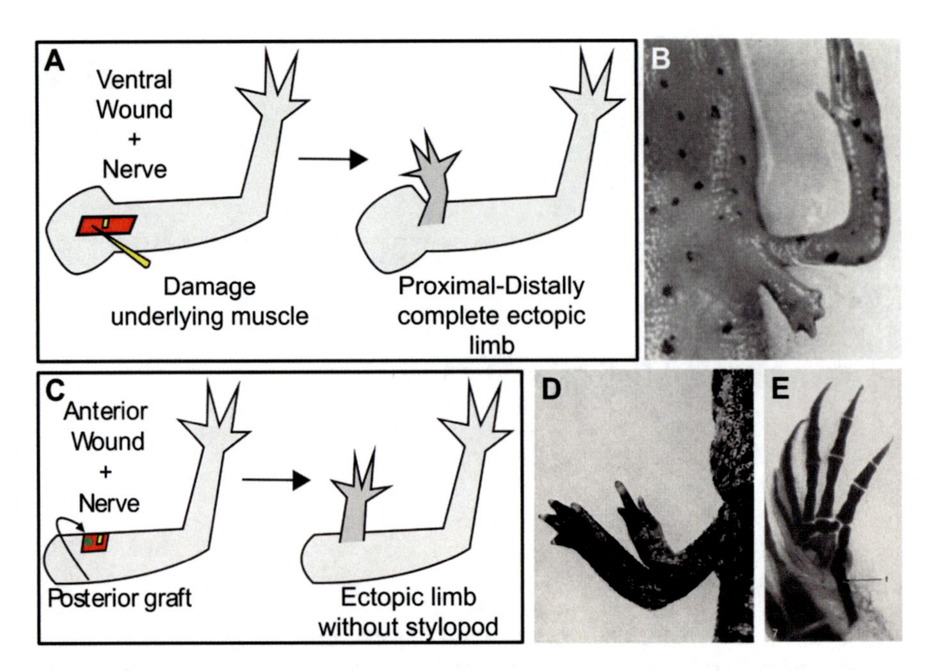

Fig. 1 Historical examples of accessory limbs: (**a**) Accessory limbs can be generated by removing a flap of skin from the ventral side of the arm overlapping the pectoral muscle. The underlying muscle was surgically damaged, and a nerve was deviated to the wound site (Bodemer 1958). (**b**) Accessory limb generated through nerve deviation and muscle damage in *Tritus viridescens* (Bodemer 1958). (**c**) Accessory limbs can also be generated through nerve deviation and grafting of contralaterally positioned skin into the wound (Lheureux 1977). (**d**) Anterior positioned accessory limb generated through grafting of contralateral skin in *Pleurodeles waltlii* (Lheureux 1977). (**e**) Victoria blue staining of accessory limb induced in axolotl using the contralateral graft method (Maden and Holder 1984)

blastemas and structures resembled normal regeneration, including skeletal elements, vasculature, and muscle, though ectopic cartilage was noted in the most proximal regions [3].

Shortly after, Bodemer published a follow-up study in which various types of tissue grafts were utilized as a "replacement" for damaging pectoral muscle to generate ectopic limbs [4]. He demonstrated that while either nerve diversion or tissue grafting alone was insufficient for ectopic pattern development, grafts from limb tissue or internal organs (such as liver and lung) in conjunction with a deviated nerve led to the formation of limb structures [4]. Grafts of epidermal, muscle, or peripheral nervous tissue, in contrast, were insufficient to induce ectopic growth when a deviated nerve is present; thus, some tissues appeared to be able to induce the formation of ectopic limbs, while others do not [4]. Last, he observed that the timing of wound innervation was also an important aspect of ectopic limb induction, since delaying the nerve deviation led to decreased structure formation, and after 2 weeks, completely prevented this outcome [4]. This demonstrated that there was a window of time that the wound epithelium, and potentially other cells in the wounded environment, is competent to receive nerve-based signals to generate a regeneration permissive environment.

In 1960, Bodemer next sought to determine if there was a correlation between the quantity of nerve fibers that deviated into a wound site and the ability to form ectopic structures [5]. This study was motivated by the mutual interdependence of deviated nerves and tissue grafts seen previously [4] and the works of Marcus Singer, which demonstrated there is a quantitative relationship between the number of nerve fibers present in the amputation stump and the ability to regenerate a limb [6, 7]. By deviating nerve bundles with differing nerve abundance, he was able to show that the nerve bundles that had the highest nerve content (the brachial nerves) resulted in a larger number of ectopic structures. In contrast, deviations of smaller nerve bundles (the caudal extensor nerve) resulted in far fewer ectopic limbs, and of these, many formed simple structures such as cartilage mounds or rods. Thus, the ability to induce ectopic structures with complex pattern was dependent on a threshold of nerve-dependent signals [5].

With the advent of the Polar Coordinate Model in 1976, which posited that interactions between cells with discrete positional identities around the limb circumference lead to the generation of new cells with the missing identities during limb regeneration [8], researchers looked beyond tissue identity and began to focus on the positional identity of the grafted tissues. In 1977, Emile Lheureux demonstrated skin grafts with oppositional positional information from the innervated wound site lead to ectopic structure formation in the newt *Pleurodeles waltlii* by generating either anterior-posterior or dorsal-ventral positional discontinuities (Fig. 1d, e) [9]. Lheureux further demonstrated that the orientation of the ectopic structures was controlled by the orientation of the grafted tissues within the wound site [9]. This important observation was also demonstrated in the hindlimb of the Italian Crested Newt *Triturus cristantus* [10] and in the hindlimbs of the Mexican axolotl [11]. Maden and Holder also noted that although the wound sites were located in the stylopod, the stylopod elements failed to form, and the ectopic skeletal structures did not integrate into the host limb (Fig. 1f) [11].

In 1988, Margaret Egar investigated ipsilateral grafting in the forelimb, hindlimb, and trunk wound sites in the axolotl [1]. The forelimb was found to be more conducive to accessory limb development as diverting the sciatic nerve in the hindlimb lead to host limb regression and the formation of simple structures, while deviations to the trunk tissue did not lead to accessory limbs [1]. Egar determined that for accessory limbs to form, both a large limb wound site and the presence of the nerve directly under the wound epithelium were required [1]. Egar determined that grafts were unnecessary for ectopic limb formation [1], which stood in contrast to multiple other contemporary studies, though it is possible that the surgical techniques used, such as widening the wound site subsequently for nerve deviation [9–11], may have led to damage to the underlying muscle tissue or introduced some positional disparity to the wound site.

1.2 The Establishment of the ALM and Its Use in Research

Endo, Bryant, and Gardiner recognized that the generation of accessory limbs was advantageous because they provided an avenue to study the mechanisms behind each step of the regenerative process [1, 3–5, 9–12]. Traditional studies on amputated limbs were limited mostly to loss-of-function approaches, which made it slow and difficult to distill the molecular and cellular contributions that play a direct role in regeneration. Moreover, limb amputation is a traumatic injury, which results in tissue death and sustained inflammatory response. In contrast, nerve-induced accessory limbs did not require a large traumatic injury, allowing researchers to more directly study the regenerative response [1, 3–5, 9–11]. However, the previous methods of ectopic limb formation varied among researchers, so a straightforward surgical method with a robust and reproducible regeneration phenotype was needed.

In standardizing the ALM, Endo, Bryant, and Gardiner showed that limb regeneration occurred in three basic steps, wound healing, dedifferentiation, and growth and pattern formation [12]. Focusing on the anterior and posterior positions of the mid-stylopod regions of axolotl forearms they observed that lateral limb wounds were unable to make ectopic blastemas unless the branchial nerve bundle has deviated into the wound. These blastemas would eventually regress without the formation of ectopic limbs unless skin from the contralateral side of the limb was grafted into the wound. After further showing that the ALM blastemas were indistinguishable from amputation blastemas, Endo, Bryant, and Gardiner proposed this system as a tool to study the three steps of regeneration [12] (Fig. 2).

In the Stepwise Model, regeneration begins with a wound healing response that leads to the formation of the wound epithelium [12]. Interactions between the wound epithelium and the deviated nerve lead to the formation of the apical-epithelial cap, which acts as a signaling hub, attracting connective tissue cells and influencing them to proliferate and de-differentiate [12, 13]. Fibroblasts from the underlying stump tissue and the tissue graft then provide positional cues that drive pattern formation in the blastema, leading to an accessory limb [12]. Thus, the ALM can be used to study the latter two steps of regeneration by either supplementing a lateral wound site to test for factors that can replace the presence of the deviated nerve or by supplementing an innervated wound site to test for the factors that are used to communicate positional information.

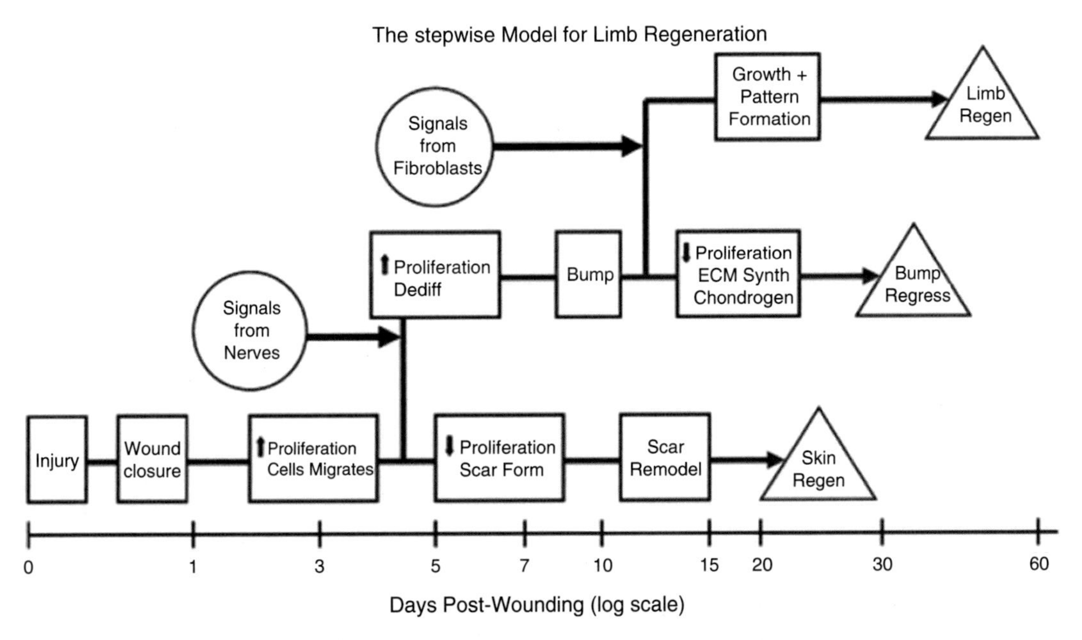

Fig. 2 The three basic steps of regeneration as illustrated by the ALM surgery. The formation of a limb requires (1) an injury, (2) signaling from the nerves, and (3) positional signals from fibroblasts. The ALM surgery allows for researchers to study the molecular and cellular contributions that are sufficient to push the injured tissue to the next step of regeneration. Figure obtained from (Endo et al. 2004)

Since the advent of the ALM, multiple variant procedures have also been developed. The most straightforward of which is the induction of ectopic structures elsewhere from the limb itself. For example, the Satoh research team has induced accessory limbs on the flank of axolotl by performing an ALM-like surgery where both anterior and posterior limb tissues were grafted into an innervated flank wound [14]. We have replaced tissue grafts in anterior and dorsal positioned ALMs by treating the animal with RA, which generates a positional disparity by reprogramming cells in those locations to posterior-ventral identities (Fig. 3c) [15]. The Tanaka research team has utilized the ALM to generate accessory limbs by substituting contralateral grafts by inducing ectopic *Shh* and *Fgf8* expression in anterior and posterior positioned blastemas, respectively, through drug treatment and baculovirus delivery (Fig. 3e) [16]. The Satoh research team has created aneurogenic ALMs by substituting brachial nerve deviations with beads soaked in combinations of BMPs and FGFs [17–19]. This has been shown to lead to well-patterned ectopic limbs and can be utilized with either a contralateral skin graft [17–19] or RA treatment (Fig. 3d) [20]. In addition to inducing accessory limbs, Satoh and his team have demonstrated the Accessory Tail Model [19] and the Accessory Gill Model [21] in the axolotl system. Additionally, the ALM and its variants have been established in other model organisms, such as *Xenopus laevis* [18, 22], chicken embryos [23], and *Pleurodeles*

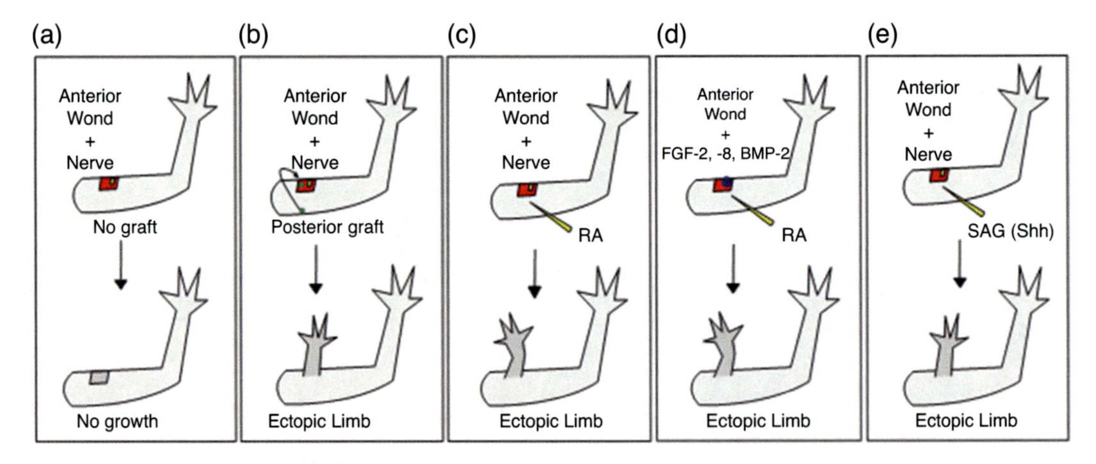

Fig. 3 Ectopic limb phenotypes reported using the ALM. (**a**) Anterior limb wounds with a deviated nerve bundle will generate an ectopic blastema, which will regress unless (**b**) skin or periskeletal tissue from the posterior side of the limb is grafted into the wound site (Endo 2004; McCusker 2016). (**c**) Ectopic limb structures can also be generated in innervated anterior wounds injected with RA (rather than a skin graft) (McCusker 2014; Vieira 2019). (**d**) Gelatin beads soaked in FGF2, FGF8, and BMP2 can substitute for the nerve deviation (Makanae 2014), and wounds treated with these growth factor-soaked beads, and later treated with retinoic acid (RA) will grow complete limb structures (Vieira 2019). (**e**) Anterior located innervated wounds also grow ectopic limb structures when *Shh* signaling is activated (Nacu 2016)

waltl [19, 24] demonstrating the conservation of the basic underlying biology of ectopic limb formation in tetrapods.

Using the ALM and its variants, multiple research groups have made headway in answering questions that would be more difficult in a traditional amputation assay. The Satoh research team demonstrated that regenerating nerves act to induce de-differentiation and reprogramming of invading dermal fibroblasts from the underlying stump tissue [25–27] and the molecular mechanisms of limb nerves in regeneration [25, 28–30]. Additionally, the ALM was critical to better characterization of the wound epithelium and the apical epithelial cap in terms of keratinocyte populations [25, 26, 29, 31, 32]. Both the Gardiner and the Satoh groups leveraged this technique to understand which cell populations contribute to the regenerate [26, 27, 33]. Multiple groups have used these methods to characterize gene expression [16, 20, 22, 34–38] and epigenetic regulation [39] during regeneration. The role of collagen deposition [18, 22, 28], how proximal-distal intercalation is regulated [40], healing of bone defects [41] the determination of at which stage of regeneration blastema cells lose positional plasticity [42], and the role of ECM in providing positional cues [43] have all also been investigated using the ALM. These studies highlight the versatility of the ALM in regeneration research.

2 Materials

2.1 Solution Preparation

Concentrated Holfreter's Stock Solution

1. Add 56 grams of CaCl2, 4 tablespoons of MgSO4-7H2O, 2 tablespoons of KCl, and 640 grams of NaCL to 3.5 L of MilliQ water (alternatively diH2O) ensuring each salt is dissolved before adding the next.
2. Backfill to 4 L with MilliQ water (alternatively diH2O).

40% Holfreter's solution (alternately animal housing water)

1. Dilute 250 Ml of Concentrated Holfreter's Stock Solution in 19.75 L of diH20 to pH 6.5–8.0

0.1% MS222 in Holtfreter's Solution

1. Add 2 gram of MS222 Ethyl 3-aminobenzoate methanesulfonate salt) powder to 1900 mL of 40% Holfreter's solution.
2. Add 2 mL of 1 M Tris–HCl and 250 μL of phenol red to the solution on the stir bar.
3. Adjust pH to 7.0–7.4 using NaOH and fill to 2 L with 40% Holtfreter's solution.
4. Store in darkness at 4 °C.

2.2 Surgical Supplies

Surgical Tools

Watchmaker forceps (number 5).

Small spring scissors.

Spray bottle.

Surgical Tray

Use a rigid plastic tray (roughly 9.5 cm wide by 20.5 cm long) that can be sterilized and fit the entire length of the animal. Depending upon animal size, larger or smaller trays may be used. Alternatively, plastic lids from animal housing can be used as well.

Stereoscope and Surgical Lighting

Surgeries are performed using a ZEISS Stemi 305 stereoscope with a 5:1 adjustable optical zoom, or an adequate substitute. Lighting is provided by 24VDC SCHOTT MLS (model 1,308,378) goose headlamps, or equivalent.

Surgical Consumables

KIMTEECH Science Brand Kimwipes (11x21cm code 34155).

Absorbent bench pads (VWR Absorbent Bench Underpads 56,617–018).

3 Methods

3.1 Standard ALM

Preparing for Surgery

1. Prepare the surgical area by spraying it down with 70% ethanol. Surgical tools should also be sharp and sterilized, either by autoclaving in advance, using a bead sterilizer, or spraying the tools with 70% ethanol.
2. Submerge the animal(s) in diluted MS222 until they do not respond to any stimuli; about 15–20 minutes for 7 cm animals, more time for larger animals. Once anesthetized, the animals will be placed on its side on a surgical tray covered in Holtfreter's-moistened Kim wipes, such that the limb to be manipulated is facing up. The animal is then covered with a moistened Kim wipe and is placed on a bed of ice. The animals are sprayed gently with Holtfreter's as needed to prevent drying out (approximately every 10 minutes). Animals are typically unconscious for 2 hours on ice.

   The ALM Surgery (Fig. 4).

3. Place the animal under the stereomicroscope and angle the goose head surgical lamps such that the animal forelimb is in focus and illuminated. Remove a square of skin on the anterior side of the upper arm, making sure that the underlying muscle is not damaged (Fig. 4, **step 1**). To do this, pinch the skin with forceps and pull it gently outwards from the underlying tissue, before cutting through the pinched region of the skin. (*see* **Note 1**).
4. Using spring-loaded scissors, make a ventral incision from the most proximal end of the limb to the elbow, dissect the brachial nerve bundle from the surrounding tissue, and sever the nerve at the elbow (Fig. 4, **step 2**). Avoid pinching the nerve proximal to where you will sever it as this causes damage, which will prevent the nerve from inducing a regenerative response. Additional comments on identifying and dissecting the nerve bundle are located in the notes section (*see* **Note 2**).
5. Create a tunnel for the nerve between the epidermis and the underlying mesenchymal tissue by pushing forceps in the closed position from the nerve dissection wound to the anterior-located wound site. The location of this channel should be proximal to the location of the anterior wound so that the length of the nerve bundle is long enough to reach across the wound site after it is channeled. Channel the nerve through the tunnel that you have made with your forceps to the center of the wound site. The nerve is pulled gently to extend across the wound site. Care must be taken to not pull the nerve too tight, however, because this tension may pull the

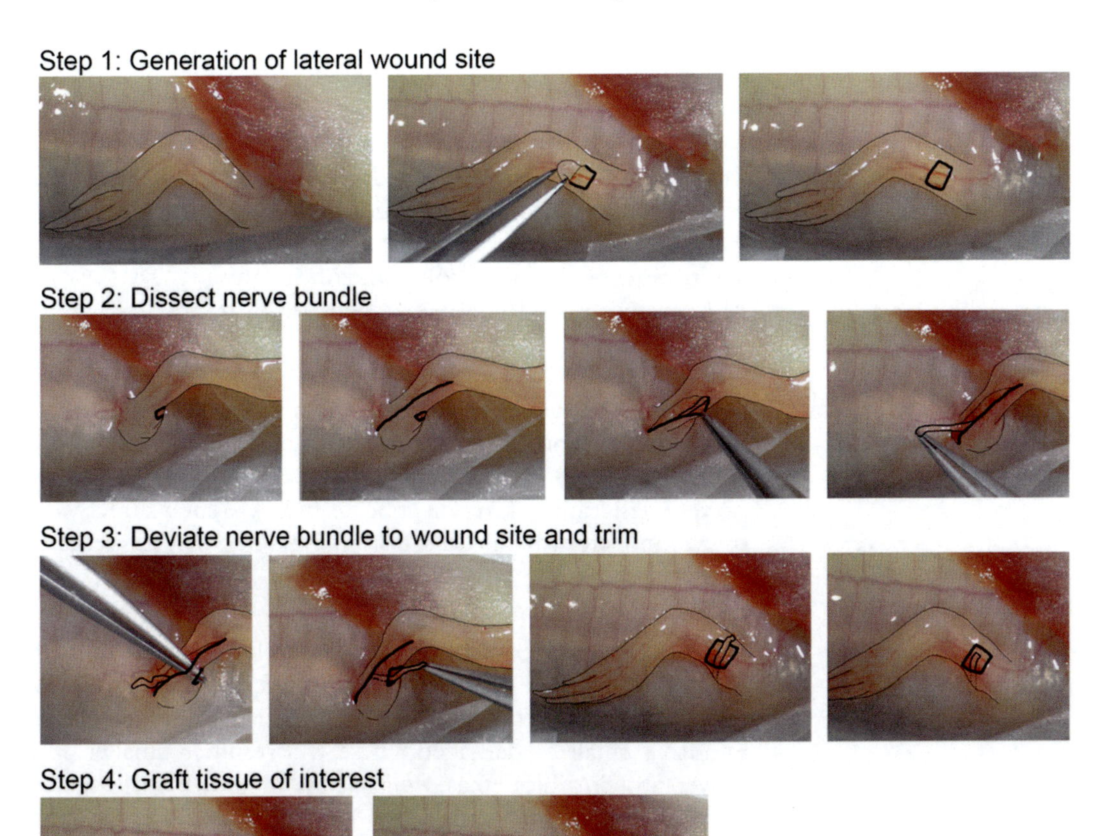

Fig. 4 Step by step standard ALM procedure: Live images of the four basic steps of the ALM surgery have been marked such that the important tissues are highlighted. *Step 1* consists of preparing the animal for surgery and generating a full-thickness skin wound on the anterior side of the limb. In *Step 2*, the limb is positioned such that the ventral limb surface is present, and a vertical incision is made from the proximal ventral side of the limb to the "elbow." Once the skin flabs are separated, the nerve bundle is located, severed distally, and dissected away from the surrounding tissues. In *Step 3*, a channel is generated through the dermal layer from the ventral incision to the anterior wound site, the nerve bundle is pulled through the channel, placed in the center of the wound site, and the damaged end of the nerve is trimmed. During *Step 4*, posterior skin tissue from a transgenic animal expressing GFP is grafted into the wound site, partially tucking graft under a corner of the mature tissue in the wound site

nerve out of the wound site when the animal wakes and begins to move. Once the nerve is placed, trim the damaged tip (Fig. 4, **step 3**). If the nerve needs to be repositioned after trimming, gently push to the new location using the side of closed forceps.

6. Graft in a rectangular piece of posterior skin from a GFP+ transgenic animal into the wound site, tucking a corner of the

graft under the mature skin surrounding the wound (Fig. 4, **step 4**). We recommend using fluorescently labeled (GFP+ or otherwise) tissue as a graft source to keep track of graft survival over time. Ensure that the grafted tissue is placed into the wound with the dermal side down and to the side of the nerve (not on top of it). Additional comments on the placement of graft are located in the notes section (*see* **Note 3**). A description of the protocol deviation in retinoic acid-treated ALMs is located in the notes section (*see* **Note 4**).

After Surgery

7. Once the surgery is completed, the animal is kept on ice for 1.5–2 hours to promote healing. The animal is sprayed gently with Holtfreter's frequently (approximately every 10 mins), to prevent desiccation. Care must be taken not to spray directly on the wound site, which could dislodge the nerve or tissue graft (*see* **Note 5**).
8. One week following the ALM surgery, the wound site is monitored for (1) the formation of an early blastema and (2) the presence of the graft. We do not continue monitoring ALMs that are missing either of these aspects. If the deviated nerve bundle is small or damaged, these will result in smaller and slower-growing blastemas, which adds unknown variables to the experiment. We typically see EB-MB staged blastemas by 1 week following surgery. Additional comments regarding surgery success rates are located in the notes section (*see* **Note 6**). ALM sites are imaged weekly until the generation of an ectopic limb has completed, or the wound site has healed over and mature skin has reformed. We often then harvest the entire limb (host limb with ectopic limb) and perform whole-mount skeletal preparations to better view the skeletal pattern.

3.2 ALMs in Different Limb Positions

3.2.1 Identification of Limb Axes

1. Holding the limb such that its dorsal/ventral axis aligns with that of the body axis will help with the identification of the limb axes (dorsal/ventral and anterior/posterior) (Fig. 5a–c). When the limbs are held this way, the surgeon can trace from the body axis to the limb and identify the apex of each side of the axes. Even without the visual landmarks describe below, this method will allow for the appropriate identification of the axes.
2. Identification of the anterior axis: The major landmark that can be used to identify the anterior axis is the branchial artery that runs anteriorly and is usually visible just under the skin in the stylopod. When the limb is held as described above, it is evident that this artery, although on the anterior side of the axis, is shifted slightly dorsal. Thus, this landmark indicates the dorsal edge of the anterior side of the limb (Fig. 5d, e).

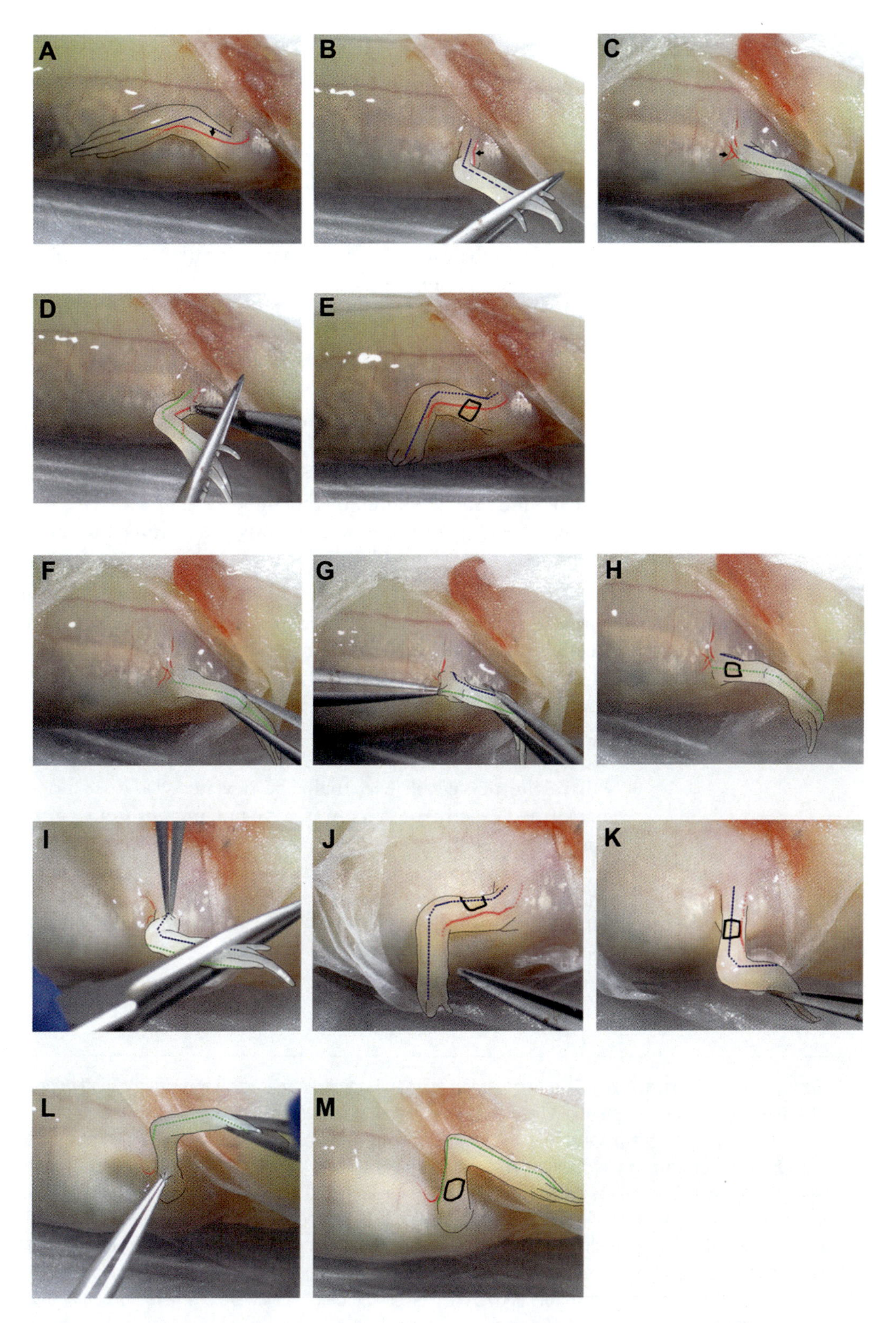

Fig. 5 Visual landmarks for making limb wounds on the circumferential axes: (**a**) When the limb is placed to the side of the animal, both the anterior and dorsal axes are prominent. The brachial artery (highlighted in red with

3. Identification of the posterior axis: The landmark of the posterior side of the limb is visualized by a cluster of blood vessels, along with white-toned nerve bundles, which are located just proximal to the limb, and enters the limb through the posterior side (Fig. 5f–h).
4. Identification of the dorsal and ventral locations: The identification of the dorsal and ventral locations is performed by aligning the limb axis with the body axis, and tracing from the body to the most dorsal or ventral location of the limb, as described above (Fig. 5i–m).

3.2.2 Location-Specific Nerve Deviations.

1. Wound site generation must factor in two major components: location and size. The appropriate location of the axes is described above. However, if the wounds are very large, then cells from the neighboring axes will also contribute to the blastema. Our general rule is to aim for a wound that is approximately one-fourth the circumference of the limb or smaller. The wounds are generated by pinching the skin with forceps in the most A/P/D or V limb locations and cutting the skin as described above (Fig. 5d, g, i, l).
2. Deviation of the nerves is most difficult in the limb positions furthest from where the nerve is dissected because the nerve needs to extend a further distance to reach the wound. Thus, it is common for dorsally located ALM surgeries to have a lower success rate [15]. In our experience, channeling the nerve through the posterior side of the limb to the dorsally located wound results in a slightly higher success rate. Regardless of where the nerve will eventually be deviated, its dissection is performed exactly the same way to maintain consistency in the surgical procedure. We recommend freeing as much of the nerve as possible to ensure that the nerve will reach each of the wound positions. After the nerve bundle is dissected, it is channeled just under the skin on the shortest route toward the wound.

Fig. 5 (continued) black arrow) is visible through the skin and demarks the anterior (shifted slightly dorsal) side of the limb. (**b**) The dorsal side of the limb (blue line) is best visualized by holding out the limb so that the dorsal/ventral (D/V) limb axis is parallel to the D/V body axis. (**c**) The posterior axis (green line) is also better visualized by holding the arm in this position. A cluster of blood vessels (highlighted in red), along with white-toned nerve bundle (dotted lines) is located just proximal to the limb and enters the limb through the posterior side (black arrow). (**d, e**) The generation of an anterior located limb wound. (**f–h**) The generation of a posterior-located limb wound. (**i–k**) The generation of a dorsally located limb wound. (**l, m**) The generation of the ventral located limb wound is achieved by holding the limb out (as described above) and making a wound in the limb region that aligns with the ventral side of the body axis. It is also helpful to be aware of where the center of the posterior and anterior sides is located

3. Tissue grafts are performed as described in the standard ALM procedure above. For consistency, we always place the tissue grafts on the most proximal edge of each wound site.

4 Notes

1. The size of the wound depends on the size of the animal that the ALM is being performed on. In general, wound sites are approximately 1/4 of the circumference of the limb or smaller. Thus, the size of the wound on large animals is larger in size than those on small animals. Under some circumstances, experimental conditions require the same size wound site for all animals (large and small). In these conditions, we recommend using biopsy punchers to generate the wound sites. We have used 2.5 mm biopsies with good outcomes. However, if a biopsy punch is used, care must be taken to avoid damage to the internal structures. When using a biopsy punch, we gently push the sharp end of the puncher onto the preferred wound location and twist it back and forth (as if twisting a dial). This will generally not cut completely through the skin but will make an impression in the epidermis, which can then be used as a template to cut with scissors. This method maintains that the wound is superficial (only skin deep), and a reproducible size.
2. In our experience, the most difficult part of the ALM for trainees is finding and dissecting the limb nerve bundle with minimal damage to the internal limb tissues. We recommend those that are new to the ALM to practice nerve dissections prior to performing the complete surgery so that they can get a feel for where the nerve bundle is located (which is very consistent), and how to dissect it without breaking the nerve bundle or extensively damaging the surrounding tissues. In our experience, small animals (5–7 cm long) that have never had their limbs amputated are the easiest to perform the surgery on because the limb skin is almost transparent, and the white-toned nerve bundle can usually be seen through the skin. The nerve bundle accompanies a large blood vessel that traverses from the proximal, posterior, and ventral side of the limb to the anterior ventral side in the more distal regions close to the "elbow" joint, just under the skin. An incision is made with scissors from the most proximal limb position and extends upward to the elbow joint, being careful not to cut the tissue underlying the skin by pulling the scissors gently upward as the skin is cut. Once the skin flaps are pulled open (separating the dermis from the underlying tissue), the nerve bundle is often visible immediately. If not, identify the largest blood vessel, and

gently push around the tissues with closed forceps until you find the nerve. Grasp the nerve bundle in the most distal location possible, sever the bundle on the distal side, and gently pull on the nerve bundle such that it detaches from the surrounding connective tissue. On occasion, the nerve bundle will be bifurcated, in which case, both nerve bundles are dissected simultaneously. If increased tension is observed while pulling on the nerve to separate it from the surrounding connective tissues, release the nerve from the forceps, and separate the connective tissues surrounding the bundle before continuing. While dissecting the nerve, make sure that the only region of the nerve bundle that is grasped is the most distal end to minimize damage. The damaged end will later be removed once the bundle has deviated into the middle of the wound site.

3. Regarding graft placement relative to the nerve bundle: When grafting mature tissue (for example, mature skin) care must be made to not cover the end of the nerve bundle with the grafted tissue, as this will inhibit the regenerative response. However, if grafting blastemas, it is preferable to graft over the nerve to promote reinnervation of the regenerating tissue as soon as possible.

4. RA-induced ectopic limbs do not require grafting tissue into the ALM [15, 20]. We prefer the injection method as opposed to the soaking method of RA treatment because we can more closely control the dosage of RA that the experimental animals receive. In RA-treated ALMs, nerve deviations into wound sites are performed as described in the protocol above. No tissue is grafted into the wound sites. 7–10 days following the nerve deviation, when a MB blastema is evident, the animal is treated with RA by injecting the animal in the flank with a solution of RA dissolved in DMSO at a dosage of 150 ug/g of body weight as described in *Niazi* et al and *Crawford* et al [44, 45]. Take care to keep the injection volume small to minimize the stress on the animals. For example, we will inject 50ul of solution per animal in animals that are 7 cm long. Upon injection, we often see the RA recrystallize below the skin. Because RA is photo-inactivated, we keep the animals in the dark for 2 days following injection. Animals should be monitored closely, and the water is changed more frequently in the tanks of recovering animals. Animals will typically exhibit redness and eventually some tissue death at the site of injection, but are generally in good health, and will eat and swim as usual within days following the injection.

5. Axolotls generally recover very well from surgery and do not require post-surgical analgesic treatment. Once returned to their 40% Holfreter's solution, animals should be fasted for 24 hours and monitored for signs of stress. Stressed animals

should be moved into fresh housing solution and closely monitored. Signs of stress include lethargy, lack of appetite, skin shedding, and shrunken gills.

6. Success rates of the ALM surgery depend greatly on experience, wound location, the nature of the grafted tissues, and whether the animal has regenerated previously. The success rate in anterior wounds reported by *Endo* et al in the original paper that established the ALM was reported to be 73% [12]. With a lot of practice, we now commonly have success rates between 90% and 100% for the standard ALM surgery. When performing ALMs on the different limb axes, our success rate is lower, particularly in dorsal and posterior located wound sites. Dorsal-located wounds require that the nerve transverses a longer distance than the other wound locations, and the most common source of failure in these wounds is due to the nerve slipping outside of the wound before the blastema forms. For example, in *McCusker* et al, the success rate (formed a blastema) of nerve deviations into dorsal wounds was only 53% [15]. The most common cause of failure in posterior wounds is loss of graft, which is likely a result of the rubbing of the wound site on the flank of the animal (which does not occur with the other wound locations). Thus, increasing the number of biological replicates to account for these lower success rates should be considered. Last, we have found that one of the best ways to ensure ALM success is to perform the surgery on virgin limbs that have never been amputated. We have found that the nerves in regenerated limbs are often highly branched and break more easily. Even if the nerve can be deviated, the abundance of innervation is highly variable among animals that have previously regenerated. Thus, performing the ALM surgery on virgin limbs will not only increase the success rate of the surgery but will also decrease variability from animal to animal.

References

1. Egar MW (1988) Accessory limb production by nerve-induced cell proliferation. Anat Rec 221:550–564. https://doi.org/10.1002/ar.1092210111
2. Locatelli, P. Formation de Membres Surnumeraires. C. R (1925) Assoc des Anatomistes. 20e Reun Turin 1:279–282
3. Bodemer CW (1958) The development of nerve-induced supernumerary limbs in the adult newt, Triturus viridescens. J Morphol 102:555–581. https://doi.org/10.1002/jmor.1051020304
4. Bodemer CW (1959) Observations on the mechanism of induction of supernumerary limbs in adult Triturus viridescens. J Exp Zool 140:79–99. https://doi.org/10.1002/jez.1401400105
5. Bodemer CW (1960) The importance of quantity of nerve fibers in development of nerve-induced supernumerary limbs in Triturus and enhancement of the nervous influence by tissue implants. J Morphol 107:47–59. https://doi.org/10.1002/jmor.1051070104
6. Singer M (1946) The nervous system and regeneration of the forelimb of adult Triturus. V. the influence of number of nerve fibers, including a quantitative study of limb innervation. J Exp Zool 101:299–337. https://doi.org/10.1002/jez.1401010303

7. Singer M (1947) The nervous system and regeneration of the forelimb of adult triturus. VI. A further study of the importance of nerve number, including quantitative measurements of limb innervation. J Exp Zool 104:223–249. https://doi.org/10.1002/jez.1401040204
8. French V, Bryant P, Bryant S (1976) Pattern regulation in epimorphic fields. Science (80) 193:969–981. https://doi.org/10.1126/science.948762
9. Lheureux E (1977) Importance des associations de tissus du membre dans le developpement des membres surnumeraires induits par deviation de nerf chez le Triton Pleurodeles waltlii Michah. J Embryol Exp Morphol 38: 151–173
10. Reynolds BS, Holder N (1983) The form and structure of supernumerary hindlimbs formed following skin grafting and nerve deviation in the newt Triturus cristatus. J Embroyl Exp Morph 77:221–241
11. Maden M, Holder N (1984) Axial characteristics of nerve induced supernumerary limbs in the axolotl. Wilhelm Roux's Arch. Dev. Biol. 193:394–401. https://doi.org/10.1007/BF00848230
12. Endo T, Bryant SV, Gardiner DM (2004) A stepwise model system for limb regeneration. Dev Biol 270:135–145. https://doi.org/10.1016/j.ydbio.2004.02.016
13. McCusker C, Bryant SV, Gardiner DM (2015) The axolotl limb blastema: cellular and molecular mechanisms driving blastema formation and limb regeneration in tetrapods. Regeneration 2:54–71. https://doi.org/10.1002/reg2.32
14. Hirata A, Makanae A, Satoh A (2013) Accessory limb induction on flank region and its muscle regulation in axolotl. Dev Dyn 242: 932–940. https://doi.org/10.1002/dvdy.23984
15. McCusker C, Lehrberg J, Gardiner D (2014) Position-specific induction of ectopic limbs in non-regenerating blastemas on axolotl forelimbs. Regeneration 1:27–34. https://doi.org/10.1002/reg2.10
16. Nacu E, Gromberg E, Oliveira CR, Drechsel D, Tanaka EM (2016) FGF8 and SHH substitute for anterior–posterior tissue interactions to induce limb regeneration. Nature 533:407–410. https://doi.org/10.1038/nature17972
17. Makanae A, Hirata A, Honjo Y, Mitogawa K, Satoh A (2013) Nerve independent limb induction in axolotls. Dev Biol 381:213–226. https://doi.org/10.1016/j.ydbio.2013.05.010
18. Mitogawa K, Hirata A, Moriyasu M, Makanae A, Miura S, Endo T, Satoh A (2014) Ectopic blastema induction by nerve deviation and skin wounding: a new regeneration model in Xenopus laevis. Regeneration 1:26–36. https://doi.org/10.1002/reg2.11
19. Makanae A, Mitogawa K, Satoh A (2016) Cooperative inputs of bmp and Fgf signaling induce tail regeneration in urodele amphibians. Dev Biol 410:45–55. https://doi.org/10.1016/j.ydbio.2015.12.012
20. Vieira WA, Wells KM, Raymond MJ, De Souza L, Garcia E, McCusker CD (2019) FGF, BMP, and RA signaling are sufficient for the induction of complete limb regeneration from non-regenerating wounds on Ambystoma mexicanum limbs. Dev Biol 451:146–157. https://doi.org/10.1016/j.ydbio.2019.04.008
21. Saito N, Nishimura K, Makanae A, Satoh A (2019) Fgf- and Bmp-signaling regulate gill regeneration in Ambystoma mexicanum. Dev Biol 452:104–113. https://doi.org/10.1016/j.ydbio.2019.04.011
22. Mitogawa K, Makanae A, Satoh A, Satoh A (2015) Comparative analysis of cartilage marker gene expression patterns during axolotl and Xenopus limb regeneration. PLoS One 10: e0133375. https://doi.org/10.1371/journal.pone.0133375
23. Satoh A, Makanae A, Wada N (2010) The apical ectodermal ridge (AER) can be re-induced by wounding, wnt-2b, and fgf-10 in the chicken limb bud. Dev Biol 342:157–168. https://doi.org/10.1016/j.ydbio.2010.03.018
24. Makanae A, Mitogawa K, Satoh A (2014) Co-operative bmp- and Fgf-signaling inputs convert skin wound healing to limb formation in urodele amphibians. Dev Biol 396:57–66. https://doi.org/10.1016/j.ydbio.2014.09.021
25. Satoh A, Graham GMC, Bryant SV, Gardiner DM (2008) Neurotrophic regulation of epidermal dedifferentiation during wound healing and limb regeneration in the axolotl (Ambystoma mexicanum). Dev Biol 319:321–335. https://doi.org/10.1016/j.ydbio.2008.04.030
26. Satoh A, Bryant SV, Gardiner DM (2008) Regulation of dermal fibroblast dedifferentiation and redifferentiation during wound healing and limb regeneration in the axolotl. Develop Growth Differ 50:743–754. https://doi.org/10.1111/j.1440-169X.2008.01072.x
27. Hirata A, Gardiner DM, Satoh A (2010) Dermal fibroblasts contribute to multiple tissues in the accessory limb model. Develop Growth

Differ 52:343–350. https://doi.org/10.1111/j.1440-169X.2009.01165.x

28. Satoh A, Hirata A, Makanae A (2012) Collagen reconstitution is inversely correlated with induction of limb regeneration in Ambystoma mexicanum. Zool Sci 29:191–197. https://doi.org/10.2108/zsj.29.191
29. Satoh A, Bryant SV, Gardiner DM (2012) Nerve signaling regulates basal keratinocyte proliferation in the blastema apical epithelial cap in the axolotl (Ambystoma mexicanum). Dev Biol 366:374–381. https://doi.org/10.1016/j.ydbio.2012.03.022
30. Satoh A, Makanae A, Nishimoto Y, Mitogawa K (2016) FGF and BMP derived from dorsal root ganglia regulate blastema induction in limb regeneration in Ambystoma mexicanum. Dev Biol 417:114–125. https://doi.org/10.1016/j.ydbio.2016.07.005
31. Ferris DR, Satoh A, Mandefro B, Cummings GM, Gardiner DM, Rugg EL (2010) Ex vivo generation of a functional and regenerative wound epithelium from axolotl (Ambystoma mexicanum) skin. Develop Growth Differ 52: 715–724. https://doi.org/10.1111/j.1440-169X.2010.01208.x
32. Makanae A, Satoh A (2012) Early regulation of axolotl limb regeneration. Anat Rec Adv Integr Anat Evol Biol 295:1566–1574. https://doi.org/10.1002/ar.22529
33. McCusker CD, Diaz-Castillo C, Sosnik J, Phan QA, Gardiner DM (2016) Cartilage and bone cells do not participate in skeletal regeneration in Ambystoma mexicanum limbs. Dev Biol 416:26–33. https://doi.org/10.1016/j.ydbio.2016.05.032
34. Iwata R, Makanae A, Satoh A (2020) Stability and plasticity of positional memory during limb regeneration in Ambystoma mexicanum. Dev Dyn 249:342–353. https://doi.org/10.1002/dvdy.96
35. Satoh A, Gardiner DM, Bryant SV, Endo T (2007) Nerve-induced ectopic limb blastemas in the axolotl are equivalent to amputation-induced blastemas. Dev Biol 312:231–244. https://doi.org/10.1016/j.ydbio.2007.09.021
36. Satoh A, Makanae A (2014) Conservation of position-specific gene expression in axolotl limb skin. Zool Sci 31:6–13. https://doi.org/10.2108/zsj.31.6
37. Moriyasu M, Makanae A, Satoh A (2012) Spatiotemporal regulation of keratin 5 and 17 in the axolotl limb. Dev Dyn 241:1616–1624. https://doi.org/10.1002/dvdy.23839
38. Satoh A, Makanae A, Hirata A, Satou Y (2011) Blastema induction in aneurogenic state and Prrx-1 regulation by MMPs and FGFs in Ambystoma mexicanum limb regeneration. Dev Biol 355:263–274. https://doi.org/10.1016/j.ydbio.2011.04.017
39. Aguilar C, Gardiner DM (2015) DNA methylation dynamics regulate the formation of a regenerative wound epithelium during axolotl limb regeneration. PLoS One 10:e0134791. https://doi.org/10.1371/journal.pone.0134791
40. Satoh A, Cummings GMC, Bryant SV, Gardiner DM (2010) Regulation of proximal-distal intercalation during limb regeneration in the axolotl (Ambystoma mexicanum). Develop Growth Differ 52:785–798. https://doi.org/10.1111/j.1440-169X.2010.01214.x
41. Satoh A, Cummings GMC, Bryant SV, Gardiner DM (2010) Neurotrophic regulation of fibroblast dedifferentiation during limb skeletal regeneration in the axolotl (Ambystoma mexicanum). Dev Biol 337:444–457. https://doi.org/10.1016/j.ydbio.2009.11.023
42. McCusker CD, Gardiner DM (2013) Positional information is reprogrammed in Blastema cells of the regenerating limb of the axolotl (Ambystoma mexicanum). PLoS One 8:e77064. https://doi.org/10.1371/journal.pone.0077064
43. Phan AQ, Lee J, Oei M, Flath C, Hwe C, Mariano R, Vu T, Shu C, Dinh A, Simkin J et al (2015) Positional information in axolotl and mouse limb extracellular matrix is mediated via heparan sulfate and fibroblast growth factor during limb regeneration in the axolotl (Ambystoma mexicanum). Regeneration 2:182–201. https://doi.org/10.1002/reg2.40
44. Niazi IA, Pescitelli MJ, Stocum DL (1985) Stage-dependent effects of retinoic acid on regenerating urodele limbs. Wilhelm Roux's Arch Dev Biol 194:355–363. https://doi.org/10.1007/BF00877373
45. Crawford K, Stocum DL (1988) Retinoic acid coordinately proximalizes regenerate pattern and blastema differential affinity in axolotl limbs. Development 102:687–698

Chapter 16

Embryonic Tissue and Blastema Transplantations

Maritta Schuez, Thomas Kurth, Joshua D. Currie,
and Tatiana Sandoval-Guzmán

Abstract

Embryo grafts have been an experimental pillar in developmental biology, and particularly, in amphibian biology. Grafts have been essential in constructing fate maps of different cell populations and migratory patterns. Likewise, autografts and allografts in older larvae or adult salamanders have been widely used to disentangle mechanisms of regeneration. The combination of transgenesis and grafting has widened even more the application of this technique.

In this chapter, we provide a detailed protocol for embryo transplants in the axolotl (*Ambystoma mexicanum*). The location and stages to label connective tissue, muscle, or blood vessels in the limb and blood cells in the whole animal. However, the potential of embryo transplants is enormous and impossible to cover in one chapter. Furthermore, we provide a protocol for blastema transplantation as an example of allograft in older larvae.

Key words Axolotl Embryos, Lateral Plate Mesoderm, Presomitic Mesoderm, Connective Tissue, Blood Vessels, Muscle, Blood Cells

1 Introduction

The large size of the axolotl's eggs has facilitated their broad use in developmental biology research for more than 200 years. An axolotl egg is double the size of an anuran egg (*Xenopus*). Furthermore, axolotls in the lab can breed all year long, and one breeding yields a clutch size of approximately 500 eggs. Added to this, their development can be modulated by temperature changes [1–3]. Thus, the availability and the egg size, make the axolotl a dominant model for embryonic grafts, while the lack of allograft rejection makes possible also postembryonic grafting. Both embryonic and postembryonic grafting have been seminal to study regenerative biology. It is probably safe to say that every major discovery in salamander regenerative biology has relied on groundwork arising from a transplantation experiment.

Ashley W. Seifert and Joshua D. Currie (eds.), *Salamanders: Methods and Protocols*,
Methods in Molecular Biology, vol. 2562, https://doi.org/10.1007/978-1-0716-2659-7_16,

Preceding the era of genome editing in axolotls, grafts of dark and white wild-type mutant embryos were used [4, 5]. In older larvae, allografts from diploid and triploid animals were convenient, for example, to determine the contribution of dermal cells to regeneration [6, 7], regeneration of skeletal structures [8], and cellular contribution to the regenerate [9–11]. These techniques uncovered seminal principles of regeneration, although the possibility to faithfully follow cells in vivo was still limited. With the introduction of transgenic axolotls expressing ubiquitously a green fluorescent protein (GFP) [12], new possibilities opened. Grafting, in combination with transgenesis, proved at once very useful to reveal migratory patterns and the embryonic origin of tissues or structures [13–15].

Genomic editing is nowadays the best option not only to genetically label a specific cell type or a specific tissue in a reliable and durable manner but also to modify or control the local and temporal expression of genes or for targeted modifications (for a review see [16]). Nevertheless, transgenesis is not always suitable due to multiple reasons: a lack of tissue-specific genetic marker, positional marker, difficulty to introduce multiple transgenes, slow development, mosaicism in F0 animals, long time for the creation of germline transgenics, and the challenges that represent working with a large genome (32 Gbps) [17, 18].

Numerous examples exist where grafting has circumvented different challenges; the lack of a genetic marker for positional identity [19], lack of a tissue-specific marker [20–22], an alternative to creating a triple transgenic [23], to disrupt positional cues [24, 25], or as an alternative to mutagenesis otherwise lethal for the organism [26]. Alone or in combination with advanced genome editing techniques and the endless creativity of scientists, grafting has been part of ground-breaking discoveries in regenerative biology. Recent examples include: determining the tissue-specific contribution to regeneration [20, 27, 28], determining the essential components and steps for regeneration using the Accessory Limb Model (ALM) [29], and discovering that blastema cells possess positional memory of their axial level [30–34]. Transplantation of limb blastema into limb buds determined that cells within the blastema could respond to signals from the limb bud and contribute to the formation of the limb [35]. Transplantations of healthy non-irradiated tissue into irradiated limbs informed the plasticity of dermal cells [36, 37]. Correspondingly, examples of embryonic transplants to address regeneration in older animals are abundant, such as the contribution of cells in circulation by transplanting the blood anlage from GFP+ to white embryos [12], and the contribution of muscle stem cells by using a presomitic mesoderm embryo transplant [21].

In many of these studies, even the availability of transgenic animals would not have been sufficient to address the unanswered question. Thus, transplantations still provide an alternative for marking tissues during embryogenesis and regeneration, disrupting positional identity, and even increasing organ size [38]. Protocols for embryo and larvae grafts have been previously published [1, 39]. In this work, we provide further optimization and transplantation techniques not yet published.

2 Materials

2.1 General Tools

1. Axolotl embryos
   Host embryos: white mutant (d/d)
   Donor embryos: white mutant (d/d) GFP+ (axolotls with ubiquitous expression of green fluorescent protein).
2. Plastic containers.
3. Petri dishes (94/16 mm and 35/10 mm).
4. Glass beaker.
5. Metal sieve.
6. Incubator preset to 16 °C.
7. Sterile plastic transfer pipettes (2 mL).
8. Dissecting microscope (Olympus SZ×10).
9. Fluorescence microscope (Olympus SZ×10).
10. Sharp forceps (two pairs Dumont #55).
11. Angled ophthalmic scissors.
12. Preparation needles:
    - 2 Pin Holders (F.S.T. No. 26018–17, jaw opening: 0–1 mm, 17 cm).
    - 2 Insert Pins (F.S.T. No. 26007–02, diameter:0.6 mm, 5 cm).
13. Scalpel.
14. 6-well plates.

2.2 Reagents

1. Agar (SERVA powder analytical grade, No.: 11393.04).
2. 5× Steinberg's Solution 1 L: Steinberg, 1957.

   NaCl.

   KCL

   Ca(NO3)2 . H2O.

   MgSO4. 7H2O.

   Tris.
3. 100X Antibiotic-Antimycotic, Gibco by Life Technologies

2.3 Reagent Preparation

1. Stock solution 5× Steinberg's

   NaCl: 17.0 g.

   KCL: 0.25 g.

   Ca(NO3)2. H2O: 0.4 g.

   MgSO4. 7H2O: 1.02 g.

   Tris: 2.8 g.

Fill up to 1 L with distilled water. Adjust pH 7.4 with HCL and autoclave.

2. Working solution 1× Steinberg's plus Antibiotic-Antimycotic

   100 mL 5× Steinberg's solution.

   400 mL distilled water.

   5 mL Antibiotic-Antimycotic.

   (Gibco by Life Technologies, No:1524062)

   In the protocol description, 1× Steinberg always refers to 1× Steinberg + Antibiotic-Antimycotic.

3. Coating of petri-dishes with agar gel solution

   Prepare a 2% Agar gel solution with autoclaved animal holding water and pour 1–2 mL into the desired number of 35 mm Petri dishes to form a thin layer of agar on the bottom of the dish. Make enough dishes to allocate 1 dish per transplanted embryo.

3 Methods

3.1 Embryo Preparation

1. Keep the freshly laid eggs in tap water in plastic containers (*see* **Note 1**).
2. Stage embryos according to the table from (Schreckenberg & Jacobson 1975), or https://ambystoma.uky.edu/education1/embryo-staging-series. For this, transfer embryos using a plastic pipette into a petri dish and screen them under a stereomicroscope.
3. When they have reached stages 12–13 (late gastrula/early neurula stage), transfer embryos into a metal sieve by using a plastic pipette and rinse with 70% Ethanol (approximately 20 seconds) to sterilize the surface of the jelly coat. Wash with sterile water and pour embryos into a glass beaker with 1× Steinberg's solution.
4. Place embryos with the help of a plastic transfer pipette into a 94 mm Petri dish filled with 1× Steinberg's solution. Dejelly embryos manually using forceps under a stereomicroscope (*see* **Note 2**).

5. Transfer the dejellied embryos into a fresh petri dish (94 mm) filled with 1× Steinberg's solution using a wide-bore sterile transfer pipette (*see* **Note 3**) and keep them at 7 °C until the desired stage is reached.

3.2 Limb Blood Vessels and Limb Muscle

Blood vessels and muscle transplantations are performed at stages 14–15 (Fig. 1). Transplantations are carried out with preparation needles. Donor and host embryos should be at the same stage. Keep Steinberg's solution and prepared embryos cold during transplantation

1. Choose embryos at stages 13–15 and sort fluorescent and nonfluorescent embryos under a fluorescent microscope, store them at 7 °C (*see* **Note 4**). Place an agar-coated 35 mm Petri dish under a microscope and fill with 1× Steinberg's Solution.
2. Cut two small wells (or one small channel) with a scalpel close together in the middle of the dish into the agar layer (*see* **Note 5**).
3. Transfer a donor embryo (GFP+) to the left side of the dish and a host embryo (d/d) to the right side. If you are not using a fluorescent microscope, care should be taken not to mix them up. Otherwise, use a stereomicroscope for steps 4–12.
4. Remove the vitelline membrane by using two pairs of forceps. Grab it first on one side of the embryo then on the opposite side and tear it apart.
5. Arrange the embryos in the wells with the neural folds facing up (toward the viewer) and the cranial side away from the researcher; donor into the left well and host into the right well (*see* **Note 6**).
6. Cut the ectoderm (epidermal layer) of the donor with preparation needles dorsally along the left neural fold in the anterior trunk region. Use gentle scratching strokes along the line to cut (*see* **Note** 7). Steps 6–11 are performed with preparation needles.
7. Make a cranial and caudal incision through the neural fold towards the ventral region and carefully flip the ectoderm back (*see* **Note 8**).
8. For blood vessel progenitors, cut out a rectangular piece of lateral mesoderm (Fig. 1a). For muscle, cut a fragment of presomitic mesoderm, roughly beneath the neural fold (Fig. 1c) and leave it in the donor while preparing the host.
9. In the host embryo, cut the ectoderm dorsally along the left neural fold **avoiding cutting through the neural fold** (*see* **Note 9**). Figure 3c, d show a micrograph of a host embryo with the neural fold opening.

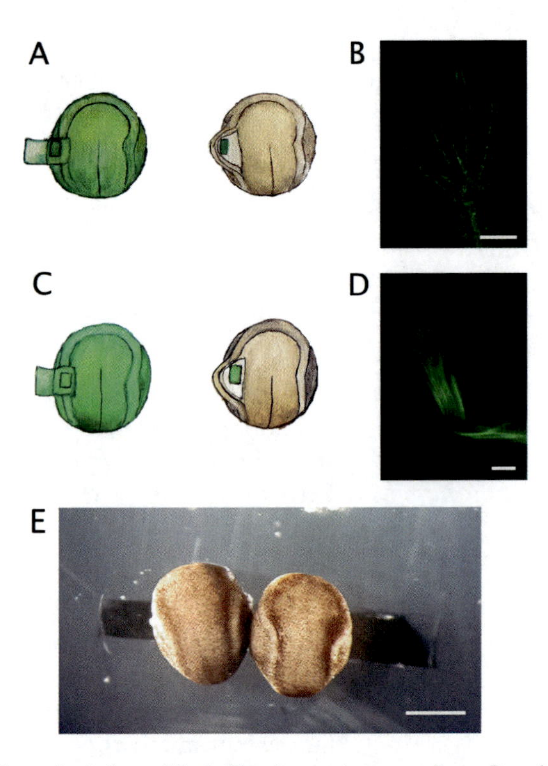

Fig. 1 (**a**) Transplantation of limb blood vessels progenitors. Drawing of donor embryo with an opened flap of ectoderm and underlying lateral mesoderm, next to a host embryo with a transplanted piece of GFP^{+} mesoderm placed into a created pocket. (**b**) Limb of an axolotl larva transplanted as in (**a**) with blood vessels expressing GFP. Scale bar 2 mm. (**c**) Transplantation of limb muscle progenitors. Drawing of donor embryo with an opened flap of ectoderm and underlying presomitic mesoderm, next to a host embryo with a transplanted piece of GFP^{+} mesoderm placed into a created pocket. D) Limb of an axolotl larva transplanted as in (**c**) with muscle expressing GFP. Scale bar 1 mm. (**e**) Photomicrographs of eggs at stage 14. The eggs are resting on a channel made in the agar plate with a scalpel. The channel holds the eggs in place to allow manipulation. Scale bar 1 mm

10. Lift the ectoderm to create a pocket and remove a fragment of mesoderm in equivalent size as the prepared graft (Fig. 1a, c).
11. Transfer the graft from the donor to the host, place it orthotopically and close back the ectoderm. Figures 3a, b show a micrograph of a donor embryo with the vacant space left after the fragment to be transplanted has already been removed.
12. Gently push the host out of the well from underneath using preparation needles, but leave it in the dish and dispose of the donor embryo. Close the lid of the dish and allow the embryo to heal overnight at 16 °C. After 1–2 hours, the opening in the host already shows signs of healing (Fig. 3e).

13. The next day, the embryos should be at the early tailbud stage (Stage 21) (Fig.3f).
14. Transfer the transplanted embryos into 6 well plates (one embryo per well) filled with 1× Steinberg's. Keep the plates at 16 °C – 18 °C for one to two weeks or until the larvae start to eat.Screen larvae for fluorescence to determine the success of the transplant (Fig. 1b, d). The success of the transplant can be fully assessed once the limb has been formed (developmental stage 54, or around 25 days post-hatching).

3.3 Limb Connective Tissue and Blood Cells

Connective tissue and blood cell transplantations are performed at stages 16–18.

1. Choose embryos in stages 13–17 and sort fluorescence and non-fluorescence embryos under a fluorescence microscope, store them at 7 °C.
2. Cut two small wells (or one small channel) with a scalpel close together in the middle of the dish into the agar layer (*see* **Note 5**).
3. Transfer a donor embryo (GFP+) to the left side of the dish and a host embryo (d/d) to the right side. If you are not using a fluorescence microscope care should be taken not to mix them up. Otherwise, use a stereomicroscope for **steps 3–10**.
4. Remove the vitelline membrane by using two pairs of forceps. Grab it first on one side of the embryo then on the opposite side and tear it apart.
5. Place the embryos sideways and horizontally into the wells, head to the right side, tail to the left, neuro folds facing up (towards the viewer).
6. Cut a flap into the ectoderm of the donor embryo with preparation needles:
   (a) For limb connective tissue, cut lateral to the anterior trunk neural fold, (lateral plate mesoderm) (Fig. 2a).
   (b) For limb blood cells, cut in the posterior ventrolateral mesodermal area (Fig. 2b).
7. Cut out a piece of underlying mesoderm and leave it in the donor while the host is being prepared.
8. Open the host embryo in the same way and remove an equivalent fragment of mesoderm.
9. Transfer the graft from the donor embryo to the host embryo, place it orthotopically and fold the ectoderm back.
10. Gently push the host out of the well but leave it in the dish and trash the donor. Close the lid of the dish and allow the embryo to heal overnight at 16 °C.

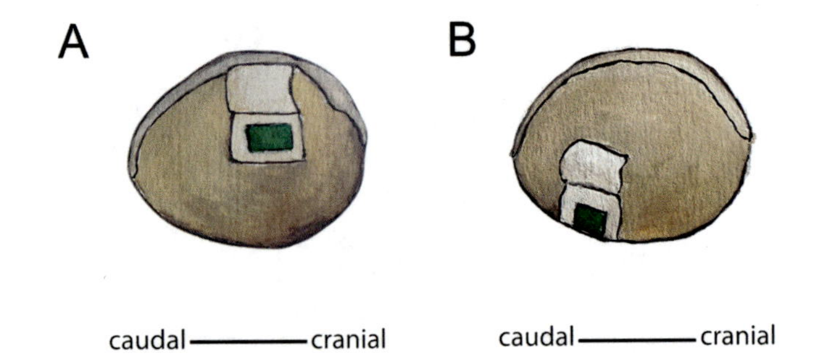

Fig. 2 Transplantation of progenitors of the limb connective tissue and blood. (**a**) Transplantation of limb connective tissue. Drawing of the side view of a white host embryo at Stage 17 with a transplanted piece of GFP^{+} mesoderm. The donor embryo is opened in the same way. (**b**) Transplantation of blood anlage. Drawing of white host embryo at Stage 17 with a transplanted piece of GFP^{+} mesoderm. The donor embryo is opened in the same way

11. The next day transfers the transplanted embryos into six well plates (one embryo per well) filled with 1x Steinberg. Keep the plates at 16 °C–18 °C for 1 or 2 weeks or until the larvae start to eat.

3.4 Limb Blastema Transplantations

The donor animal should be amputated prior to transplantation to create a blastema that is the appropriate age for the desired experiment. This protocol is adapted for axolotls of 4–6 cm total length. For bigger animals, adjust the anesthesia immersion time, size of the instruments, and recovery time.

The day of the transplant:

1. Anesthetize host and donor animals by immersion in 0.01% benzocaine for 10–15 minutes or until the full cessation of movement.
2. Transfer both host and donor axolotls to a Petri dish covered with a paper tissue soaked in anesthetic.
3. Amputate the host animal at the desired level of the proximodistal axis of the limb. With a pair of fine forceps, push gently the soft tissue back to expose approximately 1 mm of limb skeleton and trim the exposed skeletal elements with a pair of scissors. Trimming the skeleton of the host will facilitate the attachment of the donor tissue.
4. Place the animal on its back allowing the host limb to face up.
5. Place a benzocaine-soaked paper tissue under the limb to keep it raised up (Fig. 4) and to keep the animal from dehydration.
6. The axolotl should be kept moist during the entirety of the transplantation procedure and recovery time by periodically dripping anesthetic solution over the animal with a transfer pipet.

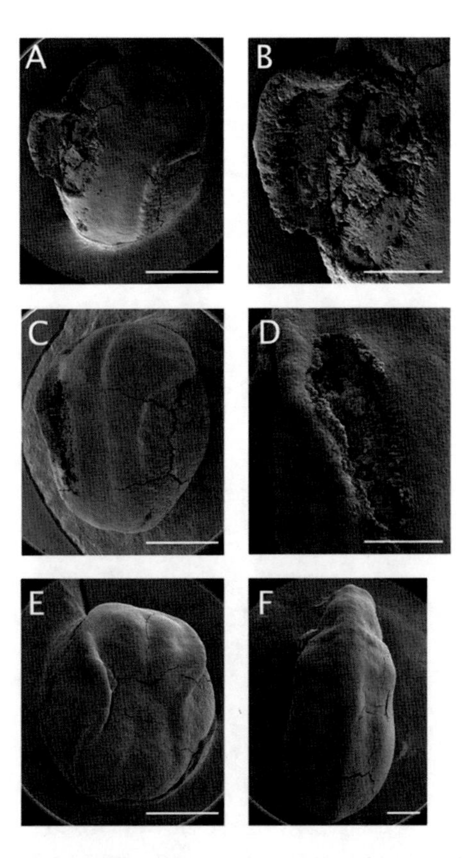

Fig. 3 Scanning Electron Microscope micrographs of host and donor embryos during and after the graft. Samples were fixed in modified Karnovsky (2% formaldehyde and 2% glutaraldehyde in 50 mM HEPES), postfixed in 1% osmium tetroxide in water, dehydrated in a graded series of ethanol/water up to pure ethanol on a molecular sieve, critical-point-dried using the Leica CPD300, mounted to aluminum stubs, and sputter-coated with gold. Analysis occurred on a Jeol JSM7500F at 5 kV acceleration voltage. (**a**) Donor embryo after a fragment of presomitic mesoderm has been removed. Scale bar 1 mm. (**b**) Higher magnification of (**a**). Scale bar 500 microns. (**c**) Host after ectoderm is open to receive the donor fragment. Scale bar 1 mm. (**d**) Higher magnification of (**c**). Scale bar 500 microns. (**e**) Host after 1–2 hours of transplant. Scale bar 1 mm. (**f**) Dorsal view of the host embryo 1 day after transplantation. Scale bar 500 microns

7. With a pair of scissors cut the blastema of the donor and gently transfer it with forceps on top of the host stump (Fig. 4). To avoid positional discontinuity between blastema and stump, position both host and donor animals equally and transfer the blastema avoiding unnecessary maneuvering.
8. Leave the host in this position for 15–20 min before returning it to its water tank. Handling the host animals with fishnets could cause the graft to detach. Avoid excessive handling for at least 3 days after the surgery.

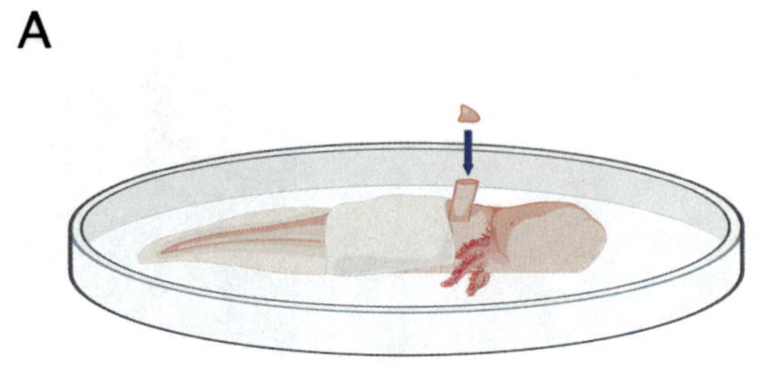

Fig. 4 Depiction of a blastema transplant. The animal is laying on its side while a wet tissue is placed under the limb to maintain it straight up. This facilitates positioning the blastema in the right place and tissue adherence. Figure created with BioRender

4 Notes

1. Two to three days after laying, the embryos have reached stages 14–15 when kept at room temperature. It is possible to delay the development by storing embryos at colder temperatures. At 4 °C development almost stops; at 11 °C it is slowed down 2 1/2 times. However, embryos at stages 10–12 should not be kept at 7 °C or colder for longer than a few hours since this will result in impaired gastrulation.
2. Embryos are enclosed in a jelly coat and a capsule. Both have to be removed before transplantation: Grab the jelly coat with forceps and pierce at the same time with one tip of the forceps through the capsule, take the second pair of forceps and tear the capsule apart.

   Cut the tip of a plastic transfer pipette to a diameter of 3–4 mm. Use always this wide-bore pipette to transfer dejellied embryos.
3. It is possible to sort the embryos earlier than stage 13 but usually, the fluorescence is not clearly recognizable at the early stages. However, if the transplantation is done under a fluorescent microscope, it is also possible during transplantation to check for fluorescence of the donor animal and differentiate between host and donor embryos.
4. To prevent the embryos from moving or dodging during the transplantation they are placed into small wells which are cut into the agar layer. The size of the wells should be a bit smaller than the embryos. We recommend a small horizontal channel, 1 mm wide and 1 cm long on which embryos are placed transversely side by side (Fig. 1e). Keep the cut-out piece of agar on the side, this can be placed back in the channel after the operation to avoid deformities on the embryo caused by the edges of the agar. Cutting out a channel or depression works better when the dish is filled with solution.

5. It is advisable to establish a routine method for arranging host and donor embryos to avoid mixing them. For example, always place the donor on the left side and the host on the right, and always perform transplantations on one side of the embryo.
6. Try to transplant quickly and always open the donor first to make sure the wound of the host has closed again as fast as possible. If left open for too long, the cut edges will fail to stick back together.
7. To separate the ectoderm from the mesoderm, insert the tip of the preparation needle between these two layers along one side of the incision and move gradually forward.
8. With practice and ensuring the procedure is fast, it is also possible to open the host embryo in the same way as the donor embryo. But always keep in mind that the wound heals much better and faster when the neural fold is not severed.

Acknowledgments

We are thankful to Beate Gruhl, Anja Wagner, and Dr. Judith Konantz for their dedication to the axolotls. We thank all members of the Sandoval-Guzmán Lab current and past, for their unconditional support. This work was supported by the Center for Regenerative Therapies Dresden and the German Research Council (DFG).

References

1. Nacu E, Knapp D, Tanaka EM, Epperlein HH (2009) Axolotl (Ambystoma mexicanum) embryonic transplantation methods. Cold Spring Harb Protoc 2009:pdb.prot5265. https://doi.org/10.1101/pdb.prot5265
2. Khattak S, Murawala P, Andreas H, Kappert V, Schuez M, Sandoval-Guzmán T, Crawford K, Tanaka EM (2014) Optimized axolotl (Ambystoma mexicanum) husbandry, breeding, metamorphosis, transgenesis and tamoxifen-mediated recombination. Nat Protoc 9:529–540. https://doi.org/10.1038/nprot.2014.040
3. Voss SR, Epperlein HH, Tanaka EM (2009) Ambystoma mexicanum, the axolotl: a versatile amphibian model for regeneration, development, and evolution studies. Cold Spring Harb Protoc 2009:pdb.emo128-9. https://doi.org/10.1101/pdb.emo128
4. Dalton CH (1953) Relations between developing melanophores and embryonic tissues in the mexican axolotl. In: Pigment cell growth. pigment cell growth, pp. 17–27.
5. Keller RE, Löfberg J, Spieth J (1982) Neural crest cell behavior in white and dark embryos of Ambystoma mexicanum: epidermal inhibition of pigment cell migration in the white axolotl. Dev Biol 89:179–195. https://doi.org/10.1016/0012-1606(82)90306-2
6. Namenwirth M (1974) The inheritance of cell differentiation during limb regeneration in the axolotl. Dev Biol 41:42–56. https://doi.org/10.1016/0012-1606(74)90281-4
7. Gardiner DM, Muneoka K, Bryant SV (1986) The migration of dermal cells during blastema formation in axolotls. Dev Biol 118:488–493
8. Pescitelli MJ, Stocum DL (1980) The origin of skeletal structures during Intercalary regeneration of larval ambystoma limbs'. Dev Biol 79:255
9. Muneoka K, Holler-Dinsmore GV, Bryant SV (1985) A quantitative analysis of regeneration from chimaeric limb stumps in the axolotl. J Embryol Exp Morphol 90:1–12

10. Steen TP (1970) Origin and differentiative capacities of cells in the blastema of the regenerating salamander limb. Am Zool 10:119–132. https://doi.org/10.1093/icb/10.2.119
11. Steen TP (1968) Stability of chondrocyte differentiation and contribution of muscle to cartilage during limb regeneration in the axolotl (Siredon mexicanum). J Exp Zool 167:49–78. https://doi.org/10.1002/jez.1401670105
12. Sobkow L, Epperlein HH, Herklotz S, Straube WL, Tanaka EM (2006) A germline GFP transgenic axolotl and its use to track cell fate: Dual origin of the fin mesenchyme during development and the fate of blood cells during regeneration. Dev Biol 290:386–397. https://doi.org/10.1016/j.ydbio.2005.11.037
13. Taniguchi Y, Kurth T, Medeiros DM, Tazaki A, Ramm R, Epperlein H-H (2015) Mesodermal origin of median fin mesenchyme and tail muscle in amphibian larvae. Nat Pub Group 5:1–14. https://doi.org/10.1038/srep11428
14. Epperlein HH, Selleck MAJ, Meulemans D, McHedlishvili L, Cerny R, Sobkow L, Bronner-Fraser M (2007) Migratory patterns and developmental potential of trunk neural crest cells in the axolotl embryo. Dev Dyn 236:389–403. https://doi.org/10.1002/dvdy.21039
15. Soukup V, Epperlein HH, Horácek I, Cerny R (2008) Dual epithelial origin of vertebrate oral teeth. Nature 455:795–798. https://doi.org/10.1038/nature07304
16. Tilley L, Papadopoulos S-C, Pende M, Fei J-F, Murawala P (2021) The use of transgenics in the laboratory axolotl. Dev Dyn 251:942. https://doi.org/10.1002/dvdy.357
17. Nowoshilow S, Schloissnig S, Fei J-F, Dahl A, Pang AWC, Pippel M, Winkler S, Hastie AR, Young G, Roscito JG, Falcon F, Knapp D, Powell S, Cruz A, Cao H, Habermann B, Hiller M, Tanaka EM, Myers EW (2018) The axolotl genome and the evolution of key tissue formation regulators. Nature 554:50–55. https://doi.org/10.1038/nature25458
18. Smith JJ, Timoshevskaya N, Timoshevskiy VA, Keinath MC, Hardy D, Voss SR (2019) A chromosome-scale assembly of the axolotl genome. Genome Res 29:317–324. https://doi.org/10.1101/gr.241901.118
19. Mchedlishvili L, Epperlein HH, Telzerow A, Tanaka EM (2007) A clonal analysis of neural progenitors during axolotl spinal cord regeneration reveals evidence for both spatially restricted and multipotent progenitors. Development 134:2083–2093. https://doi.org/10.1242/dev.02852
20. McCusker CD, Diaz-Castillo C, Sosnik J, Phan AQ, Gardiner DM (2016) Cartilage and bone cells do not participate in skeletal regeneration in Ambystoma mexicanum limbs. Dev Biol 416:26–33. https://doi.org/10.1016/j.ydbio.2016.05.032
21. Sandoval-Guzmán T, Wang H, Khattak S, Schuez M, Roensch K, Nacu E, Tazaki A, Joven A, Tanaka EM, Simon A (2013) Fundamental differences in dedifferentiation and stem cell recruitment during skeletal muscle regeneration in two salamander species. Cell Stem Cell 14:174–187. https://doi.org/10.1016/j.stem.2013.11.007
22. Gerber T, Murawala P, Knapp D, Masselink W, Schuez M, Hermann S, Gac-Santel M, Nowoshilow S, Kageyama J, Khattak S, Currie J, Camp JG, Tanaka EM, Treutlein B (2018) Single-cell analysis uncovers convergence of cell identities during axolotl limb regeneration. Science 20:eaaq0681-19. https://doi.org/10.1126/science.aaq0681
23. Currie JD, Kawaguchi A, Traspas RM, Schuez M, Chara O, Tanaka EM (2016) Live imaging of axolotl digit regeneration reveals spatiotemporal choreography of diverse connective tissue progenitor pools. Dev Cell 39:411–423. https://doi.org/10.1016/j.devcel.2016.10.013
24. Crawford K, Stocum DL (1988) Retinoic acid coordinately proximalizes regenerate pattern and blastema differential affinity in axolotl limbs. Development 102:687–698
25. Maden M (1982) Vitamin-a and pattern-formation in the regenerating limb. Nature 295:672–675. https://doi.org/10.1038/295672a0
26. Sanor LD, Flowers GP, Crews CM (2020) Multiplex CRISPR/Cas screen in regenerating haploid limbs of chimeric Axolotls. eLife 9:762–718. https://doi.org/10.7554/elife.48511
27. Kragl M, Knapp D, Nacu E, Khattak S, Maden M, Epperlein HH, Tanaka EM (2009) Cells keep a memory of their tissue origin during axolotl limb regeneration. Nature 460:60–65. https://doi.org/10.1038/nature08152
28. Iwata R, Makanae A, Satoh A (2019) Stability and plasticity of positional memory during limb regeneration in Ambystoma mexicanum. Dev Dyn 52:343–312. https://doi.org/10.1002/dvdy.96
29. Endo T, Bryant SV, Gardiner DM (2004) A stepwise model system for limb regeneration. Dev Biol 270:135–145. https://doi.org/10.1016/j.ydbio.2004.02.016

30. Wolpert L (1971) Positional information and pattern formation. Curr Top Dev Biol 6:183–224. https://doi.org/10.1016/s0070-2153(08)60641-9
31. Carlson BM (1975) The effects of rotation and positional change of stump tissues upon morphogenesis of the regenerating axolotl limb. Dev Biol 47:269–291. https://doi.org/10.1016/0012-1606(75)90282-1
32. Brockes JP (1997) Amphibian limb regeneration: rebuilding a complex structure. Science 276:81–87. https://doi.org/10.1126/science.276.5309.81
33. Pescitelli MJ, Stocum DL (2003) The origin of skeletal structures during intercalary regeneration of larval Ambystoma limbs. Dev Biol 79:255–275. https://doi.org/10.1016/0012-1606(80)90115-3
34. Echeverri K, Tanaka EM (2005) Proximodistal patterning during limb regeneration. Dev Biol 279:391–401. https://doi.org/10.1016/j.ydbio.2004.12.029
35. Muneoka K, Bryant SV (1982) Evidence that patterning mechanisms in development and regenerating limb are the same. 1–3.
36. Lheureux E (1983) Replacement of irradiated epidermis by migration of non-irradiated epidermis in the newt limb: the necessity of healthy epidermis for regeneration. J Embryol Exp Morphol 76:217–234
37. Carlson BM (1974) Morphogenetic interactions between rotated skin cuffs and underlying stump tissues in regenerating axolotl forelimbs. Dev Biol 39:263–285. https://doi.org/10.1016/S0012-1606(74)80029-1
38. Zarzosa A, Grassme K, Tanaka E, Taniguchi Y, Bramke S, Kurth T, Epperlein H (2014) Axolotls with an under- or oversupply of neural crest can regulate the sizes of their dorsal root ganglia to normal levels. Dev Biol 394:1–18. https://doi.org/10.1016/j.ydbio.2014.08.001
39. Kragl M, Tanaka EM (2009) Grafting axolotl (Ambystoma mexicanum) limb skin and cartilage from GFP+ donors to normal hosts. Cold Spring Harb Protoc 2009:pdb.prot5266. https://doi.org/10.1101/pdb.prot5266

Chapter 17

Retinoic Acid–Induced Limb Duplications

Malcolm Maden and Trey Polvadore

Abstract

Retinoic acid (RA) and the family of molecules based on vitamin A known as retinoids have remarkable effects on limb regeneration in salamanders and newts and cause whole limb duplications in a concentration-dependent manner. They respecify all three axes of the limb—the proximodistal, the anteroposterior, and the dorsoventral axis. As a result, complete limbs can be induced to regenerate from distal amputation planes producing two limbs in tandem. Here, we describe the basic methods for undertaking these experiments as well as the use of new synthetic retinoids which have retinoic acid receptor-selective actions. These will be valuable tools in future studies on the molecular basis of limb duplications and thus our understanding of the nature of positional information in the regenerating salamander limb.

Key words Retinoic Acid, Limb Duplications, Limb Regeneration, Limb Development, RARs, RAR Agonists

1 Introduction

The first report of limb duplications caused by retinoids (the family of compounds derived from vitamin A) was by Niazi & Saxena [1] who treated *Bufo andersonii* tadpoles with retinol palmitate and amputated their hindlimbs at various stages. They were trying to see whether the disappearance of the regenerative ability which normally occurs in anurans as they approach metamorphosis could be delayed by vitamin A which was known to inhibit metamorphosis and produce giant tadpoles. Rather than inducing limb regeneration at later stages, this compound induced several complete limbs from a distal amputation plane which were proximodistally duplicated. This was a remarkable result because limb regeneration normally has an extremely high fidelity—what you amputate is perfectly replaced—and up until then every compound which had been tested for its effect on limb regeneration either did nothing or inhibited it to varying degrees. Here was a compound that stimulated the process and caused ectopic limbs.

Ashley W. Seifert and Joshua D. Currie (eds.), *Salamanders: Methods and Protocols*,
Methods in Molecular Biology, vol. 2562, https://doi.org/10.1007/978-1-0716-2659-7_17,

A range of retinoids was then used on the more commonly used organism for limb regeneration studies, the axolotl, *Ambystoma mexicanum* [2]. At the time, only the naturally occurring retinoids were available—retinol palmitate, retinol acetate, retinol, and retinoic acid (RA) and several of their isomers, and these were tested for relative potency with the result that RA was the most potent (approx. ten-fold) mirroring experiments in cell culture. This work together with a subsequent publication [3] characterized the basic effects of RA on inducing duplications in the regenerating axolotl limb as follows:

1. Duplication of the complete limb (i.e., the commencement of regeneration from the level of the shoulder girdle or proximal humerus) occurs from any amputation level—the hand, the forearm, the upper arm—although the duplication is more obvious from the level of the hand (Fig. 1a).
2. There is a concentration dependence of duplication—when the limb is amputated (for example, through the radius and ulna) and a low concentration of retinoid is applied then a minimal effect is seen such as the production of a radius and ulna which is too long (Fig. 1c) or excess production of carpals (not shown). When a higher concentration is administered, then regeneration commences from the elbow level resulting in a duplicated radius and ulna (not shown). At a higher concentration, regeneration commences from the distal humerus level thus including a new elbow joint (Fig. 1d). When the highest concentration is applied, then regeneration commences from the proximal humerus level and thus a whole limb is regenerated from the amputation plane through the radius and ulna (Fig. 1e). The concentration difference between the minimal effect and the maximal effect is only 2.5-fold (Fig. 1f) which is in the order of magnitude of differences in concentration of classical developmental gradients.
3. Treatment with too high a concentration of retinoid inhibits regeneration.
4. Of the naturally occurring retinoids, there is a range of potencies with RA being the most potent and retinol palmitate the least potent—about eight- to tenfold less. However, the new synthetic RAR agonists (see below) are about 100-fold more potent than RA, acting in the low nanomolar range.
5. For any concentration, increasing the time of administration increases the degree of duplication.
6. There is a period during which the amputated limb is sensitive to this duplication event—treatment too early before limb amputation will not result in duplication (the circulating retinoid has been eliminated from the system) and treatment too late will not result in duplication (the limb has completed patterning the regenerate).

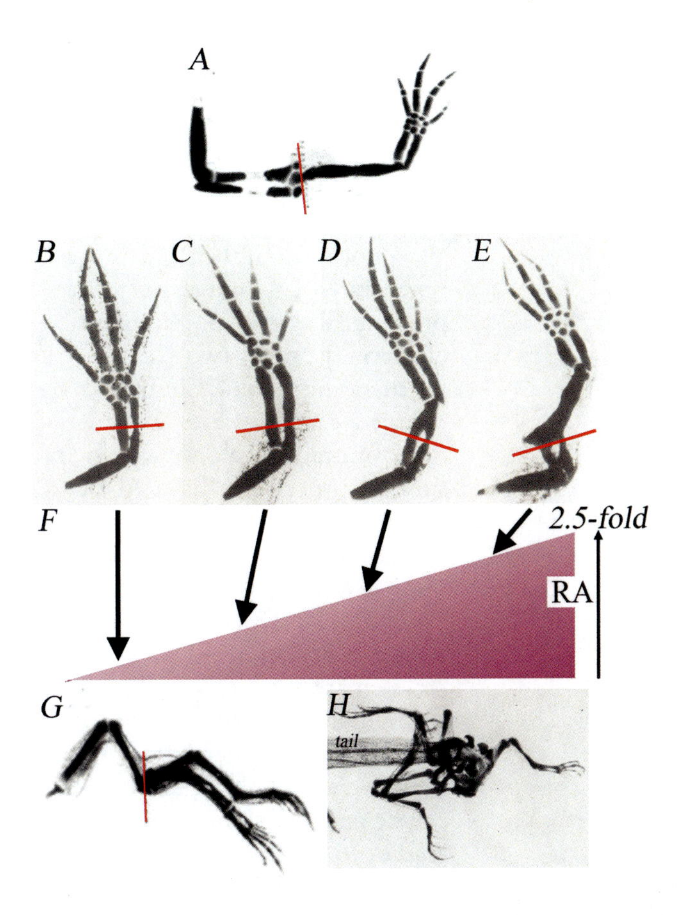

Fig. 1 (**a**) Example of an axolotl forelimb amputated through the carpals (red line) and treated with a high concentration of retinol palmitate. A complete limb beginning with the full humerus has regenerated from the hand level. (**b**) A control axolotl limb that had been amputated through the radius and ulna (red line). (**c**) A radius and ulna level amputation (red line) treated with low levels of retinol palmitate. The regenerated portion has a radius and ulna that is too long. (**d**) A radius and ulna level amputation (red line) treated with medium levels of retinol palmitate. The regenerated portion begins from the level of the distal humerus, then makes an elbow, then a new radius, and ulna, then a hand. (**e**) A radius and ulna level amputation (red line) treated with high levels of retinol palmitate. The regenerated portion begins from the proximal humerus and so regenerates a complete limb from the amputation plane. (**f**) Representation of the increasing levels of retinol palmitate or RA needed to shift the plane of regeneration in a proximal direction. The levels required are only 2.5-fold increase from the lowest effect to the highest effect. (**g**) An example of a *Rana temporaria* limb bud at the footplate stage which was amputated through the foot level (red line) and regenerated a pair of hindlimbs including the pelvic girdle from the amputation plane after treatment with a high level of retinol palmitate. (**h**) Example of the effect of retinol palmitate on tail regeneration in *Rana temporaria* tadpoles. Instead of regenerating the tail as controls do, this case has regenerated 3 pairs of hindlimbs including the pelvic girdles from the tail. (Images **a–e** from [2], image **g** from [8], image **h** from [15])

Other species of salamanders commonly used in limb regeneration studies—*Notophthalmus viridescens*, *Pleurodeles waltli*, *Triturus vulgaris*, *Hynobius leechii*—are similarly affected and produce proximodistally duplicated limbs [4–7]. However, in anurans, an additional effect on the AP and DV axis is seen in that not only are the regenerates proximodistally duplicated but they are also duplicated in the AP axis such that complete pairs of limbs are regenerated in double posterior orientation even including the pelvic girdle from foot amputation levels (Fig. 1g) [1, 8, 9]. The effect on the other two axes of the limb was then investigated in the regenerating axolotl limb because of this disparity and it was revealed that RA does indeed respecify the anteroposterior and dorsoventral axis, along with the proximodistal axis, but additional surgical procedures have to be performed to reveal the effects. Double anterior, double posterior, double ventral, and double dorsal limbs have to be constructed and then amputated whereupon it was revealed that RA posteriorizes the limb and ventralizes the limb [10–13]. Thus, RA affects all three limb axes in precise polarities in both urodeles and anurans.

Remarkably, limb duplications are also produced when the tail is amputated, but only in selective anurans and not in salamanders. This was described in *Bufo* [14] and *Rana* [15] where it was shown that several complete pairs of hindlimbs including the pelvic girdle can be produced from tail amputations (Fig. 1h). Expanding these studies to fish, it was revealed that in zebrafish and *Fundulus* fin development, AP duplications of the fin can be produced by RA when applied during gastrulation stages at very precise concentrations [16]. The same applies to mouse embryos—when RA is administered to pregnant mice at the extremely early developmental stages of the blastocyst and pre-gastrulation it has a dramatic effect on their caudal regions generating pairs of symmetrical hindlimbs and pelvic girdles and tails, hindlimb twinning and polydactylous limbs [17–19]. Thus, RA respecifies the axes of (i) the developing fins and limbs in fish and mammals; (ii) the regenerating limbs of anurans and salamanders; (iii) the regenerating tails of anurans. In the developing chick limb bud, only the AP axis is respecified [20] and the effects on lizards have not been investigated.

It was very surprising, therefore, to discover that when RA is administered to the developing limb buds of axolotls or *Xenopus* there is only a teratogenic and inhibitory effect [9, 21] and not a respecification effect, as is seen in later developing mammalian embryos where RA has long been recognized as a teratogen of the limb [22]. A developing axolotl limb bud is therefore not the same as a regenerating axolotl limb. This differing effect is clearly not due to the concentration of RA or differing animal absorption because if the distal tip of the limb bud is amputated and as a result, the developing limb bud is turned into a regenerating limb, then

RA has its duplicating effects. This has been done using the left and right limb buds of the same animal [21] so the different effects must be due to the different states of the cells induced by tip amputation. This is an interesting paradigm for elaborating the molecular action of RA in urodele development (teratogen) vs regeneration (duplication).

The mechanism of action of RA in affecting gene expression within the nucleus was subsequently elaborated and three nuclear transcription factors of the steroid hormone family were discovered—RARα, RARβ, and RARγ [23]. In newts, the mammalian RARγ is homologous to RARδ [24]. This discovery led to an explosion of interest by many researchers and by pharmaceutical companies in the retinoic acid receptors (RARs) and the genes that they regulate. RA itself has very pleiotropic effects on cells and organisms including both beneficial ones and teratogenic ones, inducing many hundreds of genes. In an effort to cut down on these unwanted side effects RAR-selective agonists were synthesized each with greater selectivity for one (sometimes two) receptor compared to the parent molecule, RA. This advance was clearly important for the development of drugs for human administration and allows us to understand the molecular basis of positional specification by retinoids. These RAR agonists do indeed regulate different sets of genes [25] so they can be used in limb reduplication experiments. However, it is important to note that all these selective binding studies have been performed with the human RAR proteins, so we have to assume that their selectivity is the same in urodeles. In support of this, the sequence similarities between human and axolotl RARs are high (87%, 90%, and 83% for RARα, RARβ, and RARγ, respectively) and axolotl RARs contain the same three specificity-determining amino acid residues that have been identified in the human RARs.

These RAR-selective agonists work at nanomolar concentrations rather than micromolar concentrations as RA does, and when the concentration of these agonists is titered down, then differential effects can clearly be seen— at 25 nM only the RARα agonist produced proximodistal duplications in the axolotl limb, and the other agonists had no effect (Fig. 2). When the concentration of agonists is increased ten-fold to 250 nM then all the agonists have a duplicating effect demonstrating the typical pharmacological loss of receptor specificity at saturating concentrations (not shown). This suggests that limb duplication, at least in axolotls is regulated by gene(s) induced/down-regulated by the RARα receptor. These RAR-selective agonists and RAR-selective antagonists are now widely available from chemical synthesis companies and will be a valuable resource for revealing more about the molecular mechanisms of limb reduplication.

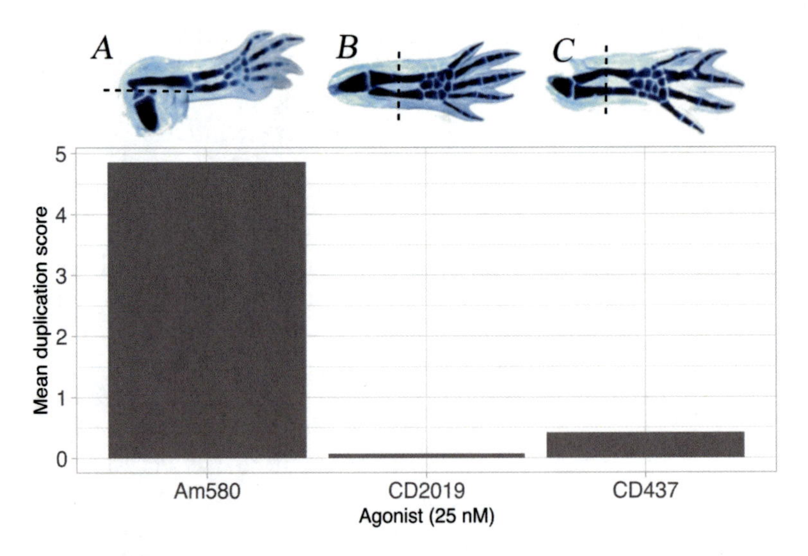

Fig. 2 Example of an experiment on axolotl hindlimbs to distinguish between the effects of different RAR agonists at 25 nM concentration on limb duplication. (**a**) Am580 (RARα agonist) produces a full limb duplication. (**b**) CD 2019 (RARβ agonist) has no effect. **C**, CD437 (RARγ agonist) has no effect. Below is the summary chart using the scoring system for the degree of duplication as elaborated in Subheading 3.3.3 and exemplified in Fig. 1. The duplication score is an average of $n = 10$ regenerates for each agonist. This demonstrates that limb duplication is a RARα-induced phenomenon

2 Materials

Naturally occurring retinoids are available from companies such as Sigma-Aldrich and include compounds such as retinyl palmitate, retinyl acetate, and retinol. Since these are used as sources of vitamin A they are measured in terms of International Units (a unit of biological activity and not a mass of the molecule), making comparisons with the other synthesized retinoids difficult. Other synthesized compounds such as all-*trans*-retinoic acid, 9-*cis* retinoic acid (the isomer of all-*trans*-RA), and retinaldehyde are pure compounds available in the all-*trans*, 9-*cis* or 13-*cis* form. Other synthetic retinoids with an emphasis on their RAR activation are available from companies such as Tocris Bioscience and below are some of their selective actions and their code names.

Pan RA agonists (activates all three receptors)

All-*trans*-retinoic acid.

TTNPB.

Receptor selective agonists

RARα agonist—AM580, AM80, BMS 753.

RARβ agonist—CD2019.

RARγ agonists—CD1530, CD437, CD1530, BMS961.

Dual agonists

RARβ and RARγ agonist—Adapalene, CD 2665.

Antagonists

Pan-RAR antagonist—AGN 193109.

RARα antagonist—ER 50891, BMS 195614.

RARβ antagonist—LE135.

RARγ antagonist—MM 1125, LY 2955303.

Also available are inverse agonists, isoform-selective agonists and fluorescent RA analogues.

2.1 Retinoid Solutions

Prepare 100 mM solution in DMSO. Store 1 ml aliquots in the -80 °C.

2.2 Anesthesia Solution

1 in 10,000 – 1 in 1000 solution (depending on the size/weight of the animal) of Ethyl 3-aminobenzoate methanesulfonate (MS-222, Tricaine) prepared in tap water. This solution is very acidic which will cause harm to the skin of the animals so the pH must be adjusted to 7.5 with NaOH.

2.3 Holtfreter's Solution

Prepare in a 44-gallon drum by adding the following to tap water: 1 teaspoon KCl, 2.5 teaspoons CaCl2, 3 teaspoons MgSO4 anhydrous, 240 cc (dry but measured in a beaker) NaCl, 20 mL AmQuel, and 20 mL NovAqua as water conditioners. The solution can be stored at room temperature for several weeks.

2.4 Victoria Blue Solution

Dissolve 1% w/v Victoria Blue powder in acid alcohol (1% conc HCl in 70% alcohol). Also needed are 10% neutral buffered formalin for fixation, 3% hydrogen peroxide solution, a graded series of ethanols, and methyl salicylate for clearing.

2.5 Animals

The protocol below is designed to work with multiple species of aquatic salamanders or newts and animals can be acquired from region-specific sources.

3 Methods

3.1 Anesthesia and Amputation of Limbs

1. Place animals in the appropriate solution of MS-222 for approximately 5 minutes until movement stops.
2. Remove the animals from anesthetic and place them on wet paper towels. With a single-sided razor blade or scissors amputate the limb at an appropriate distal level through the wrist or mid-forearm. Return the animals to Holtfreter's solution.

3.2 Administration of Retinoid

1. 24 hrs after amputating the limbs, place the animals in the appropriate concentration of retinoid. Defrost a 100 mM aliquot (1 ml) and dilute this solution in Holtfreter's solution or place it directly into 1 liter of Holtfreter's for a 100μm concentration (see **Notes 1** and **2**).

2. Change the solution that the animals are in every day for 5 days.
3. At the end of the administration period, place the animals in Holtfreter's solution until regeneration has completed (3–4 weeks depending on the size of the animals) whereupon it should be obvious due to excessively elongated growth or abnormally angled growth (for example, because of an extra elbow joint, e.g., Figs. 1e and 2 Am580).
4. Controls are treated identically except that 1 ml of DMSO is added to the water which has no effect on the animals nor any effect on regeneration.

3.3 Fixation and Cartilage Staining of Regenerates

1. At the completion of the experiment the animals are terminally anesthetized and the whole limb removed from the body and fixed in 10% neutral buffered formalin. They can then be sectioned for histology or prepared for cartilage staining using Victoria blue (Figs. 1 and 2).
2. For Victoria blue staining the fixed limbs are washed to remove the fixative, placed in 3% hydrogen peroxide overnight, 50% alcohol for 1 hr., acid alcohol for 1 hr., Victoria blue solution 45 minutes, then dehydrate and remove the stain from non-cartilaginous tissues with 70% alcohol, 90% alcohol, 95% alcohol, 100% alcohol, 100% alcohol 20 minutes each, then clear in methyl salicylate.
3. A scoring system is used [2, 3] for measuring the extent of duplication of the limbs (Fig. 1) which allows comparisons of different concentrations or comparisons between different retinoids. A control limb or no duplication scores 0; a regenerate with extra carpals scores 1; with an elongated radius and ulna scores 2 (Fig. 1c); with a complete extra radius and ulna scores 3; a part extra humerus and elbow joint scores 4 (Fig. 1d); a complete limb scores 5 (Figs. 1a, e and 2 Am580).

4 Notes

1. Other methods of administration. The method described above is the simplest and easiest method of administration and involves the least interference with the animal, but others have been used. For example, the intraperitoneal injection has been used [4] to administer RA dissolved in DMSO at the desired dose and following injection, the RA precipitates at the injection site making it visible through the skin as a yellow mass which gradually dissipates over the next 24–48 hrs. Gastric intubation has also been used for administering RA dissolved in DMSO to adult newts [26].

2. Retinoids have also been administered in Silastic rubber blocks cut to the desired size and grafted into the limb blastema. In this method, the retinoid powder at the desired concentration is mixed with one of the two silastic materials to a consistent mix before the setting agent is added and allowed to set. Based on the volume of silastic used, blocks with a known amount of retinoid are cut and grafted into the limb blastema or adjacent to the blastema underneath the skin [27]. Other nontoxic rubber materials, used in dentistry, can also be used.

Acknowledgments

Research concerning the activity of RAR agonists was supported by the National Science Foundation (IOS 1558017).

References

1. Niazi IA, Saxena S (1978) Abnormal hindlimb regeneration in tadpoles of the toad, Bufo andersonii, exposed to excess vitamin A. Folia Biol (Krakow) 26:3–8
2. Maden M (1982) Vitamin A and pattern formation in the regenerating limb. Nature 295: 672–675
3. Maden M (1983a) The effect of vitamin A on the regenerating axolotl limb. J Embryol Exp Morph 77:273–295
4. Thoms SD, Stocum DL (1984) Retinoic acid-induced pattern duplication in regenerating urodele limbs. Dev Biol 103:319–328
5. Niazi IA, Pescitelli MJ, Stocum DL (1985) Stage dependent effects of retinoic acid on regenerating limbs. Wilhelm Roux Arch Dev Biol 194:355–363
6. Lheureux E, Thoms SD, Carey F (1986) The effects of two retinoids on limb regeneration in *Pleurodeles waltl* and *Triturus vulgaris*. J Embryol Exp Morph 92:165–182
7. Ju B-G, Kim W-S (1994) Pattern duplication by retinoic acid treatment in the regenerating limbs of Korean salamander larvae, *Hynobius leechii*, correlates well with the extent of dedifferentiation. Dev Dynam 199:253–267
8. Maden M (1983b) The effect of vitamin a on limb regeneration in *Rana temporaria*. Dev Biol 98:409–416
9. Scadding SR, Maden M (1986a) Comparison of the effects of vitamin a on limb development and regeneration in *Xenopus laevis* tadpoles. J Embryol Exp Morph 191:35–53
10. Stocum DL, Thoms SD (1984) Retinoic acid-induced pattern completion in double anterior limbs of urodeles. J Exp Zool 232:207–215
11. Ludolph DC, Cameron JA, Stocum DL (1990) The effect of retinoic acid on positional memory in the dorsoventral axis of regenerating axolotl limbs. Dev Biol 140:41–52
12. Kim W-S, Stocum DL (1990) Retinoic acid modifies positional memory in the anteroposterior axis of regenerating axolotl limbs. Dev Biol 114:170–179
13. Wigmore P (1990) Serially duplicated regenerates from the anterior half of the axolotl limb after retinoic acid treatment. Rouxs Arch Dev Biol 198:252–256
14. Mohanty-Heijmadi P, Dutta SK, Mahapatra P (1992) Limbs generated at site of tail amputation in marbled balloon frog after vitamin a treatment. Nature 355:352–353
15. Maden M (1993) The homeotic transformation of tails into limbs in *Rana temporaria* by retinoids. Dev Biol 159:379–391
16. Vandersea MW, Fleming P, McCarthy RA, Smith DG (1998) Fin duplications and deletions induced by disruptions of retinoic acid signaling. Dev Gene Evol 208:61–68
17. Rutledge JC, Shourbaji AG, Hughes LA, Polifka JE, Cruz YP, Bishop JB, Generoso WM (1994) Limb and lower-body duplications induced by retinoic acid in mice. PNAS USA 91:5436–5440
18. Niederreither K, Ward SJ, Dolle P, Chambon P (1996) Morphological and molecular characterization of retinoic acid-induced limb duplications in mice. Dev Biol 176:185–198

19. Liao X, Collins MD (2008) All-*trans* retinoic acid-induced ectopic limb and caudal structures: murine strain sensitivities and pathogenesis. Dev Dynam 237:1553–1564
20. Tickle C, Alberts B, Wolpert L, Lee J (1982) Local application of retinoic acid to the limb bond mimics the action of the polarizing region. Nature 296:564–566
21. Scadding SR, Maden M (1986b) Comparison of the effects of vitamin a on limb development and regeneration in the axolotl, *Ambystoma mexicanum*. J Embryol Exp Morph 91:19–34
22. Satre MA, Kochhar DM (1989) Elevations in the endogenous levels of the putative morphogen retinoic acid in embryonic mouse limbbuds associated with limb dysmorphogenesis. Dev Biol 133:529–536
23. Chambon P (1995) The molecular and genetic dissection of the retinoid signaling pathway. In: Recent progress in hormone research. Academic Press, London, pp 317–332
24. Ragsdale CW, Gates PB, Hill DS, Brockes JP (1993) Delta retinoic acid receptor isoform is distinguished by its N-terminal sequence and abundance in the limb regeneration blastema. Mech Dev 40:99–112
25. Maden M, Chambers D, Monaghan J (2018) Chapter 7: Retinoic acid and the genetics of positional information. In: Regenerative engineering and developmental biology. CRC Press, Boca Raton, pp 163–180
26. Koussoulakos S, Sharma KK, Anton HJ (1988) Vitamin a induced bilateral asymmetries in Triturus forelimb regenerates. Biol Struct Morph 1:43–48
27. Maden M, Keeble S, Cox RA (1985) The characteristics of local application of retinoic acid to the regenerating axolotl limb. Rouxs Arch Dev Biol 194:228–235

Chapter 18

Isolation and Characterization of Peritoneal Macrophages from Salamanders

Anthony Sallese, Georgios Tsissios, J. Raúl Pérez-Estrada, Arielle Martinez, and Katia Del Rio-Tsonis

Abstract

Salamanders have been used as research models for centuries. While they exhibit a wide range of biological features not seen in mammals, none has captivated scientists like their ability to regenerate. Interestingly, axolotl macrophages have emerged as an essential cell population for tissue regeneration. Whether the same is true in other salamanders such as newt species *Notophthalmus viridescens*, *Cynops pyrrhogaster*, or *Pleurodeles waltl* remains to be seen. Unfortunately, regardless of the species, molecular tools to study macrophage function in salamanders are lacking. We propose that the readily available, terminally differentiated peritoneal macrophages from newts or axolotls could be used to validate molecular reagents in the study of macrophage function during tissue regeneration in salamanders.

Key words Peritoneal Macrophage, Isolation, Detection, CSF1R, Newts, Axolotls

1 Introduction

The requirement for macrophages during tissue regeneration has been documented in several species including zebrafish, Xenopus, axolotls, and African spiny mice [1–6]. In salamanders, we still know very little about the ontogeny of macrophages or their polarization states associated with these remarkable outcomes. Peritoneal macrophages represent an ideal macrophage population for isolation and characterization in salamanders because they are easy to collect and are present in relatively large numbers. Also, since most methods and reagents used to study macrophages are designed for use in mammals, peritoneal macrophages from salamanders represent an excellent model for an in vitro validation of reagents before their use in vivo. Here, we describe a method for the collection of naïve, tissue-resident, peritoneal macrophages from newts and axolotls (Fig. 1). This protocol follows closely those previously used in mammals and amphibians. This source of

Ashley W. Seifert and Joshua D. Currie (eds.), *Salamanders: Methods and Protocols*,
Methods in Molecular Biology, vol. 2562, https://doi.org/10.1007/978-1-0716-2659-7_18,

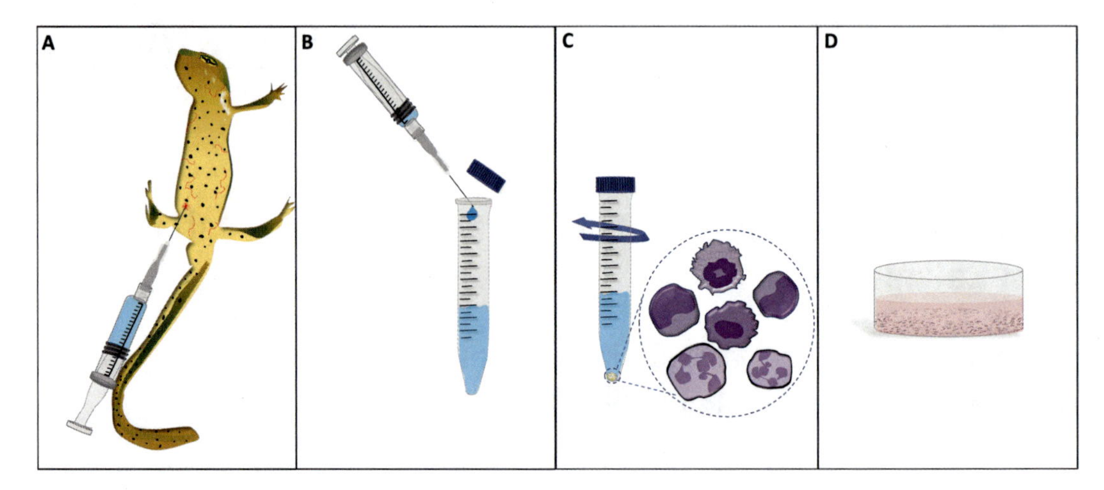

Fig. 1 Graphical illustration of peritoneal macrophage isolation in newts. (**A**) Newts are sedated and then APBS & 0.5 mM EDTA is injected into the peritoneal cavity. (**B**) The fluid is collected into a conical tube and the process is repeated 2 more times. (**C**) The peritoneal cell suspension is centrifuged to form a pellet, the supernatant removed, and the cellular pellet resuspended in tissue culture media. (**D**) Peritoneal cells are plated on tissue culture plastic or on treated coverslips and allowed to adhere for 1–4 hours. Non-adherent cells are washed off with APBS and purified peritoneal macrophages remain adhered to the surface

macrophages can be used for validation of molecular tools such as antibodies for cell detection (Figs. 2 and 3) or phenotyping (Fig. 2) and testing primers (Fig. 4). Here, we used the purified macrophages to validate a commercially available antibody that recognizes the colony-stimulating factor 1 receptor (CSF1R) for monocyte/macrophage identification in both species. Furthermore, we demonstrate these purified macrophages can be used to screen antibody markers used for M1 vs M2 phenotyping, commonly used in mammals, such as iNOS and Arginase 1 or to test primers for macrophage-specific genes. The CSF1R antibody or primers could be useful for characterizing the kinetics of macrophage accumulation following injury or for validating methods to deplete macrophages in salamanders in vivo. Missing from this description is the intraperitoneal injection (IP) of inflammatory stimulants such as heat-killed E. coli or Brewer's thioglycollate to trigger a localized inflammatory response that increases peritoneal macrophage yield. Based on the work in Xenopus, this would be expected to increase the yield of cells approximately 10 fold [7]. It should be noted, however, that the increased yield comes at the expense of obtaining macrophages that are no longer in a homeostatic state and instead are polarized to an inflammatory one. Pending the individual user's objectives, this could be advantageous or disadvantageous.

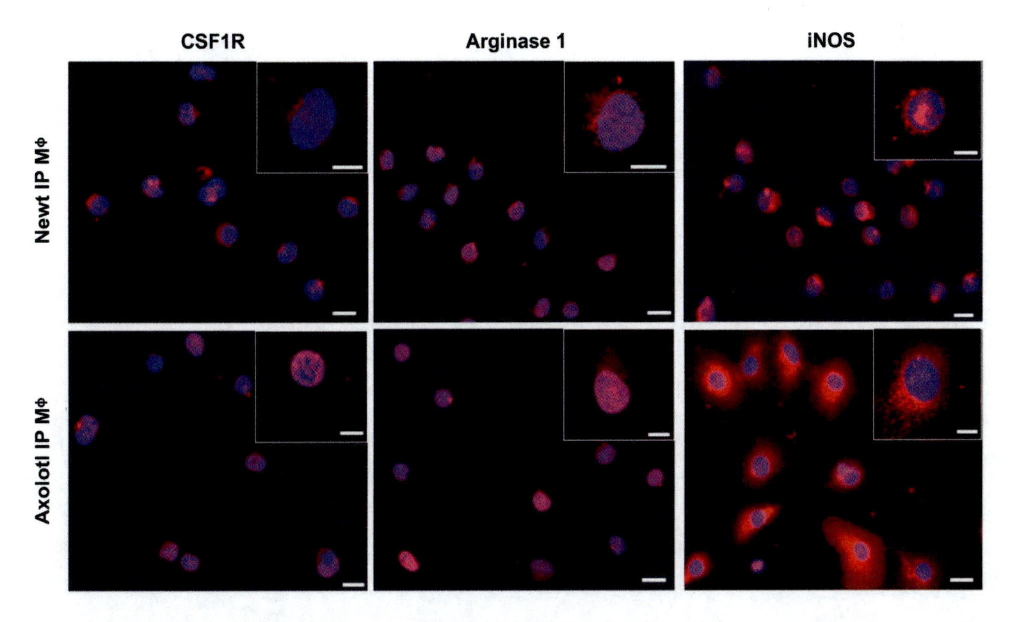

Fig. 2 Immunohistochemistry of purified peritoneal macrophages from salamanders. Peritoneal macrophages from newts (top panel) and axolotls (bottom panel) were collected as described in the protocol, adhered to poly-lysine coated coverslips for 3 hours, fixed with 4% PFA for 10 min and stained with antibodies for CSF1R, Arginase 1 and iNOS as indicated at a 1:200 dilution. Nuclei were detected with DAPI. Scale bars: For large image 10 μm and for small inset 5 μm

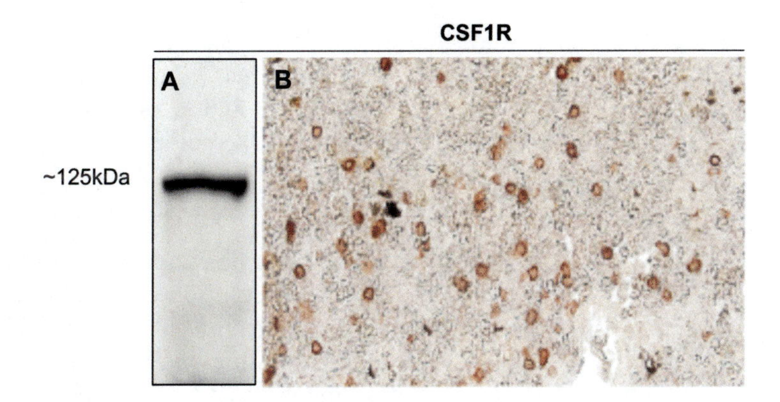

Fig. 3 Detection of total and phosphorylated CSF1R protein. (**A**) Protein was extracted from newt lungs and used for western blot detection of CSF1R. The antibody detected a single band around approximately 125 kDa. (**B**) Newt spleens were fixed, paraffin embedded, and stained with phosphorylated CSF1R antibody at 1:200 dilution. To avoid auto fluorescence of the newt spleen, the primary antibody was detected with an HRP-conjugated secondary antibody. The reddish-brown staining indicates positive phosphorylated CSF1R detection

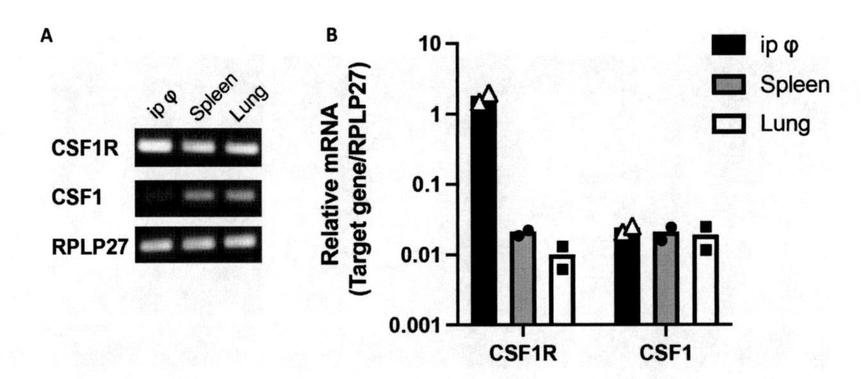

Fig. 4 Determination of CSF1R and CSF1 gene expression levels in newts. RNA was extracted from newt peritoneal macrophages, spleens and lungs in duplicate. (**A**) RT-PCR for CSFR1 (167 bp), CSF1 (124 bp) and RPLP27 (133 bp) genes using specific primers. (**B**) Relative mRNA levels of CSF1R and CSF1 quantified by RT-qPCR. Note the significant enrichment of CSF1R expression in peritoneal macrophages (black bars). Unsurprisingly, as the spleen and lung are macrophage-rich organs, CSF1R was also detected, albeit at a lower level. CSF1 was detected at comparable levels across all tissues. ip φ: IP macrophages

2 Materials

2.1 Animals

1. Wild-caught adult *Notophthalmus viridescens* (newts).
2. 10–12 cm *Ambystoma mexicanum* (axolotls).

2.2 Reagents

1. Peritoneal Macrophage Isolation.
   (a) MS222 (Ethyl 3-Aminobenzoate methanesulfonic acid salt): Weigh 0.5 g of MS222 and add 500 ml of 1× modified APBS solution.
   (b) Lugol's iodide solution: Add 5 ml of Lugol's Iodine solution and 95 ml of 70% ethanol. (*See* **Note 1**).
   (c) APBS (amphibian phosphate-buffered saline): Prepare 20× APBS solution by weighing 120 g sodium chloride, 4.5 g potassium chloride, 0.525 g sodium phosphate, 0.3 g potassium phosphate, and 0.5 mM EDTA. Add 900 ml water and adjust pH to 7.4 with HCl. Bring final volume to 1 L. Dilute to 1× with water and sterilize by filtration.
   (d) Modified L-15 media: Weigh 4.5 g of Leibovitz's L-15 powder and 2.35 g of glucose. Add 5 ml of 1 M HEPES, 5 ml of antibiotic/antimycotic, 50 ml of fetal bovine serum. Adjust pH to 7.4 with HCl and add water to 500 ml. Sterilize by filtration.
   (e) 1 ml size slip tip syringe (*see* **Note 2**).
   (f) 18 G needle (*see* **Note 3**).
   (g) Conical centrifuge tube (15 ml).

2. Immunocytochemistry (ICC).
   (a) Wash buffer: 1×APBS solution with 0.1% bovine serum albumin.
   (b) Blocking buffer: 1×APBS solution with 10% normal donkey serum, 0.3% TritonX-100, and 0.1% bovine serum albumin.
   (c) Twelve well tissue culture treated plates.
   (d) Poly-D-Lysine coated 12 mm round coverslip (Corning #354086).
   (e) DAPI (Life technologies #D1306) at 1:1000.
   (f) Antibodies:
      (i) CSF1R antibody (Cell signaling #3152 s) at 1:200 dilution.
      (ii) iNOS antibody (Abcam #ab3523) at 1:200 dilution.
      (iii) Arginase 1 antibody (Abcam #ab91279) at 1:200 dilution.
      (iv) IgG (H + L) Highly Cross-Adsorbed Donkey anti-Rabbit, Alexa Fluor® 546.
   (g) Fluoromount aqueous mounting medium (Sigma#4680).
3. Western Blot.
   (a) TBST: Prepare 10×TBST solution by weighing 24 g tris-HCl, 5.6 g tris-base and 88 g NaCl. Add 900 ml water and add 10 ml of Tween 20. Adjust pH to 7.6 with HCl and bring final volume to 1 L. Dilute to 1× with water.
   (b) 2× Laemmli buffer: Prepare solution by adding 950 μl of 2× Laemmli sample buffer (Bio-Rad #1610737) and 50 μl of beta-mercaptoethanol.
   (c) Running buffer: 10× Tris/Glycine/SDS buffer (Bio-Rad #1610772). Dilute to 1× with water.
   (d) Transfer buffer: Tans-Blot Turbo 5X transfer buffer (Bio-Rad #10026938).
   (e) Blocking buffer: 5% BSA in 1×TBST.
   (f) Mini-PROTEAN® TGX™ 12% Precast Protein Gels, 10-well, 30 μl (Bio-Rad #4561043).
   (g) Pierce BCA protein assay kit (Thermo #23227).
   (h) SuperSignal West Pico Plus chemiluminescent substrate (Thermo #34580).
   (i) RIPA lysis buffer system (Santa Cruz# SC24948).
   (j) Antibodies:
      (i) CSF1R antibody (Cell signaling #3152 s) at 1:200 dilution.

(ii) Anti-Rabbit IgG HRP-linked (Cell Signaling#7074S) at 1:1000 dilution.

4. Immunohistochemistry (IHC).
   (a) Blocking buffer: 0.5% BSA, 20% goat serum, in 1×TBST.
   (b) Peroxidase substrate ImmPACT DAB (Vector labs #SK-4105).
   (c) Antibodies:
      (i) Phospho-MCSFR (Cell signaling #3154S) at 1:200 dilution.
      (ii) Signal stain boost IHC Detection, rabbit HRP (Cell Signaling #8114S).
   (d) Permount mounting medium (Fisher #SP15–500).
5. RT-qPCR.
   (a) TRIzol reagent (ThermoFisher #15,596,018).
   (b) Tissue homogenizer (Pellet Pestle Motor, Fisher Scientific #12–141-362).
   (c) RNase-Free disposable pellet pestles (Fisher Scientific #12–141-364).
   (d) Direct-zol RNA Micropep Kit (Zymo Research #R2060).
   (e) QuantiTect Reverse Transcription kit (Qiagene #205,310).
   (f) RT2 SYBR Green Master Mix (Qiagene #204,076).
   (g) mL Corbet Type Strip Tubes & Caps (LabForce #1139Y44).
   (h) DNase/RNase-Free water.
   (i) Primers (*see* **Note 6**).
      (i) CSF1R: (Forward: CAAGCTGTTTAACCCGGTGC, reverse: GCGTTGGGTCGATGAAGGTA , product length: 167 bp).
      (ii) CSF1: (Forward: TGTCTCCCAGGGACATGCTA , reverse: TCGAGGTAATCCGGGTCCTT , product length:124 bp).
      (iii) RPLP27: (Forward: GCTGGTCGATACTCTG GACG , reverse: TGCCCATTGTGGCTGTTACT , product length: 133 bp).

2.3 Equipment

1. Centrifuge for 15 ml conical tubes.
2. Tissue culture incubator at 27 °C (without CO_2 attached).
3. Vacuum filter units (Fisher #SCGP00525).
4. Branson model 250/450 Sonifier (#100–413-016).
5. Thermo cycler Rotor-Gene Q 5plex (Qiagen).

6. Trans-Blot® Turbo™ transfer system (Bio-Rad #1704150).
7. ChemiDocMP imaging system (Bio-Rad #12003154).
8. Bioanalyzer 2100 D3300 (Agilent).

3 Methods

1. Peritoneal macrophage isolation.
   (a) Anesthetize the newts by immersion in 0.1% MS222.
   (b) Apply Lugol's solution to the newt's abdomen to sterilize the injection site.
   (c) Fill a 1 ml syringe with ice-cold 1×APBS and 0.5 mM EDTA, and connect it to an 18 G needle (*see* **Note 4**).
   (d) Insert the needle into the lower quadrant of the newt's abdomen, and once the needle is inserted, slowly inject 1 ml of the 1×APBS and 0.5 mM EDTA.
   (e) Slowly withdraw half the solution and reinject, repeat two times (*see* **Note 5**).
   (f) Once completed, leave the needle in the abdomen and remove the syringe from the needle. You should be able to recover approximately 75–80% of the injected fluid.
   (g) Use the plunger of the syringe to eject the peritoneal exudate into a conical tube and store it on ice.
   (h) Repeat the peritoneal lavage two more times with new 1×APBS and 0.5 mM EDTA each time, and pool the peritoneal exudate into the same conical tube.
   (i) Store conical tube on ice as peritoneal exudate is collected if the procedure is being performed on multiple animals.
   (j) Once all of the peritoneal lavages are complete, centrifuge the peritoneal exudate in the conical tube at 300 g for 10 min to pellet the cells.
   (k) Aspirate the supernatant and resuspend the cell pellet in modified L-15 media.
   (l) Plate the cells on tissue culture dishes or poly-lysine coated coverslips, place in the incubator, and wait 1–4 hrs for macrophages to adhere. Plating density will depend on the downstream application and the size of the tissue culture well or coverslip. We have had good results following recommendations for cell numbers based on established protocols in mice for immunohistochemical staining, RNA isolation, and protein isolation.
   (m) Wash three times with 1×APBS to remove non-adherent cells; purified adherent peritoneal macrophages will

remain attached to the surface. Proceed immediately with fixation, RNA or protein extraction depending on downstream application.

2. Macrophage validation by ICC.
   (a) Plate isolated peritoneal macrophages on poly-lysine coated coverslips.
   (b) Fix cells in 4%PFA for 10 min.
   (c) Wash cells 3 times with wash buffer for 10 min.
   (d) Incubate cells with cell IHC blocking buffer for 2 hrs at room temperature.
   (e) Incubate cells with primary antibody (1:200 dilution in cell IHC blocking buffer) overnight at 4 °C.
   (f) Wash cells 3 times with wash buffer for 10 min.
   (g) Incubate cells with secondary antibody for 1 hr (1:100 dilution in blocking buffer) at room temperature. For nuclear staining, add DAPI to the blocking buffer with the secondary antibody at 1:1000 dilution.
   (h) Wash cells 3 times with wash buffer for 10 min.
   (i) Mount coverslip into a glass slide by using fluoromount aqueous mounting medium.
3. Antibody validation by western blot.
   (a) After anesthetizing the newt, collect entire lung tissues (or tissue of interest) in ice cold APBS.
   (b) Place each lung in a 1.5 ml Eppendorf tube and add 200 μl RIPA lysis buffer mix.
   (c) Vortex briefly and proceed for sonication.
   (d) Sonicate 3 times for 5 s at low power and 40% pulse. Let samples cool on ice for 1 min in between each sonication.
   (e) Centrifuge at 14,000 rpm for 10 min at 4 °C.
   (f) Remove supernatant and use immediately or store at −20 °C.
   (g) Quantify protein using a BCA assay kit.
   (h) Dilute protein sample at 1:1 ratio with 2× Laemmli sample buffer.
   (i) Heat the samples at 100 °C for 5 min, put them on ice for 5 min, and centrifuge at 14,000 rpm for 5 min at 4 °C.
   (j) Load the desired amount of protein into the wells of a 12% precast polyacrylamide gel, along with a molecular weight marker (depending on the application, gel percentage may vary).

(k) Run the gel at 120v for approximately 2 hrs at room temperature (depending on the application, time and voltage might require optimization).

(l) Transfer the gel into a PVDF membrane for 7 min at 25 V using the trans-blot turbo transfer kit (depending on the application, time, and voltage may vary).

(m) Wash membrane with 1×TBST buffer three times for 10 min with shaking.

(n) Block with WB blocking buffer for 1 hr at room temperature with shaking.

(o) Incubate with primary antibody overnight at 4 °C with shaking.

(p) Wash membrane with 1×TBST buffer three times for 10 min with shaking.

(q) Incubate with secondary antibody for 2 hrs at room temperature with shaking.

(r) Wash membrane with 1×TBST buffer three times for 10 min with shaking.

(s) Develop signal by incubating the membrane with chemiluminescent substrate.

(t) Image membrane using ChemiDocMP imaging system.

4. Antibody Validation by IHC.

(a) After anesthetizing the newts, collect entire spleen tissues (or tissue of interest) in ice-cold APBS.

(b) Fix spleens in 4% PFA at 4 °C overnight.

(c) After paraffin embedding, section the tissue at 10 μm thickness.

(d) Deparaffinize and gradually hydrate sections.

(e) Perform antigen retrieval by submerging the sections in 0.01 M sodium citrate and incubating at sub-boiling temperature (95 °C–98 °C) for 15 min.

(f) Let sections to cool and then wash with Di water three times for 5 min.

(g) Incubate sections in 3% hydrogen peroxide for 10 min.

(h) Wash sections with Di water three times for 5 min.

(i) Wash sections with 1×TBST buffer three times for 5 min.

(j) Block sections with tissue IHC blocking buffer for 1 hr at room temperature.

(k) Incubate sections with primary antibody overnight at 4 °C.

(l) Wash sections with 1×TBST buffer three times for 5 min.

(m) Incubate sections in secondary antibody for 30 min.

(n) Wash sections with 1×TBST buffer three times for 5 min.

(o) Incubate sections with Peroxidase Substrate ImmPACT DAB solution for 10 min according to manufacturer's instruction (depending on the application, incubation time may vary). Monitor the reaction using a microscope and once a light brown signal is observed stop the reaction.

(p) Wash sections with Di water, gradually dehydrate, and clear sections.

(q) Coverslip sections with permount mounting media.

5. Macrophage Validation by RT-qPCR.

(a) Total RNA isolation and cDNA synthesis.

(i) Collect tissue and cell samples (IP macrophages, spleen, and lung) in 1 mL of TRIzol reagent. Use 2×10^5 cells for IP macrophages, and 10–50 mg of tissue for spleen and lung.

(ii) Homogenize the tissue samples (lung and spleen) mechanically using a pellet pestle motor and RNase-Free disposable pellet pestles. Vortex samples vigorously for 1 min. The samples can be used immediately for RNA isolation or stored at −20 °C until needed.

(iii) Isolate total RNA using Direct-zol RNA Microprep kit, following manufacturer's instructions, including DNase I treatment for DNA removal.

(iv) Determinate RNA quality and quantity using Agilent 2010 Bioanalyzer, following the manufacturer's instructions.

(v) Use 200 ng of total RNA for cDNA synthesis using QuaniTect Reverse Transcription Kit, following manufacturer's instructions. Use 20 μl of total volume reaction.

(b) qPCR Reaction (*see* **Note 7**).

(i) Prepare each qPCR reaction according to the following table:

(ii) Run the qPCR reaction on a thermocycler Rotor-Gene Q 5plex (Qiagen) according to the following conditions: Activation: 95 °C for 5 min. PCR amplification: 94 ° C for 20s, 60 °C for 20s, and 72 °C for 20s, for 40 cycles. Melting curve: 72–95 °C.

(iii) Calculate the gene expression levels as the ratio between the Cq value of the target and housekeeping gene (Rplp27) (Table 1).

Table 1
List of reagents and volumes for qPCR

Component	Volume
RT2 SYBR Green master mix 2×	10 μl
Forward primer, 10 μM	1 μl
Reverse primer, 10 μM	1 μl
cDNA (1:10 dilution)	2 μl
DNase/RNase-free water	6 μl

4 Notes

1. In mice, this procedure is performed after soaking the fur with 70% ethanol and surgically removing the outer skin and exposing the inner skin layer of the peritoneal cavity. In this manner, the needle punctures through the sterile inner skin layer of the peritoneal cavity and the collected peritoneal fluid should be free of bacterial contamination supporting the long-term culture of peritoneal cells. Since this is not feasible in salamanders, if the user wants to culture peritoneal cells, it is recommended to soak the abdomen with Lugol's solution in an effort to disinfect the skin as much as possible before the needle is inserted into the abdomen.
2. Slip tip syringes are preferred over Luer lock because it makes performing the repeated lavage easier and decreases the risk of accidentally pulling the needle from the abdomen as the Luer lock is being removed.
3. Other gauges may be used but beware the smaller the gauge, the easier the needle will clog.
4. Roughly 1 ml of fluid was used to lavage adult *Notophthalmus viridescens* weighing approximately 3–4 g. This volume will need to be adjusted based on the size of the animal. We recommend slowly injecting the fluid as the peritoneal cavity will start to inflate like a balloon. Stop injecting once the skin appears tight. We have noticed several times in axolotls that a smaller injection volume may be required. The bladder of the axolotls appears more fragile than the one in newts, and if too large a volume is injected into the peritoneal cavity of the axolotls, the fluid will start leaking from its urethra.
5. After injecting the fluid into the peritoneal cavity, the internal organs, especially the intestines, are free-floating and will easily clog the needle once suction from the syringe is applied. The plunger of the syringe must be withdrawn very slowly to

maintain a low suction force. If too much suction is applied, the needle will immediately clog. It can also be helpful to keep the needle horizontal to the lab bench and slightly raised, almost lifting the animal off the bench, creating a tent-like shape "∧" with the skin of the abdomen and the needle at the top. This can help create a void or pocket where you can withdraw the fluid without getting clogged on internal organs. Finally, with this technique, it is also important to have the bevel of the needle facing down so the skin does not clog it.

6. Primer blast (https://www.ncbi.nlm.nih.gov/tools/primer-blast/index.cgi) was used for primer design. Specific primers for *Notophthalmus viridescens* genes were tested and selected if the Tm = 60 °C. Sequences were obtained from a previously published paper [8].
7. The RT-qPCR method validation described in this chapter complies with the qPCR MIQE guidelines [9].

Acknowledgments

This work was supported by National Eye Institute Grant EY027801 and by the John W. Steube Professor endowment to KDRT. Financial support from Fight for Sight is gratefully acknowledged (AS). The authors thank the laboratory animal facility at Miami University and Erika Grajales-Esquivel and Jared Tangeman for the critical reading of the protocol.

References

1. Aztekin C et al (2020) The myeloid lineage is required for the emergence of a regeneration-permissive environment following Xenopus tail amputation. Development 147:3
2. Var SR, Byrd-Jacobs CA (2020) Role of macrophages and microglia in zebrafish regeneration. Int J Mol Sci 21:13
3. Godwin JW et al (2017) Heart regeneration in the salamander relies on macrophage-mediated control of fibroblast activation and the extracellular landscape. NPJ Regen Med 2
4. Godwin JW, Pinto AR, Rosenthal NA (2013) Macrophages are required for adult salamander limb regeneration. Proc Natl Acad Sci U S A 110(23):9415–9420
5. Cavone L et al (2021) A unique macrophage subpopulation signals directly to progenitor cells to promote regenerative neurogenesis in the zebrafish spinal cord. Dev Cell 56(11): 1617–1630.e6
6. Simkin J et al (2017) Macrophages are necessary for epimorphic regeneration in African spiny mice. elife 6
7. Grayfer L (2018) Elicitation of xenopus laevis tadpole and adult frog peritoneal leukocytes. Cold Spring Harb Protoc 2018:7
8. Sousounis K et al (2013) Transcriptome analysis of newt lens regeneration reveals distinct gradients in gene expression patterns. PLoS One 8(4):e61445
9. Bustin S, Nolan T (2017) Talking the talk, but not walking the walk: RT-qPCR as a paradigm for the lack of reproducibility in molecular research. Eur J Clin Investig 47(10):756–774

Part IV

Bioinformatics and Genomics

Chapter 19

Navigation and Use of Custom Tracks within the Axolotl Genome Browser

Sergej Nowoshilow and Elly M. Tanaka

Abstract

The availability of the chromosome-scale axolotl genome sequences has made it possible to explore genome evolution, perform cross-species comparisons, and use additional sequencing data to analyze both genome-wide features and individual genes. Here, we will focus on the UCSC genome browser and demonstrate in a step-by-step manner how to use it to integrate different data to approach a broad question of the *Fgf8* locus evolution and analyze the neighborhood of a gene that was reported missing in axolotl – *Pax3*.

Key words Genome Browser, Synteny, Custom Tracks, Genome Evolution, Annotation, Gene Expression

1 Introduction

The advances in sequencing technologies and assembly techniques in the past three decades allowed for a giant leap from a handful of 1–3 kb long *expressed sequence tag* (ESTs) sequences to complete genome sequences of *C. elegans* (100 Mb) [1], human (3.2Gb) [2, 3] and finally the Mexican axolotl, *Ambystoma mexicanum* (32Gb) [4–6].

Despite the tremendous progress, no sequencing platform can currently produce a read as long as a eukaryotic chromosome. Therefore, the chromosomes must be reconstructed from much shorter reads in a process called assembly (reviewed in [7]) (Fig. 1). In general, one needs to identify all overlapping reads and re-build longer *contiguous consensus sequences*, referred to as contigs (Fig. 1a). In an ideal world, this would be enough to reconstruct the full DNA sequence as it is; enough to find all matching pieces to reconstruct a puzzle. However, in reality, the genomic sequences are often repetitive, which complicates the reconstruction of contigs [8] (Fig. 1b). Since all the exact overlaps cannot be determined, one can reconstruct the sequential order of the contigs in a

Ashley W. Seifert and Joshua D. Currie (eds.), *Salamanders: Methods and Protocols*,
Methods in Molecular Biology, vol. 2562, https://doi.org/10.1007/978-1-0716-2659-7_19,

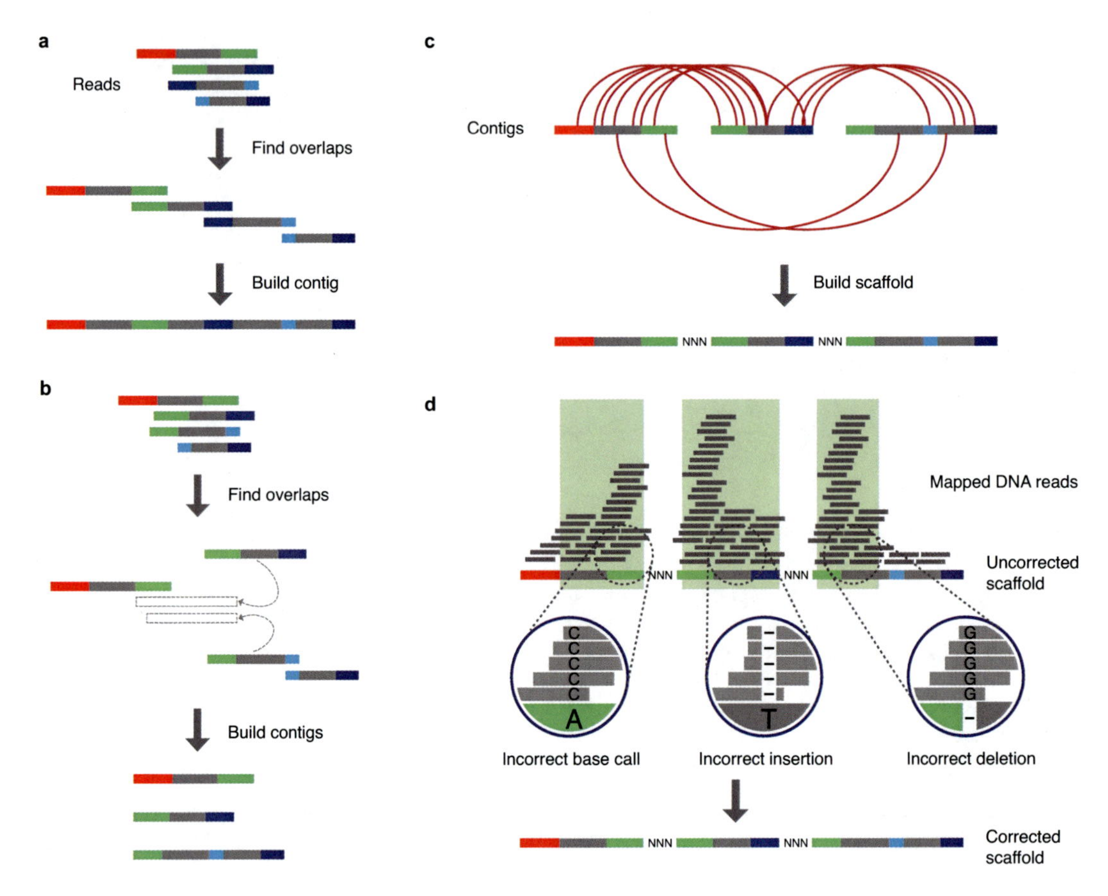

Fig. 1 Assembly approaches and sequence correction. (**a**) Unique overlaps between the reads allow to reconstruct a much longer contig. (**b**) Repetitive sequences complicate the assembly as the unique overlapping reads cannot be identified. (**c**) Chromosome conformation capture allows to determine of the spatial orientation of the contigs within the scaffold but cannot fill the missing sequences between the contigs. (**d**) Simple mistakes such as incorrect base calls, insertions, and deletions can be corrected using short reads in the areas with sufficiently deep coverage (shaded in green)

structure called a *scaffold* using high-throughput *chromosomal conformation capture* (Hi-C) [9–12]. In brief, it captures interactions between genomic regions, while regions that are close to one another in the genome interact more frequently than distant regions. Thus, the linear order of the contigs in a scaffold can be determined based on the interaction frequency [6, 13] (Fig. 1c). Long read sequencing, such as PacBio technology [14], used for genome assembly is characterized by a relatively high error rate. Therefore, after assembly, we sequenced the genome at 30× coverage with short Illumina reads which have a high accuracy rate to correct simple mistakes such as incorrect base calls, insertions, and deletions (Fig. 1d). The reads were aligned to the genome and for each position, which was overlapped by at least 5 reads, we generated a consensus base call based on the read sequences. This allowed to reduce the error rate to less than 1%. However, it is important to emphasize that did not eliminate every possible mistake in the genome due to the fact that some positions did not have enough sequencing depth due to technical reasons (e.g., high local GC content).

As is often the case with much sequencing data, the correction data can be re-used for quite different purposes to get additional insights. For example, one can inspect polymorphic sites across the genome. Additional types of versatile data such as RNA-seq of different issues that shed light on different cellular events in a variety of conditions can now be produced at a relatively low cost and be integrated with the raw genomic sequences to facilitate further analyses. Those data can be viewed in a number of interactive graphical applications such as the *Integrative Genome Viewer* (IGV) [15] from the Broad Institute or the *UCSC genome browser* [16].

2 Methods

In this chapter, we will introduce the instance of the UCSC genome browser containing the axolotl genome data, which can be accessed at https://genome.axolotl-omics.org [17]. We will show how multiple available data such as gene annotation, RNA-seq, Hi-C, and correction data can be integrated and used to approach complex questions. To demonstrate the potential of the graphical viewers, we will first examine the neighborhood of a gene (*Fibroblast Growth Factor 8*, *Fgf8*) and compare it to the homologous human locus. Then we will use the same methods to look at the neighborhood of a gene that was shown to be missing in axolotl—*Paired box 3* (*Pax3*) [4] and use the repeat annotation and mapped RNA-seq data to further investigate the fate of one of those neighboring genes—*Phenylalanyl-TRNA Synthetase Subunit Beta* (*Farsb*), which was found on a different chromosome. Note that the examples below only touch the tip of the iceberg, since a number of different sequencing data such as methylation data [18], *Assay for Transposase-Accessible Chromatin* data (ATAC [19]), and *chromatin immunoprecipitation sequencing* data (ChIP-seq) [20] as well as non-sequencing annotation data such as *CpG islands* [21], which are often associated with promoter regions, or multiple sequence alignment data [22] that is useful for cross-species comparisons are generated at a high rate and can be dynamically added to the genome browser.

2.1 A Practical Guide to Exploring the Axolotl Genome Using the UCSC Browser

The UCSC genome browser offers three releases of the axolotl genome (Fig. 2a). It is always advisable to use the latest release (currently, v6.0-DD) unless the additional data mentioned above was prepared using an older genome release.

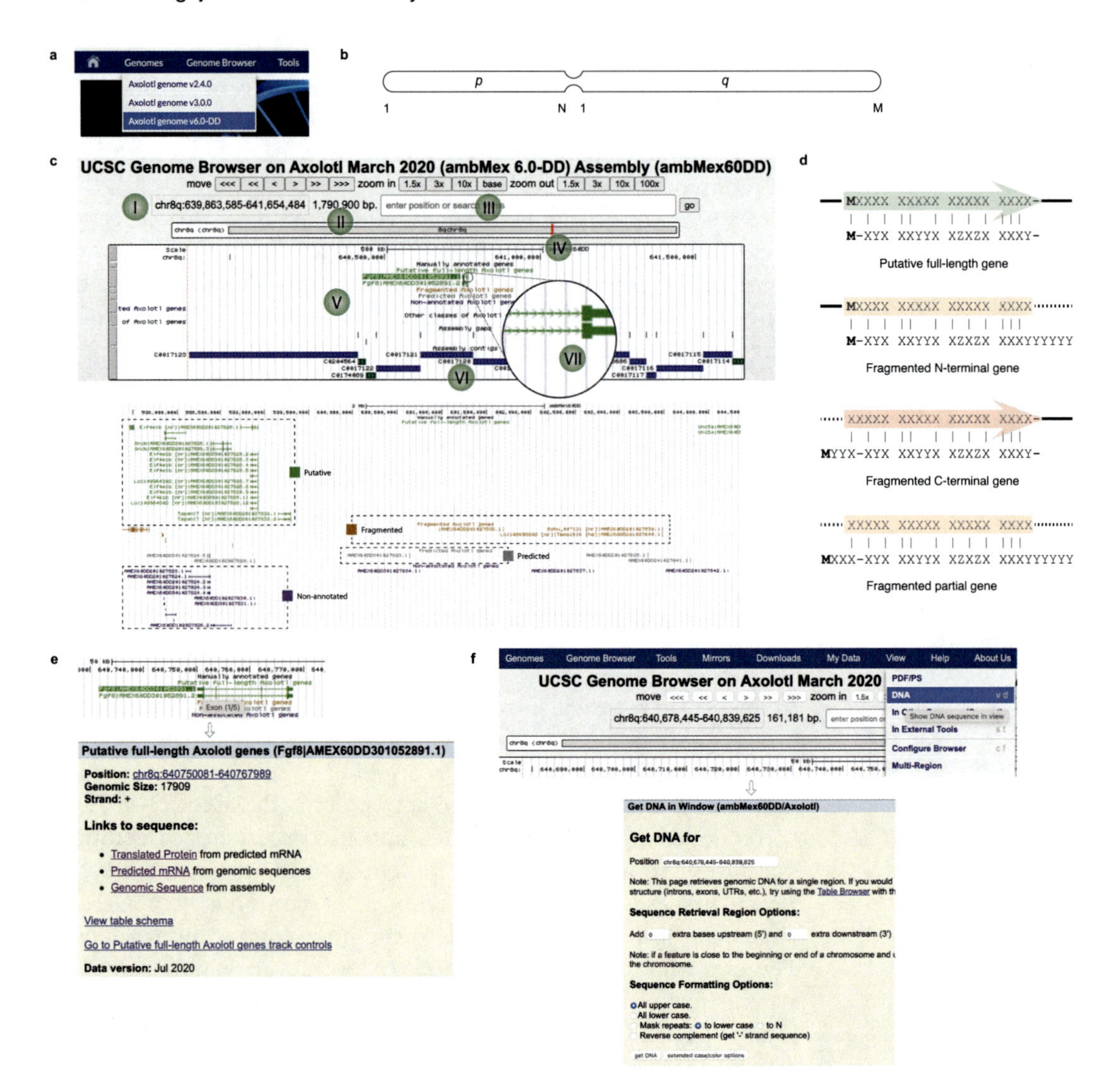

Fig. 2 Genome browser overview and naming conventions. (**a**) Currently, three genome releases are available. (**b**) Chromosome arm naming convention–the shorter arm is called *p*, while the longer *q*. The coordinates within an arm always start with 1. The total chromosome length is the length of both arms (p: 1-N and q: 1-M – N + M) combined. (**c**) **top**. Standard genome browser view displays the genomic coordinates (I), the length of the displayed region (II), the search/navigation box (III), and the position within the chromosome arm (IV). Additionally, one or more tracks are shown, e.g., the annotated genes (V) and the contigs that constitute the displayed region of the scaffold (VI). The orientation of the gene with respect to the chromosome arm is indicated by arrows (VII) **bottom**. Different classes of genes. Green, putative; orange, fragmented; gray, predicted; blue, non-annotated. See the main text for details. (**d**) Different classes of genes that have homologs. Genes that have both the starting methionine and a terminal codon are labeled as putative full length. Those that lack either of those or both are labeled as fragmented. Solid lines indicate parts of the transcript contained in the transcript contig, while the dashed lines indicate the portions of the real transcript missing from the contig. Colored rectangles indicate open reading frames, while the letters exemplify alignments to putative homologous sequences in other species (vertical bar – match, space – mismatch). (**e**) Fetching the gene sequence. (**f**) Fetching the DNA sequence of a user-defined region

In the default configuration, the UCSC genome browser only displays the current position within the genome at the top, the outline of the chromosome arm, identified genes in the middle, and the contigs that constitute the scaffold at the bottom. Shorter chromosome arms (*p*) are oriented telomere-to-centromere, while longer chromosome arms (*q*) are oriented in a centromere-to-telomere fashion (Fig. 2b).

The genes are displayed as five differentially colored tracks below the chromosomal representation (Fig. 2c). The classes are defined based on whether the genes therein have homologs in other organisms and on the properties of the *Open Reading Frame* (ORF) yielded by the homologous alignment. The most abundant class (green, e.g., *Fgf8* in Fig. 2b) contains the genes that have homologs and a full-length ORF (Fig. 2d). The genes that have homologs, but lack the starting methionine, the stop codon, or both (Fig. 2d) are classified as fragmented (orange). Predicted genes (gray) do not have any homologs in other species but have an ORF that is at least 150 amino acids long. Non-annotated genes (dark blue) have neither a homolog nor a long enough ORF. Finally, all other genes, for instance, those that have a disrupted or an incomplete ORF and, therefore, are likely pseudogenes, are collectively called "Other" and are black.

All five categories can be searched by putting either a gene symbol or the transcript ID from www.axolotl-omics.org into the position box (Fig. 2c, III) at the top of the page. The genome browser offers the possibility to zoom in and out to see a more detailed representation of the selected region or to explore the neighborhood of the gene of interest, respectively.

2.1.1 Retrieving the DNA Sequence

If you need to retrieve the sequence of a gene, simply click on it and select the type of the sequence you want to get (Fig. 2e). If you need to fetch the genomic sequence outside of the gene, select the region you are interested in and click **View** > **DNA** in the top menu. This will return the DNA sequence of the region you are currently viewing (Fig. 2f). In the example shown in the figure, this will extract 161,181 bp from the genome.

2.1.2 Exploring the Gene Neighborhood

One particularly important gene in the field of limb development and regeneration is the *fibroblast growth factor 8*, *Fgf8*. It has been reported that the *FGF8* locus in humans is inverted with respect to its homologs in other vertebrates [23]. The following example demonstrates how to use the genome browser to compare the *Fgf8* loci in axolotl and humans.

1. Navigate to https://genome.axolotl-omics.org and click **Genomes** > **Axolotl genome v6.0-DD** in the top menu (Fig. 2a).

2. Search for the gene by typing **Fgf8** into the position box (Fig. 2c, III). This will bring you to the list of genes that match the query. Click the most suitable entry (*see* **Note 1**) to navigate to that gene.
3. The gene page (Fig. 2c) displays the outline and the orientation (arrows, Fig. 2c, VII) of the gene within the chromosome arm.
4. Drag with the mouse cursor along the top margin of the viewer box to highlight the gene region (blue highlight in Fig. 3a). After that, a dialog is displayed that allows to either zoom into the selected region or highlight it with a color (Fig. 3a). Select a color, click "**Choose**" and "**Add highlight.**"
5. Zoom out approximately 1000-fold to see the neighboring genes and highlight *Npm3* and *Poll* in yellow and orange, respectively (Fig. 3b, top). In case your view is different, please refer to **Note 2**.
6. Do the same for the human homologs at https://genome.ucsc.edu (Fig. 3b, bottom).
7. Subsequent analysis of the locus reveals that the locus between *Slf2* and *Pprc1* is inverted in humans with respect to their order in axolotl and some fish (Fig. 3c), while the relative orientation of the genes within the locus did not change.

2.1.3 Missing Genes and Hi-C Data

Imagine, you need to analyze the neighborhood of the gene that was shown to be missing in axolotl – *Paired Box 3* (*Pax3)* [4]. First, you need to find the syntenic locus in the axolotl genome. To do so, open the human genome browser (https://genome.ucsc.edu) and navigate to the *PAX3* gene. Since the gene is missing in axolotl, one needs to look for the neighboring genes of the human *PAX3* in the axolotl genome. Ten of thirteen neighboring genes in humans are located in close proximity on the axolotl chromosome arm 10q (Fig. 4a). However, the axolotl locus appears reversed with respect to the human as is indicated by the relative position of the genes *EPHA4*, *ACSL3*, *AGFG1*, and *KIF1A1* (Fig. 4a, yellow, pink, blue, and green, respectively).

1. Open the UCSC genome browser and locate the human *PAX3*.
2. Select some 10 neighboring genes that would allow you to confidently define the syntenic region in the axolotl and search for the homologs of those neighboring genes in the axolotl genome browser.
3. You will find that *Epha4* exists in the axolotl (Fig. 4a, yellow). Highlight *Epha4* as described in Subheading 2.1.2 and continue with the next human gene – *SGPP2*.

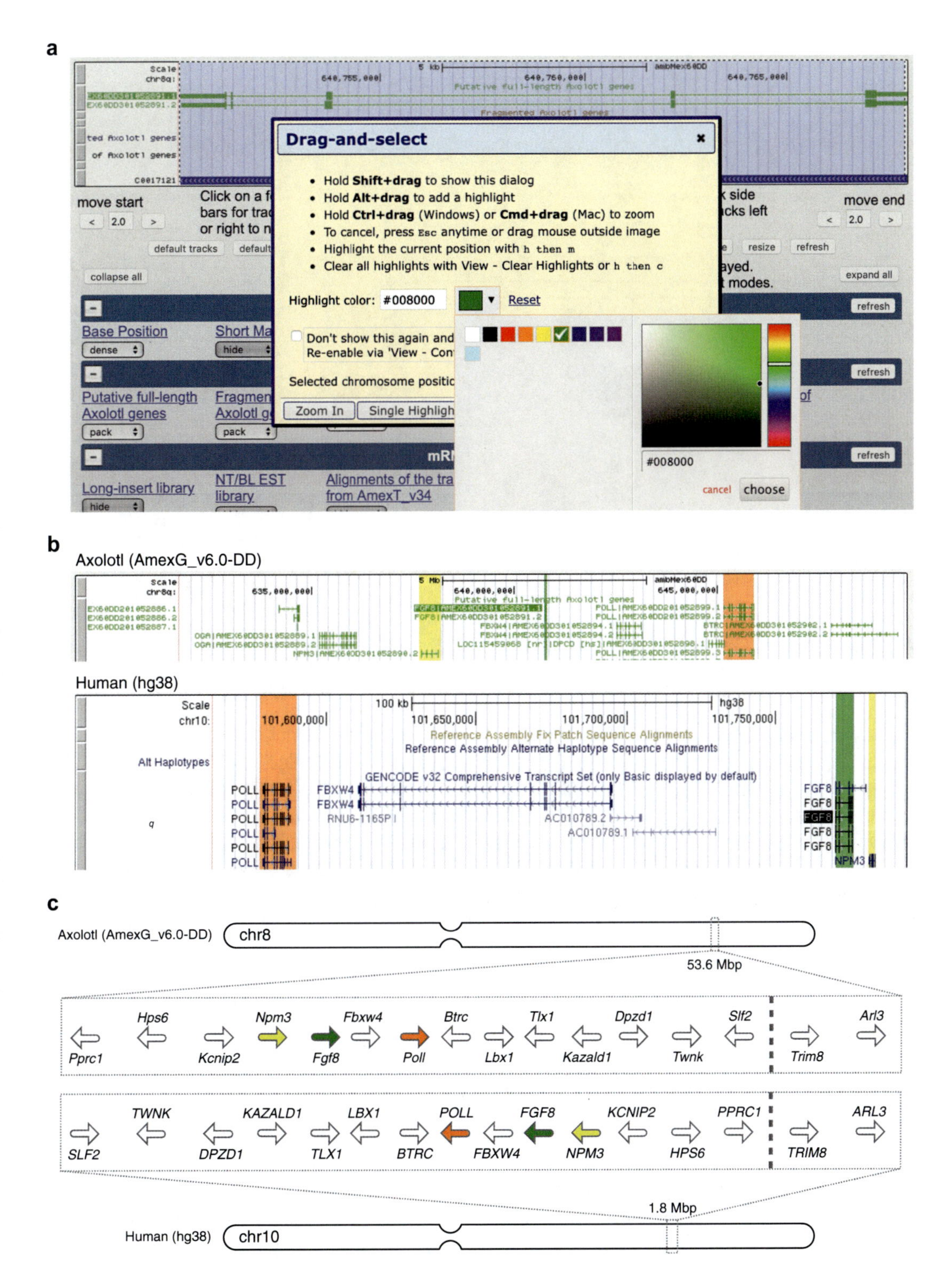

Fig. 3 Synteny analysis. (**a**) Highlighting a region. (**b**) Highlighting syntenic regions in axolotl and humans. (**c**) Differences in the orientation of the loci in axolotl and human. The dashed line indicates the start of the region where the gene order is the same again

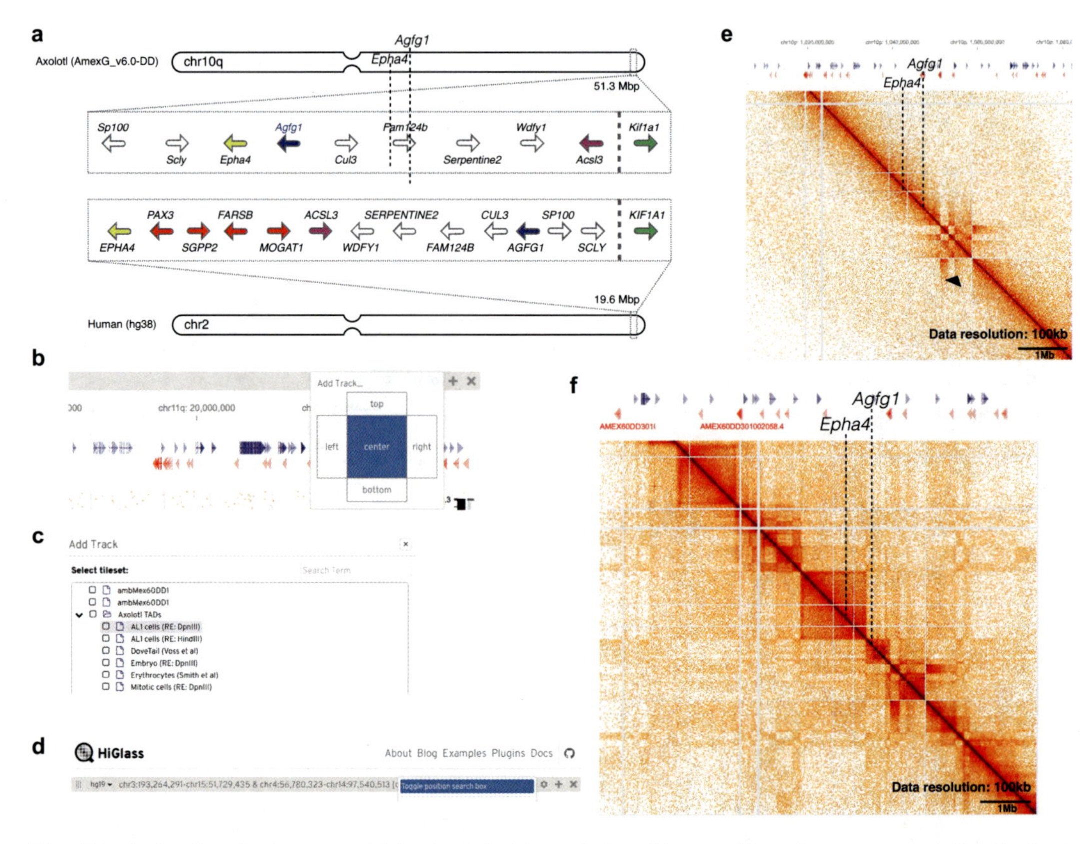

Fig. 4 Analysis of a missing gene. (**a**) Syntenic loci in axolotl and human. Homologous genes *Epha4* (yellow) and *Agfg1* (blue) flank the missing region in the axolotl. Human genes highlighted in red are missing from the syntenic locus in axolotl. ACSL3 (pink) and KIF1A1 (green) are neighbors in the axolotl, but far apart in humans. (**b**) Adding a track to HiGlass. (**c**) Selecting a Hi-C dataset. (**d**) HiGlass position search box. (**e**) Hi-C map of a mitotic dataset. The thick bright continuous diagonal line indicates the contacts between adjacent genomic loci. The dashed lines indicate the positions of *Epha4* and *Agfg1*. The arrowhead shows a potential misassembled position. Data resolution: 100 kb per pixel, scale bar – 1 Mb. (**f**) Hi-C map of an interphase AL1 dataset. Data resolution: 100 kb per pixel, scale bar – 1 Mb

4. *Sgpp2* (Fig. 4a, red) is located on a different chromosome in axolotl. Keep this in mind and do the same search with the remaining genes.
5. The axolotl locus that contains most of the neighboring human genes is most likely the true syntenic region.
6. Ultimately, you will find that *Epha4* and *Agfg1* are neighbors in the axolotl, while they are quite far away from one another in the human genome (Fig. 4a).
7. The whole locus between *Agfg1* and *Acsl3* is flipped in the axolotl with respect to the human (Fig. 4a, blue and pink, respectively), while *Kif1a1* (Fig. 4a, green) is still in place in the axolotl. Thus, the inversion must have happened upstream of it as indicated by the dashed line in Fig. 4a.

8. In the axolotl genome browser, highlight *Agfg1* and zoom out so that the region flanked by *Epha4* and *Agfg1* is visible.Click on the box with the coordinates (Fig. 2c, I) and copy the coordinates to the clipboard.

As it was shown above, the axolotl locus appears to be flipped. Since the ancestral order of the whole locus is not known, you can check other species. A comparison with the chicken genome reveals that the gene order and orientation in the human genome correspond to the ancestral, while in the axolotl, the locus underwent an internal inversion. To check if that inversion is an assembly artifact or a biological fact, one should look at the Hi-C data [9, 11, 12, 24, 25], which are meanwhile available for axolotl.

1. Navigate to the web-based Hi-C data viewer HiGlass [26] at https://higlass.axolotl-omics.org.
2. Click on the plus symbol in the top right corner and select the central panel in the pop-up window (Fig. 4b).
3. In the displayed dialog (Fig. 4c), click on the triangle next to "Axolotl TADs" to expand the list, select **"AL1 cells (RE: DpnII)"** and hit **"Submit"** (*see* **Note 3** for more details).
4. Additional data such as chromosome labels or gene annotations can be added similarly by clicking the plus symbol, selecting one of the side panels (usually, top or left), and choosing the appropriate dataset, for example, the latest gene annotations – "AmexT_v47".
5. Navigate to the locus of interest either by copying the coordinates from the genome browser as described above (**step 9** of previous instructions, Fig. 2c,l)) or by entering the gene symbol into the position search box. If the box is not displayed, click on the gear symbol and select **"Toggle position search box"** (Fig. 4d).

As can be seen in Fig. 4e, there is a thick contiguous diagonal line indicating that the genome in the region of interest was assembled correctly along the chromosome because genomic regions only contact their neighbors and not the distant regions, which would appear as the off-diagonal signal. Such an assembly anomaly can be seen in Fig. 4e (arrowhead) downstream of *Agfg1*, while the region around *Epha4* and *Agfg1* is correct.

Fig. 4f shows the real biological data (AL1) in the same region. *Epha4* and *Agfg1* belong to two different *topologically associating domains* (TADs).

2.2 Looking for Missing Genes

There is no evidence of a major mis-assembly, almost all genes are conserved between the axolotl and the human. The genomic region between *Acsl3* and *Scly* (see Subheading 2.1.3 and Fig. 4a) seems to be inverted in the axolotl such that *Acsl3* ended up very close to the end of the chromosome. It is likely that the four genes *Pax3*, *Sgpp2*, *Farsb*, and *Mogat1* were lost or translocated during this event.

1. Search for each of the neighboring genes that are missing from the locus (*Sgpp2*, *Farsb*, and *Mogat1*) by gene symbol to check whether they can be found on a different chromosome.
2. Since the genes appear to be missing, it is likely that the search does not yield any results. However, in order to make sure the gene is indeed missing and is not simply mis-annotated or non-annotated, you need to use the protein sequences of the respective human genes to run a BLAST search.
   (a) Navigate to https://www.axolotl-omics.org/tblastn and paste the protein sequence into the query box.
   (b) You can run the blast search against the transcriptome (preferred option), set of contigs, or chromosome scaffolds (Fig. 5a). Due to the large size of the genome, the search against the scaffolds is likely to fail.

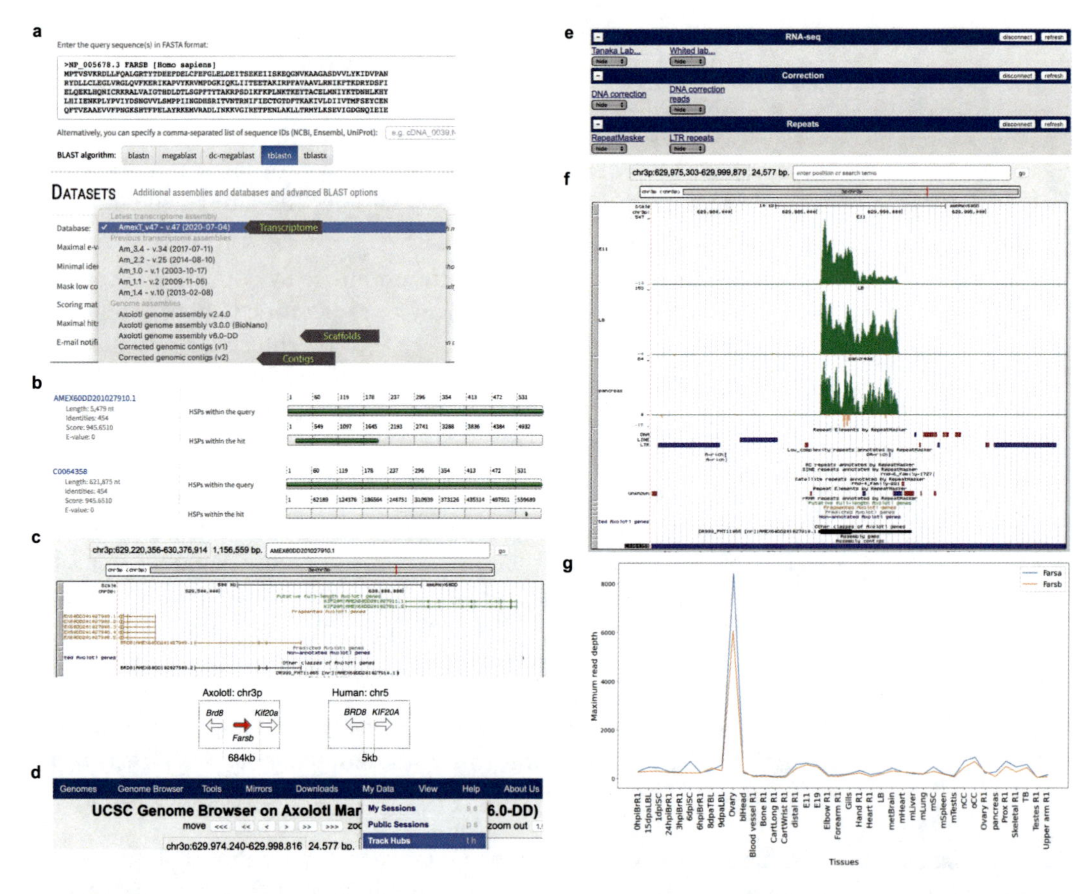

Fig. 5 Verification of the gene identity. (**a**) Blast options and databases. (**b**) Blast hits. Top: transcript hit, bottom: contig. (**c**) Location of the *Farsb* integration in the axolotl genome and the structure of the homologous human locus. (**d**) Adding a track hub. (**e**) Track hubs view. (**f**) Expression of *Farsb* demonstrated by RNA-seq data. (**g**) Expression comparison of *Farsa* and *Farsb*

(c) If the search against the transcriptome does not find any hit, it is recommended to run a blast against the set of genomic contigs (also *see* **Note 4** for more details).

3. The name of the identified homologous transcript (Fig. 5b, top) or the contig (Fig. 5b, bottom) can be pasted into the position box of the genome browser to further explore its neighborhood.

In the case of *Farsb*, a homologous transcript AMEX60DD201027910.1 was found, which is annotated as DR999_PMT11065. However, the detailed information on this ID on NCBI lists the *phenylalanyl-tRNA synthetase beta chain* as the region, which is the full name of *Farsb* (Fig. 5c). A further examination of its neighboring genes reveals that the insertion happened in a large 684-kb fragment between the genes *BrdB* and *Kif20a*, which otherwise retained their relative orientation (Fig. 5c). Interestingly, the human gene contains 17 exons, while the axolotl homolog is a monoexonic gene suggesting that this is not a true homolog of the human *FARSB*, but rather a retrotransposition event.

2.2.1 Track Hubs

In order to gain further insights into what happened to *Farsb* in the axolotl, it is necessary to take a closer look at the annotated repeats and the mRNA expression profiles. Both of those can be added to the genome browser as *track hubs* (*see* **Note 5**) – web-accessible genomic data that can be viewed in the genome browser [16].

1. Navigate to **My Data** > **Track Hubs** in the menu at the top of the window and further to the tab **My Hubs** (Fig. 5d).
2. For the further steps, we will need the RNA-seq expression data and the repeat annotation. Paste the URL https://www.axolotl-omics.org/trackhubs/AmexG_v6.0-DD/RNAseq/hub.description into the URL box and hit "Add hub."
3. To add the repeat annotation track, repeat the same procedure using the URL https://www.axolotl-omics.org/trackhubs/AmexG_v6.0-DD/repeats/hub.description

Afterward, return to the genome view. The added track hubs are displayed at the bottom of the screen. You can choose individual categories that should be displayed, e.g. only the RepeatMasker [27] data, but not the *Long Terminal Repeat* (LTR) or correction data (Fig. 5e). You can also select individual datasets, i.e., tissues, by clicking on the group name in the RNA-seq category, e.g., "Tanaka Lab."

In Fig. 5f, you can see that the minigene sequence is highly expressed as mRNA in a number of tissues such as the stage 11 embryo [28], limb bud, and pancreas. Moreover, the locus is flanked by LTRs and *Long Interspersed Nuclear Elements* (LINEs), which can be the potential driving force behind the retrointegration. Although the integrated cDNA does not have its original promoter, it is likely that it uses the promoter of the nearby LINE or LTR element. Moreover, the overall structure of the transcript – a very short *5' Untranslated Region* (UTR) and a long 3' UTR is very similar to that of the human homolog. Interestingly, *Farsb* has been previously shown to be up-regulated in the limb during regeneration [29]. The FARSB protein comprises the beta subunits of a tetramer that also contains two catalytic alpha subunits. Thus, if the retrotransposed version of *Farsb* is indeed carrying out the same function as the original gene, one would expect that both subunits are up-and down-regulated similarly. One can use the maximum read depth (Fig. 5f, Y-axis) in each sample as a very approximate proxy for the expression level. The expression profiles of both *Farsa* and *Farsb* are strikingly similar suggesting that the expression of both is co-regulated (Fig. 5g). Although the described approach is only valid as a rough estimate, a proper statistical analysis using the FPKM values [30] (fragments per kilobase of exon model per million reads mapped) shows the same pattern with the Pearson correlation of 0.87.

Although it is possible to study many aspects of genome evolution and organization with the current version of the genome, it is important to keep in mind that the genomic sequence is being constantly improved and may still contain residual errors, especially insertions and deletions, that might cause frameshifts and, thus, have great implications on the analysis results. In the case shown in Fig. 6a, the genomic sequence was corrected using the approach outlined in Fig. 1d in a first-round where approximately 5x coverage of Illumina reads was available, but still had a frameshift in the coding sequence. The frameshift occurred at the junction between exon 4 and exon 5. Although at the first glance, the intron recognition sites (GT – AG) are correct and each position is covered by 5 reads, it first remained unnoticed that a G was missing at the very beginning of exon 5 due to an unfortunate property of that piece of sequence. All correction reads ended in the stretch of 3 Gs (Fig. 6a, dashed rectangle) and, thus, the missing G (Fig. 6a, red) could not be spotted. However, the alignment to a homologous protein revealed the missing G that was required to restore the ORF (Fig. 6a, bottom). This example illustrates that manual inspection of the region of interest and its correction data is crucial.

1. Navigate to **My Data** > **Track Hubs** in the menu at the top of the window and further to the tab **My Hubs** (Fig. 5d).
2. Add the correction data track from https://www.axolotl-omics.org/trackhubs/AmexG_v6.0-DD/correction/hub.description.

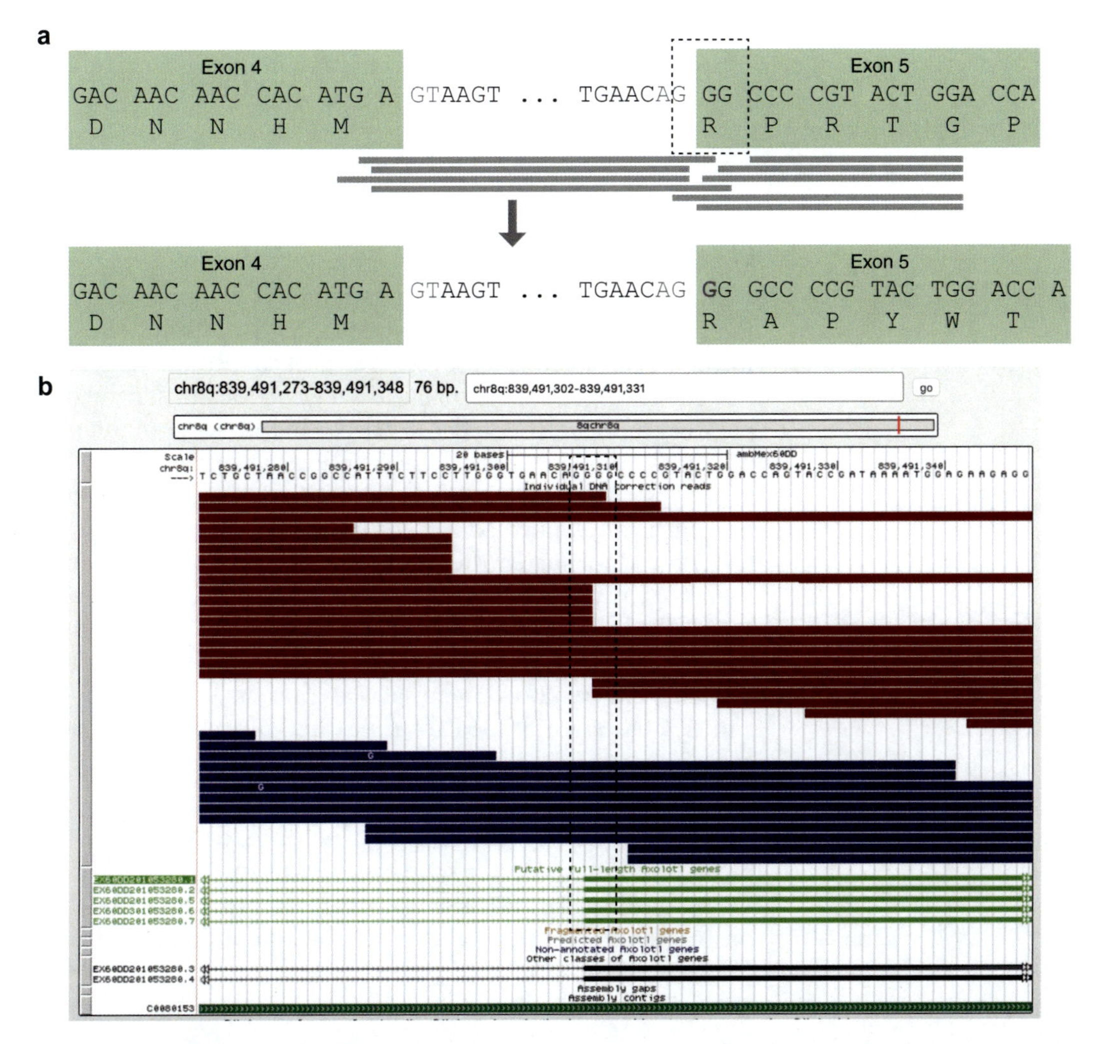

Fig. 6 Frameshift visualization. (**a**) **top**. Frameshift and mapped correction data for axolotl *Farsb*. The dashed rectangle indicates the region that should contain an additional G in exon 5 (green rectangle) in order to maintain a valid open reading frame, but could not be corrected because none of the correction reads spans the entire region. Blue GT and AG indicate the intron recognition sites, gray bars represent correction reads. **Bottom**. Corrected sequence. Red G is still missing in the corrected sequence due to the lack of reads properly covering the region. (**b**) Correction data track displaying the reads on the leading strand in dark blue and those on the lagging strand in dark red. Yellow arrowheads indicate bases in the reads that differ from those in the reference. Dashed rectangle indicates the corrected region.

The track displays individual reads that overlap the currently viewed region and visually separates them into forward, i.e. mapping to the leading strand in dark blue, and reverse in dark red. With the availability of a greater depth of the correction data (Fig. 6b), one can see that the locus was corrected properly (also *see* **Note 6**), and the missing base was inserted. Any mismatches in the reads with respect to the reference are indicated by the bases in the reads track (Fig. 6b, yellow arrowheads).

Mismatches in individual reads are most likely due to sequencing errors, while if the same mismatch is present in multiple reads, this may indicate either an assembly artifact or a polymorphic site.

3 Notes

1. Search results.

   If the query matches the annotation of multiple genes or the matching gene has multiple annotated isoforms, the isoforms are displayed as a list (Fig. 7a) broken into the respective categories (Fig. 2c,d). Usually, it is a good idea to start with a putative full-length entry if it exists. The isoform annotation may contain two different gene symbols with either "**[nr]**" or "**[hs]**" as suffix. This is the case if the annotation of the overall best hit in the entire NCBI non-redundant BLAST database (NR) does not match the gene symbol of the closest human homolog. If the suffix "[hs]" is not present, then there is no human homolog, while if none of the suffixes is present, then the gene symbol is the same in both NR and human. The human gene symbols are displayed since many other tools and databases are primed to using the human annotation in case you need to use them.

2. Adjusting the View.

   If your view is different from what is described in the procedure, make sure the scale bar is 5 Mb and that the display mode for the gene track is set to "pack." In order to do so, click on the track with the right mouse button and select "pack" from the context menu (Fig. 7b).

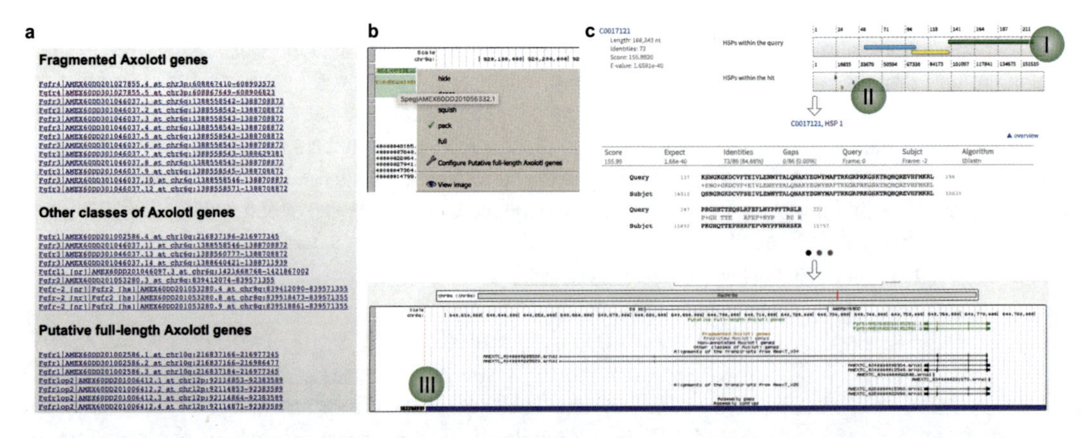

Fig. 7 Additional information. (**a**) Search results – if multiple genes match the query, they are displayed in their respective annotation category. (**b**) Switching the track view mode and using the tooltips to see the full gene name. If the gene is annotated, then the gene symbol (*Speg* in this example) is separated from the transcript ID by a vertical line. If the gene is not annotated, only a transcript ID is displayed. (**c**) tBLASTn results. The BLAST output displays the set of color-coded aligned regions in both the query (I) and the hit (II). In the genome browser, the genomic contig is displayed at the bottom (III) and the arrowheads indicate the orientation.

Additionally, if the gene name does not fit into the view completely, it is truncated and displayed at the left margin. Hover with the mouse cursor to see the full name (Fig. 7b).

3. Choosing the Hi-C Dataset.

   In the case of a gene regulation question, one of the AL1 [31] samples or the embryonic sample is usually the right choice. The difference between them is the *restriction enzyme* (RE) that was used to generate the fragments for sequencing. Usually, a 4-cutter gives a better resolution than a 6-cutter. However, to check for assembly artifacts, one should use the mitotic sample as it does not contain any TADs that would bias the result.

4. Searching for a Gene of Interest.

   The annotation of the genome was done using the transcriptome data. Therefore, tissue-specific, lowly expressed or otherwise problematic transcripts may either be incomplete (Fig. 2d) or missing. In this case, BLAST will not yield any hits in the transcriptome, either. A more laborious and yet often successful approach is to run a tBLASTn search against the set of genomic contigs (Fig. 5a) using the protein sequence of interest as a query. In this case, one would need to carefully examine the BLAST output and find the regions. For example, assume you are interested in finding *Fgf8* (UniProt ID P55075). There are multiple tBLASTn hits (Fig. 7c), while the first one (Fig. 7c, top panel) shows several consecutive aligned regions within the query and are in the same order (as can be seen by color-coding) in the genomic contig. Thus, they likely represent individual exons. The alignment (Fig. 7c, middle panel) also looks good. You can then copy the contig name and paste it into the search box in the genome browser. This brings you to the genomic contig and you can see that the contig is oriented backward within the scaffold (Fig. 7c, bottom panel) with *Fgf8* being on the right-hand side, which fits well to the relative orientation of the contig and *Fgf8* on the BLAST output.

5. Available Track Hubs.

   Currently, available track hubs for each genome assembly are listed in the respective assembly section at https://www.axolotl-omics.org/assemblies. Note that the URLs are only accessible to the genome browser and, thus, cannot be opened in a browser.

6. Advanced Usage of the Correction Data.

   The use of the correction reads track is not limited to frameshift detection. It can also be employed to check if, for instance, premature terminal codons are real or just assembly artifacts or even to see if certain sites are polymorphic. For the latter case, it is important to bear in mind that both primary sequencing reads and the correction reads come from the same animal and, thus, by no means reflect the full repertoire of possible polymorphisms.

References

1. Elegans Sequencing C (1998) Consortium, genome sequence of the nematode C. elegans: a platform for investigating biology. Science 282:2012–2018
2. Mcpherson JD, Marra M, Hillier L et al (2001) A physical map of the human genome. Nature 409:934–941
3. Venter JC, Adams MD, Myers EW et al (2001) The sequence of the human genome. Science 291:1304–1351
4. Nowoshilow S, Schloissnig S, Fei J-F et al (2018) The axolotl genome and the evolution of key tissue formation regulators. Nature
5. Smith JJ, Timoshevskaya N, Timoshevskiy VA et al (2019) A chromosome-scale assembly of the axolotl genome. Genome Res 29:1–8
6. Schloissnig, S., Kawaguchi, A., Nowoshilow, S., et al., The giant axolotl genome uncovers the evolution, scaling and transcriptional control of complex gene loci, PNAS under review (2021)
7. Rice ES, Green RE (2019) New approaches for genome assembly and scaffolding. Annu Rev Anim Biosci 7:17–40
8. Staden R (1980) A new computer method for the storage and manipulation of DNA gel reading data. Nucleic Acids Res 8:3673–3694
9. Imakaev M, Fudenberg G, Mccord RP et al (2012) Iterative correction of hi-C data reveals hallmarks of chromosome organization. Nat Methods 9:999–1003
10. Van Berkum NL, Lieberman-Aiden E, Williams L et al (2010) Hi-C: a method to study the three-dimensional architecture of genomes. J Visual Exp JoVE
11. Lieberman-Aiden E, Van Berkum NL, Williams L et al (2009) Comprehensive mapping of long-range interactions reveals folding principles of the human genome. Science 326: 289–293
12. Dekker J, Rippe K, Dekker M et al (2002) Capturing chromosome conformation. Science 295:1306–1311
13. Meyer A, Schloissnig S, Franchini P et al (2021) Giant lungfish genome elucidates the conquest of land by vertebrates. Nature
14. Eid J, Fehr A, Gray J et al (2009) Real-time DNA sequencing from single polymerase molecules. Science 323:133–138
15. Robinson JT, Thorvaldsdottir H, Winckler W et al (2011) Integrative genomics viewer. Nat Biotechnol 29:24–26
16. Kent WJ, Sugnet CW, Furey TS et al (2002) The human genome browser at UCSC. Genome Res 12:996–1006
17. Nowoshilow S, Tanaka EM (2020) Introducing www.axolotl-omics.org – an integrated -omics data portal for the axolotl research community, Experimental Cell Research 394
18. Macdonald WA (2012) Epigenetic mechanisms of genomic imprinting: common themes in the regulation of imprinted regions in mammals, plants, and insects. Genet Res Int 2012: 585024
19. Buenrostro JD, Giresi PG, Zaba LC et al (2013) Transposition of native chromatin for fast and sensitive epigenomic profiling of open chromatin, DNA-binding proteins and nucleosome position. Nat Meth 10:1213–1218
20. Johnson DS, Mortazavi A, Myers RM et al (2007) Genome-wide mapping of in vivo protein-DNA interactions. Science 316: 1497–1502
21. Hartl DL, Jones EW (2005) Genetics : analysis of genes and genomes. Jones and Bartlett Publishers, Sudbury
22. Blanchette M, Kent WJ, Riemer C et al (2004) Aligning multiple genomic sequences with the threaded blockset aligner. Genome Res 14: 708–715
23. Marinić M, Aktas T, Ruf S et al (2013) An integrated holo-enhancer unit defines tissue and gene specificity of the Fgf8 regulatory landscape. Dev Cell 24:530–542
24. Belton J-M, Mccord RP, Gibcus JH et al (2012) Hi-C: A comprehensive technique to capture the conformation of genomes. Methods (San Diego, Calif.) 58:268–276
25. Dekker J, Marti-Renom MA, Mirny LA (2013) Exploring the three-dimensional organization of genomes: interpreting chromatin interaction data. Nat Rev Genet 14:390–403
26. Kerpedjiev P, Abdennur N, Lekschas F et al (2018) HiGlass: web-based visual exploration

and analysis of genome interaction maps. Genome Biol 19:125

27. Tarailo-Graovac, M., Chen, N., Using Repeat-Masker to identify repetitive elements in genomic sequences, Curr Protoc Bioinformatics Chapter 4 (2009) Unitas 4 10
28. Bordzilovskaya NP, Dettlaff TA, Duhon ST (1989) Developmental-stage series of axolotl embryos
29. Monaghan JR, Athippozhy A, Seifert AW, et al. (2012) Gene expression patterns specific to the regenerating limb of the Mexican axolotl, Biology Open BIO20121594
30. Mortazavi A, Williams BA, Mccue K et al (2008) Mapping and quantifying mammalian transcriptomes by RNA-Seq. Nat Methods 5: 621–628
31. Roy S, Gardiner DM, Bryant SV (2000) Vaccinia as a tool for functional analysis in regenerating limbs: ectopic expression of Shh. Dev Biol 218:199–205

Chapter 20

Chromosome Conformation Capture for Large Genomes

Akane Kawaguchi and Elly M. Tanaka

Abstract

The gigantic 32Gb Axolotl genome inspires fascinating questions such as: how such a big genome is organized and packed in nuclei and how regulation of gene transcription can happen over such large genomic distances. Currently, there are many technical challenges when we investigate chromatin architecture in axolotl. For example, probing promoter–enhancer interactions in such a large genome. Chromatin capture methods (e.g., Chromatin Conformation Capture) have been used in a variety of species. The large size of the axolotl nuclei and its genome requires the adaptation of such methods. Here, we describe a detailed protocol for high-throughput genome-wide conformation capture (Hi-C) using axolotl limb cells. This Hi-C library preparation protocol can also be used to prepare libraries from other nonmodel organisms such as Lungfish and Cephalopods. We believe that our protocol could be useful for a variety of animal systems including other salamanders.

Key words Huge chromosome, Chromatin interaction capture, Chromatin assembly

1 Introduction

Axolotl is a well-known animal model for development, evolution, and regenerative biology for which well-established methods like animal husbandry, electroporation, and genome editing are used to probe the genetic and cell biological basis of these processes [1–3]. In the field of genomics, the model poses challenges because the size of an axolotl genome is around 10 times larger than the human genome and contains highly repetitive sequences [4, 5]. However, it has recently been possible to adopt chromosome interaction capture strategies in order to scaffold the 2D linear sequence and decipher the 3D architectural features of the huge axolotl genome [6].

High-throughput genome-wide chromatin conformation capture (Hi-C) reveals All-vs-All genome-wide chromosome conformation contacts which output both "short-range" and long-distance contacts occurring in the genome. The basis of the method starts from the genomic sequences which present proximal

Ashley W. Seifert and Joshua D. Currie (eds.), *Salamanders: Methods and Protocols*,
Methods in Molecular Biology, vol. 2562, https://doi.org/10.1007/978-1-0716-2659-7_20,

distances in nuclear space via protein DNA interactions that are fixed with a fixative solution (e.g., formaldehyde). The chromatin interactions are digested by restriction enzyme and the remaining digested sites are filled in with biotin-NTP. The newly created blunt ends are ligated by DNA ligase. These ligation events are favored between cross-linked fragments regardless of their actual distances in the chromosome, creating chimeric DNA fragments that indicate the physical proximity between two chromosome fragments in the 3D nuclear space. The ligated fragments are randomly sheared into an appropriate size for high throughput sequencing. Following an immunoprecipitation step based on biotin–streptavidin interaction, only the fragments which contain physical interaction between one locus and the other one are collected and applied to high-throughput sequence determination. After sequencing on next-generation sequencing platforms, the reads reflect the contact frequency between one genome locus to the other. The frequency of short-range contacts (seen on the diagonal of Hi-C contact maps) shows a power-law decay with respect to distance from any given genomic location. This feature is used to take genomic contig sequences and scaffold them to chromosome-level contiguous genome assemblies [7, 8]. The long-range contacts reveal the existence of Topological Association Domains (TADs) which are largely cell-type invariant domains of increased association, seen as triangles off of the diagonal on Hi-C maps. Various work has associated the coincidence of TADs with regions where long-distance gene regulation, namely enhancer-promoter interactions are occurring [9].

In order to assemble the 32Gb Axolotl genome and investigate their gigantic TAD structures, we constructed three different Hi-C libraries from the Axolotl culture cell line in the interphase and mitotic phases, and from larval tissue respectively [6]. We collected data from mitotic cells because the density of short-range contacts that are relevant for genome assembly are far higher in mitotic chromosome datasets compared to those from interphase cells. Regarding mitotic chromosomes, it has been known that interphase chromatin interactions are canceled and chromosome structures highly condensed chromatin during mitotic phase and it show only short-range contacts. To synchronize Axolotl cells in the mitotic phase efficiently, we established a two-step cell cycle arrest protocol-taking into account the long cell cycle (4–7 days) [10, 11]. Moreover, we generated a Hi-C library from larvae to correct for chromosome rearrangements that may be associated with cells in long-term culture. Beyond providing short-range contacts for genome assembly, these datasets allowed us, for the first time, to assess the length scales of interphase TADs and mitotic loops in such a gigantic genome [6].

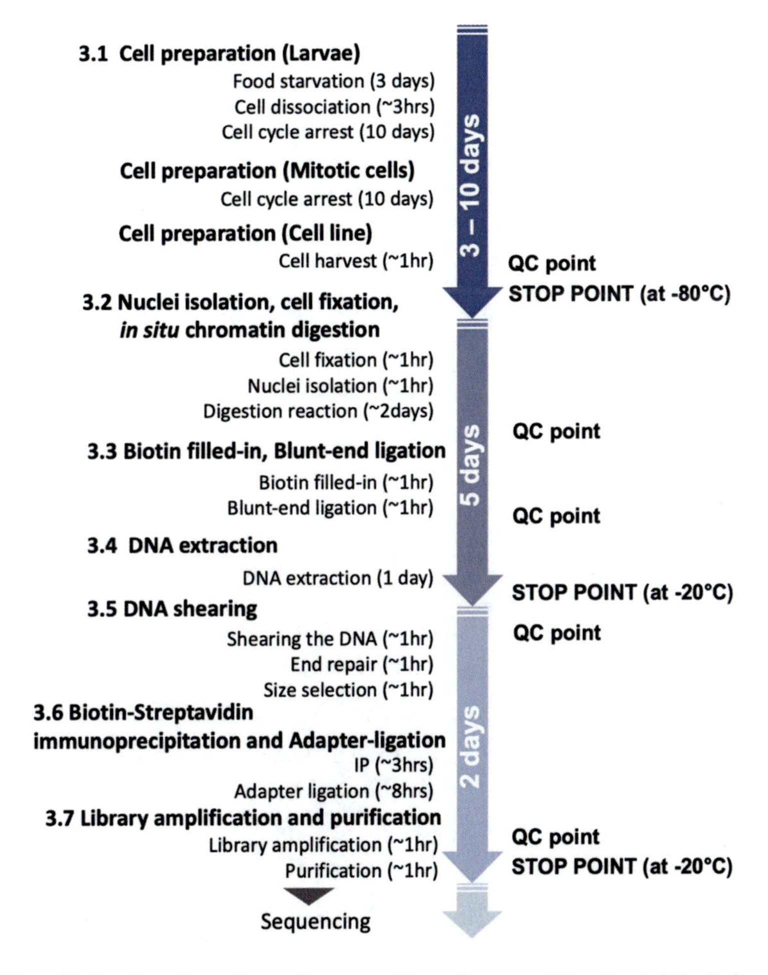

Fig. 1 Estimation date of sample preparations steps, and the stop points during procedures

Here, we describe our experimental experiences pertaining to the preparation of Hi-C libraries from these three different starting samples [Fig. 1]. Our protocol was also successfully used to generate Hi-C libraries for the Lungfish genome assembly that has a genome size of 42Gb [12] and the investigation of 3D chromatin architectures of Squid. We would suggest that our Hi-C protocol is not only useful for many nonanimal Hi-C preparations, but also all types of Chromosome Capture methods.

2 Materials

2.1 Materials for Hi-C Preparations

1. 1 M Tris–HCl (pH 8.0): Mol. Biol. Grade solutions can be used.
2. 5 M NaCl: Mol. Biol. Grade solutions can be used.
3. 10 mM Tris–HCl (pH 8.0): Mol. Biol. Grade solution can be used.

4. 0.5 M EDTA (pH 8.0): Mol. Biol. Grade solution can be used.
5. Phosphate-buffered solution (PBS): Mol. Biol. Grade solution can be used.
6. 37% Formaldehyde: MERCK Millipore, Ca.# 104003.
7. IGEPAL CA-630: SIGMA, Ca.# I8896.
8. Methyl green-pyronin solution: MERCK Millipore, Ca.# HT70116.
9. Trypan blue solution: Thermo Fisher Scientific, Ca.# 15250061.
10. Protease inhibitor Cocktail EDTA free: SIGMA, P8340.
11. 10× NEB2: NEW ENGLAND BioLabs, Ca.# B7002S.
12. 10× NEB2.1: NEW ENGLAND BioLabs, Ca.# B7202S.
13. 1.25× NEB2: e.g., 0.25 mL of 10×NEB2.1 with 1.75 mL of ddH_2O, Beforeprepare fresh before the start of the experiment or make aliquot and store at −80 °C.
14. Phenol:Chloroform:Isoamyl Alcohol: SIGMA, Ca.# P3803.
15. Chloroform: SIGMA, Ca.# C2432.
16. dNTPs: Thermo Fisher Scientific, Ca.# 10297118.
17. Biotin-14-dATP: Thermo Fisher Scientific, Ca.# 19524016 (*See* **Note 1**).
18. Restriction enzyme (*See* **Note 2**)
19. DpnII: NEW ENGLAND BioLabs, Ca.# R0523M.
20. T4 DNA ligase: NEW ENGLAND BioLabs, Ca.# R0202M.
21. T4 DNA polymerase: NEW ENGLAND BioLabs, Ca.# M0203S.
22. T4 DNA polynucleotide kinase: NEW ENGLAND BioLabs, Ca.# M0201S.
23. Klenow Fragment (3′–5′ exo-): NEW ENGLAND BioLabs, Ca.# M0212S.
24. RNase A: MERCK Millipore, Ca.# 11119915001.
25. Glycogen: Roche, Ca.# 10901393001.
26. AMPure XP: BECKMAN COULTER, Ca.# A63880.
27. Dynabeads MyOne Streptavidin C1: NEW ENGLAND Bio-Labs, Ca.# 65001.
28. NEB Next Ultra II Q5 Master Mix: NEW ENGLAND Bio-Labs, Ca.# M0544S.
29. Bovine Serum Albumin (BSA): SIGMA, Ca.# A3294.
30. Dithiothreitol (DTT): Thermo Fisher Scientific, Ca.# R0861.

31. APBS:
    (a) 70% PBS with 30% of ddH_2O.
    (b) e.g., 350 mL of PBS with 150 mL of ddH_2O, and sterilized with 0.45 μm filter and store at room temperature.
32. 10% FCS/AMEM:
    (a) 70% volume of MEM (Gibco, Cat.# 21090-022), 10% volume of Fetal Cow Serum (Sigma, Cat.# F7524-500ML), 100 U of Penicillin-Streptomycin (Sigma, Cat. # P0781-100ML), Glutamine (Gibco, Cat.# 25030-024), Insulin (Sigma, Cat.# 91077C) and sterilized with 0.45 μm filter.
    (b) e.g., 280 mL of MEM, 40 mL of Fetal Cow Serum, 4 mL of Glutamine, Penicillin/Streptomycin and Insulin. Sterilized with a 0.45 μm filter. Store at 4 °C.
    (c) 0.1% Trypsin: Dissolve Ttrypsin in APBS and store at 4 °C for up to 2 weeks.
    (d) Protease K: Prepare 10 mg/mL of proteinase K in glycerol, make an aliquot, and store at −80 °C.
    (e) 2 M Glycine: Dissolve Gglycine in ddH_2O, freshly prepared.
    (f) 20% SDS: Dissolve SDS in ddH_2O and store at room temperature.
    (g) 20% Triton X-100: Dissolve Triton X-100 in ddH_2O and store at room temperature.
    (h) 10% BSA solution: Weigh 1 g BSA and dissolve 10 mL ddH_2O and store in −20 °C.
    (i) 0.5 M DTT solution: Weigh 771.25 mg DTT and dissolve ddH_2O and fill up to 10 mL. Store at −20 °C.
33. Cell Lysis Buffer (nuclei isolation buffer):
    (a) 10 mM Tris–HCl (pH 7.5), 10 mM NaCl, 0.1% IGEPAL, 1% Triton X-100, 100× Protease inhibitor EDTA free, freshly prepared, and keep it on ice (*See* **Note 3**).
    (b) e.g., 500 μL of Tris–HCl (pH 7.5), 100 μL of 5 M NaCl, 500 μL of 10% IGEPAL, 5 mL of 10% Triton X-100, 500 μL of Protease inhibitor EDTA free and fill up to 50 mL with ddH_2O.
34. SDS Buffer:
    (a) 1% SDS, 50 mM Tris–HCl (pH 8.0), 150 mM NaCl, 10 mM EDTA in ddH_2O and store at room temperature.
    (b) e.g., 2.5 mL of 20% SDS, 2.5 mL of Tris–HCl (pH 8.0), 1.5 mL of NaCl, 1 mL of 0.5 M EDTA, and fill up to 50 mL with ddH_2O.

35. 10× Ligation Buffer:
    (a) 10× concentration: 500 mM Tris–HCl (pH 7.5), 100 mM $MgCl_2$, 1 mg/mL BSA, 100 mM DTT, 10 mM ATP, make aliquot and store at −80 °C.
    (b) e.g., 25 mL of 1 M Tris–HCl (pH 7.5), 5 mL of 1 M $MgCl_2$, 5 mL of 10% BSA solution, 10 mL of 0.5 M DTT solution, 5 mL of 100 mM ATP solution, and mix in 50 mL tube.
36. 2× NTB Buffer:
    (a) 10 mM Tris–HCl (pH 7.5), 2 M NaCl, 1 mM EDTA, and store at room temperature.
    (b) e.g., 500 μL of 1 M Tris–HCl (pH 7.5), 20 mL of 5 M NaCl, 100 μL of 0.5 M EDTA, and fill up to 50 mL with ddH_2O.
37. TB Buffer:
    (a) 5 mM Tris–HCl (pH 7.5), 1 M NaCl, 0.5 mM EDTA, and store at room temperature.
    (b) e.g., 250 μL of 1 M Tris–HCl (pH 7.5), 10 mL of 5 M NaCl, 50 μL of 0.5 M EDTA, and fill up to 50 mL with ddH_2O.
38. 2× Oligo Ligation buffer:
    (a) 20 mM Tris–HCl (pH 7.5), 2 mM EDTA, 100 mM NaCl, and store at room temperature.
    (b) e.g., 40 μL of 1 M Tris–HCl (pH 7.5), 8 μL of 0.5 M EDTA, 40 μL of 5 M NaCl, and fill up to 2 mL with ddH_2O.
39. Illumina Tru-seq Universal Adapter: (*See* **Note 4**)
    (a) 5′- AATGATACGGCGACCACCGA GATCTACACTCT TTCCCTACACGACGCTCTTCCGATC(S)T-3′
    (b) Illumina Tru-seq index primer
    (c) 5′-(P) GATCGGAAGAGCACACGTCTGAACTCCAGT CAC (index:NNNNNN) ATCTCGTATGCCGTCTTCT GCTTG-3′
    (d) Illumina Tru-seq Universal primer Fw (12.5 μM working stock concentration and store at −20 °C)
    (e) 5′-CAAGCAGAAGACGGCATACGAGAT-3′
    (f) Illumina Tru-seq Universal primer Rv (12.5 μM working stock concentration and store at −20 °C)
    (g) 5′-AATGATACGGCGACCACCGAGAUCTACAC-3′

2.2 Equipment for all Hi-C Preparations

1. Covaris S2 Focused-Ultrasonicator: Covaris, S220 series (*See* **Note 5**).
2. Covaris microtube AFA Fiver Pre-Slit Snap Cap: Covaris, Ca.# PN520045.
3. QIAquick PCR purification kit: QIAGEN, Ca.# 28104.
4. NGS Fragment Kit (1-6000bp): Agilent, Ca.# DNF-473-0500.
5. Qubit Fluorometer: Thermo Fisher Scientific.
6. 15 mL conical tube: Any 50 mL tube can be used.
7. 50 mL conical tube: Any 50 mL tube can be used.
8. 70 μm cell strainer: Corning, Ca.# 352350 (100 μm cell strainer (Corning, Ca.# 431752) also works).
9. Nunc™ EasYFlask™: Thermo Fisher Scientific, Ca.# 159910 (any kind of flask works).
10. DNA LoBind tube: Eppendorf, Ca.# EP0030108051.

2.3 Additional Materials for Larval Hi-C Preparations

1. Benzocaine: SIGMA, Ca.# E1501-100G.
2. 10× TBS:
   (a) Weigh 24.2 g of tris, which is powder and 90 g NaCl. Add MilliQ water to 990 mL. Mix and adjust pH to 7.0 with HCl. Fill up to final volume 1 L with MilliQ water. Sterilize with autoclaving and store at room temperature.
3. 40× Holtfreter's Solution:
   (a) Weigh 158.4 g of NaCl, 2.9 g KCl, 5.4 g $CaCl_2{\cdot}2H_2O$ and 11 g $MgSO_4{\cdot}7H_2O$. Fill up to 1 L with MilliQ water. Sterilize with Autoclaving and store at room temperature.
4. 10% Benzocaine Solution
   (a) Dissolve 10 g Benzocaine in 100 mL of 100% EtOH and store at room temperature.
5. 0.03% Benzocaine Solution
   (a) Mix 500 mL of 10× TBS and 50 mL of 40× Holtfreter's solution. Fill up to 9970 mL with de-ionized water. Add 30 mL of 10% Benzocaine and store at room temperature.
   (b) Heparin sodium: SIGMA, Ca.# 9041-08-1.
6. 0.04% Heparin with APBS
   (a) Diluted Heparin in APBS.
   (b) e.g. Weight 200 mg of Heparin sodium, dissolve in 50 mL of APBS, fleshly prepared.
7. Liberase TM: Roche, Ca.# 5401119001.
   (a) Reconstruct 26 Unit/mL in ddH_2O and store at −80 °C.

8. Liberase TM Cell dissociation buffer:
   (a) 1:100 dilution in APBS, freshly prepared.
   (b) e.g. 500 μL Liberase TM stock in 50 mL APBS.

2.4 Additional Materials for Mitotic Hi-C Preparations

1. Thymidine: SIGMA, Ca.# T9250.
2. Nocodazole: SIGMA, Ca.# M1404.
3. Propidium Iodide: SIGMA, Ca.# P4170.
4. Prepare 50 μg/mL stock in ddH_2O and store at −20 °C (Only for the cell cycle arresting test).
5. Nunc™ EasYFlask™ (225 cm^2 size): Thermo Fisher Scientific Ca.# 159934.
6. FACS Aria™ III Cell Sorter: BD Biosciences (Any type of Cell sorter with the 488 nm excited laser can be used. Only for the cell cycle arresting test).

3 Method

3.1 Cell Preparation

3.1.1 Cell Preparation of Cycling Cultured Axolotl Cells

Here we describe cell preparation from a cultured cell line (e.g., AL1 cell line) [13, 14].

1. Prepare cells inappropriate conditions in the flask. Cells should be healthy and 80–90% confluent.
2. Discard the medium and add 10 mL of APBS for washing the cells.
3. Harvest 1 million cells by treating with 5 mL of 0.1% trypsin solution for 3–5 min at room temperature and quench the trypsin by adding 10 mL of 10% FCS/AMEM.
4. Collect the cell suspension in a 15 mL conical tube, and spin the cells at 200 × g for 5 min at room temperature (*see* **Note 6**).
5. Discard the supernatant and resuspend the pellet in 5 mL of 10% FCS/AMEM.
6. To discard cell debris and make single-cell suspension, the cell suspension from Subheading 3.1.1, **step 5** is filtrated using a 70 μm cell strainer in a 50 mL conical tube. After the staining, the strainer is washed with 15 mL of 10% FCS/AMEM (now the total volume is 20 mL) (*see* **Note** 7).
7. Take 10 μL cell solution and mix with 3–5 μL of Trypan blue solution and count the number of living cells (*see* Fig. 2).
8. Collect 1–2 million cells in a 50 mL conical tube (*See* **Note 8**) and fill up to 36 mL with 10% FCS/AMEM (now 36 mL of cell suspension is placed in the tube).

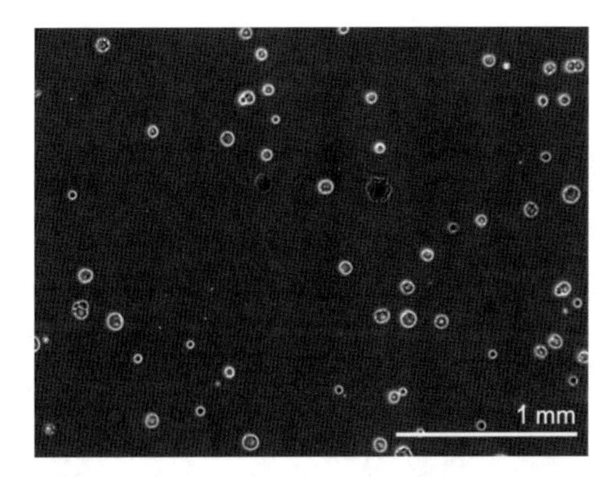

Fig. 2 The cell vitality test by Trypan Blue staining. With any kind of cell dissociation method, the viability of cells is need to be checked. The figure shows the Trypan blue stained cells in Subheading 3.1 Cell preparation with ×4 Objective. For the library preparation of Chromatin conformation capture method, the cell viability should be >95% with the cell line. Although in vivo cell harvest would be more difficult than the cell line, the cells need to be intact condition, and the viability should be >90%. If you see less viabilities, you need to optimized your cell harvest conditions. If the cells are needed to be sorted, Trypan blue staining should be done after the sorting procedures

3.1.2 Protocol for Mitotic Synchronization of Cultured Axolotl Cells

In this passage, we describe the adapted protocol for synchronizing the cell cycles in the mitotic phase with the steps of drug treatment [11].

1. Split the cells at 30% confluency in 5–10 flasks and culture for 5 days in normal culture conditions.
2. Change media to 10% FCS/AMEM with supplemental thymidine (final concentration: 2 mM) and culture for 72 h.
3. Every 24 h, the culture medium with Thymidine needs to be changed through the incubation time.
4. After 72 h of thymidine treatment, the culture medium is replaced by 10% FCS/AMEM with supplemental nocodazole (final concentration: 4 nM) and cultured for 18 h (*see* Fig. 3 and Fig. 4).
5. After 18 h from nocodazole treatment, the cell cycle arrested cells are detached by shaking and the supernatant is collected into a 50 mL tube.
6. Spin the cells at 200 × g for 10 min at room temperature.
7. Resuspend pellets in 10 mL of 10% FCS/AMEM.
8. Take 10 μL cell solution and then mix with 3–5 μL of Trypan blue solution, and count the number of living cells (*see* Fig. 2).
9. The number of cells is counted at this step, and minimum 1 million cells should be corrected in 36 mL of 10% FCS/A-MEM (*See* **Note 8**).

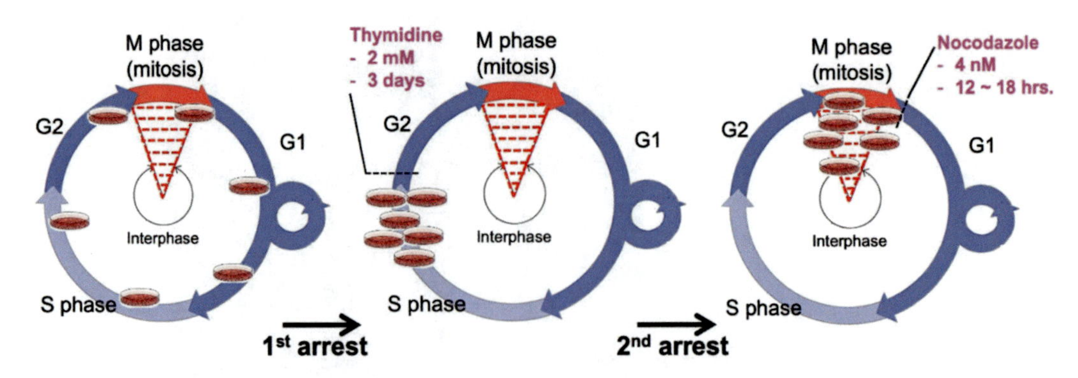

Fig. 3 Schema of the cell cycle arresting with two-step chemical treatment with the cultured cell line. (**a**) Before the arresting cell cycles, the cells (pink round dishes) stay in the interphase which includes 4 major cell cycle phases, known as G0, G1, S, and M phase. (**b**) In the 1st thymidine treatment, the cells are arrested in front of the G2 phase. The 1st arresting needs some time (~72 h). (**c**) Soon after 1st cell arrest treatment, the culture media need to be changed with supplemental Nocodazole. Then the cells are arrested in the M phase after 12–18 h culture

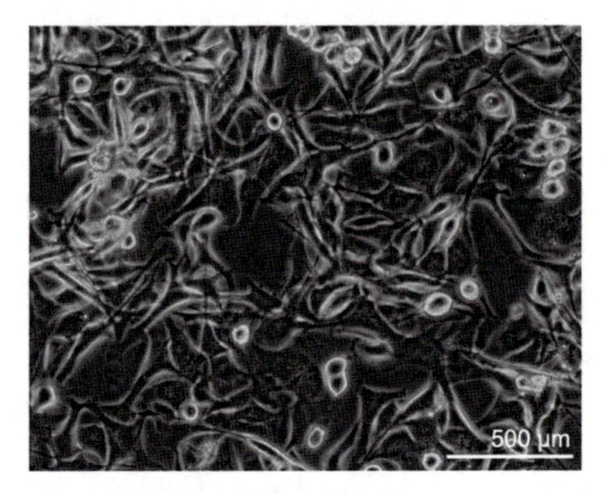

Fig. 4 The cell shape of the mitotic AL1 cell line in M phase. The image shows AL1 cells in the interphase and M phase. There two different types of cell shapes. One take is large, flat, and spindle-shaped forms, and the others are a spherical shape (known as mitotic cell rounding). Because of the spherical shape, the mitotic cells are loosely attached to the bottom. To harvest mitotic cells selectively, the tapping or shaking of culture flasks is sufficient. Then the supernatant contains only mitotic cells

10. If you can have an extra 10 K cells, you can check the proportion of the arrested cells in the G2/M phases with PI staining and flow cytometry (*See* Fig. 5, and **Note 10**).
11. Check the cell vitality by Trypan Blue staining after harvesting the cells (*See* Fig. 2).

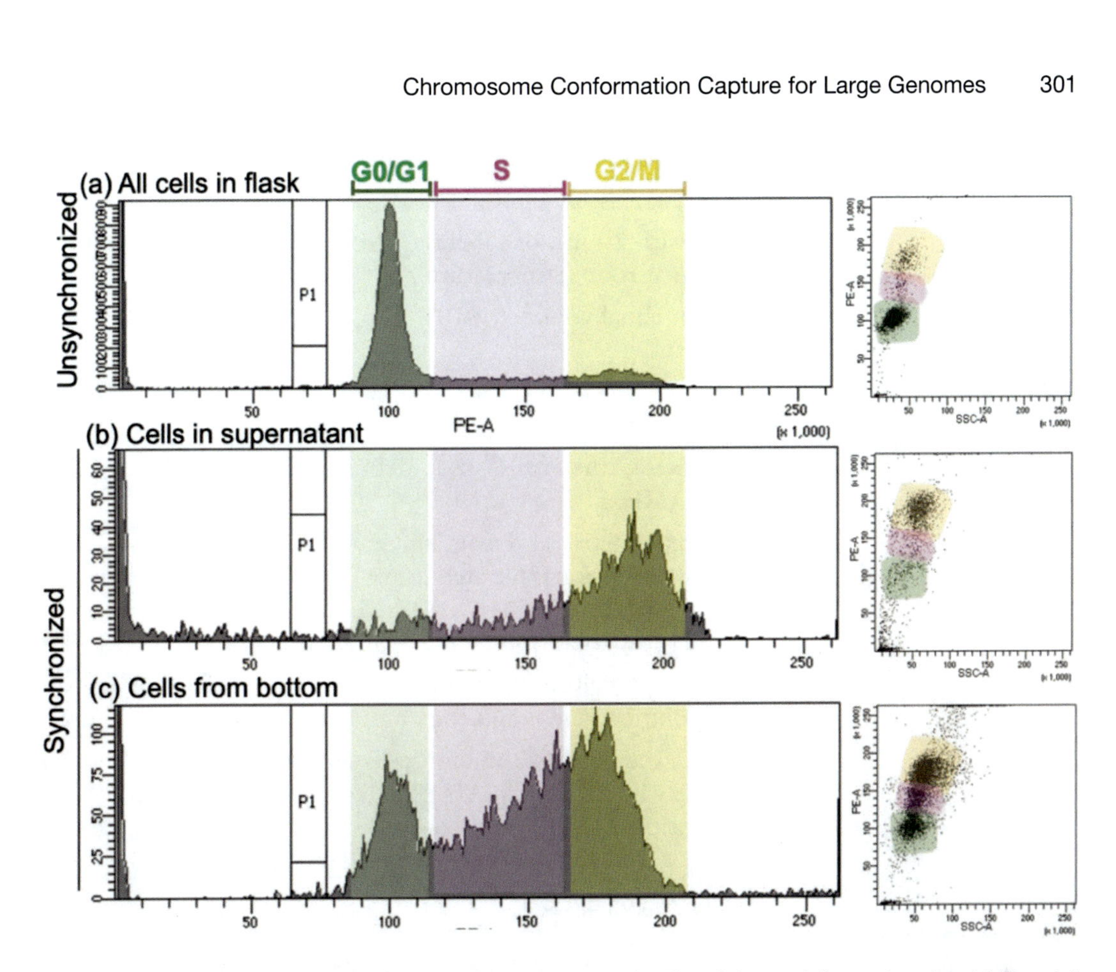

Fig. 5 Validation of the cell cycle arresting with AL1 cell line by PI staining and flow cytometry. (**a**) The plot shows Un-synchronized AL1 cells. The tested cells were harvest from the supernatant and trypsin treatment. (**b**) The plot shows the cell cycle synchronized AL1 cells from the supernatant population. The cells were treated by two-step cell arresting and harvested from the supernatant after shaking off the flask. (**c**) After harvesting the cells of (**b**), the cells were harvested by trypsin treatment. Green: G0/G1 phases, Pink: S phase, Yellow: G2/M phases. In plot (**a**), there huge G0/G1 populations exist ~85%, S phase, and G2/M phase populations are ~10%. Compare to (**a**), the plots from (**b**) and (**c**) contain the large populations of G2/M phases. Plot (**b**) shows that the cells in G2/M phases (yellow marked area) are selectively collected by shaking of the flask compare to plot (**c**)

3.1.3 Cell Preparation (from Larvae)

Here we describe cell isolation from the whole larvae body expect blood and intestine organs (*see* **Note 11**).

1. 3 cm (the body length from nose to tail) of Axolotl larvae are kept under starving conditions for 3 days to avoid genomic DNA contamination from their diets.
2. Put 10 larvae in 0.03% benzocaine solution and leave it for 5 min.
3. After anesthesia, intestines and guts are removed by surgical knife the remaining parts are dissected.
4. Wash the dissected body parts with heparin-sulfate /APBS solution twice.

5. Briefly sterilize the dissected body (~3 s) with 70% EtOH.
6. Wash the dissected body with APBS twice.
7. Treat with 20 mL of Liberase TM Cell dissociation buffer for 30 min at room temperature.
8. Add an equal volume of 10% FCS/AMEM and mix the solution gently.
9. Spin the cells and debris at 200 × g for 10 min at room temperature.
10. Resuspend the cells and debris into 20 mL of 10% FCS/AMEM.
11. To discard debris and make single-cell suspension, the cell and debris suspension is filtrated using 70 μm cell strainer. After the Cell Strain, the strainer is washed with 10 mL 10% FCS/A-MEM (now total volume is 20–30 mL).
12. Take 10 μL cell solution and then mix with 3–5 μL of Trypan blue solution, and count the number of living cells (*see* Fig. 2).
13. Collect the cells 1–2 million cells (*see* **Note 8**) and fill up to 36 mL with 10% FCS/AMEM (Now you have 36 mL of cell suspension).
14. Check the cell vitality by Trypan Blue staining after harvesting the cells (*see* Fig. 2).

3.2 Nuclei Isolation, Fixation, In Situ Chromatin Digestion

You have 36 mL of cell suspension in the culture media.

1. Add 1 mL of 37% formaldehyde (final concentration: 1%) and mix the solution by inversion and rotate for 10 min at room temperature (*see* **Note 12**).
2. Fixation is stopped by adding 2.5 mL of 2 M glycine (final concentration: 125 mM) and mix the solution by inversion until the whole solution turns evenly yellow.
3. Incubate for 5 min at room temperature, and incubate the tube on ice for 15 min.
4. Spin the tube at 200 × g for 5 min at 4 °C (*see* **Notes 13** and **15**).
5. Resuspend cell pellet into 50 mL of pre-ice-cold APBS with Protease inhibitor by gently pipetting.
6. Repeat Subheading 3.2, **steps 4** and **5**.
7. Optimal stop point: You can keep the fixed cells pellet at −80 °C.
8. Resuspend cells in 50 mL of pre-ice-cold Cell lysis buffer (nuclei isolation buffer) with gently pipetting, and incubate on ice ~1 h with occasional mixing by inversion (*see* **Note 3**).
9. Collect the nuclei at 200 × g for 10 min at 4 °C.

10. Add 50 mL of ice-cold APBS and mix gently.
11. Collect the nuclei at 200 × g for 10 min at 4 °C.
12. Carefully discard APBS supernatant with aspiration as much as possible.
13. Add 50 mL of ice-cold APBS and mix gently. Collect the nculei at 200 × g for 10 min at 4 °C.
14. Carefully discard APBS supernatant with aspiration as much as possible.
15. Resuspend nucleus in 500 μL of 1.25× NEB2 by gently pipetting with a cut-off P1000 tip.
16. Transfer 125 μL of nuclei suspension into several 1.5 mL DNA LoBind tubes by cut-off P1000 tip (0.25 million cells/tube) (*See* **Note 14**).
17. Spin the tubes at 100 × g, 30 s at room temperature (*See* **Note 15**).
18. Resuspend nucleis in 450 μL of 1.25× NEB2 by gently pipetting with a cut-off P1000 tip.
19. Add 14 μL of 20% SDS (final concentration: ~ 0.6%, now the volume is 465 μL), by cut-off P1000 tip, gently pipetting.
20. Incubate tubes for 2 h at 37 °C with 900 rpm agitation.
21. Add 91 μL of 20% Triton-X (final concentration: ~3.3%, now the volume is ~560 μL), by cut-off P1000 tip, gently pipetting.
22. Incubate tubes for 2 h at 37 °C with 900 rpm agitation.
23. Collect the nuclei at 100 × g for 30 s at room temperature.
24. Resuspend nucleus in 450 μL of 1× NEB2 by cut-off P1000 tip, gently pipetting.
25. Put aside 20 μL of Nuclei solution into another tube in order to check the digestion efficiency. Keep it at −20 °C until Subheading 3.2, **step 27**.
26. Add 15 μL of DpnII (1500 U) and gently mix by cut-off P1000 tip then incubate for 8–10 h at 37 °C with 900 rpm agitation.
27. Take 20 μL of the digestion nuclei suspension and undigested nuclei solution (from Subheading 3.2, **step 25**) for checking the digestion efficiency (Fig. 6).
28. Keep ~550 μL of the remaining reaction at 37 °C.
29. Add 180 μL of SDS buffer and 2 μL Protease K into 20 μL of the nuclei solutions (from Subheading 3.2, **steps 25** and **27**) and incubate at 65 °C for 2 h.
30. Add 200 μL of Phenol/Chloroform, and vortex 1 min.
31. Spin the tubes for 2 min at 4 °C with maximum speed.

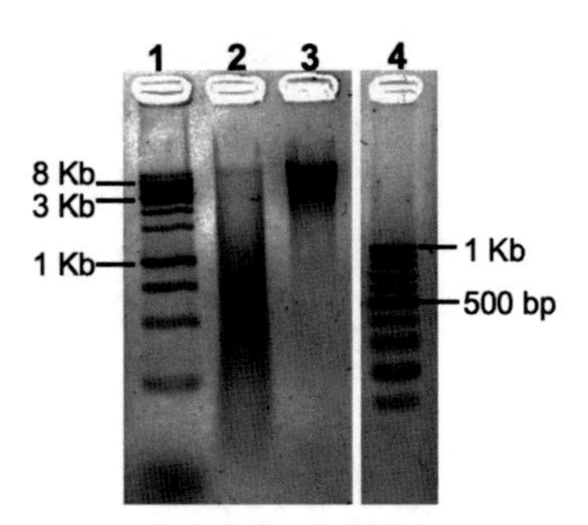

Fig. 6 Efficiency check of In situ chromatin digestion by gel electroporation. To check the efficiency of in situ chromatin digestion, the digested genomic DNA is able to run on the 2% agarose gel. The gel picture shows that Lne 1: 1 kb ladder, Lane 2: DpnII digested genomic DNA (500 ng), Lane 3: in efficient chromatin digestion with DpnII (500 ng), Lane 4: 500 bp ladder. With 30 h digestion reaction, the main peak is of DpnII digestion shown around 500 bp to 1 Kb. In efficient digestion sample shows strong specific band more than 10 Kb, and smear band exist between 3 and 10 Kb

32. Collect supernatant carefully and transfer it into new tubes.
33. Add 200 μL of Chloroform, and vortex 1 min.
34. Spin the tubes for 2 min at 4 °C with maximum speed.
35. Correct supernatant carefully and transfer into new tubes.
36. Add 500 μL of 100 EtOH (×2.5 volume), 20 μL of 3 M NaoAC (1/10 volume) and vortex 1 min.
37. Spin the tubes for 10 min at 4 °C with maximum speed.
38. Discard the supernatant and add 500 μL of 70% EtOH (×2.5 volume) and mix gently.
39. Spin the tubes for 5 min at 4 °C with maximum speed.
40. Discard the supernatant and let the pellet dry at room temperature.
41. DNA pellet is dissolved in 10 μL of 0.1 M Tris–HCl (300–500 ng of gDNA in total).
42. Run all the gDNA suspension on the 2% agarose gel.
43. Check the digestion efficiency (*See* Fig. 6).
44. If the digestion of chromatin is not sufficient, add 10 μL of DpnII (1000 U) to the reaction (from **Step 28**) and incubate for at least 6 h at 37 °C with 900 rpm agitation.

3.3 Biotin Filled-In, Blunt-End Ligation

1. Incubate the tubes (~550 μL of reaction from Subheading 3.2, **steps 28** or **44**) for 20 min at 65 °C without any agitation or mixing (*See* **Note 16**).
2. Replace the tubes at room temperature and leave it until the tube temperature reaches slowly room temperature.

3. Collect the nuclei mix into one tube by cut-off P1000 tip.
4. Pellet down the nuclei at 100 × g for 30 s at room temperature.
5. Resuspend nuclei in 400 μL of 1× NEB2, by cut-off P1000 tip with gently pipetting.
6. Mix the nuclei suspension with the 50 μL of fill-in mix:

10 × NEB2	5 μL
dCTP (10 mM) dGTP (10 mM) dTTP (10 mM)	1.5 μL 1.5 μL 1.5 μL
Biotin-14-dATP (0.4 mM)	37.5 μL
H_2O	3 μL
	50 μL

7. Add 10 μL (50 U) Klenow then mix well, by cut-off tip, gently pipetting.
8. Incubate for 1 h at 37 °C with shaking program in Thermal cycler (Program: 700 rpm, 10 s on and 30-s breaks. Repeat the program for 1 h).
9. Collect the nuclei at 100 × g for 30 s at room temperature.
10. Resuspend the nuclei pellet in 500 μL of T4 Ligation buffer gently by cut-off tip.
11. Repeat Subheading 3.3, **steps 9** and **10**.
12. Resuspend the nuclei pellet in 200 μL of T4 Ligation by cut-off tip, gently pipetting.
13. Put aside 10 μL of Nuclei solution into another tube in order to test the fill-in and ligation efficiency. Keep it in −20 °C until the quality checkpoint.
14. Add 10 μL of T4 DNA ligase and mix by cut-off tip gently.
15. Incubate at 16 °C for 8–12 h.
16. Add 10 μL of T4 DNA ligase and mix by cut-off tip gently and incubate for at least 2 h.
17. Take 20 μL of nuclei suspension and isolate gDNA extraction (*see* Subheading 3.2, **steps 29–41**).
18. Check the ligation product on the 0.6% agarose gel (*see* Fig. 7).
19. If the ligation efficiency is not sufficient, add 10 μL of T4 DNA ligase and incubate at 16 °C at least for 6 h.

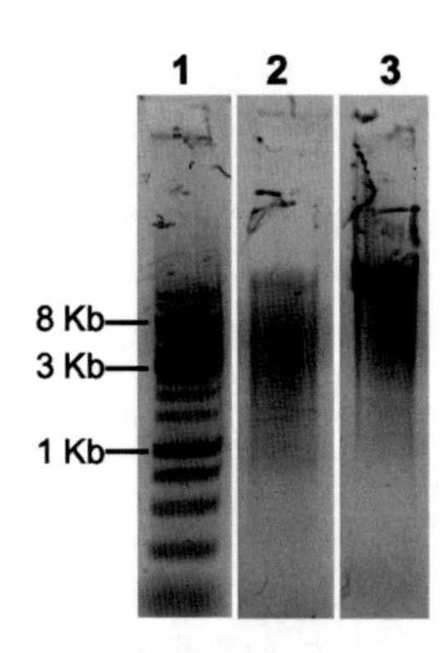

Fig. 7 Efficiency check of fill-in and ligation efficiency by gel electroporation. To check the efficiency of biotin fill-in and blunt-end ligation, the blunt-end ligation genomic DNA is checked by running on the 0.6% agarose gel. The gel picture shows (Lane 1: 1 kb ladder, Lane 2: blunt-end ligation genomic DNA (350 ng), Lane 3: inefficient Biotin filled in ligation product (500 ng). With 30 h ligation reaction, efficient biotin fill-in and blunt-end ligation products should be smear band 3–10 Kb. Lane 3 shows the band mainly bigger than 10 Kb, it causes by a sticky end from inefficient biotin fill-in ends

3.4 DNA Extraction

1. Collect the nuclei at 100× g for 30 s at room temperature.
2. Resuspend the nuclei pellet in 750 μL of SDS buffer and 30 μL Protease K, and incubate at 65 °C for 8–12 h.
3. Add extra 10 μL Protease K and incubate for at least 2 h.
4. Split 400 μL of Protease K treatment solution in two DNA LoBind tubes.
5. Add 400 μL of Phenol/Chloroform, and vortex 1 min.
6. Spin the tubes for 2 min at 4 °C with maximum speed.
7. Collect supernatant carefully and transfer it into new tubes.
8. Repeat Subheading 3.4, **steps 5–7**.
9. Add 400 μL of chloroform and vortex for 1 min.
10. Spin the tubes for 2 min at 4 °C with maximum speed.
11. Collect the supernatant carefully and transfer it into new tubes.
12. Prepare EtOH precipitation (*See* Subheading 3.2, **steps 35–40**) by adding 1 μL glycogen.
13. DNA pellet is dissolved in 100 μL of 0.1 M Tris–HCl.
14. Add 1 μL RNase A and incubate for 2 h at 37 °C.
15. Add 400 μL of phenol/chloroform and vortex for 1 min.
16. Spin the tubes for 2 min at 4 °C with maximum speed.
17. Collect the supernatant carefully and transfer it into new tubes.
18. Add 400 μL of chloroform, and vortex 1 min.
19. Spin the tubes for 2 min at 4 °C with maximum speed.

20. Collect supernatant carefully and transfer it into new tubes.
21. Prepare EtOH precipitation (*see* Subheading 3.2, **step 35–40**) by adding 1 μL Glycogen (twice for 70% EtOH wash).
22. Add 50 μL of 0.1 M Tris–HCl to each tube and genomic DNA pellets are dissolved.
23. Combine both 50 μL of genomic DNA solution into one DNA LoBind tube.
24. Check the DNA yield by Qubit or Pico-green plate reader.
25. *Optimal Stop Point*: genomic DNA can be stored at −20 °C.

3.5 DNA Shearing

1. Take 20 μg of genomic DNA from Subheading 3.4, **step 25**, and split into 4 individual tubes (5 μg DNA per tube).
2. To remove the excess biotin-14-dATP, add the mix into each tube following (Volume × μL can be changed by the concentration of genomic DNA):

genomic DNA (for 5 μg)	× μL
10× NEB2	10 μL
dATP (2.5 mM)	0.5 μL
dGTP (2.5 mM)	0.5 μL
T4 DNA polymerase (3 U/μL)	3 μL
10% BSA	0.5 μL
H_2O	up to 50 μL

3. Incubate for 3 h at 20 °C.
4. The reaction is stopped by adding 2 μL of 0.5 M EDTA and mix by pipetting.
5. Add 80 μL of H_2O in each tube (Now, 5 μg DNA/130 μL in 4 tubes).
6. Transfer 130 μL of genomic DNA solution to Covaris micro tubes (each 5 μg of genomic DNA solution into individual 4 Covaris micro tubes).
7. Shear genomic DNA to obtain 300–400 bp DNA fragments using Covaris S2 following the settings (*See* **Note 5**):

   Temperature (°C): 4

   Duty Factor (%): 10

   Peak incident power (w): 5

   Cycles per burst: 200

   Time (Sec): 65

8. DNA Shearing Check: Pick 10 μL from the sonicated DNA (approximately 130 ng of genomic DNA).
9. Apply the DNA on 2% Agarose gel and run short time (135 V, 10–15 min).
10. The peak of sheared DNA size should be around 350–400 bp.
11. After the sonication, the sheared DNA is purified by QIAGEN purification column (each 5 μg DNA of sheared DNA is purified on an individual column).
12. Elute with 65 μL of ddH_2O.
13. The fragment ends of sheared DNA are repaired by mixing with the solution:

		(×4.5 mastermix)
T4 DNA ligation buffer (×10)	10 μL	45 μL
dNTP mix (2.5 mM)	18 μL	81 μL
T4 DNA polymerase (3 U/μL)	4 μL	18 μL
T4 DNA polynucleotide kinase	4 μL	18 μL
Klenow	0.8 μL	3.6 μL

14. Incubate the End repair mixture for 30 min at room temperature (Still with 4 tubes).
15. Purify the DNA by QIAGEN purification Colum and elute in 100 μL of ddH_2O.
16. Add 60 μL of pre-warmed Ampure Beads and incubate on Rotator for 10–20 min at room temperature.
17. Move tubes onto the magnetic stand, then leave for 5 min.
18. During Subheading 3.5, **steps 14–17**, take 120 μL of pre-warmed Ampure Beads in a tube and move it to the Magnetic stand, and leave for 5 min. Discard 90 μL of cleared supernatant, then resuspend the beads in 30 μL of remaining supernatant: now the Ampure beads are prepared at 4 times concentration).
19. Collect the supernatant from Subheading 3.5, **step 17** into new tubes.
20. Add 30 μL of high concentrated Ampure beads (from Subheading 3.5, **step 18**) and incubate on a rotator for 10–20 min at room temperature.
21. Move tubes onto the Magnetic stand, then leave for 5 min.
22. Discard supernatant.

23. Add 500 μL of 75% (or 80%) EtOH, then suck EtOH up immediately. In that step, don't disturb the beads which are on the tube wall.
24. Repeat Subheading 3.5, **step 23**.
25. Take tubes from the magnetic stand and dry the beads.
26. Add 50 μL of 10 mM Tris–HCl, then resuspend the beads and leave for 10 min at room temperature.
27. Replace the tubes on the Magnetic stand and collect the supernatant.
28. Check the DNA concentration with Qubit Fluorometer.
29. Optimal Stop Point: DNA can be kept at −20 °C up to several months.

3.6 Biotin-Streptavidin Immunoprecipitation and Adapter-Ligation

1. The pre-washing of Dynabeads MyOne Streptavidin C1 (from Subheading 3.6, **steps 2–8**) should be done before you start immune precipitation.
2. Transfer 20 μL of Dynabeads MyOne Streptavidin C1 in a new tube and add 500 μL TB buffer, then rotate for 5 min at room temperature.
3. Replace the tube on the Magnetic stand, then discard the supernatant carefully.
4. Repeat Subheading 3.6, **steps 2** and **3**.
5. Add 500 μL of 2× NTB buffer and rotate tubes for 5 min at room temperature.
6. Move the tubes to the Magnetic stand and discard the supernatant.
7. Repeat Subheading 3.6, **steps 5** and **6**.
8. Resuspend pre-cleaned Dynabeads MyOne Streptavidin C1 in 150 μL 2× NEB buffer.
9. The DNA samples are filled up to 150 μL with ddH_2O.
10. Add 150 μL of pre-cleaned Dynabeads MyOne Streptavidin C1 to DNA suspension and rotate for 1 h at room temperature.
11. Replace the tube on the magnetic stand and discard the supernatant carefully.
12. Add 500 μL of 1× NTB buffer and rotate the tube for 5 min at room temperature.
13. Replace the tube on the magnetic stand and discard the supernatant carefully.
14. Repeat Subheading 3.6, **steps 12** and **13**.
15. Add 500 μL of 1× NEB2 buffer and rotate the tube for 5 min at room temperature.
16. Replace the tube on the Magnetic stand and discard supernatant carefully.

17. Repeat Subheading 3.6, **steps 15** and **16**, and resuspend the beads in 94.75 μL of 1× NEB2 buffer.
18. Add A-tailing reaction mix following:

dATP (10 mM)	5 μL
Klenow Fragment (3′–5′ex-)	1.25 μL

19. Incubate tube for 1 h at 37 °C.
20. Move it on the Magnetic stand, then collect beads and discard supernatant carefully.
21. Add 500 μL of 1× T4 Ligation buffer Kienow and rotate the tube for 5 min at room temperature.
22. Replace the tube on the magnetic stand and discard the supernatant carefully.
23. Add 200 μL of 1× T4 Ligation buffer and rotate the tube for 5 min at room temperature.
24. Replace the tube on the magnetic stand and discard the supernatant carefully.
25. Repeat Subheading 3.6, **steps 23** and **24** and resuspend the beads in 50 μL of 1× T4 Ligation buffer.
26. Add Adapter ligation reaction mix the following:

Pre-annealed Illumina index adapter (15 μM)	5 μL
T4 DNA ligase	1 μL

27. Incubate overnight at 18 °C.
28. Replace the tubes with magnetic stand and discard supernatant carefully.
29. Add 200 μL of 1× NEB2 buffer and rotate for 5 min at room temperature.
30. Repeat Subheading 3.6, **steps 28** and **29**.
31. Add 200 μL of 1× NEB2 buffer and rotate for 5 min at room temperature.
32. Replace the tubes with the magnetic stand and discard the supernatant carefully.
33. Add 200 μL of 1× NEB2 buffer and rotate for 5 min at room temperature.
34. Repeat Subheading 3.6, **steps 32** and **33**.
35. Resuspend the beads into 20 μL of 1× NEB2 buffer.
36. Optimal stop point: You can keep your DNA solution at −20 °C.

3.7 Library Amplification and Purification

Test PCR (*See* **Note 17**).

1. Mix the 10 μL of individual test PCR reactions for 4 different PCR cycles following:

		Master mix
DNA with beads	1 μL	4.1 μL
2× Q5 UltraII	5 μL	20.5 μL
Primer F (2.5 μM)	0.4 μL	1.64 μL
Primer R (2.5 μM)	0.4 μL	1.64 μL
H_2O	3.2 μL	13 μL
	10 μL	

PCR Program

98 °C	30
[5, 7, 9, 11 cycles]	
98 °C	10
65 °C	75
65 °C	7 min
4 °C	∞

2. After PCR, apply PCR reactions that contain the 10 μL of reactions from 4 different PCR cycles on the 2% agarose gel and check the band.
3. Define the minimum PCR cycles in which you observe smear product (*See* Fig. 8).
4. Based on defined PCR cycles, prepare PCR reaction (actual Hi-C library amplification) as follow (*See* **Note 17**):

DNA with Beads	2 μL
2× Q5 UltraII	10 μL
Primer F (2.5 μM)	0.8 μL
Primer R (2.5 μM)	0.8 μL
H_2O	6.4 μL
	20 μL

5. After PCR, transfer the PCR reaction into a new 1.5 mL tube.
6. Add 1.8× volume of pre-warm (at room temperature) Ampure beads and rotate for 10 min at room temperature.
7. Place the tube on the Magnet stand and leave 5 min.
8. Discard the supernatant carefully.

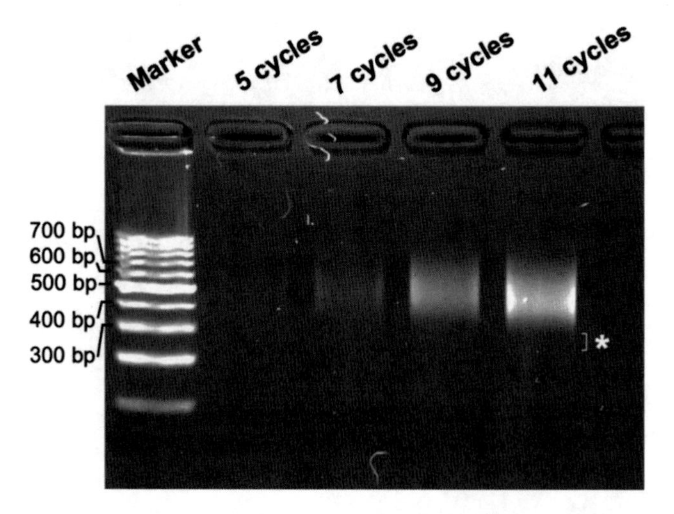

Fig. 8 Hi-C library after PCR cycles. 10 μL of marker (100 bp ladder) is applied with samples. 5 PCR cycles shows weak smear band between 400–600 bp. With more cycles, the smear band is more clear regarding the PCR cycles. The library PCR cycle number should be defined by the PCR cycles of weak smear band plus 1 PCR cycle or the clear band minus 1 cycles. It means 6 cycles are suitable from the picture. In 11 cycles, the products start to show PCR bias product below 300 bp (*see* asterisk in figure)

9. Add 300 μL of 70–80% EtOH and discard (do not disturb the beads).
10. Repeat Subheading 3.7, **step 9**.
11. Air-dry the tubes for 5 min.
12. Add Add 20 μL of 0.1 M Tris–HCl and elute the PCR fragment.
13. *Optimal Stop Point*: DNA can be stored at −80 °C until the sequencing.
14. *Quality checkpoint*: Check fragment size and DNA concentration by Fragment analyzer before starting to run the Illumina sequencing (*see* Fig. 9 and **Note 18**).
15. Send Hi-C library for sequencing (*see* **Notes 19** and **20**).

4 Notes

1. Biotin-ATP: If the restriction site of the desired enzyme does not contain a T, alternatively it can be Biotin-14-dCTP (Thermo Fisher SCIENTIFIC, Ca.# 19518018). In this case, the restriction enzyme site should contain a G. Although the recent paper reported that the incorporation efficiency of Biotin-14-dATP could be better than Biotin-14-dCTP in the

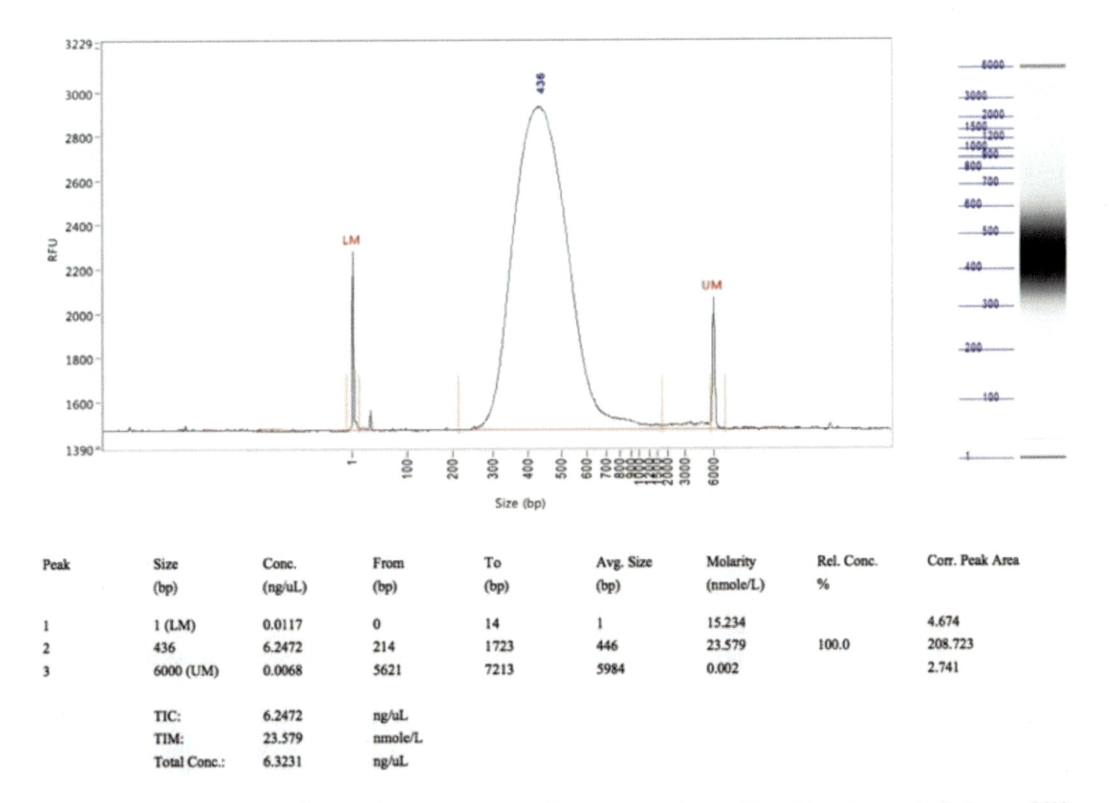

Peak	Size (bp)	Conc. (ng/uL)	From (bp)	To (bp)	Avg. Size (bp)	Molarity (nmole/L)	Rel. Conc. %	Corr. Peak Area
1	1 (LM)	0.0117	0	14	1	15.234		4.674
2	436	6.2472	214	1723	446	23.579	100.0	208.723
3	6000 (UM)	0.0068	5621	7213	5984	0.002		2.741

TIC:	6.2472	ng/uL
TIM:	23.579	nmole/L
Total Conc.:	6.3231	ng/uL

Fig. 9 Example of our Hi-C library proportion with Fragment analyzer. The data shows Hi-C library DNA patterns by NGS HS fragment analysis kit (1–6000 bp). After library purification with Ampure beads, the library can submit to sequencing facility. However, we highly recommend checking the library proportion and the ratio of adapter contamination yourself using Fragment analyzer or Bioanalyzer (both from Agilent) if there is the expected peak (350–600 bp) and no un-expected peak from adaptor (~140 bp)

restriction enzyme site [15], we did not see big differences of the incorporation efficiency between Biotin-14-dATP to Biotin-14-dCTP.

2. Restriction enzyme: Restriction enzymes need to be carefully chosen for an experimental purpose. 4-base cutters give smaller interaction fragments than 6-base cutter enzymes. Many reports use MboI, NheIII, DpnII as 4-base cutters and HindIII, BglII, and EcoRI s 6-base cutters [16]. In our experience, HindIII (NEW ENGLAND BioLabs, Ca.# R0104M) also works nicely. Any other known enzyme which works with 3C or 4C experiments can also be an appropriate option. Besides, glycerol concentration in the digestion reaction should be kept in low concentration. In this regard, we prefer to use a highly concentrated restriction enzyme to keep a low glycerol concentration.
3. Cell Lysis Buffer: The detergent components would depend on the target cell type. With the cell lysis buffer, we mainly use axolotl fibroblast cells. When we tried to prepare Hi-C library

from axolotl red blood cells, it required a more stringent cell lysis buffer with a higher concentration of non-ionic detergents because of the different cell cytoskeleton (e.g., Spectrin) maintaining the shape of cells.

4. Illumina Tru-seq Universal Adapter: Annealing reaction could be done before you start the experiment. Dissolved Universal Adaptor and index Adaptor to 30 μM, respectively. Mix 10 μL of Universal and Index primer into 20 μL of 2× 2×Oligo ligation buffer and transfer to PCR reaction tube. Run the program following 95 °C for 2 min, and cool down to 25 °C for over 1 h. Make aliquots into PCR tubes, then store at −20 °C.
5. Shearing DNA: We use Covaris S2 for shearing the DNA, although there are many options for the machine (e.g. Covaris E200, Bioruptor sonication, Epishear from Active Motif). The DNA shearing program may vary with each piece of equipment. we recommend optimizing the programs with the purified gDNA that include the time and power using an available machine. The first optimization would be the time of shearing.
6. Collect/Spin the cells: Given that the size and volume of Axolotl (Salamander) cells are bigger than other vertebrate animals, the spinning speed (rpm or ×g) is an important parameter to optimize between different cell types. After collecting the cells, it is important to check whether the cells are alive or damaged by Trypan blue staining (*See* Fig. 2). Cell viability should be >90%. With our experiments, the fibroblasts are not damaged at <300 × g for <10 min, while red blood cells were not damaged at <600 × g for 10 min.
7. Making the single-cell suspension would be important for fixing the cells equally by formaldehyde. A cell strainer would be helpful, while you need to choose a cell strainer that has an appropriate pore diameter size. We recommend using a strainer of 70–100 μm pore size for fibroblasts. If you start your experiment with erythroid cells, there is no need to use a cell strainer.
8. The number of cells for start material: Minimum 1 million cells are good for starting your initial experiment. Since Axolotl genome size is 10 times larger than the mouse and human, all the enzymatic reactions such as the in situ chromatin digestion, the polymerize activity for biotin fill-in, and the end ligation activity for blunt-end ligation need to be split into multiple tubes (for example, 1 million cells could be split 0.25 million cells each into 4 tubes).
9. To increase the resolution of the Axolotl genome assembly, we prepared the Hi-C library from the cell cycle arrested AL1 cells in Mitotic phases as well as the interphase Hi-C.

10. In the mitotic phase, the chromosomes start to take a condensed structure and all TAD structures disappear. Following the chromosome structural change, only the short-range contacts appear to the contact frequency which suggests the helical condensed chromosome structure and arraigned as a uniformly looped structure [17]. Our Axolotl mitotic Hi-C gave us the idea to investigate the feature of the gigantic axolotl mitotic chromosome structure compared to other mitotic chromosomes with small genome models [17, 18].
11. To collect mitotic cells selectively, the cells need to be detached by shaking off [11]. However, there is the possibility to harvest the apoptotic cell debris along with mitotic cells. There are two critical points: one is the appropriate thymidine and nocodazole titrations because both chemicals are toxic for cells. Second is the variability check by Trypsin staining after collecting the cells. In our experience, if we change the medium every 24 h during the Thymidine treatment, the contaminations of debris cells become a small population.
12. We highly recommend using the culture cell line or the type of tissue in which you can get a homogenous population (e.g., sorted limb fibroblasts). To dissociate and isolate the specific cell type, the protocol is needed to optimize appropriate methods such as FACS sorting with the cell lineage labeled cells. It is important to design the experiment and get better results. If you start your experiment with internal organs (e.g., liver and kidney), you need to design your plan very carefully as to how to collect your target cells. From our experiences, Axolotl liver was one of the most difficult organs to start for Hi-C preparation, because the liver contains multiple cell types (e.g., lipid cells, red blood cells, hepatocytes, and fibroblasts). It creates difficulty to fix the cells evenly.
13. The condition which is mentioned in the protocol is the maximum time, and formaldehyde concentration with Axolotl fibroblasts, however, it depends on the cell type. When we tested longer fixation, the amount of undigested gDNA would be significantly increased. When the first trial is started, you need to adjust the fixation condition.
14. If possible, it's better to use a swing rotor rather than a fixed angle rotor to collect the cells and nuclei in Subheading 3.1 Cell preparation procedures.
15. We recommend using a low-binding tube and tips for all pipetting. In the step of the nuclei pipetting and mixing, using the cut-off low binding tip is recommended.
16. The step of adding SDS into nuclei suspension is important to permeabilize nuclei allowing the restriction enzyme to enter into nuclei, while the nuclei become fragile and weak. To

maintain the intact nuclei structure as much as possible, collecting the nuclei should be done with a lower speed and quick spin, and mixing the nuclei should be done with cut-off tips and extremely gentle pipetting (e.g. no harsh agitation and making bubbles).

17. Since DpnII recognition site creates new DpnII site after fill-in and ligation, heat-inactivation (Subheading 3.3, **step 1**) is required. If you use HindIII for digesting chromatin, it is not required as the HindIII site after those reactions creates NheI site. Check if the restriction enzyme site does create a new enzyme site after fill-in and ligation [16].
18. To keep library complexity high, the number of PCR cycles for library amplification should be minimal. In our experience, 2 μL of Biotin-bind DNA in 20 μL of PCR reaction needs to amplify 6–9 cycles, but should not be more. If you need to apply a large amount of prepared library (e.g., loading the samples into Nova-seq flow cell), increase the number of PCR reaction tubes.
19. Check fragment size and DNA concentration by Fragment analyzer before starting to run the Illumina sequencing. The steps are depending on the policy of your NGS facility. We check the fragment size by Fragment Analyzer. Simply ask your facility what is required. Our FA plot is here (*see* Fig. 9).
20. We recommend testing your sample quality with a small sequencing platform. There are several options with Illumina devices that produce several million sequencing reads (e.g., Next-seq, Mi-seq). Given that the genome sizes of salamander species are bigger than other vertebrate species, the sequencing reads need a minimum of 5–10 million reads for testing the quality.
21. After sequencing, there are several options for informatics analysis regardless of whether your sequence reads are high quality enough. We highly recommend using the HiCUP program [11]. Assuming the library quality is good, the quality check with HiCUP gives 70–90% of varied pairs and more than 40% of unique mapped reads. Based on our experiments, Hi-C libraries from a cultured cell line show approximately 95% of varied pairs and 40% of unique mapped reads. While the library from in vivo samples might be difficult to handle compared to the cultured cell ones, these scores should be more than 70% and 30%, respectively.

Acknowledgment

We would like to thank Dr. Takuji Sugiura, Dr. Gordana Wutz and Dr. Kota Nagasama (IMP in Vienna Bio-Center), and Dr. Chang Liu (ZMBP, University of Tubingen) for their help giving technical advice. AK was supported by a JSPS Postdoctoral Fellowship for Overseas Researchers. ET was supported by ERC AdG 742046.

References

1. Voss SR, Epperlein HH, Tanaka EM (2009) Ambystoma mexicanum, the axolotl: a versatile amphibian model for regeneration, development, and evolution studies. Cold Spring Harb Protoc 2009(8):pdb emo128. https://doi.org/10.1101/pdb.emo128
2. Khattak S, Murawala P, Andreas H, Kappert V, Schuez M, Sandoval-Guzman T, Crawford K, Tanaka EM (2014) Optimized axolotl (Ambystoma mexicanum) husbandry, breeding, metamorphosis, transgenesis and tamoxifen-mediated recombination. Nat Protoc 9(3): 529–540. https://doi.org/10.1038/nprot.2014.040
3. Fei JF, Schuez M, Tazaki A, Taniguchi Y, Roensch K, Tanaka EM (2014) CRISPR-mediated genomic deletion of Sox2 in the axolotl shows a requirement in spinal cord neural stem cell amplification during tail regeneration. Stem Cell Reports 3(3):444–459. https://doi.org/10.1016/j.stemcr.2014.06.018
4. Nowoshilow S, Schloissnig S, Fei JF, Dahl A, Pang AWC, Pippel M, Winkler S, Hastie AR, Young G, Roscito JG, Falcon F, Knapp D, Powell S, Cruz A, Cao H, Habermann B, Hiller M, Tanaka EM, Myers EW (2018) The axolotl genome and the evolution of key tissue formation regulators. Nature 554(7690):50–55. https://doi.org/10.1038/nature25458
5. Jeramiah J, Smith NT et al (2018) A chromosome-scale assembly of the axolotl genome. Genome Res 29(2):317–324
6. Schloissnig S, Kawaguchi A, Nowoshilow S, Falcon F, Otsuki L, Tardivo P, Timoshevskaya N, Keinath MC, Smith JJ, Randal Voss S, Tanaka EM (2021) The giant axolotl genome uncovers the evolution, scaling and transcriptional control of complex gene loci. Proc Natl Acad Sci USA 118(15):e2017176118
7. Burton JN, Adey A, Patwardhan RP, Qiu R, Kitzman JO, Shendure J (2013) Chromosome-scale scaffolding of de novo genome assemblies based on chromatin interactions. Nat Biotechnol 31(12):1119–1125. https://doi.org/10.1038/nbt.2727
8. Kaplan N, Dekker J (2013) High-throughput genome scaffolding from in vivo DNA interaction frequency. Nat Biotechnol 31(12): 1143–1147. https://doi.org/10.1038/nbt.2768
9. Spitz F, Furlong EEM (2012) Transcription factors: from enhancer binding to developmental control. Nat Rev Genet 13(9):613–626. https://doi.org/10.1038/nrg3207
10. Rodrigues AR, Yakushiji-Kaminatsui N, Atsuta Y, Andrey G, Schorderet P, Duboule D, Tabin CJ (2017) Integration of Shh and Fgf signaling in controlling Hox gene expression in cultured limb cells. Proc Natl Acad Sci U S A 114(12):3139–3144. https://doi.org/10.1073/pnas.1620767114
11. Whitfield ML, Sherlock G, Saldanha AJ, Murray JI, Ball CA, Alexander KE, Matese JC, Perou CM, Hurt MM, Brown PO, Botstein D (2002) Identification of genes periodically expressed in the human cell cycle and their expression in tumors. Mol Biol Cell 13(6): 1977–2000. https://doi.org/10.1091/mbc.02-02-0030
12. Meyer A, Schloissnig S, Franchini P, Du K, Woltering JM, Irisarri I, Wong WY, Nowoshilow S, Kneitz S, Kawaguchi A, Fabrizius A, Xiong P, Dechaud C, Spaink HP, Volff JN, Simakov O, Burmester T, Tanaka EM, Schartl M (2021) Giant lungfish genome elucidates the conquest of land by vertebrates. Nature 590(7845):284–289. https://doi.org/10.1038/s41586-021-03198-8
13. Groell AL, Gardiner DM, Bryant SV (1993) Stability of positional identity of axolotl blastema cells in vitro. Rouxs Arch Dev Biol 202: 170–175. https://doi.org/10.1007/BF00365307
14. Roy S, Gardiner DM, Bryant SV (2000) Vaccinia as a tool for functional analysis in regenerating limbs: ectopic expression of Shh. Dev Biol 218(2):199–205. https://doi.org/10.1006/dbio.1999.9556
15. Nagano T, Varnai C, Schoenfelder S, Javierre BM, Wingett SW, Fraser P (2015) Comparison

of Hi-C results using in-solution versus in-nucleus ligation. Genome Biol 16:175. https://doi.org/10.1186/s13059-015-0753-7

16. Belton JM, McCord RP, Gibcus JH, Naumova N, Zhan Y, Dekker J (2012) Hi-C: a comprehensive technique to capture the conformation of genomes. Methods 58(3): 268–276. https://doi.org/10.1016/j.ymeth.2012.05.001

17. Natalia Naumova MI, Fudenberg G, Zhan Y, Lajoie BR, Mirny LA, Dekker J (2013) Organization of the mitotic chromosome. Science 342(6161):948–953

18. Gibcus JH, Samejima K, Goloborodko A, Samejima I, Naumova N, Nuebler J, Kanemaki MT, Xie L, Paulson JR, Earnshaw WC, Mirny LA, Dekker J (2018) A pathway for mitotic chromosome formation. Science 359(6376): eaao6135. https://doi.org/10.1126/science.aao6135

Part V

Transgenics and Lineage-Tracing

Chapter 21

Axolotl Transgenesis via Injection of I-SceI Meganuclease or Tol2 Transposon System

Maritta Schuez and Tatiana Sandoval-Guzmán

Abstract

The axolotl (*Ambystoma mexicanum*) has been widely used as an animal model for studying development and regeneration. In recent decades, the use of genetic engineering to alter gene expression has advanced our knowledge on the fundamental molecular and cellular mechanisms, pointing us to potential therapeutic targets. We present a detailed, step-by-step protocol for axolotl transgenesis using either I-SceI meganuclease or the mini Tol2 transposon system, by injection of purified DNA into one-cell stage eggs. We add useful tips on the site of injection and the viability of the eggs.

Key words Axolotl eggs, Transgenesis, Egg injection, I-SceI meganuclease, mTol2 transposase

1 Introduction

The use of transgenic axolotls has been crucial to further understand their development and mechanisms of regeneration. The introduction of plasmid and Bacterial Artificial Chromosomes (BAC) constructs, randomly integrated into the genome, has been applied in the last 15 years successfully for studying biological processes in living animals.

I-SceI meganuclease has been a powerful tool in fish, newt, and Xenopus, with efficient transgenesis resulting in a high percentage of F0 animals displaying strong expression throughout the body [1–3]. These genetic models have paved the road for transgenesis in axolotls. The *Tol2* transposon system has also been used successfully to complement conventional BAC transgenesis in fish and mammalian organisms [4, 5]. The minimal sequences required for the transposition by the *Tol2* are used to deliver large cargo into the genome. Tol2-mediated transgenesis effectively transmits to F1.

In axolotls, injections of DNA and I-SceI meganuclease or DNA and *Tol2* transposase mRNA into the one-cell stage axolotl embryo have resulted in stable germline transgenic lines [6–9].

Ashley W. Seifert and Joshua D. Currie (eds.), *Salamanders: Methods and Protocols*,
Methods in Molecular Biology, vol. 2562, https://doi.org/10.1007/978-1-0716-2659-7_21,

Especially, in axolotl a minimal *cis*-sequence of the *Tol2* element has been used to successfully establish germline transgenic lines [7, 10, 11].

Good knowledge of axolotl mating behavior is an important step for successful egg injections. In our colony, a pair of sexually mature axolotls is placed in a mating aquarium (different from the holding aquarium) with plenty of artificial floating leaves. The temperature is set at around 15–16 degrees Celsius and the water-flow is turned down. The male deposits the spermatophores on the bottom of the aquarium where the female takes them to her cloaca. In most cases, spermatophores are laid within 24 h. If after 48 h, there are no signs of spermatophores replacing the male or the female proves successful. Approximately 18–24 h after the female has taken the spermatophores, she will lay the eggs on the artificial leaves from where the researcher collects the eggs. While the female lays her eggs, she should be left undisturbed. Disturbances can cause her to interrupt the laying for some s. It is advisable to remove the male when he is a cause of disturbance. The eggs are then scrapped off the leaves with a scalpel and the injecting protocol can proceed.

Since transgenesis efficiency can vary, animals must be screened to select the progeny with successful genome integration and uniform transgene expression. Our suggestion is to introduce a fluorescent reporter that allows easy identification. The molecular mapping of insertion sites in the genome by PCR is also recommended.

2 Materials

2.1 Equipment and Materials

1. Freshly laid eggs from axolotl.
2. Scalpel.
3. Plastic container for collecting eggs.
4. A metal sieve.
5. Glass beaker.
6. Sterile plastic transfer pipettes, (2 mL). Adjust the borehole size by cutting the tip.
7. Petri dishes, different sizes.
8. Dissecting microscope (Olympus SZX10).
9. Light source (Olympus KL 1500 compact).
10. Sharp forceps (two pairs Dumont #55).
11. DNase-/RNase-free sterile pipette filter tips (Fisher Scientific): 0.1–10 μL, cat. no. 02-707-439; 2–20 μL, cat. no. 02-707-432; 10–100 μL, cat. no. 02-707-431;20–200 μL, cat. no. 02-707-430.

12. PCR tubes, 0.2 mL (Bio-Rad, cat. no. TWI-0201).
13. Microloader tips (Eppendorf).
14. Slide with a micrometric scale of 0.1 mm to measure the size of the injection droplet (Bresser).
15. PV830 Pneumatic Pico Pump (WPI).
16. Micromanipulator, Narishige MN-153 (Science Products).
17. Magnetic stand GJ (Science Products).
18. Steel plate IP for use with magnetic stand (Science Products).
19. M/F pressure-monitoring/injection line, 5 feet (Argon Medical Devices, cat. no. 040105005A).
20. Microelectrode holder (WPI, cat. no. MPH6S10).
21. Injection needles, borosilicate glass capillaries GC100F-15, 1 mm outer diameter × 0.58 mm inner diameter (Harvard Apparatus, cat. no. 30-0020).
22. Flaming/Brown micropipette puller P-97 (Sutter Instrument Company).
23. Box filament FB255B 2.5 × 2.5 mm.
24. Mold, custom-made to prepare injection plates (*see* **Note 1**).
25. 24-well plates to keep embryos before hatching.
26. Small plastic containers (10 cm × 10 cm) for keeping hatched animals.

2.2 Reagents

1. Ethanol, absolute diluted in H_2O to 70%, to rinse the eggs (VWR, cat. no. 20821-330).
2. I-SceI enzyme and 10× SceI buffer (NEB, cat. no. R0694) (*see* **Note 2**).
3. Plasmid DNA.
4. Ficoll PM 400 (Sigma-Aldrich, cat. no. 46327-500G-F).
5. Benzocaine, to anesthetize the animals (Sigma-Aldrich, cat. no. E1501).
6. NaCl (Merck, cat. no. 106404).
7. $MgCl_2$ (Merck, cat. no. 105833).
8. KOH (Merck, cat. no. 814353).
9. EDTA (Merck, cat. no. 324506).
10. HEPES (Merck, cat. no. 391338).
11. KCl (Merck, cat. no. 104936).
12. $CaCl_2$ (Merck, cat. no. 102382).
13. $MgSO_4$ (Merck, cat. no. 105886).
14. HCl, fuming, 37% (vol/vol) (Merck, cat. no. 100317).
15. Tris (Carl Roth, cat. no. 4855.2).

16. Penicillin-streptomycin (PAA, cat. no. P11-010).
17. DMSO (Sigma, cat. no. D2650).
18. BspEI (New England Biolabs (NEB), cat. no. R0540).
19. Plasmids: pBSII-SK-mTol2-MCS (Addgene #51817) and T7-TPase (Addgene #51818).
20. RNaseZap (Ambion-Life Technologies, cat. no. AM9780).
21. Qiagen plasmid maxi kit (Qiagen, cat. no. 12163).
22. UltraPure DNase-/RNase-free distilled water (Life Technologies, cat. no. 10977-023).
23. DPBS (Life Technologies, cat. no. 14190).
24. DNA-grade agarose (Life Technologies, cat. no. 75000-500).
25. Agar to prepare injection plates (SERVA, cat. no. 11393.04).

2.3 Reagent Preparation

1. *Marc's Modified Ringer's Solution (MMR), 10×*

   For 2 L of 10× MMR, mix 400 mL of 5 M NaCl, 40 mL of 1 M KCl, 20 mL of 1 M $MgCl_2$, 40 mL of 1 M $CaCl_2$, 4 mL of 0.5 M EDTA and 100 mL of 1 M HEPES. Add deionized water to a volume of 1.5 L. Mix and adjust pH to 7.8 with 10 M KOH. Makeup to 2 L with deionized water. Autoclave and store at room temperature (up to 12 months).
2. *MMR 1×/pen-strep*

   For 1 L of 1× MMR/pen-strep, mix 100 mL of 10× MMR with 13 mL of penicillin-streptomycin and add deionized water to a volume of 1 L. Sterile-filter the solution. Store at room temperature (up to 2 months).
3. *MMR 0.1×/pen-strep*

   For 2 L, mix 20 mL of 10× MMR and 25 mL of penicillin-streptomycin. Add deionized water to a volume of 2 L. Sterile-filter the solution. Store at room temperature or (up to 2 months).
4. *MMR 0.1×/5% (wt/vol) Ficoll/pen-strep*

   For 1 L, mix 10 mL of 10× MMR and 50 g of Ficoll, and add 500 mL of deionized water. Dissolve at 50 °C on a magnetic hot plate stirrer. Add 13 mL of penicillin-streptomycin and bring the volume up to 1 L with deionized water. Sterile-filter the solution. Store at 4 °C (up to 2 months).
5. *MMR 1×/20% (wt/vol) Ficoll/pen-strep*

   For 250 mL, mix 25 mL of 10× MMR and 50 g of Ficoll. Add 150 mL of deionized water. Dissolve at 50 °C on a magnetic hot plate stirrer. After cooling down to room temperature add 7 mL of penicillin-streptomycin and fill up to 250 mL with deionized water. Sterile-filter the solution. Store at 4 °C (up to 2 months).

6. *Holtfreter's Solution, 400% (wt/vol) Concentration*

   For 10 L of Holtfreter's solution, mix 2.875 g of KCl, 5.36 g of $CaCl_2 \cdot 2\ H_2O$, 11.125 g of $MgSO_4 \cdot 7\ H_2O$, and 158.4 g of NaCl. Add distilled water to a volume of 10 L. Store at room temperature for up to 6 months.

7. *Benzocaine 10% (wt/vol)*

   For 500 mL of 10% (wt/vol) benzocaine, dissolve 50 g of benzocaine in 500 mL of 100% ethanol. Store at room temperature (up to 12 months).

8. *TBS, 10×*

   For 1 L, add 24.2 g of Tris and 90 g of NaCl to 990 mL distilled water, mix well on a plate stirrer, add 10 mL of 12 M HCl, and adjust the pH to 8.0. Store at room temperature (up to 6 months).

9. *Anesthesia for Axolotls, 0.03% (vol/vol) Benzocaine*

   For 10 L of anesthetic, mix 500 mL of 10× TBS, 500 mL of 400% Holtfreter's solution, and 30 mL of 10% (wt/vol) benzocaine. Add distilled water to a volume of 10 L. Store at room temperature (up to 6 months).

2.4 Injection Needles

Prepare injection needles by pulling borosilicate glass capillaries on the micropipette puller. They should have a very long thin tip. We use the following settings: Pressure = 500; heat = 530; pull = 100; velocity = 120; time = 150.

2.5 Injection Plate

Prepare a 2% Agar gel solution with autoclaved animal holding water and pour about 10 mL into a 94 mm Petri-dish. Put in the custom-made plastic mold (*see* **Note 1** and Fig. 1a) upside down and take it out when the agar has solidified. The agar plate will have channels that hold the eggs during the injection. Close the lid and seal the dish with Parafilm. Keep it at 4 °C until use (*see* **Note 3**).

3 Methods

3.1 Preparation of I-SceI and Tol2-Compatible Transgenic DNA Construct

1. Clone the expression cassette into the multiple cloning site (MCS) of the pBSII-SK-mTol2-MCS plasmid such that the I-SceI nuclease sites and miniTol2 recognition sites surround the cloned expression cassette. The resulting DNA construct can be used for both I-SceI– and Tol2- mediated transgenics [7, 11].
2. Purify the plasmid DNA with a plasmid maxiprep kit according to the manufacturer's instructions.
3. Dissolve the purified DNA pellet in 200–300 μL of DNase-/RNase-free water (*see* **Note 4**).

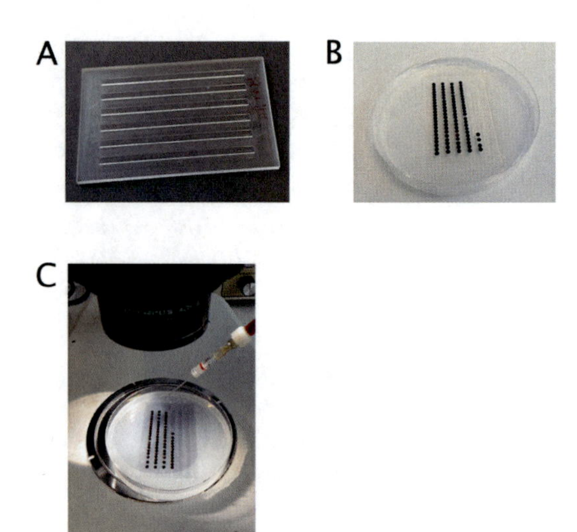

Fig. 1 (**a**) Plastic mold for creating channels into an agar layer as an injection plate for injecting Axolotl eggs. (**b**) Axolotl eggs arranged in the injection plate with the animal pole upwards. (**c**) The injection plate under the microscope and the injection needle

4. Verify the integrity of plasmid DNA by restriction analysis and sequencing of the plasmid DNA to make sure that there are no mutations in the cloned cassette (promoter, gene of interest, and polyadenylation signal). This is achieved by designing multiple primers for sequencing that cover the entire cloned cassette.
5. Divide the plasmid into small aliquots (about 10 μL). Store at −20 °C until use (*see* **Note 5**).

3.2 In Vitro Transcription of the Tol2 Transposase mRNA

Linearize 5–10 μg of T7-TPase plasmid DNA with BspEI restriction enzyme with an overnight digestion reaction at 37 °C:

1. Mix 5–10 μg of T7-Tpase plasmid DNA, 10 μL of NEBuffer 3.1, and 2 μL of BspEI enzyme. Make the volume up to 100 μL with DNase-/RNase-free H_2O.
2. Run 1–2 μL of the digested DNA in a 1% (wt/vol) agarose gel to check if all of the plasmids have been linearized. The linearized plasmid will show one sharp band at 5.3 kb (*see* **Note 6**).
3. Use the mMESSAGE mMACHINE T7 ultra kit to transcribe the Tol2 RNA in vitro, follow the manufacturer's instructions. But keep 2.5 μL before proceeding with polyA tailing (*see* **Note** 7).
4. Dissolve the RNA pellet after precipitation in 20 μL of DNase-/RNase-free water and measure the concentration (*see* **Note 8**).

5. To confirm the addition of a polyA tail run 500 ng of the RNA and the 2.5 μL RNA saved before polyA tailing on 1% (wt/vol) agarose gel. Both samples should run as single sharp bands without smears. The band of the RNA after polyA tailing should be bigger in size than that of the RNA before polyA tailing.
6. Divide in aliquots in 0.2-mL tubes (10 μL). Keep the tubes on ice while aliquoting. Store at −80 °C (up to 2 years).

3.3 Injection of DNA or RNA-DNA Mix

An injection is carried out in freshly laid fertilized eggs at one-cell stage. The Axolotl eggs are enclosed by a jelly coat and a capsule. Both have to be removed manually before injection.

1. Collect freshly laid eggs by removing them from the plastic leaves with the help of a scalpel. Put them in a plastic container with fresh tap water.
2. Pour the eggs into a metal sieve and rinse them with 70% Ethanol for about 20 s to sterilize the surface of the jelly coat. Then wash with sterile water and pour them into a glass beaker with 1× MMR/pen-strep solution.
3. Put some of the eggs in a 94 mm petri dish containing 1× MMR/pen-strep solution by using a sterile transfer pipette. Keep the glass beaker with the remaining eggs at 4 °C (*see* **Note 9**).
4. Place the Petri dish under a dissecting microscope and dejelly them manually by using two pairs of forceps (*see* **Note 10**).
5. Cut the tip of a transfer pipette to get a bigger diameter and transfer the dejellied eggs into a fresh petri dish (94 cm) filled with 1× MMR/pen-strep solution. Keep only healthy-looking eggs and discard damaged ones (*see* **Note 11**).
6. Pour some 1× MMR/20% Ficoll/pen-strep solution into an injection plate and transfer the dejellied eggs and arrange them with the brown animal pole facing up in the channels (*see* **Note 12**) (Fig.1b). Place the plate under the microscope for injections (Fig. 1c).
7. Prepare the DNA/RNA injection mix on ice.
   (a) *For I-SceI–Mediated Transgenesis*:

   Concentration: 0.5 ng DNA + 0.005 U I-SceI in 5 nL

   1 μg Plasmid DNA

   1 μL 10× SceI buffer

   2 μL I-SceI enzyme

   DNase-/RNase-free H_2O to 10-μL final volume

   (b) *mTol2 Transposase RNA-Mediated Transgenesis:*

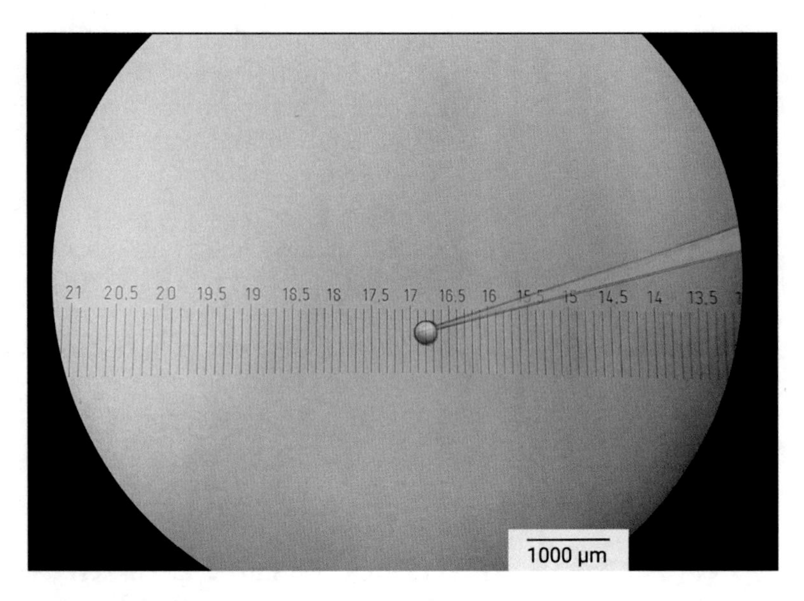

Fig. 2 A scale bar slide as seen from the microscope. A droplet is attached to the needle showing the approximate diameter that would contain 5 nL

Concentration: 0.05 ng DNA + 0.25 ng Tol2-RNA in 5 nL

μg Plasmid DNA

0.5 μg mTol2 RNA

DNase-/RNase-free H_2O to 10-μL final volume

8. Fill the glass needle (Subheading 3.2) with the injection mix and insert the needle into the micromanipulator.
9. Measure the droplet size with the help of a slide with a micrometric scale under the microscope.

   Break off the very tip of the needle with forceps.
10. Make one injection into the air above the scale bar slide to check the droplet volume by measuring the diameter (Fig. 2). Adjust the droplet size to 5 nanoliters by either altering the injection time (start with 200 msec) or by breaking off more of the needle tip. Avoid leaving the loaded needle for a long time in the air, this will dry the contents and clog the needle.
11. Inject the eggs in the dark brown animal pole region (*see* **Note 13**); this is where the cytoplasm and pronucleus are located.
12. Transfer the injected eggs into a fresh Petri dish filled with 1× MMR/20% Ficoll/pen-strep solution. Close the lid and leave the dish for 2 h at room temperature (up to 24 °C).
13. Transfer the eggs into a Petri dish filled with 0.1× MMR/5% Ficoll/pen-strep and leave them for 24 h at room temperature (up to 24 °C) (*see* **Note 14**).

14. The next day, transfer the properly developed embryos into sterile 24-well plates filled with 0.1× MMR/pen-strep solution; one embryo per well. Leave the embryos in these plates for about 1 week without changing the solution at room temperature (up to 24 °C) (*see* **Note 15**).

3.4 Screening

1. After 1 week, transfer the larvae into bigger well plates (e.g., 6-well plates) or Petri dishes filled with fresh 0.1× MMR/pen-strep solution.
2. About 2 weeks after injection, screen the larvae for transgenic expression under a fluorescence microscope (*see* **Note 16**).
3. Prepare a 0.01% anesthetic by diluting the 0.03% Benzocaine solution withholding water.
4. Gently transfer the larvae with a transfer pipette one by one and place them into a Petri dish filled with 0.01% Benzocaine solution and wait until they are anesthetized (*see* **Note 17**).
5. Transfer one larva with the transfer pipette and place it together with a drop of the Benzocaine solution in a Petri dish under the fluorescence microscope to check transgenic expression (*see* **Note 18**).
6. Keep only strong founders (fluorescence gene expression in 95–100% of the body) and place them into small plastic containers. Feed them daily with Artemia (*see* **Note 19**).
7. Characterization and validation of the transgenic lines is an important step. We suggest tissue sectioning and the use of specific antibodies when available, a PCR-based genotype, or tamoxifen conversion, and excision when a reporter with floxed allele is used.
8. After 9–12 months when the animals are sexually mature check for germline transmission by mating them with a non-transgenic animal and screening F1 progeny for transgene expression.

4 Notes

1. Our plastic mold was custom made at the Max Planck Institute for Cell Biology and Genetics workshop. It should fit into a 94 mm petri dish (about 4 cm × 6.5 cm, thickness 2.5 mm) and the channels should be 2 mm wide and 0.5 mm deep, the distance between the channels 3 mm.
2. Divide the I-SceI enzyme in 2 μL aliquots in 0.5 mL tubes and freeze at −80 °C. When preparing the injection mix, add the other components to this tube.

3. Prepare more injection plates, seal them with Parafilm and keep them in the fridge. Take a new one when you change the construct.
4. We dilute the plasmid to a concentration of 1 μL/μg with DNase-/RNase-free water for easier calculation of the injection mix.
5. You can store the plasmid at −20 °C for an unlimited time.
6. Run 500 ng of undigested plasmid together with the linearized plasmid as a control.
7. Wipe pipettes and the bench surface with RNaseZap (or equivalent reagents) and use DNase-RNase-free pipette tips. In vitro transcription and all RNA, handling should be performed under RNase-free conditions.
8. We dilute the concentration to 500 ng/μL with DNase-/RNase-free water for easier calculation of the injection mix.
9. The Axolotl eggs start to divide 6–7 h after laying. When kept at 4 °C, this timepoint will be postponed. But they should not be kept at 4 °C for more than 2–3 h, as this would result in poor development.
10. Grab the jelly coat with forceps and pierce at the same time with one tip of the forceps through the capsule, take the second pair of forceps and tear the capsule apart. Keep dejellied eggs at 4 °C.
11. Always use this prepared wide-bore pipette for dejellied eggs since they are very sensitive.
12. If there is enough liquid in the channels, the brown animal pole points upwards automatically. The eggs should be covered with liquid, but too much liquid would cause the eggs to move or dodge during the injection. Thus, when all eggs are arranged, remove excess liquid leaving just enough to cover the eggs.
13. Inject only eggs in a one-cell stage. Try to hit an imaginary line going through the middle, from the animal to the vegetal pole of the egg. In freshly laid eggs (Fig. 3a), aim to inject below the pole. In older eggs, aim toward the mid-height of the animal pole (Fig. 3b). Very fresh eggs are uniformly brown (Fig. 4a). Later the brown color fades and becomes irregular (Fig. 4d), this is also a good time to inject, but do not inject if you can already see a cleavage.
14. Do not place more than 60 eggs in a 94 mm Petri dish and 100 in a 145 mm Petri dish. Distribute them evenly over the entire surface.
15. We recommend to check the embryos under the microscope the next day of the injection and remove the damaged or unfertilized eggs (Fig. 4b, c), and before transferring them into 24 well plates. Unhealthy eggs are better identified under the microscope.

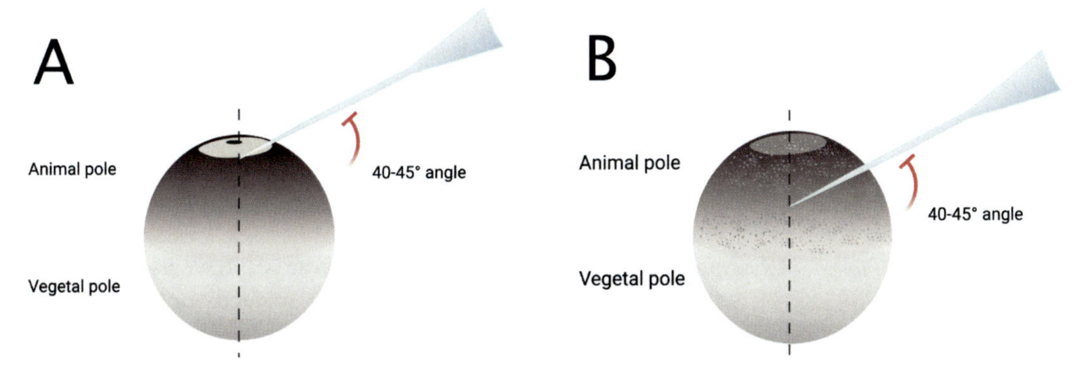

Fig. 3 (**a**) Graphic representation of a one-cell stage egg that has been freshly laid. At an approximate angle of 40–45°, the needle is positioned just below the upper pole of the egg. (**b**) Graphic representation of a one-cell stage egg that has been laid for a few s, where the brown color starts to fade and becomes irregular. At an approximate angle of 40–45°, the needle is positioned at mid-height of the animal pole

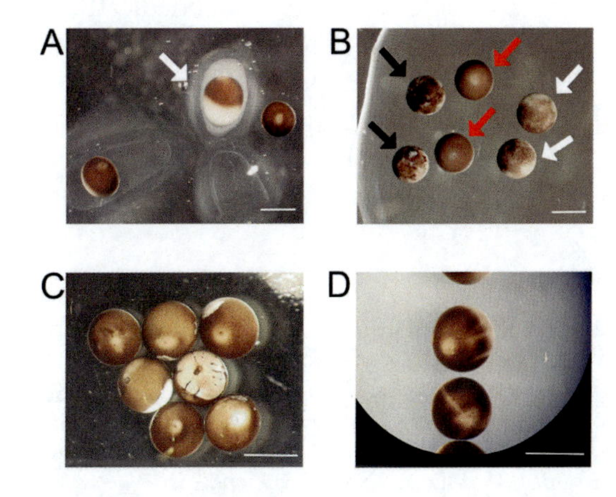

Fig. 4 (**a**) Freshly laid eggs result in higher efficiency of injections; these look uniformly brown in the animal pole. The arrow shows an egg still encapsulated in jelly, with the two poles visible. Scale bar 2 mm. (**b**) One day after injection, the egg batch will contain eggs with a different appearance. The eggs to the left of the image are damaged eggs and should be removed (black arrows). The eggs in the middle are unfertilized and should be removed (red arrows). Only keep the eggs that look like those marked with the white arrows. Scale bar 2 mm. (**c**) Examples of damaged eggs; these should be removed and not be injected. Scale bar 2 mm. (**d**) Example of eggs laid 4 h before. The brown color fades and becomes irregular. These eggs can also be injected, following the Fig. 3 indications. Scale bar 2 mm

16. Poor survival or very low expression can be due to the DNA amount injected. When poor survival is the case, inject only 25–50% of the initial DNA concentration. When there is weak or no expression, inject 25–50% more DNA. If you experience nonuniform expression with Tol2 RNA-injected animals, inject 25–50% less transposase mRNA.

17. Anesthetize no more than 4–5 larvae at once to avoid prolonged exposure to benzocaine. They are sensitive to benzocaine.
18. It is not possible to screen the floating larvae in a Petri dish filled with Benzocaine solution. Take just enough Benzocaine solution to prevent the larvae from drying out. Put them back into freshwater as soon as possible.
19. About 2 weeks after injection, the larvae will have used up their yolk supply and have to be fed. Start to feed them when the previously black belly turns transparent.

Acknowledgments

We are thankful to Beate Gruhl, Anja Wagner, and Dr. Judith Konantz for their dedication to the axolotls. We would like to thank all members of the Sandoval-Guzmán Lab for their unconditional support. This work was supported by the Center for Regenerative Therapies Dresden and the German Research Council (DFG).

References

1. Thermes V, Grabher C, Ristoratore F, Bourrat F, Choulika A, Wittbrodt J, Joly J-S (2002) I-SceI meganuclease mediates highly efficient transgenesis in fish. Mech Dev 118: 91–98. https://doi.org/10.1016/s0925-4773(02)00218-6
2. Ueda Y, Kondoh H, Mizuno N (2005) Generation of transgenic newt Cynops pyrrhogaster for regeneration study. Genesis 41:87–98. https://doi.org/10.1002/gene.20105
3. Pan FC, Chen Y, Loeber J, Henningfeld K, Pieler T (2006) I-SceI meganuclease-mediated transgenesis in Xenopus. Dev Dyn 235:247–252. https://doi.org/10.1002/dvdy.20608
4. Suster ML, Abe G, Schouw A, Kawakami K (2011) Transposon-mediated BAC transgenesis in zebrafish. Nat Protoc 6:1998–2021. https://doi.org/10.1038/nprot.2011.416
5. Suster ML, Sumiyama K, Kawakami K (2009) Transposon-mediated BAC transgenesis in zebrafish and mice. BMC Genomics 10:477–477. https://doi.org/10.1186/1471-2164-10-477
6. Sobkow L, Epperlein HH, Herklotz S, Straube WL, Tanaka EM (2006) A germline GFP transgenic axolotl and its use to track cell fate: dual origin of the fin mesenchyme during development and the fate of blood cells during regeneration. Dev Biol 290:386–397. https://doi.org/10.1016/j.ydbio.2005.11.037
7. Khattak S, Schuez M, Richter T, Knapp D, Haigo SL, Sandoval-Guzmán T, Hradlikova K, Duemmler A, Kerney R, Tanaka EM (2013) Germline transgenic methods for tracking cells and testing gene function during regeneration in the axolotl. Stem Cell Reports 1:90–103. https://doi.org/10.1016/j.stemcr.2013.03.002
8. Gerber T, Murawala P, Knapp D, Masselink W, Schuez M, Hermann S, Gac-Santel M, Nowoshilow S, Kageyama J, Khattak S, Currie J, Camp JG, Tanaka EM, Treutlein B (2018) Single-cell analysis uncovers convergence of cell identities during axolotl limb regeneration. Science 20:eaaq0681-19. https://doi.org/10.1126/science.aaq0681
9. Currie JD, Kawaguchi A, Traspas RM, Schuez M, Chara O, Tanaka EM (2016) Live imaging of axolotl digit regeneration reveals spatiotemporal choreography of diverse connective tissue progenitor pools. Dev Cell 39: 411–423. https://doi.org/10.1016/j.devcel.2016.10.013
10. Sandoval-Guzmán T, Wang H, Khattak S, Schuez M, Roensch K, Nacu E, Tazaki A, Joven A, Tanaka EM, Simon A (2013)

Fundamental differences in dedifferentiation and stem cell recruitment during skeletal muscle regeneration in two salamander species. Cell Stem Cell 14:174–187. https://doi.org/10.1016/j.stem.2013.11.007

11. Khattak S, Murawala P, Andreas H, Kappert V, Schuez M, Sandoval-Guzmán T, Crawford K, Tanaka EM (2014) Optimized axolotl (Ambystoma mexicanum) husbandry, breeding, metamorphosis, transgenesis and tamoxifen-mediated recombination. Nat Protoc 9:529–540. https://doi.org/10.1038/nprot.2014.040

Chapter 22

A Practical Guide for CRISPR-Cas9-Induced Mutations in Axolotls

Konstantinos Sousounis, Katharine Courtemanche, and Jessica L. Whited

Abstract

Clustered regularly interspaced short palindromic repeats (CRISPR) is a powerful tool that enables editing of the axolotl genome. In this chapter, we will cover how to retrieve gene sequences, confirm annotation, design CRISPR targets, analyze indels, and screen for Crispant axolotls. This is a comprehensive guide on how to use CRISPR on your favorite gene and gain insights into its function.

Key words CRISPR, Axolotls, CRISPR design, Axolotl screening, Genome targeting, CRISPR-Cas9

1 Introduction

CRISPR was initially found as a bacterial defense mechanism against viruses [1, 2]. This RNA molecule contains sequences required for its three-dimensional structure and targeting specificity. In the laboratory, CRISPR is usually broken into two pieces for convenience: the crRNA, which contains the guide RNA (gRNA) sequence and provides the specificity and targeting, and the tracrRNA which is a supporting sequence and common in all CRISPR experiments. The resulting RNA sequence forms a complex with the Cas9 nuclease which is responsible for cleaving the target DNA. Indels—insertions and deletions which deviate from the original DNA sequence—are introduced by the innate cellular machinery in an attempt to repair the damage [3, 4]. These indels may be predicted to cause loss-of-function alleles or even to eliminate gene function, thus creating mutant alleles at a target locus. CRISPR has become increasingly popular in understanding the role of genes in axolotl biology due to the ease of using the technique and widespread technical literature [5–13].

Designing a CRISPR experiment involves several steps while quality controls are critical to ensure that the intended gene is targeted, off-target effects are minimized, and reared Crispant

Ashley W. Seifert and Joshua D. Currie (eds.), *Salamanders: Methods and Protocols*,
Methods in Molecular Biology, vol. 2562, https://doi.org/10.1007/978-1-0716-2659-7_22,

axolotls have the best chance of propagating an indel of interest. Given that axolotls may require 9 months to reach adulthood and have a chance for F1-edited offspring, all these steps are essential, and this guide provides considerations on how to tackle them. Even by taking all the necessary precautions, a fully edited Crispant axolotl may not have the intended phenotype because the gene in question has no role for what it is studied for, truncated versions and isoforms of the gene or even other related genes are sufficient to replace its function [14]. In addition, resulting phenotypes may hinder important downstream analysis. For instance, mutations that cause embryonic lethality or significant organ deformities during development would not allow for proper functional analysis during tissue regeneration. Taken together, CRISPR is a powerful tool that may allow for significant advances in understanding axolotl biology, but researchers should take every precaution possible to fully take advantage of this technology.

In this chapter, we will provide a guide on how to design CRISPR experiments, prepare materials, inject embryos, and screen Crispant axolotls for mutations in the target locus.

2 Materials

2.1 Molecular Biology Materials

1. Agarose (IBI #IB70072).
2. Phusion High-Fidelity DNA polymerase (NEB #M0530).
3. DNA Extraction Kit (Qiagen #69504).
4. PCR purification (Beckman Coulter AMPure XP #A63881).
5. Penicillin-Streptomycin 10,000 U/mL (PenStrep, Gibco #15140148).
6. Ficoll 400 (Millipore Sigma #F4375).
7. UltraPure DNA/RNase-Free Distilled Water (Invitrogen #10977015).
8. Alt-R CRISPR-Cas9 crRNA (IDT Custom design).
9. Alt-R CRISPR-Cas9 tracrRNA (IDT #1072534).
10. Alt-R S.p. Cas9 Nuclease V3 (IDT #1081058).
11. 500 mL 0.22 μm PES Filter System (Celltreat #229707).

2.2 Bioinformatics

1. SnapGene (https://www.snapgene.com) [15].
2. NCBI Blast (https://blast.ncbi.nlm.nih.gov/Blast.cgi) [16].
3. NCBI Gene (https://www.ncbi.nlm.nih.gov/gene) [17].
4. NCBI primer Blast (https://www.ncbi.nlm.nih.gov/tools/primer-blast/) [18].
5. Axolotl Genome Blast (https://ambystoma.uky.edu/genome-resources) [19, 20].

6. Benchling (https://www.benchling.com) [21].
7. IDT CRISPR-Cas9 guide RNA design checker (https://www.idtdna.com/site/order/designtool/index/CRISPR_CUSTOM) [22].

2.3 Injections

1. Stereoscope (Linitron #113723).
2. Microinjector (Warner Instruments #PLI-90A).
3. Manual Control Micromanipulator (Warner Instruments #MM-33).
4. Micropipette puller (Sutter Instrument #P-97).
5. Capillary Tubing Length 10 cm, Borosil 1 mm OD × 0.75 mm ID/Fiber (FHC #30-30-0).
6. Petri Dishes 10 cm (Celltreat #229693).
7. Petri Dishes 3.5 cm (Celltreat #229635).
8. Microloader tips (Eppendorf).
9. Well mold 6.5 cm × 4 cm × 0.2 cm with 6 well strips 5 cm × 0.2 cm × 0.1 cm, 0.2 cm apart.

2.4 Solutions

1. CRISPR-Cas9 solution recipe: Mix 2 μl 100 μM crRNA and 2 μl 100 μM tracrRNA. Heat solution at 95 °C for 5 min. Let the solution cool to room temperature. Add 7 μl UltraPure water and 1.2 μl 10 μg/μl Cas9 protein. Mix solution with gentle pipetting. Heat at 37 °C for 10 min and store on ice until same-day injection or at −80 °C for long-term.
2. 10× Marc's modified Ringer's (MMR): For 2 L solution mix 400 mL 5 M NaCl, 40 mL 1 M KCl, 20 mL 1 M $MgCl_2$, 40 mL 1 M $CaCl_2$, 4 mL 0.5 M EDTA, 100 mL 1 M HEPES. pH solution to 7.8 with 10 M KOH and adjust volume to 2 L with milliQ water. Autoclave.
3. 1× MMR PenStrep: For 1 L solution mix 100 mL 10× MMR and 13 mL PenStrep. Adjust volume to 1 L with milliQ water and sterilize with 0.22 μm filter.
4. 1× MMR 20% Ficoll PenStrep: For 250 mL solution mix 25 mL 10× MMR, 3.125 mL PenStrep, and 50gr Ficoll. Adjust volume to 250 mL with milliQ water. Stir until Ficoll is completely dissolved. Adjust volume again to 250 mL with milliQ water if necessary. Sterilize the solution with 0.22 μm filter.
5. 0.1× MMR 5% Ficoll PenStrep: For 500 mL solution mix 50 mL 10× MMR, 6.25 mL PenStrep and 25gr Ficoll. Adjust volume to 500 L with milliQ water. Stir to completely dissolve Ficoll. Sterilize the solution with 0.22 μm filter.

6. 0.1× MMR PenStrep: For 2 L solution mix 200 mL 10× MMR and 26 mL PenStrep. Adjust volume to 2 L with milliQ water. Sterilize the solution with 0.22 μm filter.

3 Methods

3.1 Bioinformatics and Guide RNA Design (Fig. 1)

1. *Find the Complete Open Reading Frame (ORF) of the Target Gene Using SnapGene.*

   This can be accomplished by data mining available axolotl transcriptomes and the process will depend on how the target gene was selected (*see* **Note 1**). The overall process of recovering gene sequences will involve downloading files linked to a particular axolotl transcriptome: typically, a file containing the sequences (usually in .fasta format) and the associated annotation (file usually in .txt format) are useful starting points. The two files are usually connected by shared identifiers (ID). Select the sequence of your target gene by using the ID or the gene name/annotation.

2. *Analyze Sequence by Verifying Its Annotation Using NCBI blastx.*

   This step is important as the axolotl transcriptome and genome are not curated and thus it is imperative to properly distinguish among paralogous genes (*see* **Note 2**). If your sequence contains 5′ and/or 3′ untranslated regions, copy and paste the target sequence into a software that can identify ORFs, such as SnapGene. Start by copying the sequence of the biggest ORF available with a 5′–3′ orientation and +1 frame, and paste it on the NCBI blastx software to verify that the original annotation is accurate. Make sure that the frame shown is +1 indicating that the translation of the ORF started from the first three nucleotides. The following would provide a strong indication that the gene is the vertebrate ortholog: very low E-value ($<1E^{-20}$, with 0 being the best), high query coverage (>80%, with 100% being the best), and high identity (>75% with 100% being the best). Take care to ensure you have recovered the entire ORF (starting with a methionine, and that the methionine that appears first maps to the beginning M in orthologous sequences recovered by the blast).

3. *Find the Target Locus in the Axolotl Genome and Identify Exons, Introns, and Functional Motifs in Both the Locus and the ORF Using the Axolotl Genome Blast.*

   This step will provide the canvas that will inform your decisions for CRISPR targeting. Compare the target ORF against the axolotl genome. One way to accomplish that is using the axolotl genome blast software or browser from Sal-Site (https://ambystoma.uky.edu). The correct locus

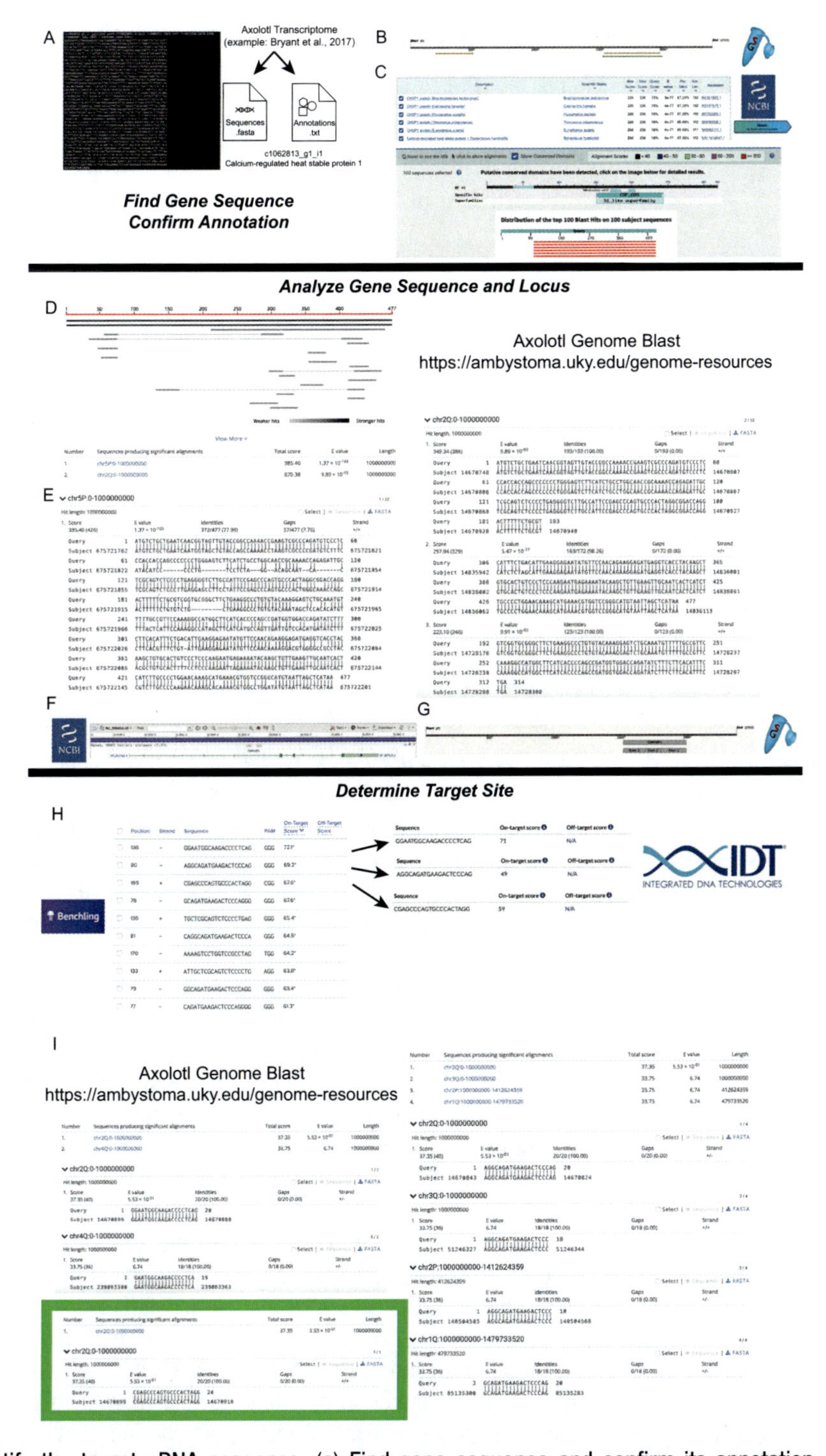

Fig. 1 Identify the target gRNA sequence. (**a**) Find gene sequence and confirm its annotation. (**a1**) As an example, the generation of *calcium-regulated heat-stable protein 1* (*carhsp1 or chsp1*) CRISPant axolotls will be shown. The gene was found as blastema-enriched during axolotl limb regeneration and its contig ID is c1062813_g1_i1 with CHSP1_MOUSE annotation [26]. (**a2**) Copy the whole sequence as shown from the de novo assembled transcriptome and paste it into SnapGene. Find the available ORFs (yellow and green arrows). (**a3**) Copy the largest ORF; in our case, this is represented by the green arrow on the reverse and complementary side. Make sure that the sequence is copied at 5′ to 3′ orientation and beginning with the

should have 100% coverage of the query, low e-value, and high identity. Based on the number of exons the gene contains, the results would be split into different sections. In most cases the number of exons matches that of other vertebrates, providing strong evidence of orthology. The different sections provide a rough estimate of the intron/exon junctions in your ORF (*see* **Note 3**).

4. *Identify the Best Possible Site to Target.*

   Copy and paste a particular exon, part of the sequence, or the whole ORF as a New DNA Sequence into Benchling. Identify all possible targetable sites by utilizing the CRISPR function of Benchling with default parameters (single guides, guide length: 20, Genome: AmbMex13_14.1.0 (Ambystoma mexicanum) and PAM: NGG (SpCas9, 3'side)). Rank the sequences based on their on-target score and select them based on their location on the ORF (*see* **Note 4**). Copy and paste the sequences in the IDT gRNA checker without selecting any target genome and rank them based on their on-target score alone. Select the top10 and paste them in the Axolotl

Fig. 1 (continued) starting codon. Paste the sequence in the NCBI blastx software to validate the annotation. The sequence shows high identity and coverage, and the low e-value against the CHSP1 protein provides strong evidence that the annotation is correct. In this example, we note that the beginning of the gene lacks coverage against the target protein, so caution should be exercised, and additional sequence analysis may be necessary. (**b**) Analyze gene sequence and target locus. (**b1**) Copy the ORF and paste it into the Axolotl Genome Blast against the newest genome assembly available. The results show two good hits on the genome at chromosomes 2 and 5 (black lines covering the whole red query). (**b2**) By looking at the sequence alignment we can determine with high confidence that our sequence is on chromosome 2 and has three exons based on the three sections our gene has been broken down into. Note that for the hit result from chromosome 2 almost all sequence alignments are identical. In this example, the sequence in chromosome 5 has a lower e-value due to the continuity of the sequence alignment so all parameters should be taken into consideration before identifying the correct locus on the genome. (**b3**) In order to double-check that our gene has actually three exons, the human CHSP1 gene is searched on the NCBI gene database. The results show that the human *Chsp1* gene also has three exons (dark green boxes), providing strong evidence that the locus located in chromosome 2, on the q arm from 14,670,748 bp until 14,836,113 bp contains the axolotl *Chsp1* gene. Phylogenetic tree analysis would further prove the correct orthology of the gene. (**b4**) Based on the Axolotl Genome blastn result, the Snapgene file can be updated containing the exon information. Using the available sequence information, the location of functional motifs, potential alternative start codons, and potential alternative splicing variants, determine the best exon to target with CRISPR. In our example, exon 1 would be the best. (**c**) Determine target site. (**c1**) Copy the exon 1 sequence and paste it into Benchling. Determine the available targetable sites using the available features in the software and rank them based on their on-target score. Copy the top10 sequences and paste them into the IDT gRNA checker. Rank them based on their on-target score to get the final best-to-worst sequence list. (**c2**) Copy the top 3 sequences and paste them into the Axolotl Genome blast to check their specificity against the axolotl genome. The results show that gRNA sequence #3 is very specific (Green box). In our example, extra precautions should be taken that the gRNA sequence chosen is very dissimilar to the previously identified, highly similar, genomic area in chromosome 5 (*see* **b2**)

Genome Blast to test their specificity. Select only gRNA sequences that have a single hit to the intended locus. Order crRNA from IDT.

3.2 CRISPR-Cas9 Injections (Fig. 2)

1. *Inject single-cell axolotl embryos with CRISPR-Cas9 complex* (*see* **Note 5**): Make CRISPR-Cas9 solution using the crRNA against the gene of your choice.
2. Fill 10 cm petri dishes with 2% agarose solution in milliQ water and gently float the well molds on top, raised ridges down. Let them solidify and then remove the mold.
3. Use the following program on the micropipette puller to generate fine needles from capillary tubing: HEAT Ramp+10, PULL 90, VEL 70, DEL 90, and PRESSURE 200.
4. Breed axolotls and provide plastic plants for the female to lay eggs on. When eggs start to appear, collect them every 30 min in a container with pre-chilled axolotl water.
5. Remove the jelly coat using fine forceps in chilled 1× MMR and place them in fresh 1× MMR at 4 °C (*see* **Note 6**).
6. Add 1× MMR 20% Ficoll to the agarose plates until molded space is completely submerged and use a disposable plastic transfer pipet (with the tip trimmed to be slightly wider than the embryos) to align the eggs in the wells with the animal pole (black hemisphere) facing upwards.
7. Count the total number of eggs to be injected.
8. Fill the needle with 5 μl CRISPR-Cas9 solution using a microloader tip (*see* **Note 7**).
9. Connect the needle to the microinjector and place it on a micromanipulator.
10. Using a stereoscope, bring the tip of the needle into your field of view. Break the tip of the needle using forceps until it is firm enough that does not bend when touching the surface of an axolotl embryo. Be careful not to break too much lest the diameter of the opening be too wide to safely inject the embryo without destroying it (*see* **Note 8**).
11. Adjust the injection volume to 5 nL (approximately 1/10 of the egg's diameter).
12. Inject the eggs at a 45° angle on the side of the egg so that the tip of the needle is below the top when inserted under the membrane.
13. Transfer injected embryos to a petri dish filled with fresh 1× MMR 20% Ficoll. Let them sit undisturbed for 2 h. Spread the embryos out across the dish.

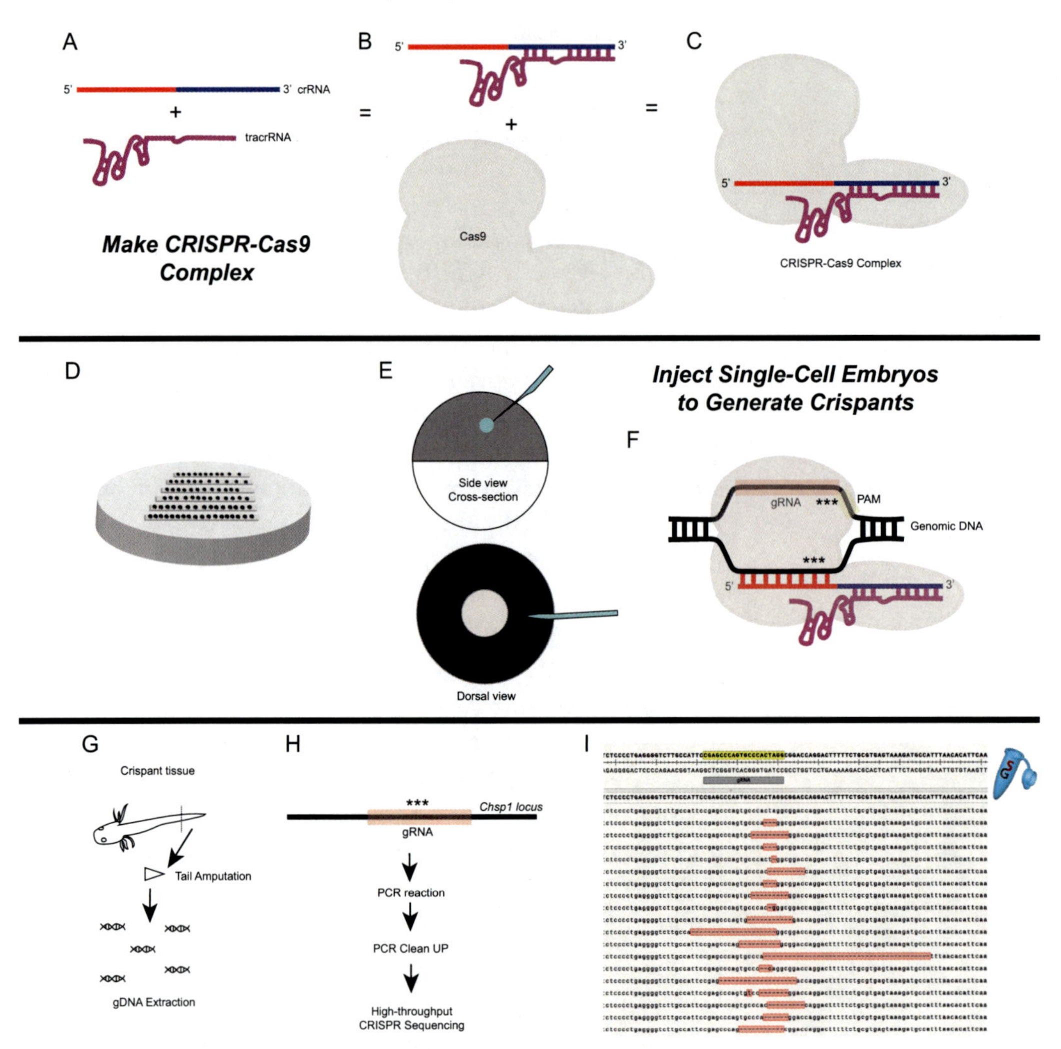

Fig. 2 Generate CRISPants. (**a**) Make the CRISPR-Cas9 Complex. (**a1**) An equimolar mixture of crRNA (red/blue line) and tracrRNA (pink line) is initially made. Red-line indicates the gRNA sequence. (**a2**) The crRNA-tracrRNA complex is then incubated with the Cas9 protein (gray). (**a3**) The completed CRISPR-Cas9 complex. (**b**) Inject single-cell embryos to generate CRISPants. (**b1**) Prepare an agarose petri dish with carved wells. Place the single-cell embryos in the wells containing 1× MMR 20% Ficoll PenStrep. (**b2**) Inject the CRISPR-Cas9 complex at a 45° angle on the side of the egg as shown from the side and dorsal view. Light-gray indicates the characteristic spot at the dorsal side of the egg. Cyan sphere indicates the relative to egg size of the injection volume. (**b3**) The CRISPR-Cas9 complex will cleave the target genomic locus and indels will subsequently be generated. The red shaded area is the gRNA sequence that was previously designed. The yellow shaded area is the PAM sequence required for CRISPR targeting. The black lines indicate the gDNA strands. (**c**) Genotype CRISPants. (**c1**) Anesthetize CRISPants and remove a small piece of their tail using a sterile razor blade. Take care not to use too much tissue. Digest the tissue and perform genomic DNA extraction according to the kit-specific protocol or using standard lab reagents. (**c2**) PCR amplify the genomic locus containing the gRNA target site and purify the amplicon before sending for high-throughput sequencing. (**c3**) Analyze the data by aligning the reads to the target locus using Snapgene. Our results indicate that the designed CRISPR against the axolotl *Chsp1* gene locus generates many different indels at the target site. Note that frame-shifting or stop codon generating mutations are preferred when loss-of-function is the end goal. Western blot and downstream functional analysis would further confirm whether gene knockout has been successfully achieved

14. Transfer embryos to a petri dish filled with 0.1× MMR 5% Ficoll and take care to spread them out. Incubate them overnight in a location where they will not be disturbed.
15. Transfer embryos to a petri dish filled with 0.1× MMR solution the following morning.
16. Transfer and count surviving embryos to fresh plates with 0.1× MMR every other day or as needed. Discard and count dead embryos daily (*see* **Note 9**).
17. Approximately 2 weeks following the injection, resume normal animal care based on their size and age. Axolotl larvae are typically ready to start feeding on live brine shrimp once the rostral-most aspect of their heads, when viewed from top, have become flattened rather than pointed.

3.3 Phenotype Crispants

1. Mutations caused by CRISPR-Cas9 may cause lethality during embryonic development. Compare the ratio of dead to surviving injected embryos to that of the *tyr* injected control. Discrepancies may indicate that the targeted gene is required for embryo survival.
2. Mutations caused by CRISPR-Cas9 may cause significant developmental abnormalities. Compared the development of the injected embryos to that of *tyr* controls under a stereoscope and note significant differences in patterning. Continue to observe the axolotls as they age until all appendages have fully matured.
3. Evaluate the correct development of internal organs and tissues of juvenile Crispants using Hematoxylin and Eosin of histological sections: Euthanize Crispants including *tyr* controls and fix them in 4% paraformaldehyde 0.7× PBS overnight at 4 °C. Dehydrate the specimens in an increasing ethanol series (25%, 50%, 75%, 80%, 85%, 90%, 95%, 100%) for 2 h each at room temperature, and then wash them two times in 100% xylene for 2 h at room temperature under a chemical hood. Replace xylene with melted paraffin and keep samples in an incubator set at 65 °C for 2 h. Replace with fresh paraffin and incubate samples at 65 °C for 2 h or until paraffin is liquid. Place specimen in tissue block and fill it with paraffin. Let the sample solidify overnight. Using a microtome, acquire representative cross-sections of the whole body by placing 1 section in every 10 cuts on a microscope slide. Align sections next to each other for easier observation later. Slides containing sections are then washed with 100% xylene and 100% ethanol before hematoxylin and eosin stains are performed. Compare the morphology of internal tissues and organs between experimental and control samples to determine whether your Crispant axolotls show any developmental phenotypes.

3.4 Genotype Crispants

1. Use NCBI primer blast to design primers flanking the gRNA location based on the axolotl genome (*see* **Note 10**).
2. Design more than one primer pair and test them on naïve axolotl genomic DNA. The best primer pair should produce a single bright band (*see* **Note 11**).
3. Test the overall efficiency of the CRISPR-Cas9 complex to produce indels by randomly euthanizing 5–10 injected embryos following neurulation [23].

   Collect tissue from individual Crispant larvae or older axolotls while numbering all animals and their respective specimens (*see* **Note 12**).

   Extract genomic DNA using a Qiagen DNA extraction kit.
4. Perform PCR using Phusion High-Fidelity DNA polymerase and check the band size and specificity in an agarose gel following electroporation (*see* **Note 13**).
5. Purify PCR product using AMPure XP beads (*see* **Note 14**).
6. Sequence amplicon using a high-throughput platform (*see* **Note 15**).
7. Align the reads to the target locus using SnapGene to identify indels generated by CRISPR.

Considerations for Breeding Schemes When Generating CRISPant Axolotls

How the generated F_0 CRISPants are bred will determine the effects of their genetic background on the downstream experiments. The editing efficiency, potential of CRISPR-Cas9 off-targets, and contributions from the genetic background are listed below based on the experimental setup.

1. No Breeding: Using F_0 animals
   1. Editing efficiency: Gradient (0–100%) – will be based on the ability of the CRISPR-Cas9 complex to generate indels. Note that the editing efficiency, as determined using the means explained here, cannot give information about the exact breakdown of how many individual cells are heterozygous versus homozygous at a given locus; it only gives conglomerate information about the frequency of alleles across a population of cells.
   2. Off-targets specific to gRNA – If any, likely to be the same in all animals, though there is an element of stochasticity to this issue.
   3. Contributions from their genetic background – If siblings, the same in all animals.
2. $F_0 \times F_0$ sibling breeding: Using F_1 inbred animals
   4. Editing efficiency: Binary (0%, 50%, 100%) with multiple indels – will be based on the parental indel complexity.

5. Off-targets specific to gRNA – If any, the same in all animals except if the off-target location is on the same chromosome and very close to the target locus (because low chance of crossover between these loci).
6. Contributions from their genetic background – The same in all animals but homozygosity from inbreeding may contribute to increased variability during data collection.

3. $F_0 \times F_0$ non-sibling breeding: Using F_1 animals
7. Editing efficiency: Binary (0%, 50%, 100%) with multiple indels – will be based on the parental indel complexity.
8. Off-targets specific to gRNA – If any, the same in all animals except if the off-target location is in the same chromosome and very close to the target locus.
9. Contributions from their genetic background – The same in all animals.

4. $F_1 \times F_1$ Breeding: Using F_2 animals is the Best but Time-Consuming
10. Assumes each F_0 animal was outcrossed to a *wild-type* or *white* mutant mate to generate F_1, and that the individual F_1 animals were raised to maturity and bred to one another.
11. Editing efficiency: Binary (0%, 50%, 100%) with specific indels so animals could be homozygous.
12. Off-targets specific to gRNA – If any, the same in all animals except if the off-target location is in the same chromosome and very close to the target locus but with higher chances to be diluted.
13. Contributions from their genetic background – The same in all animals.

4 Notes

1. If the sequence is unknown, select a sequence (contig) from the axolotl transcriptome using the available annotation. There are many resources for axolotl transcriptomic data [24–28]. If the target gene was selected based on the gene expression data from a published paper, and in order to avoid discrepancies in gene annotation and potentially recover the wrong gene sequences, it is important to work with the same transcriptome files that the original paper was using. This is because expression values and annotation are sequence-dependent and sometimes annotations are redundant in non-curated transcriptomes like that of the axolotl. Please refer to chapters in this book on efforts to streamline genomic and transcriptomic data in axolotls.

2. To confirm the exact orthology of the gene, phylogenetic tree and synteny analysis should be used. This is important in order to properly name the gene and subsequently the locus targeted using CRISPR.

   If the gRNA sequence is located near the junction, identify its exact location. This is important because the gRNA sequence will be determined based on the sequence as shown in the axolotl transcriptome/ORF since currently this has the least amount of sequence discrepancies which may affect CRISPR specificity.

3. The location on the locus where CRISPR-Cas9 will introduce DNA damage is important in generating CRISPant animals. Alternative start sites and intron splicing may render the resulting CRISPants to have close to normal protein activity. Targeting the functional domain may increase the odds of nullifying the function even if some form of truncated protein is being translated. Functional motifs and their location can be found by utilizing the human orthologue annotation and sequence alignment or available literature.

   Again, make sure the identification of the gRNA sequences is based on the ORF from an axolotl transcriptome because these have the best sequence fidelity, and keep in mind that the CRISPR target cannot span intron/exon junctions. If there are no targetable areas in the exon and the gRNA sequence spans to the nearby intron, it is recommended that the target is sequenced before the initiation of the experiment.

4. In the laboratory, the most common way to generate the CRISPR-Cas9 complex is by combining the tracrRNA, crRNA, and Cas9 protein. TracrRNA and Cas9 for loss-of-function experiments are the same irrespective of the target gene. CrRNA provides specificity since it contains the guide RNA sequence. All molecules can be purchased from companies like IDT. The components can also be developed in the lab.

5. The jelly coat is more easily removed from freshly collected eggs versus those that are several hours old. Eggs can remain in the fridge for up to 1 day before injection.

6. CRISPR-Cas9 against the *tyr* locus is recommended as control [6]. These animals will develop unpigmented (white) eyes, indicating that the CRISPR-Cas9 reagents and protocol, as well as the egg injection technique, were successful. In addition, the *tyr* gene is not required for many developmental, regenerative, or homeostatic biological processes, making it an ideal experimental control, especially if F_0 Crispant animals are to be utilized.

7. In practice, a useful approach is for experienced lab members to train new researchers on embryo injection and for new researchers to then practice to perfect the technique. Because every (handmade/pulled and opened) needle is microscopically different, the output of each needle should be visually inspected following opening. Some researchers recommend calibrating needle output with a dye injection or by expelling some mock solution into the media. However, practice with injection (inject beside the egg, into the MMR/Ficoll media) and correlating the visual size of the injection droplet to survivability is an effective alternative method, and it does not require the solution to be removed and changed in an individual needle, which is technically difficult. Good injections will produce few—if any—exploded embryos, and embryo survival in subsequent days will also be high. Researchers should be sure to screen embryos for evidence of transgenesis and/or targeted genome editing while they are learning to guard against developing an injection technique with too little volume being administered.
8. Dead embryos may be disintegrating or show blebbing, but they may simply look significantly different than the healthy embryos (for example, very large variation in cell size, and swirling pigmentation patterns).
9. If genotyping will be determined via high-throughput sequencing, the gRNA location should be within 100 bp off of one of the sequencing primers so the sequencer will be able to properly detect the indels. Ideally, the genotyping product should not exceed 250–300 bp.
10. Optimize the protocol by performing temperature, gDNA amount, and primer concentration gradients until the best possible PCR efficiency is achieved. It is important to use a high-fidelity polymerase so indels are accurately preserved during amplification.
11. Tissue collection could be coupled with tail or limb amputation for a planned experiment if applicable.
12. Run PCR products in 4% agarose gels alongside controls to detect potential differences in band sizes which would indicate truncations in the target locus.
13. This step is best performed when the PCR reaction produces a single product with no additional bands. If multiple bands are shown during gel electrophoresis, then primers should be further troubleshot so only the intended amplicon is made. If this is not possible with the primers selected, consider designing new primers. It is not recommended to extract amplicons/bands from agarose gels due to potential cross-contamination between samples and low throughput when hundreds of animals are to be genotyped.

14. It is recommended that the exact indels are detected using high throughput sequencing, especially for F_0 Crispant animals. Core facilities may offer services to sequence amplicons or they should be contacted for specific instructions in generating amplicon libraries that can be sequenced.

Acknowledgment

Funding for the project was provided by the Sara Elizabeth O'Brien Trust, Bank of America, NA Trustee Postdoctoral Fellowship (K S), the National Eye Institute of the National Institutes of Health under Award Number K99EY029361 (K S), NIH Office of the Director (1DP2HD087953 to J L W), and the Eunice Kennedy Shriver National Institute of Child Health and Human Development (1R01HD095494 to J L W). We thank members of Whited laboratory for the discussions. The authors declare no conflicts of interest.

References

1. Jinek M, Chylinski K, Fonfara I, Hauer M, Doudna JA, Charpentier E (2012) A programmable dual-RNA-guided DNA endonuclease in adaptive bacterial immunity. Science 337(6096):816–821. https://doi.org/10.1126/science.1225829
2. Mali P, Esvelt KM, Church GM (2013) Cas9 as a versatile tool for engineering biology. Nat Methods 10:957–963
3. Hsu PD, Lander ES, Zhang F (2014) Development and applications of CRISPR-Cas9 for genome engineering. Cell 157(6):1262–1278
4. Doudna JA, Charpentier E (2014) The new frontier of genome engineering with CRISPR-Cas9. Science 346(6213):1258096
5. Fei J-F, Schuez M, Knapp D, Taniguchi Y, Drechsel DN, Tanaka EM (2017) Efficient gene knockin in axolotl and its use to test the role of satellite cells in limb regeneration. Proc Natl Acad Sci U S A 114(47):12501–12506. https://doi.org/10.1073/pnas.1706855114
6. Fei JF, Schuez M, Tazaki A, Taniguchi Y, Roensch K, Tanaka EM (2014) CRISPR-mediated genomic deletion of Sox2 in the axolotl shows a requirement in spinal cord neural stem cell amplification during tail regeneration. Stem Cell Reports 3:444–459. https://doi.org/10.1016/j.stemcr.2014.06.018
7. Fei J-F, Knapp D, Schuez M, Murawala P, Zou Y, Pal Singh S, Drechsel D, Tanaka EM (2016) Tissue- and time-directed electroporation of CAS9 protein–gRNA complexes in vivo yields efficient multigene knockout for studying gene function in regeneration. NPJ Regen Med 1(1):1–9. https://doi.org/10.1038/npjregenmed.2016.2
8. Fei JF, Lou WPK, Knapp D, Murawala P, Gerber T, Taniguchi Y, Nowoshilow S, Khattak S, Tanaka EM (2018) Application and optimization of CRISPR–Cas9-mediated genome engineering in axolotl (Ambystoma mexicanum). Nat Protoc 13(12):2908–2943. https://doi.org/10.1038/s41596-018-0071-0
9. Flowers GP, Timberlake AT, Mclean KC, Monaghan JR, Crews CM (2014) Highly efficient targeted mutagenesis in axolotl using Cas9 RNA-guided nuclease. Development 141(10):2165–2171. https://doi.org/10.1242/dev.105072
10. Sousounis K, Bryant DM, Fernandez JM, Eddy SS, Tsai SL, Gundberg GC, Han J, Courtemanche K, Levin M, Whited JL (2020) Eya2 promotes cell cycle progression by regulating DNA damage response during vertebrate limb regeneration. Elife 9:e51217. https://doi.org/10.7554/eLife.51217
11. Sanor LD, Flowers GP, Crews CM (2020) Multiplex CRISPR/Cas screen in regenerating haploid limbs of chimeric axolotls. Elife 9:e48511. https://doi.org/10.7554/eLife.48511
12. Lou WPK, Wang L, Long C, Liu L, Fei JF (2019) Direct gene knock-out of axolotl spinal cord neural stem cells via electroporation of

cas9 protein-grna complexes. J Vis Exp 149: e59850. https://doi.org/10.3791/59850

13. Flowers GP, Sanor LD, Crews CM (2017) Lineage tracing of genome-edited alleles reveals high fidelity axolotl limb regeneration. Elife 6:e25726. https://doi.org/10.7554/eLife.25726
14. Junker JP (2019) Detouring the roadblocks in gene expression. Nat Rev Mol Cell Biol 20(4): 197. https://doi.org/10.1038/s41580-019-0107-5
15. SnapGene software (from Insightful Science; available at snapgene.com)
16. Altschul SF, Gish W, Miller W, Myers EW, Lipman DJ (1990) Basic local alignment search tool. J Mol Biol 215(3):403–410. https://doi.org/10.1016/S0022-2836(05)80360-2
17. Benson D, Boguski M, Lipman DJ, Ostell J (1990) The national center for biotechnology information. Genomics 6(2):389–391. https://doi.org/10.1016/0888-7543(90)90583-G
18. Ye J, Coulouris G, Zaretskaya I, Cutcutache I, Rozen S, Madden TL (2012) Primer-BLAST: a tool to design target-specific primers for polymerase chain reaction. BMC Bioinformatics 13(1):1. https://doi.org/10.1186/1471-2105-13-134
19. Smith JJ, Timoshevskaya N, Timoshevskiy VA, Keinath MC, Hardy D, Voss SR (2019) A chromosome-scale assembly of the axolotl genome. Genome Res 29(2):317–324. https://doi.org/10.1101/gr.241901.118
20. Priyam A, Woodcroft B, Rai V, Munagala A, Moghul I, Ter F, Gibbins MA, Moon H, Leonard G, Rumpf W, Wurm Y (2015) Sequenceserver: a modern graphical user interface for custom BLAST databases. bioRxiv. https://doi.org/10.1101/033142
21. Benchling [Biology Software] (2021) Retrieved from https://benchling.com
22. Vakulskas CA, Dever DP, Rettig GR, Turk R, Jacobi AM, Collingwood MA, Bode NM, McNeill MS, Yan S, Camarena J, Lee CM, Park SH, Wiebking V, Bak RO, Gomez-Ospina N, Pavel-Dinu M, Sun W, Bao G, Porteus MH, Behlke MA (2018) A high-fidelity Cas9 mutant delivered as a ribonucleoprotein complex enables efficient gene editing in human hematopoietic stem and progenitor cells. Nat Med 24(8):1216–1224. https://doi.org/10.1038/s41591-018-0137-0
23. Schreckenberg GM, Jacobson AG (1975) Normal stages of development of the axolotl, Ambystoma mexicanum. Dev Biol 42(2): 391–399. https://doi.org/10.1016/0012-1606(75)90343-7
24. Campbell LJ, Suárez-Castillo EC, Ortiz-Zuazaga H, Knapp D, Tanaka EM, Crews CM (2011) Gene expression profile of the regeneration epithelium during axolotl limb regeneration. Dev Dyn 240(7):1826–1840. https://doi.org/10.1002/dvdy.22669
25. Stewart R, Rascón CA, Tian S, Nie J, Barry C, Chu LF, Ardalani H, Wagner RJ, Probasco MD, Bolin JM, Leng N, Sengupta S, Volkmer M, Habermann B, Tanaka EM, Thomson JA, Dewey CN (2013) Comparative RNA-seq analysis in the unsequenced axolotl: the oncogene burst highlights early gene expression in the blastema. PLoS Comput Biol 9(3):e1002936. https://doi.org/10.1371/journal.pcbi.1002936
26. Bryant DM, Johnson K, DiTommaso T, Tickle T, Couger MB, Payzin-Dogru D, Lee TJ, Leigh ND, Kuo TH, Davis FG, Bateman J, Bryant S, Guzikowski AR, Tsai SL, Coyne S, Ye WW, Freeman RM, Peshkin L, Tabin CJ, Regev A, Haas BJ, Whited JL (2017) A tissue-mapped axolotl de novo transcriptome enables identification of limb regeneration factors. Cell Rep 18(3):762–776. https://doi.org/10.1016/j.celrep.2016.12.063
27. Nowoshilow S, Schloissnig S, Fei JF, Dahl A, Pang AWC, Pippel M, Winkler S, Hastie AR, Young G, Roscito JG, Falcon F, Knapp D, Powell S, Cruz A, Cao H, Habermann B, Hiller M, Tanaka EM, Myers EW (2018) The axolotl genome and the evolution of key tissue formation regulators. Nature 554(7690): 50–55. https://doi.org/10.1038/nature25458
28. Smith JJ, Putta S, Walker JA, Kump DK, Samuels AK, Monaghan JR, Weisrock DW, Staben C, Voss SR (2005) Sal-site: integrating new and existing ambystomatid salamander research and informational resources. BMC Genomics 6(1):1–6. https://doi.org/10.1186/1471-2164-6-181

Chapter 23

Applying a Knock-In Strategy to Create Reporter-Tagged Knockout Alleles in Axolotls

Liqun Wang, Yan-Yun Zeng, Yanmei Liu, and Ji-Feng Fei

Abstract

Tetrapod species axolotls exhibit the powerful capacity to fully regenerate their tail and limbs upon injury, hence serving as an excellent model organism in regenerative biology research. Developing proper molecular and genetic tools in axolotls is an absolute necessity for deep dissection of tissue regeneration mechanisms. Previously, CRISPR-/Cas9-based knockout and targeted gene knock-in approaches have been established in axolotls, allowing genetically deciphering gene function, labeling, and tracing particular types of cells. Here, we further extend the CRISPR/Cas9 technology application and describe a method to create reporter-tagged knockout allele in axolotls. This method combines gene knockout and knock-in and achieves loss of function of a given gene and simultaneous labeling of cells expressing this particular gene, that allows identification, tracking of the "knocking out" cells. Our method offers a useful gene function analysis tool to the field of axolotl developmental and regenerative research.

Key words Axolotl, CRISPR/Cas9, Genomic editing, Knock-in, Knockout, Pax7

1 Introduction

Axolotl is one of the most representative salamander species that has long been used as a model organism in the fields of evolutionary, developmental, and regenerative biology. In terms of tissue regeneration, unlike in mammals, axolotls show extraordinary ability to regenerate complex structures, such as the limbs and the brain, even in adulthood. Hence, studying the molecular and cellular mechanisms of axolotl tissue regeneration provides an opportunity to understand the potential limiting factors in mammals and improve the restricted mammalian tissue regeneration.

In addition to their exceptional regenerative capability, axolotl also possesses many other advantages, making it an excellent research model. Firstly, axolotls grow in freshwater. It is relatively

Liqun Wang and Yan-Yun Zeng contributed equally with all other contributors.

Ashley W. Seifert and Joshua D. Currie (eds.), *Salamanders: Methods and Protocols*,
Methods in Molecular Biology, vol. 2562, https://doi.org/10.1007/978-1-0716-2659-7_23,

easy to maintain them in the laboratory. A female axolotl can produce a large number, around 300–500 eggs from single breeding. These eggs develop externally and are accessible to all kinds of manipulation during entire embryonic stages. Remarkably, the body of larva axolotls is relatively transparent, allowing for in vivo imaging to visualize the cell behavior during a particular physiological process, e.g., tissue regeneration [1]. Secondly, perhaps due to low immune response/rejection in axolotls, it is relatively easy to carry out tissue/organ transplantation between individuals [2]. Thirdly, the recently released axolotl genome and the transcriptome [3–6], and the expansion of CRISPR/Cas9-mediated genomic engineering technologies [2, 7, 8], combined with meganuclease- and Tol2 transposon-based transgenic approaches in axolotls [9, 10], allow labeling and tracing given types of cells, overexpressing or knocking out of particular genes. All of these provide essential resources and critical tools for functional analysis of regenerative related cells/genes and help uncover crucial mechanisms underlying regeneration.

Loss of function through knocking out of targeting gene is a critical approach to study gene function. Soon after the first report showing that CRISPR/Cas9 could precisely and effectively create DNA double-strand break (DSB), further introduce "indels" at targeted genomic loci in eukaryotic cells [11–13], this technique has been applied to many species, including axolotls for the generation of knockout and knock-in animals [14]. Through injection of the mixture of guide RNA (gRNA) and Cas9 protein or mRNA into single-cell stage axolotl eggs, Crews and Tanaka labs have achieved knockout of *eGFP* reporter and endogenous genes [7, 8], respectively. Further investigation has identified the roles of *Sox2* in promoting spinal cord regeneration and *Pax7* in determining muscle and pigment cell development [3, 7, 15]. Also, when supplied with an extra targeting construct together with CAS9 protein and gRNA, either the m*Cherry* reporter gene or a relatively sizeable *Cre-ER*^{T2} recombinase expressing cassette (≈4–4.5 kb) was inserted into the desired axolotl *Pax7* and *Sox2* genomic loci. Taking advantage of the developed *Pax7: Cre-ER*^{T2} knock-in line, Fei and colleagues have labeled and traced muscle stem cells and found that they are the primary cell source of muscle regeneration [2]. In all previous work, the knockout lines lack labels for direct identification of the gene knocking out cells [7, 8]. However, the labeled cells in the knock-in lines still actively express targeted gene from the integrated rescue construct [2].

This article describes a protocol that applies a previously developed homology-independent knock-in strategy [2, 16] to create reporter-tagged knockout alleles in axolotls. Based on this approach, we have established the *Pax7: Cherry* reporter line through knock-in of the *Cherry* fluorescence reporter gene into axolotl *Pax7* genomic locus (Fig. 1). In this line, *Pax7* coding

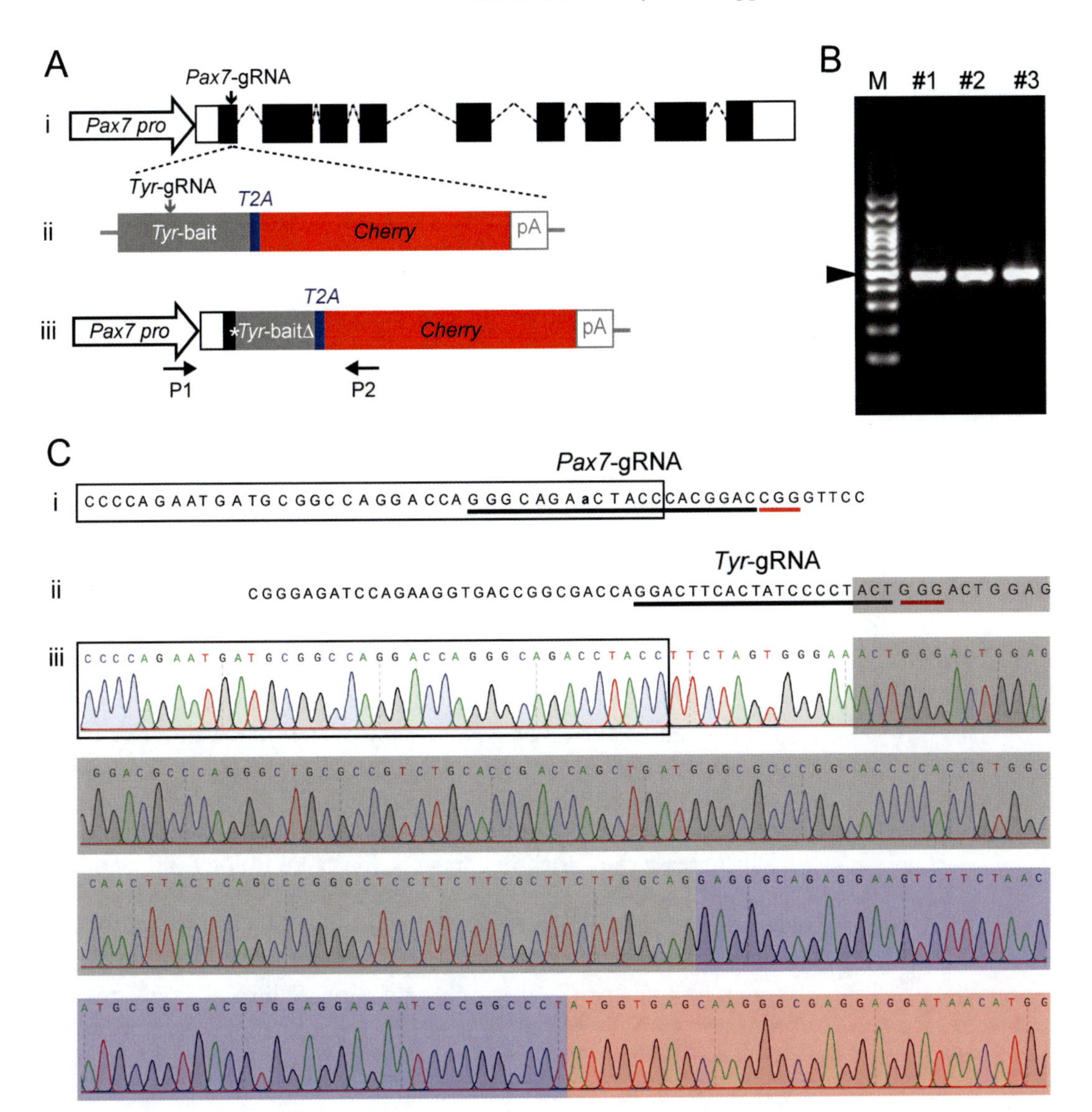

Fig. 1 Generation of *Cherry*-tagged, loss-of-function *Pax7* allele in axolotl. (**a**) The knock-in strategy for the generation of the *Pax7: Cherry* reporter/knockout axolotl line. (i) The *Pax7* genomic locus. Solid rectangles: coding regions in the exons; empty rectangles: untranslated regions; dashed lines: introns; arrow: the *Pax7-gRNA* targeting site. (ii) The targeting plasmid. It contains a piece of *Tyr* sequence as bait (*Tyr*-bait). pA: an exogenous polyadenylation sequence; arrow: the *Tyr-gRNA* targeting site. (iii) The structure of the *Pax7* genomic locus after knock-in of the targeting construct. Asterisk: integration junction between genomic and targeted construct; arrows: genotyping primers. (**b**) Genomic PCR of *Cherry*-tagged, loss-of-function *Pax7-Cherry* knock-in allele in F1 individuals. Genotyping primers are indicated in (**a**, iii). (**c**) Sequence analysis of the genomic PCR product in (**b**). Black underlines indicate the *Pax7-gRNA* targeting sequence in the axolotl genome (i) and the *Tyr-gRNA* targeting sequence in the targeting construct (ii). Red underlines indicate the PAM sequences for the *Pax7-gRNA* (i) and the *Tyr-gRNA* (ii). Empty black rectangles highlight the genomic sequences adjacent to the *Pax7-gRNA* cleavage site, before (i) and after (iii) knock-in of the targeting construct. Gray rectangles highlight the sequences adjacent to the *Tyr-gRNA* cleavage site, in the targeting construct (ii) and axolotl genome after knock-in of the targeting construct (iii), respectively. Blue and red rectangles in (iii) indicate the corresponding *T2A* and *Cherry* coding sequences in the targeting construct. The lowercase letter in (**c**, i) is a single nucleotide polymorphism

sequence is interrupted and the corresponding allele becomes loss of function, upon insertion of the *Cherry* cassette (Fig. 1). Therefore, when maintaining *Pax7: Cherry* line as heterozygous status, one *Pax7: Cherry* allele drives *Cherry* expression to label the PAX7+ cells, the other unmodified wild-type allele drives endogenous *Pax7* expression. In this situation, it could generally consider that CHERRY-labeled PAX7-positive cells function as wild-type cells. However, when making *Pax7: Cherry* allele into homozygous, both *Pax7: Cherry* alleles drive the expression of *Cherry*, instead of *Pax7* gene, which leads to the knockout of *Pax7* gene and meanwhile allows the fluorescence labeling of PAX7-knocking out cells. This is particular important for identification of gene (e.g., *Pax7*) knocking out cells, especially in the case of lacking the expression of target gene and the possibility to be detected with antibodies.

In our protocol, we used a piece of axolotl *Tyrosinase* (*Tyr*) genomic sequence as universal bait in targeting construct (Fig. 1). We used the mix of a previously identified *Tyr-gRNA* that works very efficient on targeting sequence modification, and a *Pax7-gRNA* for generation of *Pax7: Cherry* knock-in/knockouts. *Tyr-gRNA* works a lot more efficient than *Pax7-gRNA* on targeting sequence modification [2, 7]. The knock-in efficiency very much determined by the cleavage efficiency of the gRNA that targets to the genomic locus of interest (e.g., *Pax7*, Fig. 1), and the size of the insert for genomic integration. The combination of *Tyr-* and *Pax7-gRNAs*, when compared to the single *Pax7-gRNA*, may dramatically increase the knock-in efficiency in the homology-independent method [16]. The efficiency of successful *Pax7: Cherry* knock-in/knockouts in all injected F0 axolotls ranges about 5–15% [2]. There are two more advantages of using the universe *Tyr* bait to generate knock-in animals: (1) it simplifies the procedure by applying one common gRNA which does not need to be further evaluated, and a newly designed gRNA targeting to gene of interest; (2) it also provides an additional internal control to monitor Cas9 activity for an easy screening of injected F0 embryos, because introduction of *Tyr*-gRNA into eggs leads to an obvious albino phenotype [7]. Indeed, pigmentation is lost in most, if not all, of the F0 animals from a successful injection. And all selected F0 founders for the purpose of germline transmission show nearly complete albino phenotype as well. Furthermore, we could identify the presence of "indels" in *Tyr* genomic loci in all (6 out of 6) of the genotyped F1 *Pax7: Cherry* axolotls. However, although the *Tyr* gene mutation does not cause any apparent phenotypes, except for the loss of pigmentation, it is worthwhile to consider choosing other neutral sequences (e. g., *eGFP*) as bait under given situations [16].

Moreover, the selection of proper F0 *Pax7: Cherry* knock-in/knockout founder animals can be challenging. Both the introduction of "indels" and the insertion of *Cherry* cassette into *Pax7* locus (indel mutation production is generally more efficient than knock-in) lead to the loss of function of *the Pax7* gene. Homozygous *Pax7* loss-of-function mutations hamper embryonic development at very early stages and cannot be used as a founder for germline transmission. Therefore, the ideal F0 *Pax7: Cherry* knock-in/knockout founders should fulfill one of the following standards: (1) one allele harbors a *Cherry* insertion at the *Pax7* locus and the other is a wild-type allele or contains an in-frame indel that likely codes for a functional PAX7 protein; (2) F0 individual is mosaic knock-in/knockout, which contains sufficient *Cherry* knock-in and healthy alleles to balance the germline transmission, animal survival, and growth.

Overall, our protocol is valuable for researchers using axolotls, other salamanders, even other species to create transgenic lines to achieve, in parallel, the purposes of gene-knockout and labeling of the gene-knockout cells.

2 Materials

2.1 Axolotls

Axolotl (*Ambystoma Mexicanum*) strain d/d.

2.2 Targeting Construct and Other Plasmids

1. Targeting constructs: *pGEMT-Pax7bait-T2A-Cherry-pA* (Addgene, #111154) [2], *pGEMT-Tyrosinase-bait-T2A-Cherry-pA* (Addgene, #111155).
2. gRNA plasmids: *DR274* (Addgene, #42250) [17], *DR274-Pax7-gRNA* [7], *DR274-Tyrosinase-gRNA* [7].

2.3 Primers

1. *Tyr-fw*: 5′-TGAATTGTAATACGACTCACTATAGacgcgtAGAACATCGACTTCGCGCACG-3′.
2. *Tyr-rev*: 5′-TGTTAGAAGACTTCCTCTGCCCTCgcatgcCTGCCAAGAAGCGAAGAAGGAG-3′.
3. *DR274-fw*: 5′-AAATGGTCAGTATTGAGCCTCAG-3′.
4. *DR274-rev*: 5′-AAAAGCACCGACTCGGTGCCAC-3′.
5. *pGEMT-seq*: 5′-GATGTGCTGCAAGGCGATTAAGTTG-3′.
6. *Pax7-fw*: 5′-CCAACTCCTCCCAAGAACTCTG-3′.
7. *Cherry-rev*: 5′-CCCTCCATGTGCACCTTGAAGC-3′.

2.4 Preparation of Targeting Construct

1. Phusion High-Fidelity PCR Master Mix with GC Buffer (Thermo Fisher Scientific, F532S).
2. MluI-HF (NEB, R3198S).
3. SphI-HF (NEB, R3182S).
4. Agarose.
5. Gibson Assembly Master Mix (NEB, E2611L).

6. DH5α *E. Coli* competent cells.
7. Ampicillin sodium salt USP.
8. LB Agar Powder.
9. LB Broth Powder.
10. QIAGEN Plasmid Mini Kit (Qiagen, 12123).
11. EndoFree Plasmid Maxi Kit (Qiagen, 12362).
12. 6× DNA Loading Dye.
13. 100 bp plus DNA ladder.
14. 1 kb plus DNA ladder.

2.5 gRNA Synthesis and Preparation of the Mixture of CAS9 Protein, gRNA, and Targeting Construct

1. QIAquick PCR Purification Kit (Qiagen, 28104).
2. MEGAshortscript T7 Transcription Kit (Thermo Fisher Scientific, AM1354).
3. CAS9 protein (PNA Bio, CP03; or MPI-CBG protein expression facility, 5 mg/mL).
4. NaOH.
5. HEPES.
6. KCl.

2.6 Microinjection

1. Ethanol absolute.
2. Petri dishes.
3. Forceps (Fine Science Tools, 11295-51).
4. Borosilicate glass capillaries (Harvard Apparatus, 30-0020).
5. Microloader (Eppendorf, 5242956003).
6. 24-Well plate.
7. $CaCl_2 \cdot 2H_2O$.
8. EDTA.
9. KOH.
10. Ficoll PM 400 (Sigma-Aldrich, F4375-500G).
11. $MgCl_2$.

2.7 Genotyping

1. Benzocaine (Sigma-Aldrich, E1501-500G).
2. Tris.
3. 2× Taq Master Mix (Dye Plus).
4. Boric acid.
5. $MgSO_4$ $7H_2O$.
6. Fluorescence stereomicroscope (Olympus, Model SZX16).

2.8 Buffers and Other Reagents

1. 10% Benzocaine: Dissolve 10 g of benzocaine in 100 mL of ethanol absolute. Store at 25 °C for up to 1 year.
2. 10× Tris-buffered saline (TBS): Mix 500 mL of 1 M Tris–HCl (pH 7.5) and 300 mL of 5 M NaCl, and; add deionized water up to 1 L. Autoclave and store at 25 °C for up to 1 year.
3. 4000% Holtfreter: Dissolve 158.4 g of NaCl, 2.875 g of KCl, 11.125 g of $MgSO_4 \cdot 7H_2O$, and 5.36 g of $CaCl_2 \cdot 2H_2O$ in 1-L deionized water. Autoclave and store at 25 °C for up to 1 year.
4. 0.03% Benzocaine: Prepare 10 L of this solution by mixing 500 mL of 10× TBS, 50 mL of 4000% Holtfreter, and 30 mL 10% benzocaine, and add deionized water to 10 L. Store at 25 °C for up to 6 months.
5. 10× Cas9 Buffer: 200 mM HEPES, 1.5 M KCl, adjust pH to 7.5 using NaOH, and filter sterilize the solution. Store at 25 °C for up to 1 year.
6. 100 mg/mL Ampicillin (1000× stock): Dissolve 1 g of ampicillin in 10 mL deionized water and sterilize by passing 0.22 μm syringe filter. Aliquot and store at –20 °C for up to 1 year.
7. LB–Ampicillin agar plate: Mix 16 g of LB agar into 400 mL of deionized water. Autoclave and cool to 50 °C, add 400 μL 100 mg/mL Ampicillin, and pour about 20 mL to 10-cm Petri dishes. Store at 4 °C for up to 1 month.
8. LB–Ampicillin medium: Mix 10 g of LB medium powder into 400 mL of deionized water. Autoclave and store at 4 °C for up to 1 month.
9. 10× Tris/Borate/EDTA (TBE) buffer: To prepare 1 L of this solution, 108 g Tris, 9.3 g EDTA powder, and 55 g boric acid are mixed and filled up to 1 L with distilled water. This solution shall be autoclaved before used and can be stored at room temperature for up to 6 months.
10. 10× Marc's modified Ringer's solution (MMR, pH 7.8): To prepare 1 L of 10× MMR, 200 mL 5 M NaCl, 20 mL of 1 M KCl, 10 mL 1 M $MgCl_2$, 20 mL 1 M $CaCl_2$, 2 mL 0.5 M EDTA, and 50 mL 1 M HEPES (pH 7.2) are mixed and filled up with deionized water to 1-L final volume. pH is adjusted with 10 M KOH. This solution shall be autoclaved before used and can be stored at 4 °C for up to 6 months.
11. 1× MMR solution containing 1× Penicillin/Streptomycin: To prepare 1 L of this solution, 100 mL 10× MMR solution is mixed with 13 mL 100× penicillin/streptomycin (5000 U/mL penicillin and 5000 μg/mL streptomycin) and then filled up with deionized water to 1-L final volume. This solution is filtered to sterilize and can be stored at 4 °C for up to 2 months.

12. 1× MMR solution, added with 10% (wt/vol) Ficoll and 1× Penicillin/Streptomycin: To prepare 500 mL of this solution, 50 mL 10× MMR solution is mixed with 50 g Ficoll and 7 mL 100× penicillin/streptomycin, then filled up with deionized water to 500 mL final volume. Ficoll shall be stirred to dissolve fully. This solution is filtered to sterilize and can be stored at 4 °C for up to 2 months.
13. 0.1× MMR solution, supplied with 5% (wt/vol) Ficoll and 1× Penicillin/Streptomycin: To prepare 500 mL of this solution, 5 mL 10× MMR solution is mixed with 25 g Ficoll and 7 mL 100× penicillin/streptomycin, then filled up with deionized water to 500 mL final volume. This solution is filtered to sterilize and can be stored at 4 °C for up to 2 months.
14. 0.1× MMR solution with 1× Penicillin/Streptomycin: To prepare 1 L of this solution, 10 mL 10× MMR solution is mixed with 13 mL 100× penicillin/streptomycin, then filled up with deionized water to 1-L final volume. This solution is filtered to sterilize and can be stored at 4 °C for up to 2 months.

3 Methods

3.1 Preparation of Targeting Construct

1. PCR-amplify *Tyrosinase* bait sequence with a high-fidelity DNA polymerase (Phusion High-Fidelity PCR Master Mix with GC Buffer), with primer pair *Tyr-fw* and *Tyr-rev*, axolotl genomic DNA as the template (*see* **Note 1**).
2. Linearize the plasmid *pGEMT-Pax7bait-T2A-Cherry-pA* (Addgene, #111154) with restriction enzyme MluI-HF and SphI-HF.
3. Check the PCR product (from Subheading 3.1, **step 1**) and the linearized plasmid (from Subheading 3.1, **step 2**) on a 1.2% agarose gel.
4. Recover the PCR product and linearized vector from the gel using QIAquick Gel Extraction Kit, according to the manufacturer's instructions.
5. Clone the *Tyrosinase* bait sequence into the linearized vector using Gibson assembly master mix, according to the manufacturer's instructions.
6. Take 2–5 μL of ligated product to perform transformation using *E. Coli* chemical competent cells according to standard protocol.
7. Spread transformants on an LB agar plate containing ampicillin antibiotics (final concentration 100 μg/mL) and incubate the plate at 37 °C overnight.

8. Pick single colonies, in total 3–4 *E. coli* colonies, and culture them overnight at 37 °C in 4 mL LB medium supplied with ampicillin antibiotics (final concentration 100 μg/mL).
9. Extract plasmid DNA using QIAGEN Plasmid Mini Kit according to the manufacturer's instructions.
10. Sequence the targeting construct *pGEMT-Tyrosinase-bait-T2A-Cherry-pA* using primer *pGEMT-seq*; run the sequence blast to verify the cloned *Tyrosinase* bait sequence (*see* **Note 2**).
11. Prepare maxiprep for the identified correct targeting construct clone using EndoFree Plasmid Maxi Kit according to the manufacturer's instructions.

3.2 Preparation of gRNA

1. Set up separately two PCR reactions, 100 μL volume per reaction, with high-fidelity DNA polymerase (Phusion High-Fidelity PCR Master Mix with GC Buffer), primer pair *DR274-fw* and *DR274-rev*, and use plasmids *DR274-Pax7-gRNA* and *DR274-Tyrosinase-gRNA* as a template, respectively.
2. Run 3 μL of each PCR product on a 1.5% agarose gel; the expected size of the PCR product is 282 bp.
3. Purify the rest 97 μL of each PCR product using QIAquick PCR Purification Kit if the expected PCR product appears as a sharp single band on the gel.
4. Synthesize and purify *Tyr-gRNA* and *Pax7-gRNA* using MEGAshortscript T7 Transcription Kit, mix the components as listed below, and incubate the mixture at 37 °C overnight in PCR machine (*see* **Note 3**).

Components	Amount
DNA template (Subheading 3.2, **step 3**)	0.7–1 μg
10× T7 Reaction Buffer	2 μL
ATP solution (75 mM)	2 μL
UTP solution (75 mM)	2 μL
CTP solution (75 mM)	2 μL
GTP solution (75 mM)	2 μL
Enzyme mix	2 μL
Nuclease-free water	to 20 μL

5. Add 1 μL DNase and incubate the mixture at 37 °C for 15 min to remove DNA template.

6. To precipitate RNA, add 30 μL nuclease-free water and 30 μL lithium chloride solution into the reaction solution; mix and keep the solution at −20 °C for at least 30 min.
7. Pellet down RNA by centrifuging at 13,000 rpm for 15 min at 4 °C; then carefully remove the supernatant.
8. To completely remove the unincorporated nucleotides, wash the RNA pellet with 1 mL 70% ethanol followed by centrifuging at 13,000 rpm for 15 min at 4 °C, repeat this step at least twice, and then completely remove the remaining ethanol after one more brief centrifugation.
9. Air-dry RNA pellet for 5 min and dissolve it in 30 μL Nuclease-free water.
10. Measure the concentration of the RNA solution, aliquot the solution into 10 μL each, and store at −80 °C for up to 6 months.

3.3 Preparation of the Mixture of CAS9 Protein, gRNA, and Targeting Construct

1. Set up 20 μL of the mixture of CAS9 protein, gRNA, and targeting construct on ice, as described in Table 1 below.
2. Aliquot the mixture into 3–4 μL each, snap freeze in liquid nitrogen, and store the aliquots at −80 °C, for up to 2 months.

3.4 Axolotl Breeding, Egg Collection, and Microinjection

1. Set up breeding using axolotl strain d/d. A detailed axolotl breeding procedure is described in a previously published protocol [18].
2. Once the female axolotl starts to lay eggs, collect fresh eggs and briefly rinse them with 70% ethanol for about 10 s.
3. Wash the eggs 3–4 times using deionized water to remove residual ethanol; then transfer them into a 10-cm petri dish prefilled with 1× MMR solution containing 1× penicillin/streptomycin (*see* **Note 4**).
4. Dejelly axolotl eggs using forceps under a stereomicroscope.
5. Fill the injection plate (made from 2% agarose) with 1× MMR solution, supplemented with 10% (wt/vol) Ficoll and 1× penicillin/streptomycin; then transfer the dejellied eggs into the injection plate.
6. Prior to microinjection, thaw the pre-mixed solution of CAS9 protein, gRNAs, and targeting construct (injection solution, from Subheading 3.3, **step 2**); then incubate the injection solution at room temperate for 5 min (*see* **Note 5**).
7. Load 3–4 μL of the injection solution into a glass capillary (needle) with Eppendorf Microloader tip, and then sharpen the injection needle using forceps under a stereomicroscope.

Table 1
List of reagents and volumes for qPCR

Components	Amount
CAS9 protein	5 μg
Tyr-gRNA	3 μg
Pax7-gRNA	3 μg
Targeting construct: *pGEMT-Tyrosinase-bait-T2A-Cherry-pA*	0.5 μg
10× CAS9 buffer	1 μL
Nuclease-free water	To 10 μL

8. Inject 5 nl injection solution into each egg. For each construct, inject at least 100 eggs.
9. Carefully transfer the injected eggs into a 3-cm or 10-cm petri dish pre-filled with 1× MMR solution, added with 10 % (wt/vol) Ficoll and 1× penicillin/streptomycin, and then incubate for 2–3 h at room temperature.
10. Carefully transfer the injected eggs into 0.1× MMR solution, supplied with 5% (wt/vol) Ficoll and 1× penicillin/streptomycin, and then keep the eggs at room temperature overnight.
11. Transfer the injected eggs into a 24-well plate (one egg/ per well), pre-filled with 0.1× MMR solution with 1× penicillin/streptomycin; develop the eggs at room temperature for about 2 weeks (*see* **Note 6**).

3.5 Screening of F0 Transgenic Axolotls by Phenotyping

1. Anesthetize about 2-week-old F0 axolotls in 0.03% benzocaine solution (*see* **Note 7**).
2. Check the albino phenotype under stereomicroscope. It is expected that most of the injected F0 embryos (e.g., above 90%) show nearly complete loss of pigmentation in individuals, particularly in the eyes, when compared to the control (e.g., un-injected embryos). This phenotype indicates the injection itself is successful and CAS9 protein functions properly.
3. Exam further the expression of *Cherry* fluorescence reporter in 2-week-old *Pax7: Cherry* knock-in/knockout F0 axolotls under a fluorescence stereomicroscope. Successful knock-in of *Cherry* cassette into axolotl *Pax7* genomic locus leads to the expression of the CHERRY protein in the central nervous system, olfactory bulb, and muscle compartments [2] (*see* **Notes 8** and **9**).

Table 2
List of reagents and volumes for qPCR

Components	Amount
Axolotl DNA crude extract	1 μL
2× Taq Master Mix (Dye Plus)	5 μL
Pax7-fw primer (10 μM)	0.5 μL
Cherry-rev primer (10 μM)	0.5 μL
ddH_2O	To 10 μL

3.6 Screening of F0 Transgenic Axolotls by Genotyping

1. Prepare the axolotl DNA crude extract using an alkaline lysis method [19] (*see* **Notes 10** and **11**).
2. Carry out genomic PCR reaction to identify the proper integration of *Cherry* cassette in axolotl *Pax7* genomic locus using primer pair *Pax7-fw* and *Cherry-rev* (Fig. 1), as listed in Table 2 (*see* **Note 12**).
3. Run 5 μL PCR product on a 1.2% agarose gel prepared in 1× TBE. The expected size of the PCR product is around 450 bp upon proper knock-in of the *Cherry* cassette into the axolotl *Pax7* genomic locus.
4. Raise the phenotype- and genotype-identified F0 founders to the adult stage (*see* **Note 13**).

3.7 Germline Transmission and Establishing F1 Transgenic Axolotl Line

1. Breed adult *Pax7: Cherry* knock-in/knockout F0 founders with d/d axolotls (*see* **Note 14**).
2. Exam the CHERRY protein expression of F1 progeny after hatching (around 2 weeks old) under a fluorescence stereomicroscope. In F1 *Pax7: Cherry* axolotls, CHERRY protein is expressed in CNS, olfactory bulb, and muscle compartments (Fig. 2) (*see* **Note 15**).
3. Repeat the genomic PCR (*see* Subheading 3.5) and sequence the PCR product using primer pair *Pax7-fw* and *Cherry-rev* (Fig. 1).
4. Sequence and analyze the PCR product that reveals the DNA sequences, including *Tyrosinase* bait, *T2A,* and *Cherry* integrated into the *Pax7* genomic locus. Particularly, pay attention to the integration junction between the *Pax7-gRNA* and *Tyr-gRNA* cleavage sites.
5. Establish the phenotype- and genotype-identified F1 individuals as a line and raise some of them to the adult stage (*see* **Note 16**).

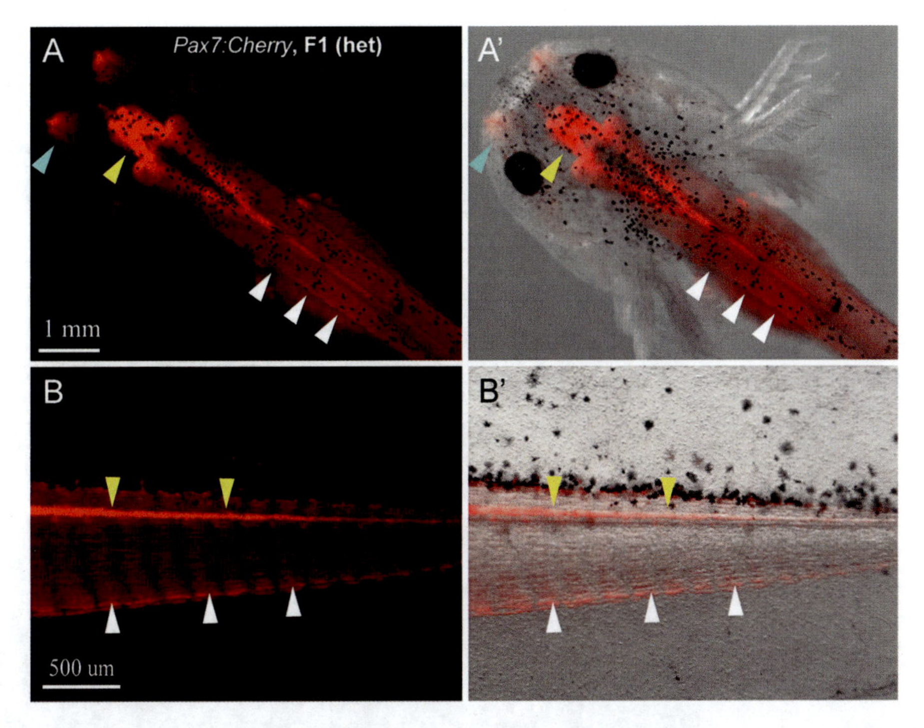

Fig. 2 CHERRY expression in heterozygous (het) F1 *Pax7: Cherry* reporter/knockout axolotls. CHERRY fluorescence (**a** and **b**) and merged (**a**′ and **b**′, with the bright field) images show the expression of the CHERRY protein, from dorsal (**a** and **a**′, head and trunk) and lateral (**b** and **b**′, tail) views, in 1.5-month-old F1 *Pax7: Cherry* axolotls. Cyan, yellow, and white arrowheads highlight CHERRY expression in olfactory bubs, central nervous system, and muscle compartments, respectively. Scale bars: 1 mm in (**a**); 500 μm in (**b**)

3.8 Mutant Phenotype in F2 Generation Transgenics

1. Breed a pair of adult F1 *Pax7: Cherry* knock-in/knockout axolotls (*see* **Note 17**).
2. Raise the progeny to various stages for phenotype examination. Homozygous *Pax7* mutant showed a broad range of phenotypes, including curved body, reduction of trunk muscle, and loss of limb muscle [3]. As expected, the homozygous *Pax7: Cherry* F2 transgenics exhibit typical *Pax7* mutant phenotypes (Fig. 3). In addition, the robust expression of CHERRY driven by the endogenous *Pax7* promoter allows the identification, tracking, and study of the "PAX7 positive" cells in the absence of PAX7.

4 Notes

1. An alternative approach is to carry out the first round of PCR to obtain *Tyrosinase* bait sequence using a primer pair lacking the Gibson assembly adaptors, then use the PCR product from the first round of PCR as template to perform another PCR reaction with primers *Tyr-fw* and *Tyr-rev.*

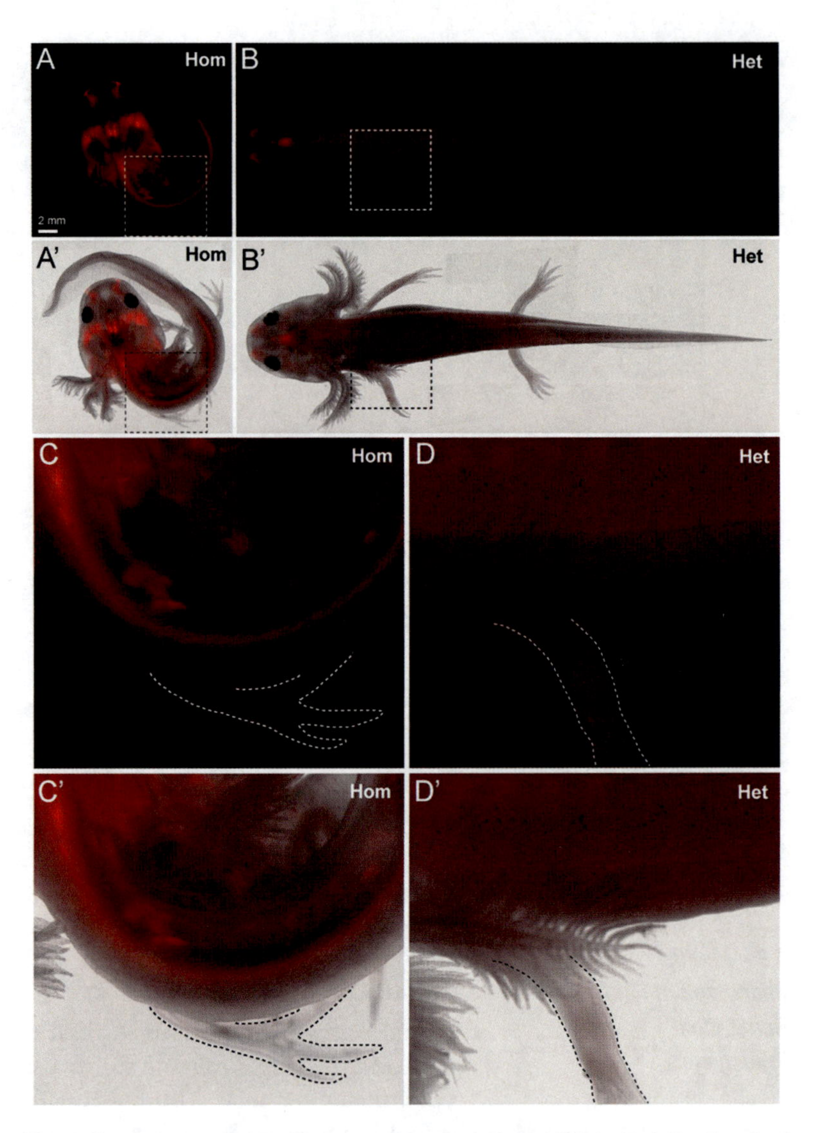

Fig. 3 Homozygous *Pax7: Cherry* reporter axolotls exhibit typical *Pax7* mutant phenotypes. CHERRY fluorescence (**a** and **b**) and merged (**a**′ and **b**′, with the bright field) images show the expression of CHERRY protein and developmental defects [2], in 6-month-old F2 generation homozygous (**a** and **a**′, hom), compared to heterozygous (**b** and **b**′, het) *Pax7: Cherry* siblings. The areas in dashed squares in (**a** and **b**′) are further shown at higher magnification (**c** and **d**′). Dashed lines in (**c** and **d**′) define the forelimbs and highlight the presence in heterozygous (**d** and **d**′), but the loss in homozygous (**c** and **c**′) of the CHERRY expression. Scale bar: 2 mm

2. Targeting construct *pGEMT-Tyrosinase-bait-T2A-Cherry-pA* used in this protocol is available at Addgene (Addgene, #111155).

3. Addition of excess T7 RNA polymerase and extension of the incubation time at 37 °C to overnight may improve the yield of *in vitro* gRNA production.
4. Axolotl eggs can be kept at a lower temperature (e.g., 8–11 °C) to slow down their development and extend the one-cell stage duration.
5. This step allows gRNA and CAS9 protein in the mixture to form a complex *in vitro*, and perhaps pre-cut of the bait sequence in the targeting construct.
6. The survival rate of the injected F0 embryos is generally more than 70% at early gastrulation stage. The following possibilities may cause a lower survival rate: (1) the targeted gene is essential for early embryonic development; (2) the batch effect, the quality of the eggs used for particular injection is too low. This potential could be ruled out by comparing to un-injected eggs; (3) the injected targeting constructs and gRNAs contain too much impurities.
7. Too long incubation in 0.03% Benzocaine kills axolotls. Therefore, if it is possible, keep axolotls as short as possible in Benzocaine solution. Early-stage axolotls are more sensitive to the toxicity of Benzocaine. In general, after anesthetization of larvae using 0.03% Benzocaine, further dilution of Benzocaine improves the survival rate of anesthetized early-stage axolotl larvae.
8. In F0 *Pax7: Cherry* founders, CHERRY expression may not be detected in all the expected PAX7-expressing tissues and may show weak fluorescence due to the mosaicism. For F0 *Pax7: Cherry*, we choose the animals at least showing 1/3 to 1/2 of the body bearing clear transgenesis in CNS and muscle, and grow up 5–10 individuals as founders. For certain transgenesis, it may be difficult to screen successful knock-in F0 by fluorescence observation, e.g., fluorescence protein expression is too low in case the transgene is driven by a weak endogenous promoter, or the cells expressing fluorescence transgene are too sparse or located in a deep organ/tissue. In this case, a genomic PCR is necessary to be carried out to determine F0 founders. In either case, if the transgenes are expressed in appendages, such as limbs and tail, mRNA in situ hybridization using probes made for transgene or immunohistochemistry using antibody against transgene can also be performed on the tissue sections to validate F0 founders (appendages of F0 axolotl founders can be collected for analysis because appendages will regenerate back and it wound not affect later breeding).
9. In general, gRNA-mediated targeting gene knockouts (no insertions) is more efficient than knock-in/knockout

(with insertions). In the case of generation *Pax7: Cherry* knock-in/knockouts, we observed about 70% of injected axolotl embryos showing obvious *Pax7* mutant phenotypes. However, only within 10–15% of injected individuals, CHERRY is expressed in the majority of the expected PAX7-expressing tissues. It suggests that most of the survived F0 *Pax7: Cherry* knock-in/knockout larvae bear biallelic *Pax7* mutations, and likely, only single allele is *Pax7-Cherry* knock-in/knockout insertion. Due to the essence of *Pax7* during early embryonic development, *Pax7* biallelic loss-of-function mutants are hard to survive. Therefore, we chose individual F0 showing less severe mutant phenotype, but higher ratio of CHERRY expression in desired tissues as founder for growing up. Generally, 1/5–1/3 of F0 *Pax7: Cherry* knock-in/knockouts fulfill this criterion and can grow up to adult, in which the non-insertion mutant allele in most of the cells likely harbors in-frame indels at the PAX7 N-terminal coding region (*Pax7-gRNA* targeting site). To obtain a good number of such founders, we normally inject at least 300–500 eggs for proper F0 *Pax7: Cherry* knock-in/knockout screening.

10. To avoid cross contamination, we place anesthetized axolotls on area-divided petri dishes and collect a tiny piece of tail tissue (1–2 mm in length) for DNA extraction. In order to remove DNA from instrument, forceps and blades were burnt throughout after collecting each animals' tissue.
11. Alkaline lysis method is carried out as follow. Add 100 μL of 50 mM NaOH to the collected tissue and keep the mixture at 95 °C for about 10–15 min, until the tissue is completely dissolved. Then 10 μL of 1M Tris (pH 8.0) is added to neutralize DNA solution. This mixture is used as the template for the genomic PCR reaction.
12. For F0 *Pax7: Cherry* knock-in/knockouts, we only choose a few, e.g., 5–6 animals for genomic PCR validation. However, if it is difficult to screen F0 knock-ins by fluorescence microscope as stated in *see* **Note 8**, we suggest to carry out genomic PCR on at least 100, or all survivals of injected F0 knock-ins.
13. Based on phenotyping and/or genotyping, if it is possible, at least 10 (generally 10–15) proper F0 founders should be kept for growing up.
14. If the selected founders (generally exhibit mild defects) have difficulties breeding naturally, in vitro fertilization is another option [20].
15. In F1 *Pax7: Cherry* axolotls, CHERRY expression should be detected in all the expected PAX7-expressing tissues.
16. In general, 4 males and 6 females are raised to adulthood for establishing a new line.

17. If the F0 animals are healthy and capable of nature mating, breeding together F0 founders may shorten the generation time, in terms of obtaining homozygous mutants. However, for *Pax7: Cherry* knock-in/knockouts, this approach is not applicable, because the F0 founders always exhibit certain levels of mutant phenotypes and the germline transmission rate of these founders is general lower than that of phenotypical healthy knock-ins from other genomic loci.

Acknowledgments

Ji-Feng Fei was supported by the National Key R&D Program of China (2019YFE0106700), the Natural Science Foundation of China (31970782), Project of Department of Education of Guangdong Province (2018KZDXM027), Key-Area Research and Development Program of Guangdong Province (2018B030332001, 2019B030335001), and Guangdong-Hong Kong-Macao-Joint Laboratory Program (2019B121205005). Yanmei Liu was supported by the Natural Science Foundation of China (32070819, 91854112, 91750203).

References

1. Currie JD, Kawaguchi A, Traspas RM, Schuez M, Chara O, Tanaka EM (2016) Live imaging of axolotl digit regeneration reveals spatiotemporal choreography of diverse connective tissue progenitor pools. Dev Cell 39: 411–423
2. Fei JF, Schuez M, Knapp D, Taniguchi Y, Drechsel DN, Tanaka EM (2017) Efficient gene knockin in axolotl and its use to test the role of satellite cells in limb regeneration. Proc Natl Acad Sci U S A 114:12501–12506
3. Nowoshilow S, Schloissnig S, Fei JF, Dahl A, Pang AWC et al (2018) The axolotl genome and the evolution of key tissue formation regulators. Nature 554:50–55
4. Smith JJ, Timoshevskaya N, Timoshevskiy VA, Keinath MC, Hardy D, Voss SR (2019) A chromosome-scale assembly of the axolotl genome. Genome Res 29:317–324
5. Caballero-Perez J, Espinal-Centeno A, Falcon F, Garcia-Ortega LF, Curiel-Quesada E et al (2018) Transcriptional landscapes of axolotl (Ambystoma mexicanum). Dev Biol 433:227–239
6. Bryant DM, Johnson K, DiTommaso T, Tickle T, Couger MB et al (2017) A tissue-mapped axolotl de novo transcriptome enables identification of limb regeneration factors. Cell Rep 18:762–776
7. Fei JF, Schuez M, Tazaki A, Taniguchi Y, Roensch K, Tanaka EM (2014) CRISPR-mediated genomic deletion of Sox2 in the axolotl shows a requirement in spinal cord neural stem cell amplification during tail regeneration. Stem Cell Reports 3:444–459
8. Flowers GP, Timberlake AT, Mclean KC, Monaghan JR, Crews CM (2014) Highly efficient targeted mutagenesis in axolotl using Cas9 RNA-guided nuclease. Development 141: 2165–2171
9. Khattak S, Schuez M, Richter T, Knapp D, Haigo SL et al (2013) Germline transgenic methods for tracking cells and testing gene function during regeneration in the axolotl. Stem Cell Reports 1:90–103
10. Sobkow L, Epperlein HH, Herklotz S, Straube WL, Tanaka EM (2006) A germline GFP transgenic axolotl and its use to track cell fate: dual origin of the fin mesenchyme during development and the fate of blood cells during regeneration. Dev Biol 290:386–397
11. Cong L, Ran FA, Cox D, Lin S, Barretto R et al (2013) Multiplex genome engineering using CRISPR/Cas systems. Science 339:819–823

12. Mali P, Esvelt KM, Church GM (2013) Cas9 as a versatile tool for engineering biology. Nat Methods 10:957–963
13. Jinek M, East A, Cheng A, Lin S, Ma E, Doudna J (2013) RNA-programmed genome editing in human cells. eLife 2:e00471
14. Barrangou R, Doudna JA (2016) Applications of CRISPR technologies in research and beyond. Nat Biotechnol 34:933–941
15. Fei JF, Knapp D, Schuez M, Murawala P, Zou Y et al (2016) Tissue- and time-directed electroporation of CAS9 protein–gRNA complexes in vivo yields efficient multigene knockout for studying gene function in regeneration. NPJ Regen Med 1:16002
16. Auer TO, Duroure K, De Cian A, Concordet JP, Del Bene F (2014) Highly efficient CRISPR/Cas9-mediated knock-in in zebrafish by homology-independent DNA repair. Genome Res 24:142–153
17. Hwang WY, Fu Y, Reyon D, Maeder ML, Tsai SQ et al (2013) Efficient genome editing in zebrafish using a CRISPR-Cas system. Nat Biotechnol 31:227–229
18. Khattak S, Murawala P, Andreas H, Kappert V, Schuez M et al (2014) Optimized axolotl (Ambystoma mexicanum) husbandry, breeding, metamorphosis, transgenesis and tamoxifen-mediated recombination. Nat Protoc 9:529–540
19. Fei JF, Lou WP, Knapp D, Murawala P, Gerber T et al (2018) Application and optimization of CRISPR-Cas9-mediated genome engineering in axolotl (Ambystoma mexicanum). Nat Protoc 13:2908–2943
20. Khattak S, Tanaka EM (2009) Axolotl (Ambystoma mexicanum) in vitro fertilization. Cold Spring Harb Protoc 2009:pdb prot5263

Chapter 24

Baculovirus Production and Infection in Axolotls

Prayag Murawala, Catarina R. Oliveira, Helena Okulski, Maximina H. Yun, and Elly M. Tanaka

Abstract

Salamanders have served as an excellent model for developmental and tissue regeneration studies. While transgenic approaches are available for various salamander species, their long generation time and expensive maintenance have driven the development of alternative gene delivery methods for functional studies. We have previously developed pseudotyped baculovirus (BV) as a tool for gene delivery in the axolotl (Oliveira et al. Dev Biol 433(2):262–275, 2018). Since its initial conception, we have refined our protocol of BV production and usage in salamander models. In this chapter, we describe a detailed and versatile protocol for BV-mediated transduction in urodeles.

Key words Baculovirus, Gene delivery, Regeneration, Axolotl

1 Introduction

Axolotl and newts have emerged as favored model organisms due to their tremendous tissue regeneration capabilities. Their rise as a model organism is complemented by the recent development of cellular, molecular, and genetic tools in these systems [1–8]. Efficient gene delivery into a somatic cell of these models has tremendous potential in developmental and regeneration biology studies. Electroporation [9, 10] and virus-mediated gene delivery are some of the most preferred techniques of gene delivery into the somatic cells. Among the virus-mediated gene delivery, a number of viruses—Vaccinia virus [11], Maloney Murine Leukemia Virus (MMLV) [12], Foamy virus [13], and Baculovirus (BV) [14]—were developed as a tool for gene delivery in salamanders.

Among all the viruses, BV has gained significant attention because it can host large DNA cassettes (>38 kb), is accessible at high titer [15, 16], and its infectivity does not depend upon mitosis, thus successfully infecting even quiescent somatic cells [17]. The BV is a rod-shaped, double-stranded DNA virus which

Ashley W. Seifert and Joshua D. Currie (eds.), *Salamanders: Methods and Protocols*,
Methods in Molecular Biology, vol. 2562, https://doi.org/10.1007/978-1-0716-2659-7_24,

is non-replicating in vertebrate cells. Their replicative infection cycle is restricted to insect cells, which enables their use in BSL-1 settings unless they harbor oncogenic genes in their cassette. Furthermore, pseudotyping of BV using vesicular stomatitis virus G glycoprotein (VSVG) or its truncated version VSV-GED [18] allows improved delivery and transduction of the vertebrate cells [19, 20]. Due to these distinct advantages, BV has emerged as a favored tool for gene delivery in salamander and other vertebrate species, including zebrafish [17] and mice [21].

Using a defective bacmid (DefBac) and a rescue plasmid (pOEM1) that harbors pseudotyped VSV-GED cassette, we have developed a BV production protocol that yields high titer of this virus [14]. In the past, we have demonstrated the use of such pseudotyped BV in gene function studies during axolotl [22] and newt [23] limb regeneration. In this chapter, we describe a detailed protocol for pseudotyped DefBac baculovirus production, its titer determination, and its use in axolotl as a method for gene delivery. Additionally, this system has also been adapted and used for protein purification; however, that aspect will not be covered in this chapter and readers are encouraged to refer to the original paper [24].

2 Materials

2.1 Bacmid and pOEM1 Transfer Vector Preparation

1. DefBac (Supplied upon request from the Protein Expression and Purification (PEP) facility, MPI-CBG, Dresden—MTA required).
2. *E. coli* DH10B.
3. Luria broth.
4. Kanamycin B sulfate.
5. 50% glycerol in deionized H_2O.
6. SbfI-HF (NEB # R3642).
7. NucleoBond Xtra Maxi Kit (Macherey-Nagel # 740414.50).
8. pOEM1-GED-CMV:GFP (Addgene # 102782).

2.2 Recombinant BV Production and Virion Amplification

1. DefBac Digest from Subheading 2.1.
2. pOEM1-GED vector of interest.
3. SF9 Cells (Oxford Expression Technologies # 600100).
4. Escort IV reagent (Sigma # L3287-1 mL).
5. ESF921 medium (Expression Systems # 96001-01).
6. Penicillin-streptomycin (Sigma # P0781-100 mL).
7. 24-well Blocks RS plate (Qiagen # 19583).
8. Breath-Easy tape (Sigma # Z380059-1PAK).

9. 500-mL Bottle Top Vacuum Filter, 0.45 μm Pore (Corning # 430512).
10. 50-mL sterile conical tube.
11. Cell culture hood.
12. Shaking incubator.

2.3 BV Purification and Concentration

1. Cell culture hood.
2. P3 supernatant.
3. 500-mL Bottle Top Vacuum Filter, 0.45 μm Pore (Corning # 430512).
4. 1-L beaker.
5. 20% Sucrose in deionized H_2O (Filter sterile), stored at 4 °C.
6. 25% Sucrose in deionized H_2O (Filter sterile), stored at 4 °C.
7. 50% Sucrose in deionized H_2O (Filter sterile), stored at 4 °C.
8. 1× Phosphate buffer saline (PBS) without Ca^{2+}, Mg^{2+} ions, adjust pH to 7.4 using hydrochloric acid or sodium hydroxide (Filter sterile), stored at 4 °C.
9. Sorvall LYNX 6000 Superspeed Centrifuge (ThermoFisher Scientific # 75006591).
10. Fiberlite™ F9-6 × 1000 LEX Fixed Angle Rotor (ThermoFisher Scientific # 096-061075).
11. Beckman Coulter Optima™ LE-80 K ultracentrifuge (Beckman Coulter Life Sciences # A95765).
12. SW 32 Ti Swinging-bucket rotor (Beckman Coulter Life Sciences # 369650).
13. Open-Top Thinwall Ultra-Clear Tube (Beckman Coulter # 344058).
14. 18G Needle (BD # 305196).
15. 5-mL sterile Syringe (BD # 309657).
16. 10% Pluronic™ F-68 non-ionic surfactant (100×) (ThermoFisher Scientific # 24040032).
17. Baculovirus storage solution (filter sterile).

1× PBS	89 mL
100% Glycerol	10 mL
10% Pluronic F-68	1 mL
Total	*100 mL*

18. Slide-A-Lyzer™ MINI dialysis device, 20K MWCO, 0.1 mL (ThermoFisher Scientific # 69590).

2.4 Titer Determination

1. Purified BV aliquot.
2. SF9-ET cells (ATCC # CRL3357).
3. 96-well flat-bottom plate (Sigma # CLS3595-50EA).

2.5 In Vitro Application of Baculovirus

1. AL1 Cells (Roy et al., 2000).
2. 150 cm^2 tissue culture flask.
3. Amphibian phosphate buffer saline (APBS), dilute 1× PBS (pH-7.4) to 0.7% using deionized H_2O.
4. 24-well plate (Sigma # CLS3527-100EA).
5. 1 mg/mL Insulin (Sigma # I5500–250 mg).
   (a) To prepare 1 mg/mL Insulin add 25 mL 0.1 M HCl to 250 mg insulin powder and swirl until dissolved. While swirling slowly add 225 mL of APBS (final solution should be clear).
   (b) Filter sterilize the solution and aliquot and store it at −20 °C.
6. Penicillin-streptomycin (Sigma # P0781-100 mL).
7. L-Glutamine (Sigma # G7513).
8. Trypsin-EDTA solution (Sigma # 59418C-100 mL).
9. MEM—growth medium.

MEM	250 mL	ThermoFisher Scientific # 11095080
Sterile MiliQ water	100 mL	
FBS	40 mL	ThermoFisher Scientific # 26140079
1 mg/mL Insulin	4 mL	Sigma # I5500
Pen/Strep solution	4 mL	Sigma # P0781
200 mM L-Glutamine	4 mL	Sigma # G7513
Total	*~400 mL*	

2.6 In Vivo Application of Baculovirus

1. BV stock aliquot.
2. Benzocaine (Sigma # E1501).
3. 30G needle (BD # 305106).
4. 10× Fast Green FCF (Sigma # F7252) in PBS.
   (a) 25 mg Fast Green FCF +10 mL PBS to make 10× stock solution.
5. GC120-10 Borosilicate glass capillary (Warner Instruments # 30-0042).
6. Microloader tip (Eppendorf # 930001007).
7. Micromanipulator (Narishige # MN-153).

8. PV830 pneumatic Pico pump (WPI # SYS-PV830).
9. Sylgard 184 (Sigma # 761036-5EA) plate.
   (a) Mix and pour Sylgard as per manufacturer's instruction in a 100-mm plate. Let it solidify. Sylgard plates are reusable indefinitely, subject to normal wear and tear.
10. Fluorescent stereoscope.

3 Methods

1. For the biosafety adherence of this protocol, *see* **Note 1**.

3.1 Bacmid and pOEM1 Transfer Vector Preparation (One Day Before Virus Infection)

1. Transform DefBac bacmid into *E. coli* DH10B using Kanamycin plate.
2. Transformed bacteria with DefBac can be stored in 50% glycerol in −80 °C freezer, whereas isolated DefBac can be stored at 4 °C for up to 6 months (*see* **Note 2**).
3. Inoculate 1 L of Luria broth containing 50 μg/mL Kanamycin with a single colony and shake at 180 rpm, 37 °C overnight.
4. Isolate DefBac DNA using NucleoBond Xtra Maxi kit for high molecular weight DNA following manufacturer's instruction.
5. Measure the absorbance at 260 nm and resuspend the DNA pellet in water to a final concentration of 200 ng/uL (*see* **Note 3**).
6. Set up DefBac digestion with SbfI restriction enzyme to linearize it (described below). Incubate at 37 °C overnight.

Bacmid Digestion (for 1× R×n)	
200 ng/μL DefBac (1.4 μg)	7 μL
DNase-free H_2O	1 μL
Cutsmart buffer	1 μL
20 units/μL SbfI-HF (NEB# R3642)	1 μL
Total	*10 μL*

7. The next day morning, halt DefBac digestion by placing the tube at 80 °C for 20 min (*see* **Note 4**).
8. DefBac is ready for transfection.
9. Make a plasmid prep of the pOEM1-GED rescue plasmid using any standard plasmid purification kit and dissolve it at 500 ng/μL final concentration (*see* **Notes 5** and **6**).

3.2 Recombinant BV Production and Virion Amplification

1. *P1 Production (Day 1–Day 4)*
   1. Label 2× 1.5-mL microfuge tubes as tube 1 and tube 2 in the cell culture hood (*see* **Note 7**).
   2. In tube 1, add 12 μL of Escort IV reagent to 100 μL of ESF921 media. Mix three times by pipetting.
   3. In tube 2, add 10 μL of DefBac digest and 600 ng-1 μg of pOEM1-transfer vector of interest to 100 μL of ESF921 media. Mix three times by pipetting (*see* **Note 8**).
   4. Incubate 10 min at room temperature in the hood.
   5. Mix contents of tube 1 and tube 2. Mix three times with pipetting and incubate 40 min at room temperature in the hood.
   6. Meanwhile, take a 24-well block plate and add 0.8–1.0 million SF9 cells in 800 μL of ESF921 media.
   7. After 40 min, apply ~200 μL of the transfection mix to the cells by pipetting three times.
   8. Cover the plate with Breath-Easy tape to seal the edges.
   9. Shake in an incubator at 27 °C and 200 rpm for 3 h in dark (*see* **Note 9**).
   10. Add 1 mL ESF921 media to the well after 3 h, and again place it back for shaking at 27 °C and 200 rpm for 96 h (4 days).
   11. On day 4, take out 10 μL cells and observe for the fluorescence (if your cassette carries fluorescence marker gene) (*see* **Note 10**).
   12. Transfer cells to microfuge tube and harvest virus by spinning down cells at 500 rcf for 5 min. Take the supernatant and pass it through a 0.45-μm syringe filter.
   13. Supernatant can be stored at 4 °C for up to 3 months or immediately used for P2 production.
2. *P2 Production (~Day 5–Day 10)*
   1. Seed 50 mL SF9 cells at 0.5–0.7 million cells/mL in a 250-mL flask.
   2. Add entire P1 supernatant (~1 mL) to the cells. Incubate flask at 100 rpm and 27 °C.
   3. Let the infection proceed for the next 4 days.
   4. On day 10 (day 4 post-inoculation), take out 10 μL of cells and observe for fluorescence (if your cassette carries a fluorescence marker gene). If >80% cells are showing fluorescence, the culture is ready for harvest. If not, add 1 mL FBS and 10 mL ESF921 media and wait for one more day to harvest (*see* **Note 11**).

5. Transfer cells to 50-mL sterile conical tube and harvest virus by spinning down cells at 500 rcf for 10 min. Take the supernatant and pass it through a 0.45-μm syringe filter.
6. Such supernatant can be stored at 4 °C for up to 3 months or immediately used for P3 production.

3. *P3 Production (~Day 10–Day 14)*
   1. Mix 2 L flask with 500 mL ESF921 medium, 10 mL FBS, and 5 mL Pen/Strep in a 2-L flask.
   2. Seed SF9 cells at 0.5–0.7 million cells/mL in the flask (total 250–350 million cells/500 mL).
   3. Add 1 mL of P2 supernatant to infect cells (*see* **Note 12**).
   4. Shake flask at 100 rpm and 27 °C for 4 days.
   5. On day 15, take out 10 μL cells and observe for fluorescence (if your cassette carries a fluorescence marker gene). If >80% cells are showing fluorescence, then the culture is ready to harvest. If not, add 5 mL FBS and 50 mL ESF921 media and wait for up to two more days for the harvest.
   6. Harvest virus by centrifuging at 500 rcf for 10 min at 4 °C (*see* **Note 13**).
   7. Filter supernatant using a 0.45-μm bottle top vacuum filter.
   8. Store at 4 °C until proceeding with purification and concentration protocol (*see* **Note 14**).

3.3 BV Purification and Concentration (Day 1–Day 3)

1. Load 500 mL of filtered P3 supernatant in a 1-L Sorvall centrifuge bottle.
2. Using a 50-mL pipet, carefully transfer 50 mL of cold 25% sucrose to the bottom of the bottle, layering a sucrose solution beneath the P3 virus preparation.
3. Place the bottle in the F9 rotor without disturbing the sucrose layer.
4. Centrifuge at 5000 rcf for 20 h at 4 °C.
5. The next day (day 2), a white virus pellet should be visible at the bottom of the tube (Fig. 1a).
6. Discard the supernatant and mark the bottle with a marker pen to remember the pellet position.
7. Add 22 mL of cold PBS to the virus pellet and leave the bottle overnight on a gentle shaker in a cold room to dissolve the pellet (*see* **Note 15**).
8. On day 3, check the marked area to ensure that the pellet is completely dissolved.
9. Transfer viral suspension to a Thinwall Ultra-Clear ultracentrifuge tube in the cell culture hood.

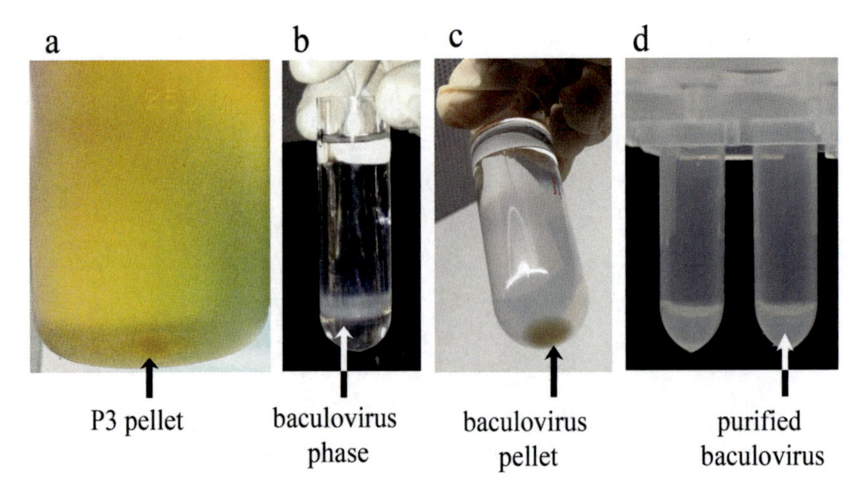

Fig. 1 BV purification and concentration. (**a**) Pelleted P3 baculovirus obtained in Subheading 3.3, **step 5**. (**b**) The sucrose gradient phase separation performed in Subheading 3.3, **step 13**, allows for the isolation of a white translucent phase containing baculovirus. (c) Baculovirus forms a translucent white and firm pellet after a PBS washing step described in Subheading 3.3, **step 18**. (d) Example of purified and concentrated baculovirus solution obtained, stored in a 2-mL microfuge tube described in Subheading 3.3, **step 19**

10. Prepare a sucrose gradient by layering two different sucrose density at the bottom of the BV solution.
    (a) First, slowly and carefully layer 5 mL of cold 20% sucrose solution underneath the base of BV solution.
    (b) Next, slowly and carefully layer 5 mL of cold 50% sucrose solution underneath the 20% sucrose without disturbing it.
11. Carefully place the tubes in the tube holder of the SW32Ti rotor without disturbing the gradient and balance the tubes using PBS.
12. Centrifuge the virus in Beckman Coulter Optima™ LE-80 K at 25,000 rcf (22,000 rpm) for 1 h at 4 °C.
13. The virus will form a dense white layer in the gradient between 20% and 50% phases (Fig. 1b).
14. Prepare a Thinwall Ultra-Clear™ ultracentrifuge tube with 20 mL of cold PBS.
15. Collect the BV containing white phase by puncturing tube using an 18G needle connected to a 5-mL syringe. The volume obtained is typically 4–5 mL.
16. While puncturing the tube after sucrose gradient step, place a 1-L beaker underneath the thin-wall tube to avoid spillage of the virus in the hood.
17. Transfer the virus on top of the PBS containing Thinwall Ultra-Clear™ ultracentrifuge tube and carefully place them in the tube holder of the SW32Ti rotor. Balance tubes using cold PBS.

18. Centrifuge the virus in Beckman Coulter Optima™ LE-80 K at 25,000 rcf (22,000 rpm) for 1 h at 4 °C.
19. Remove the supernatant and add 450 μL of 1× PBS or BV storage solution to the pellet. The pellet will be firm (Fig. 1c), and hence, leave the tube overnight in the dark at 4 °C to soften it (*see* **Note 16**).
20. The next day (day 3), transfer the BV suspension to a 2.0-mL microfuge tube (Fig. 1d).
21. Make 50 μL aliquots and snap freeze virus in −80 °C for long-term storage (*see* **Note 17**).

3.4 Titer Determination

1. Thaw one aliquot of a BV virus and dilute it 1:1000 in ESF921 medium.
2. Add 100 μL ESF921 medium to each well of a 96-well plate.
3. Add 11 μL of the diluted virus preparation to each well of the first column and mix it thoroughly by pipetting at least 5 times.
4. Using an 8-channel micropipette transfer 11 μL of the contents of the first column of wells to the second column (*see* **Note 18**).
5. Repeat this process to generate a serial dilution from the 1st to the 11th column.
6. Leave the 12th column free of the virus as a negative control.
7. Add 100 μL of SF9ET cells at 0.25 × 10^6 cells/mL to every well.
8. Incubate without shaking at 25 °C for 6 days.
9. After 6 days, observe cultured cells for GFP expression and mark each well where at least one cell is expressing GFP as positive (*see* **Note 19**).
10. Calculate the titer using the tissue culture infectious dosage 50 value ($TCID_{50}$) (Fig. 2 and Table 1) (*see* **Notes 20–24**).

3.5 In Vitro Application of BV to Axolotl Limb-Derived AL1 Cells (24-Well Format)

1. Grow axolotl limb-derived AL1 cells in a 150-cm^2 flask. When they reach 90% confluency, remove media from the starting AL1 cell culture and trypsinize them by adding 4 mL Trypsin-EDTA solution to the flask.
2. Wait for 3 min at RT and when cells start to detach, add 5 mL MEM.
3. Spin cells at 500 rcf for 3 min and remove liquid from the pelleted cells.
4. Add 5 mL fresh MEM and resuspend cells.
5. Count cell density and seed cells at 10^4 cells/well along with 500 μL MEM medium in each well of the 24-multiwell plate.

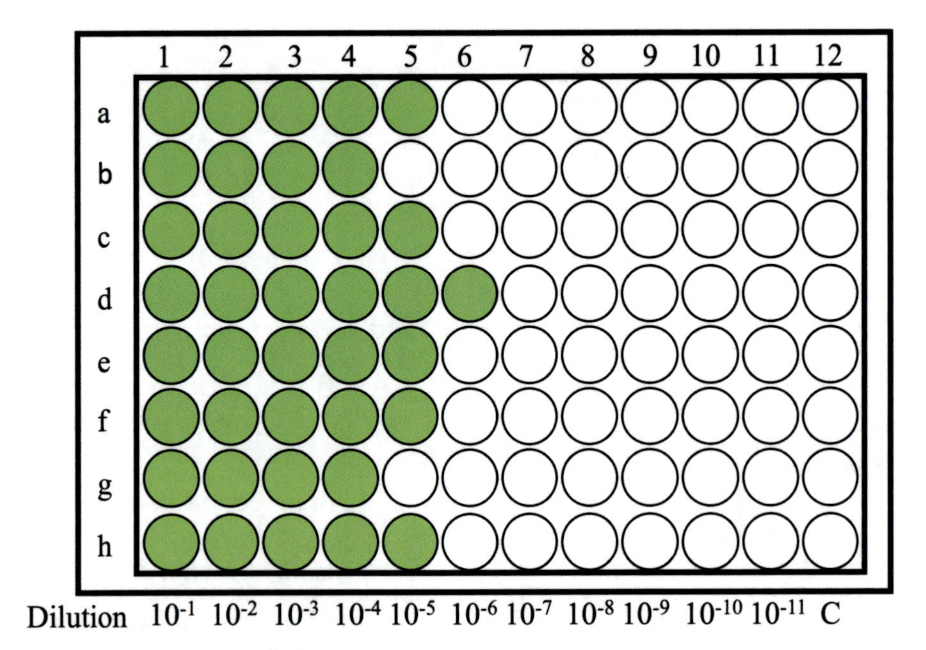

Fig. 2 A typical example of virus titration plate. Green circles indicate wells with at least one infected cell. White circles indicate wells with no infected cell. C: negative control wells

Table 1
A typical example of virus titration plate translated into a tabular form and how to derive 50% infection response

Dilution	Infected (no. of wells)	Uninfected (no. of wells)	Total Infected (No. of wells)	Total infected (No. of wells)	% infected
A	B	C	D	E	
10^{-1}	8	0	sum(B1: B11) = 39	sum(C1:C1) = 0	D1/ (D1+E1)% = 100
10^{-2}	8	0	sum(B2: B11) = 31	sum(C1:C2) = 0	D2/ (D2+E2)% = 100
10^{-3}	8	0	sum(B3: B11) = 23	sum(C1:C3) = 0	D3/ (D3+E3)% = 100
10^{-4}	8	0	sum(B4: B11) = 15	sum(C1:C4) = 0	D4/ (D4+E4)% = 100
10^{-5}	6	2	sum(B5:B11) = 7	sum(C1:C5) = 2	D5/

(D5+E5)% = 77.78$10^{-6}$17sum(B6:B11) = 1sum(C1:C6) = 9D6/(D6+E6)% = 10$10^{-7}$08sum(B7: B11) = 0sum(C1:C7) = 17D7/(D7+E7)% = 0$10^{-8}$08sum(B8:B11) = 0sum(C1:C8) = 25D8/ (D8+E8)% = 0$10^{-9}$08sum(B5:B11) = 0sum(C1:C9) = 33D9/(D9+E9)% = 0$10^{-10}$08sum(B6: B11) = 0sum(C1:C10) = 41D10/(D10+E10)% = 0$10^{-11}$08sum(B7:B11) = 0sum(C1: C11) = 49D11/(D11+E11)% = 0Control011NANANA

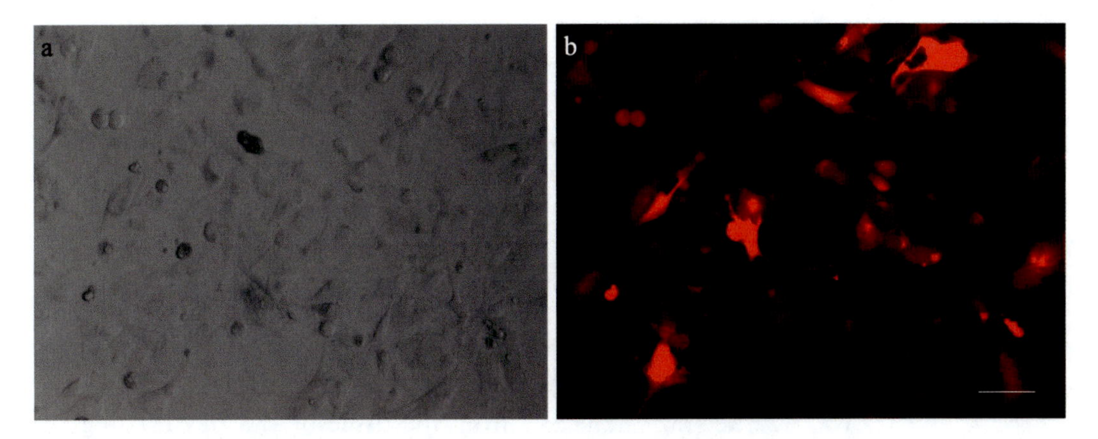

Fig. 3 In vitro limb transduction. AL-1 cell line transduction using CMV:Cherry baculovirus using multiplicity of infection (MOI) of 1000 (**a**) bright field, (**b**) fluorescence image. Scale bars = 50 μm

6. Add 1–2 μL of concentrated BV (Titer $>10^9$) or at an MOI of 1000, while seeding the cells. Mix it by pipetting 5 times (*see* **Note 25**).
7. Leave one well without BV infection as a negative control.
8. Incubate plate at 25 °C in 5% CO_2 incubator for infection to proceed.
9. Observe for gene expression in BV infected well after 3–4 days (Fig. 3) (*see* **Note 26**).

3.6 In Vivo Application of BV

1. Anesthetize animals by placing them in 0.01% Benzocaine (from 0.03% stock solution in tap water) (*see* **Note 27**).
2. Keep the virus on ice and protected from light near the working space.
3. Thaw an aliquot of frozen purified BV. Take 18 μL of BV and add 2 μL of 10× Fast Green FCF solution. Mix it by pipetting (*see* **Note 28**).
4. Use a microloader tip to load the microinjection needle with 20 μL microliters of BV preparation.
5. Insert the needle into the micropipette holder within the micromanipulator and ensure that the working range of the movable parts is appropriate.
6. Open the injection needle by breaking its tip with the help of fine forceps under stereo microscope (*see* **Note 29**).
7. Place the animal belly down in a Sylgard plate (*see* **Note 30**).
8. Stretch the arm/tail to maximize the exposure of area of interest, and to have a clear visualization of the arm/tail structure. If the tissue of interest is blastema, then amputate appendage (limb/tail) and let the blastema develop to a desired stage before injection (*see* **Note 31**).

9. Carefully puncture the skin on the designated place of injection by using a 30G needle (*see* **Note 32**).
10. Insert the glass capillary needle in the aperture created and locate its tip deep into the mesenchyme for a broad distribution of the virus within the tissue. It is possible to locate the tip beneath the skin and to introduce it into muscle tissue as well (*see* **Note 33**).
11. Press the footswitch to activate the pneumatic pump and continuously inject the BV suspension into the tissue for a couple of seconds. Each injection should transfer a total of about 1–2 μL of BV suspension into the tissue (*see* **Note 34**).
12. Remove the needle from the tissue by carefully rotating the micromanipulator knob and repeat the procedure in another area of the limb/tail/blastema if required.
13. To introduce the BV suspension to the ventral side of the limb, place the animal on its side, making sure the target structure is well visualized.
14. Select the sites to inject the virus and repeat the procedures described in **step 11**.
15. Once the injection procedure is completed—and before returning to the aquarium—place the animal in a humidified chamber (a petri dish with a moist paper towel in it) for 10 min to avoid possible leakage of the BV suspension from the tissue.
16. Return animal to the aquarium and let it recover from the anesthetic.
17. The expression of fluorescence in transduced tissue is visible after 2–7 days post-injection of virus (Fig. 4a, b). To observe fluorescence, anesthetize animal in 0.01% Benzocaine and place it under fluorescent stereoscope. Depending upon the goal of experiment, appendage can be amputated to study participation of the labeled cell population (*see* **Notes 35** and **36**).

4 Notes

1. These protocols were developed for those recombinant baculoviruses that confirm to the Biosafety Level 1 (BSL-1) norms, such as the reporter gene pseudotyped baculovirus (such as Cherry or GFP baculovirus). Since baculovirus can efficiently transduce mammalian cells, when using a transgene with oncogenic potential, the set-up must be adapted to comply with BSL-2 requirements. Accordingly, all work should be conducted using appropriate protective wear, such as gloves and lab coat.

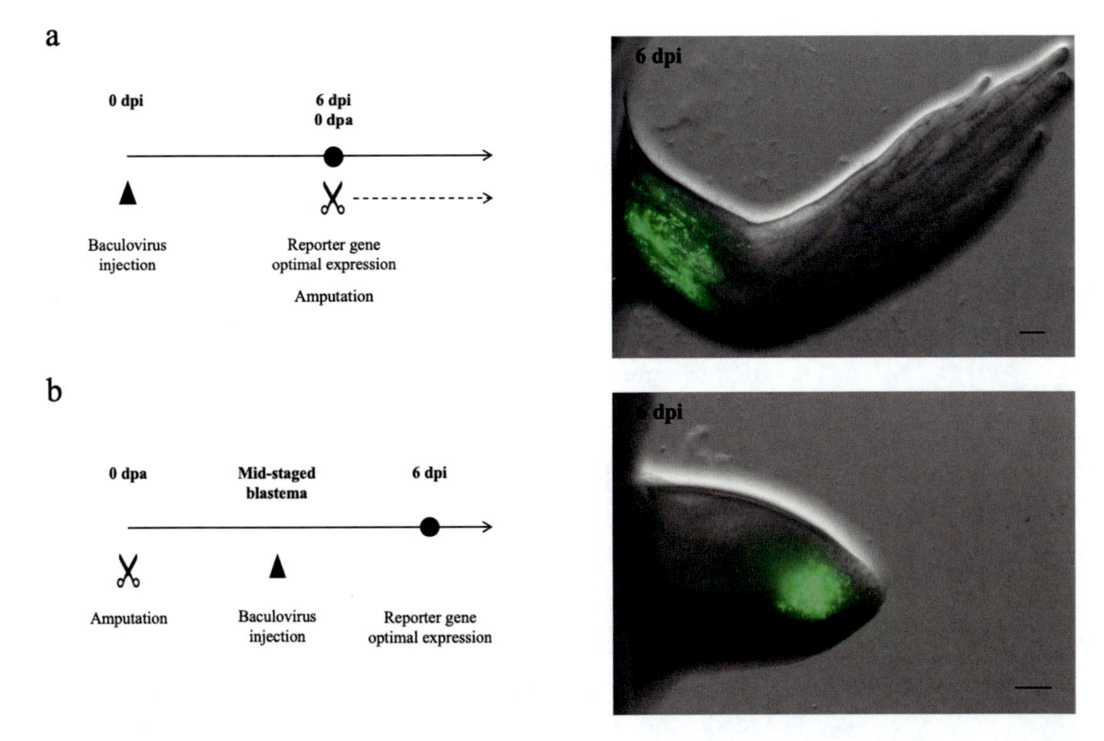

Fig. 4 In vivo limb transduction. In vivo limb transduction protocol (left) and resulting transduction (right panel) of CMV:eGFP baculovirus in (**a**) intact limb and (**b**) blastema. Transgene signal detection peaks at 6 days post-infection (dpi). Scale bars = 500 μm

4.1 Bacmid and pOEM1 Transfer Vector Preparation

2. DefBac is a bacmid which is difficult to transform. Hence, we recommend making a glycerol stock of DefBac and store it at −80 °C.
3. Avoid repeated freeze-thaw of the DefBac DNA. Repeated freeze-thaw of the bacmid significantly affects P1 production. If you intend to freeze it for a longer duration, then make aliquots and store it.
4. Always prepare a fresh digestion mix of DefBac. Storing digested DefBac for more than a week affects P1 virus production.
5. Due to the low yield of bacmid, maxiprep is preferred. pOEM1-GED rescue vector can be prepared as a miniprep or maxiprep.
6. pOEM1-GED is a rescue plasmid that upon recombination with DefBac creates BV genome which expresses cassette of your interest. To generate pOEM1-GED rescue (transfer) vector of your interest, replace CMV:GFP of the pOEM1-GED-CMV:GFP construct with your cassette of interest.

4.2 Recombinant BV Production and Virion Amplification

7. The pOEM-GED construct carries partial lef2 and ORF1629 loci of the BV. Both pseudotyping and gene expression cassettes are located between these partial lef2 and ORF1629 loci. The other half of the lef2 and ORF1629 loci are located on the DefBac bacmid and upon transfection complete lef2 and ORF1629 are recreated due to recombination. This recombination event transfers the vertebrate expression cassette to the DefBac and creates a fully functional viral genome that is capable of replicating in the SF9 cells.
8. We use 600 ng of pOEM-GED transfer vector for constructs that are smaller than 10 kb and 1 μg of pOEM-GED transfer vector for constructs that are larger than 10 kb for virus production.
9. Viruses are light sensitive. Avoid exposure to excessive light during handling.
10. Cells that are infected and actively producing virus display an increase in their diameter and irregular shape. The uninfected cells have a diameter of 16–18 μm, whereas infected cells have a diameter of 20–22 μm. If >20% cells are showing fluorescence, the culture is ready for the harvest. If not, add 500 μL ESF921 media and wait for one more day to harvest. We have produced a virus with pOEM-GED constructs that are up to 14 kb in length using this protocol. We have observed that pOEM-GED constructs that are large take a longer time for virus production, and it may be worth to keep P1 for up to 8 days. For larger constructs, when the P1 has 10–15% of total cells exhibiting fluorescence, the culture is ready for the harvest. Supplementing cells at day 4 with extra 500 μL of ESF921 media supports their growth and helps increase infection, particularly for larger constructs or when cell density is very high.
11. The presence of FBS in the cell culture media helps in virus production. For most transfer constructs that are smaller than 10 kb in size, 4 days of incubation with infection is sufficient for P1, P2, and P3 production and they do not need FBS supplement.
12. Since the CMV promoter is active in SF9 cells, gene of interest may affect virus production, particularly when a gene has a strong tumor suppressor effect. In such cases, it might be advisable to infect higher density of cells from the beginning or produce multiple P3 and pool them to obtain high titer of the virus.
13. Centrifuging the virus in 500-mL tubes results in loose pellets and more debris in the supernatant. This may result in a longer filtration time or clogging of a filter. To avoid this, split the P3 supernatant into smaller tubes, e.g., 50-mL falcon tubes.

Spinning in smaller tubes produces a more compact pellet and clearer supernatant.

14. A filtered, sterile P3 supernatant can be stored for 48 h in dark at 4 °C. However, we recommend that one should proceed to concentrate the virus as soon as possible.

4.3 BV Purification and Concentration

15. BVs are fragile, and vigorous vortexing can affect the quality of viruses for the subsequent steps. Hence, it is better to dissolve pellet by gentle shaking overnight in the dark on a table-top shaker.

16. A filtered, sterile BV storage solution can be stored for more than a year at 4 °C.

17. In our previous publication [14], we describe dialysis after the sucrose gradient separation and concentration step. Dialysis improves purity of the virus. If viruses are aliquoted and frozen immediately after purification, then dialysis can be an optional step after thawing. However, if the purity of the virus is important, then we suggest dialyzing using Slide-A-Lyzer™ MINI dialysis device after thawing and just before using it for infection. We have observed better results with such dialyzed virus on AL1 cells.

4.4 Titer Determination

18. Make sure to change pipette tips while doing serial dilution of the virus.

19. SF9ET cells carry a PH:GFP cassette (PH—polyhedrin promoter) which expresses GFP in these cells upon successful infection of the baculovirus.

20. $TCID_{50}$ calculates titer based on dilution that induces 50% infection of the cultured cells [25–27].

    (a) This method assumes that all cultures infected at a certain dilution would have been infected at lower dilutions (more concentrated virus) and that all cultures not infected at a certain dilution would not have been infected at higher dilutions (less concentrated virus). Figure 2 shows a typical example of results obtained from virus titration.

    (b) For reader's reference, we have converted titer results from Fig. 2 into a tabular form which allows to derive 50% infection response.

21. As per Table 1, the dilution that leads to a 50% response lies between 10^{-4} and 10^{-5}. This dilution is calculated by linear interpolation which allows us to calculate the proportionate distance (PD) as follows:

    PD = (X-50)/(X+Y), where X is the % response above 50%, and Y is the % response below 50%.

(a) In the above example, PD = (77.78 – 50)/(69.2 – 10) = 27.78/59.2 = 0.469.

22. The $TCID_{50}$ is calculated as follows:

Log $TCID_{50}$ = log of the dilution giving a response above 50% – the PD of that response.

(a) In this case, Log $TCID_{50}$ = – 5 – 0.469 = – 5.283.

(b) Hence, $TCID_{50} = 10^{-5.283}$.

23. The virus titer is the reciprocal of the $TCID_{50} = 1/10^{-5.283} = 1.9 \times 10^5$ virus particle/10 μL of the 1:1000 diluted virus stock:

(a) Hence, the stock virus concentration = $(1.9 \times 10^5)(10^3) = 1.9 \times 10^8$ virus particle/10 μl = 1.9×10^{10} virus particle/mL.

24. We routinely obtain viral titer in the range of 10^9 to 10^{10} using our protocol. For large constructs, the titer can be in the range of 10^8.

4.5 In Vitro Application of BV

25. More than 2 μL BV/well leads to cell death. If you notice significant cell death, you can lower the BV infection down to 0.5 μL/well.

26. Gene expression can be detected as early as 2 days after BV infection and the signal intensity of GFP seems to be stabilized after 7 days.

4.6 In Vivo Application of BV

27. Depending upon the size of the animal, the use of the anesthetic and volume of the BV needs further optimization.

28. The Fast green solution is stable at 4 °C for up to 6 months.

29. By doing this under the stereomicroscope, it is possible to identify an area of the tip that is thin enough to minimize tissue injury and to simultaneously provide enough firmness to move deep through the internal tissue. A thin needle opening will also provide a good control of liquid injection and reduce the risk of mechanical tissue disruption created by the liquid flow.

30. Sylgard 184 provides a non-reactive and soft but firm surface for handling animals during the injection. This helps avoid harming the animals. Keep animals wet during the injection procedure to avoid tissue injury.

31. For the blastema formation, bone trimming is important for forming a consistent blastema stages, because often untrimmed bone protrudes and damages the blastema.

32. The mature limb and tail can be pierced from a distal or proximal end at an angle. Puncturing of skin through needle will facilitate the insertion of glass needle tip and avoids rupturing or clogging it. For the blastema injection, we avoid

piercing through 30G needle and rather use capillary for injection to prevent the leakage of virus. It is also important not to pierce blastema from the distal end but rather capillary should penetrate from the proximal end near the amputated stump at an angle to avoid damage to the blastema.

33. Transduction through BV is highly dependent upon the site of injection, accessibility of cells, and retention of the virus in the tissue/site of injection. We have noticed efficient transduction in PRRX1$^+$ fibroblasts, MHC$^+$ muscle fibers, PAX7$^+$ satellite cells, and SOX2$^+$ neural progenitors. Contrary, skeletal cells and neurons are some of the most difficult cell types to transduce.
34. In a 2.5-cm animal, 5 μL BV injection is sufficient to achieve infection around the injection site. For larger animals (e.g., 5 cm length), this volume can be up to 20 μL.
35. It is normal to observe inflammation (which is visible as autofluorescence) at the site of injection 24 h later. Inflammation will resolve over a period of 1 week. The earliest expression in animals is detected at 2–3 days post-injection (dpi) and the expression stabilizes around 1 week post-injection (wpi). The peak of expression is seen between 4 and 7 days. BV expression can be detected up to 4 months post-injection; however, the number of cells actively expressing viral reporter genes decreases over time.
36. If studying regeneration, the participation of transduced cells in the regeneration process can be observed by light microscopy/UV light using the appropriate filter. Typically, reporter gene-expressing cells are visible throughout the entire regeneration process, i.e., for at least 40 days, and persist after regeneration is complete.

Acknowledgement

The authors thank members of the PM, MHY, and EMT lab for their inputs in improving the protocol. Work in the PM laboratory is supported by grants from NIH-COBRE (5P20GM104318-08) and DFG (429469366). CRO is supported by predoctoral grant from the Portuguese Foundation for Science and Technology (SFRH/BD/51280/2010). Work in the MHY laboratory is supported by grants from DFG (22137416 & 450807335). Work in the EMT laboratory is supported by grants from ERC (AdG 742046) and FWF (Standalone I4846).

References

1. Fei JF, Lou WP, Knapp D, Murawala P, Gerber T, Taniguchi Y, Nowoshilow S, Khattak S, Tanaka EM (2018) Application and optimization of CRISPR-Cas9-mediated genome engineering in axolotl (Ambystoma mexicanum). Nat Protoc 13(12):2908–2943. https://doi.org/10.1038/s41596-018-0071-0
2. Khattak S, Murawala P, Andreas H, Kappert V, Schuez M, Sandoval-Guzman T, Crawford K, Tanaka EM (2014) Optimized axolotl (Ambystoma mexicanum) husbandry, breeding, metamorphosis, transgenesis and tamoxifen-mediated recombination. Nat Protoc 9(3): 529–540. https://doi.org/10.1038/nprot.2014.040
3. Masselink W, Reumann D, Murawala P, Pasierbek P, Taniguchi Y, Bonnay F, Meixner K, Knoblich JA, Tanaka EM (2019) Broad applicability of a streamlined ethyl cinnamate-based clearing procedure. Development 146(3):dev166884. https://doi.org/10.1242/dev.166884
4. Pende M, Vadiwala K, Schmidbaur H, Stockinger AW, Murawala P, Saghafi S, Dekens MPS, Becker K, Revilla IDR, Papadopoulos SC, Zurl M, Pasierbek P, Simakov O, Tanaka EM, Raible F, Dodt HU (2020) A versatile depigmentation, clearing, and labeling method for exploring nervous system diversity. Sci Adv 6(22):eaba0365. https://doi.org/10.1126/sciadv.aba0365
5. Bryant DM, Johnson K, DiTommaso T, Tickle T, Couger MB, Payzin-Dogru D, Lee TJ, Leigh ND, Kuo TH, Davis FG, Bateman J, Bryant S, Guzikowski AR, Tsai SL, Coyne S, Ye WW, Freeman RM Jr, Peshkin L, Tabin CJ, Regev A, Haas BJ, Whited JL (2017) A tissue-mapped axolotl de novo transcriptome enables identification of limb regeneration factors. Cell Rep 18(3):762–776. https://doi.org/10.1016/j.celrep.2016.12.063
6. Smith JJ, Timoshevskaya N, Timoshevskiy VA, Keinath MC, Hardy D, Voss SR (2019) A chromosome-scale assembly of the axolotl genome. Genome Res 29(2):317–324. https://doi.org/10.1101/gr.241901.118
7. Nowoshilow S, Schloissnig S, Fei JF, Dahl A, Pang AWC, Pippel M, Winkler S, Hastie AR, Young G, Roscito JG, Falcon F, Knapp D, Powell S, Cruz A, Cao H, Habermann B, Hiller M, Tanaka EM, Myers EW (2018) The axolotl genome and the evolution of key tissue formation regulators. Nature 554(7690): 50–55. https://doi.org/10.1038/nature25458
8. Whited JL, Lehoczky JA, Tabin CJ (2012) Inducible genetic system for the axolotl. Proc Natl Acad Sci U S A 109(34):13662–13667. https://doi.org/10.1073/pnas.1211816109
9. Echeverri K, Tanaka EM (2003) Electroporation as a tool to study in vivo spinal cord regeneration. Dev Dyn 226(2):418–425. https://doi.org/10.1002/dvdy.10238
10. Rodrigo Albors A, Tanaka EM (2015) High-efficiency electroporation of the spinal cord in larval axolotl. Methods Mol Biol 1290: 115–125. https://doi.org/10.1007/978-1-4939-2495-0_9
11. Roy S, Gardiner DM, Bryant SV (2000) Vaccinia as a tool for functional analysis in regenerating limbs: ectopic expression of Shh. Dev Biol 218(2):199–205. https://doi.org/10.1006/dbio.1999.9556
12. Whited JL, Tsai SL, Beier KT, White JN, Piekarski N, Hanken J, Cepko CL, Tabin CJ (2013) Pseudotyped retroviruses for infecting axolotl in vivo and in vitro. Development 140(5):1137–1146. https://doi.org/10.1242/dev.087734
13. Khattak S, Sandoval-Guzman T, Stanke N, Protze S, Tanaka EM, Lindemann D (2013) Foamy virus for efficient gene transfer in regeneration studies. BMC Dev Biol 13:17. https://doi.org/10.1186/1471-213X-13-17
14. Oliveira CR, Lemaitre R, Murawala P, Tazaki A, Drechsel DN, Tanaka EM (2018) Pseudotyped baculovirus is an effective gene expression tool for studying molecular function during axolotl limb regeneration. Dev Biol 433(2):262–275. https://doi.org/10.1016/j.ydbio.2017.10.008
15. van Oers MM, Pijlman GP, Vlak JM (2015) Thirty years of baculovirus-insect cell protein expression: from dark horse to mainstream technology. J Gen Virol 96(Pt 1):6–23. https://doi.org/10.1099/vir.0.067108-0
16. Airenne KJ, Hu YC, Kost TA, Smith RH, Kotin RM, Ono C, Matsuura Y, Wang S, Yla-Herttuala S (2013) Baculovirus: an insect-derived vector for diverse gene transfer applications. Mol Ther 21(4):739–749. https://doi.org/10.1038/mt.2012.286
17. Mansouri M, Bellon-Echeverria I, Rizk A, Ehsaei Z, Cianciolo Cosentino C, Silva CS, Xie Y, Boyce FM, Davis MW, Neuhauss SC, Taylor V, Ballmer-Hofer K, Berger I, Berger P (2016) Highly efficient baculovirus-mediated multigene delivery in primary cells. Nat Commun 7:11529. https://doi.org/10.1038/ncomms11529

18. Kaikkonen MU, Raty JK, Airenne KJ, Wirth T, Heikura T, Yla-Herttuala S (2006) Truncated vesicular stomatitis virus G protein improves baculovirus transduction efficiency in vitro and in vivo. Gene Ther 13(4):304–312. https://doi.org/10.1038/sj.gt.3302657
19. Kitagawa Y, Tani H, Limn CK, Matsunaga TM, Moriishi K, Matsuura Y (2005) Ligand-directed gene targeting to mammalian cells by pseudotype baculoviruses. J Virol 79(6): 3639–3652. https://doi.org/10.1128/JVI.79.6.3639-3652.2005
20. Mangor JT, Monsma SA, Johnson MC, Blissard GW (2001) A GP64-null baculovirus pseudotyped with vesicular stomatitis virus G protein. J Virol 75(6):2544–2556. https://doi.org/10.1128/JVI.75.6.2544-2556.2001
21. Tani H, Limn CK, Yap CC, Onishi M, Nozaki M, Nishimune Y, Okahashi N, Kitagawa Y, Watanabe R, Mochizuki R, Moriishi K, Matsuura Y (2003) In vitro and in vivo gene delivery by recombinant baculoviruses. J Virol 77(18):9799–9808. https://doi.org/10.1128/jvi.77.18.9799-9808.2003
22. Nacu E, Gromberg E, Oliveira CR, Drechsel D, Tanaka EM (2016) FGF8 and SHH substitute for anterior-posterior tissue interactions to induce limb regeneration. Nature 533(7603):407–410. https://doi.org/10.1038/nature17972
23. Wagner I, Wang H, Weissert PM, Straube WL, Shevchenko A, Gentzel M, Brito G, Tazaki A, Oliveira C, Sugiura T, Shevchenko A, Simon A, Drechsel DN, Tanaka EM (2017) Serum proteases potentiate BMP-induced cell cycle re-entry of dedifferentiating muscle cells during newt limb regeneration. Dev Cell 40(6): 608–617, e606. https://doi.org/10.1016/j.devcel.2017.03.002
24. Lemaitre RP, Bogdanova A, Borgonovo B, Woodruff JB, Drechsel DN (2019) FlexiBAC: a versatile, open-source baculovirus vector system for protein expression, secretion, and proteolytic processing. BMC Biotechnol 19(1):20. https://doi.org/10.1186/s12896-019-0512-z
25. O'Reilly DR, Miller LK, Luckow VA (1994) Baculovirus expression vectors: a laboratory manual. Oxford University Press, pp 1–347. https://doi.org/10.1016/0092-8674(93)90288-2
26. Reed LJM, H. (1938) A simple method of estimating fifty percent endpoints. Am J Epidemiol 27(3):493–497. https://doi.org/10.1093/oxfordjournals.aje.a118408
27. Sung LY, Chen CL, Lin SY, Li KC, Yeh CL, Chen GY, Lin CY, Hu YC (2014) Efficient gene delivery into cell lines and stem cells using baculovirus. Nat Protoc 9(8): 1882–1899. https://doi.org/10.1038/nprot.2014.130

Chapter 25

Cell Dissociation Techniques in Salamanders

Gabriela Johnson, Nadjib Dastagir, Zachary Beal, Andrew Hart, and James Godwin

Abstract

Cell dissociation is an important technique for the study of tissue phenotypes. The method chosen to harvest cells from solid tissues profoundly influences the types of cells recovered. Methodology also shapes any biases that are introduced that can act upon cell surface protein phenotypes or gene expression. Here we describe examples of cell surface phenotypic changes and typical yields, under 4 different isolation conditions (enzymatic/non-enzymatic), using the axolotl spleen, and the regenerating limb. We describe simple methods for evaluating the liberation of viable cells and the downstream characterization of cell diversity using a live-cell flow cytometry approach. Of note, the cellular composition of dissociated cells and surface antigen detection vary with each condition. TrypLE and "no enzyme" protocols give the highest surface marker expression, but poor liberation of non-immune cells in the blastema. Liberase-DH and Liberase-TL have alternative neutral proteases and both give acceptable dissociation of diverse cell types in the blastema. Liberase-TL provides the highest yield of all cell sizes and a larger non-immune fraction. Matching dissociation times between limb blastemas and spleens, we demonstrate the effect of "over-digestion" in soft tissues. In the spleen, the Liberase enzyme cocktails produced the lowest yields, worst viability, and the greatest loss of immune cell surface markers, when compared with non-enzymatic and TrypLE dissociation. These examples provide a template for optimizing protocols for individual tissues while achieving the balance between cell recovery and the mitigation of cellular changes appropriate for downstream applications such as single-cell RNA sequencing and flow cytometry.

Key words Cell dissociation, Single cells, Phenotyping, RNAseq, Regeneration, Salamander, Axolotl

1 Introduction

1.1 The Importance of Cell Dissociation and Potential Applications

In order to really understand the biology of a tissue or organ it is important to be able to phenotype the diverse cell types of which it is composed. This phenotyping can lead to identification of rare cells types, novel gene expression, or key signaling pathways that regulate the relevant biological process that is being studied. Flow cytometry is a leading high-throughput application for phenotyping single cells but relies on either genetic/chemical labeling or immunostaining to identify different cellular subtypes. In order to

Ashley W. Seifert and Joshua D. Currie (eds.), *Salamanders: Methods and Protocols*,
Methods in Molecular Biology, vol. 2562, https://doi.org/10.1007/978-1-0716-2659-7_25,

take advantage of these phenotyping approaches, one must isolate a single-cell population from the tissue of interest. Recently, single-cell RNA sequencing (scRNAseq) approaches have become more accessible and provide non-biased genetic profiling of cellular populations. Obtaining single cells, free of doublets or debris, is critical to good sample preparation and managing costs. Of course, combining flow cytometry with either scRNAseq or bulk RNA sequencing can be a powerful approach for zooming in on particular cell population of interest. The nature of the tissue and the intended downstream application will determine the best protocol and a "one size fits all" approach is not generally applicable. The ideal tissue dissociation protocol will extract all of the target cells with high viability and with minimal changes to gene expression and cellular phenotypes.

1.2 Manual Dissociation Approaches

Manual dissociation protocols (in the absence of enzymes) rely on physical tearing of the non-cellular extracellular matrix (ECM) using forceps, scissors, or other instruments followed by manual trituration, the process by which large chunks of tissue are repeatedly passed through a pipette to produce smaller and smaller fragments. Finally, the cells are separated from the large multicellular chunks by passing through nylon mesh or a commercial nylon sieve or cell strainer. This approach releases predominantly immune cells and other cells that are only loosely attached and has the advantage that the isolation is usually fairly rapid, cells can be kept cold, and causes the least phenotypic and genetic changes. In addition, mechanical dissociation is compatible with the addition of ion chelators like EDTA to eliminate calcium-dependent adhesion and cell–cell contacts to promote the dissociation and prevent adhesion to plastics [1]. Mechanical dissociation will also maintain surface protein epitopes that may be required for immunophenotyping, magnetic purification, or by Fluorescence Activated Cell Sorting (FACS). Enzyme-free mechanical dissociation has been successfully applied to well-developed cone-shaped newt blastema's for primary culture [2] and is applicable to soft organs, like the spleen. It should be noted that we and others [3] have found that for many target cells, mechanical digestion results in very low yields depending on the ECM state and composition within the target tissue and may also introduce an extreme bias in the types of cells that are liberated (i.e., favoring immune cells or other loosely attached cell types).

1.3 Enzymatic Dissociation Approaches

Some mature differentiated tissues have well-developed desmosomes with neighboring cells anchored to each other via cell-to-cell junctions on the cell membrane [4–6]. These adhesions can prevent cell release and dissociation of target cells via mechanical dissociation. In these cases, addition of enzymes will be required to break down the ECM components containing the target cells.

Enzymatic cellular digestion is usually performed by tailoring the enzyme cocktail to match the type of ECM components present in the target tissue. Cocktails can include common ECM degrading enzymes, such as collagenases (zinc endopeptidases), elastase (serine protease) and dispase (neutral metalloprotease), thermolysin (neutral metalloprotease), hyaluronidase (glycosaminoglycan degrading), chymotrypsin (serine protease), papain (cystine protease), and trypsin (serine protease) (reviewed in refs. 7, 8). All of these have varying levels of toxicity and effectiveness in releasing cells from various tissue types. It should also be noted that some commercial collagenase preparations (types 1–7) contain varying levels of secondary proteolytic activities, such as caseinase, clostripain, aminopeptidase, and tryptic activities, depending on the grade and manufacturer.

While digestive enzymes such as collagenase, hyaluronidase, and dispase can all target components of the ECM, like collagen, fibronectin, and hyaluronan, leading to liberation of some cells from the matrix, cell–cell junctions like that found on epithelial tissues are unaffected. Enzymes capable of breaking cell–cell junctions, such as Trypsin and Papain, are commonly used to degrade proteins within cell–cell junctions but are known to severely damage cell surface epitopes and membrane proteins [9, 10]. These proteases can also cause free DNA-induced cell aggregation leading to reduced cell yields. The free DNA released from cell lysis during the dissociation can be managed by addition of DNaseI but like collagenase and several other enzymes, DNaseI requires calcium for activity which may lead to the loss of epithelial cells and macrophages by adhesion to plastics [11].

The challenges in using enzymatic approaches depend on the downstream applications. Many cell surface proteins are degraded by various enzymatic mixtures and the enzymatic cocktail may need to be adjusted to maintain specific surface epitopes if these are required for analysis or purification. However, isolation of cells for scRNAseq may be optimized to capture all of the cells in the given tissue without regard to surface proteins. Of course, the duration and concentration of enzymes also have to be carefully managed to provide optimum viability and limit large changes in mRNA or protein expression.

There are numerous cocktails available that are tailored for target cell and tissue types. Many of these detailed protocols and recipes can be found at commercial supplier websites (i.e., www.wothington.com). In addition to the traditional custom blends, many manufacturers have commercially available optimized digestion enzyme cocktails. "Liberase," is family of enzyme blends that combine collagenases with metallopeptidases capable of breaking cell–cell junctions that can maintain satisfactory cell ultrastructure [12]. Here we chose to demonstrate the specific difference that alternative metalloproteases can have on cell dissociation outcomes by comparing Liberase-DH (collagenases with a high dispase concentration) versus Liberase-TL (collagenases with a low Thermolysin concentration). Commercial Trypsin has several co-purified proteases that are known to damage surface markers and cause cell toxicity. The modified recombinant enzyme "TrypLE" mimics the proteolytic effects of trypsin with improved preservation of surface proteins and a reduction in gene expression modification [13]. In this protocol we demonstrate that TrypLE is less toxic than Trypsin for axolotl tissue dissociation, with the caveat that the length of time tissue was exposed to the different enzyme cocktails was held constant and that shorter of longer exposure times may give different results. The reagents used and duration of enzymatic digestion for a particular tissue type should be determined empirically and trade-offs managed based on experimental endpoints. Combining manual dissociation prior to enzymatic protocols enhances enzyme access to target cells and can speed up digestion times and improve yields dramatically. Many enzymes are optimally active at 37 °C and collagenase activity drops to 75% at 27 °C and 25% activity [14, 15]. We perform enzymatic digestions at 25 °C (room temperature) to prevent stress to axolotl cells and have found this to be acceptable for digestion times up to 1 h.

1.4 Monitoring the Success of Dissociation

For downstream applications, such as single-cell sequencing or flow cytometry, the critical parameters for monitoring success of a protocol are the assessment of viability, absence of cellular debris, and absence of aggregates. A simple approach for quick and easy monitoring is the use of trypan blue cell viability exclusion assay/cell counting or Acridine orange/Propidium iodide (AO/PI) [16]. This can be performed via sampling and use of a traditional hemocytometer or automatic cell counter. The cell number and live/dead ratio can be quickly evaluated and once a successful protocol obtained, this can be followed up with flow cytometry to give a clearer indication of cellular diversity and more accurate assessment of viability. An example is shown in Fig. 1 with varying levels of cell yield, viability, and debris quickly observed. Given the importance of the salamander limb in the study of regeneration, this chapter will focus on limb tissue/blastema dissociation and provide some examples of alternative protocols and tissue examples.

2 Materials

2.1 General Reagents, Equipment, and Consumables

1. Microsurgery scissors and forceps.
2. Disposable Pasteur pipettes.
3. 5-mL and 1.7-mL microfuge tubes.
4. 15-mL and 50-mL conical centrifuge tubes.

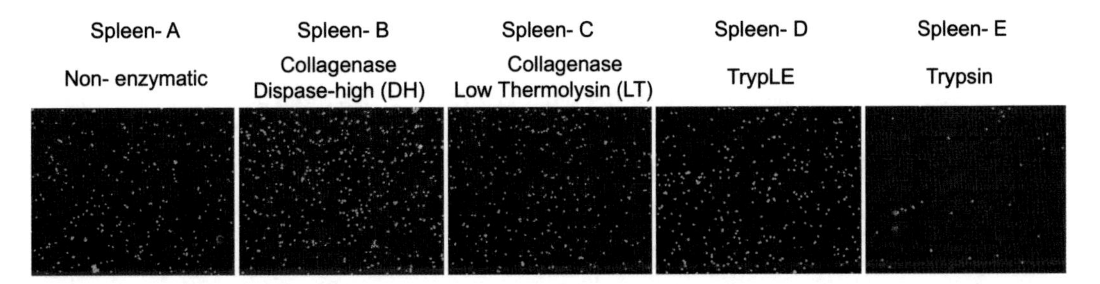

Fig. 1 AO/PI. Evaluation of cell number and viability. Live cells (AO+) shown in green and non-viable cells (PI+) shown in red/orange. 5 × 5 grid marked with black dashed line

5. Cell strainers with 70 μm (pluriSelect: 43-10070-40 and 43-10100-40 for 5 mL, and Fisher: 22-363-547 for 50 mL or equivalent).
6. 5-mL round bottom FACS tubes with 40-μm cell strainer cap (if necessary) BD:Cat. No. 35223.
7. 0.5 M Ethylenediaminetetraacetic acid (EDTA), pH 8.0.
8. 10× Hanks balanced salt solution (HBSS).
9. HBSS-EDTA solution (0.7 × HBSS, 5 mM EDTA final, no magnesium and no calcium): For example, to prepare 200 mL solution, add 14 mL 10× HBSS to 184 mL sterile water; then add 2 mLs of 0.5 M EDTA to this solution. Amount of HBSS-EDTA required varies with number of samples. Note that for Subheading 3.3. "Blastema/spleen isolation and tissue dissociation" **step 11**, one should prepare 0.7×HBSS, 10 mM EDTA by diluting 4 mL of 0.5 mL EDTA in 200 μl of 0.7×HBSS.
10. Normal goat serum.
11. Goat Block Buffer solution: To prepare 50 mL add 5 mL goat serum to 45 mL HBSS-EDTA solution.
12. FACS buffer: To prepare 50 mL add 1 mL goat serum to 49 mL HBSS-EDTA solution.
13. Holtfreter's stock solution: Dissolve 3.46 g of NaCl, 0.05 g of KCL, 0.1 g of $CaCl_2$, and 0.2 g of $NaHCO_3$ in 1 L of water.
14. Holtfreter's working solution (20%): Dilute stock Holtfreter's solution 1:5 using sterile water.
15. Stereomicroscope with fluorescence attachment (with camera preferred).

2.2 Axolotl Surgery

1. 0.2% ethyl 3-aminobenzoate methanesulfonate salt (Tricaine) Sigma: E10521-50G. For 1 L solubilize 2 g in 1 L Holtfreter's working solution. Adjust pH to 7–7.5. Store at 4 °C away from light.

2.3 Axolotl Tissue Dissociation Cocktails

1. Mechanical digestion buffer: (0.7× HBSS, 1 mM Mg^{++}, 1 mM Ca^{++} solution) with 500 μg/mL DNaseI final concentration (Roche-Sigma, Cat#4536282001). (i.e., 100 μl of 5 mg/mL DNaseI stock per 1 mL of tissue dissociation buffer.
2. Liberase DH Buffer: 0.4 mg/mL Liberase DH (Roche-Sigma, Cat#5401054001), (~2 Wünsch Units/mL final). Make by diluting 200 μl of 2 mg/mL stock (10.4 units/mL) with 800 μl of (0.7× HBSS, 1 mM Ca^{++}, 1 mM Mg^{++}). 500 μg/ mL DNaseI final concentration.
3. Liberase TL buffer: 0.4 mg/mL Liberase TL (Roche-Sigma, Cat#5401020001), (~2 Wünsch Units/mL final). Make by diluting 200 μl of 2 mg/mL stock (10.4 units/mL) with 800 μl of (0.7× HBSS, 1 mM Ca^{++}, 1 mM Mg^{++}). 500 μg/ mL DNaseI final concentration.
4. TrypLE buffer: TrypLE (Gibco-ThermoFisher, Cat# 12605010) diluted with sterile water to 0.7×.
5. Trypsin/EDTA buffer: Trypsin/EDTA (Gibco-ThermoFisher, Cat# 25300054) diluted with sterile water to 0.7× final concentration (0.035%).

2.4 Antibody Staining and Flow Cytometry

1. 5 mg/mL 4′, 6-diamidino-2-phenylindole (DAPI). Use at a 1: 2000 dilution of 5 mg/mL stock for a final concentration of 25 μg/mL (*see* **Note 1**).
2. 0.5 mg/mL Isolectin B4-Biotin (Vector Laboratories B-1205-.5). Use at a 1:50 final dilution.
3. Axolotl-specific anti-Light Chain. A kind gift from Françoise Salvadori. Conjugated to R-PE (using Biotium's Mix-n-Stain™ R-PE Antibody Labeling Kit according to the manufacturer's instructions Cat #92299). Use 0.5 mg/mL stock at 1:100 final dilution (*see* **Note 2**).
4. 2 mg/mL Streptavidin, PE-Cy7Conjugate (ThermoFisher, Cat# 25-4317-82). Use at a 1:500 final dilution.
5. Anti-CD18 clone TS1/18 (Bio Legend, Cat#302114). Use at 1:100 final dilution.
6. Standard flow cytometer and FACS machine. This study used an Invitrogen Attune NxT with the standard BRV6Y 4-laser configuration but was also validated on equivalent BD Biosciences LSRII or Symphony machines.

3 Methods

It is recommended to prepare the necessary equipment and reagents prior to performing any of the methods described in this section. An effective collagenase/dispase enzymatic method using the gentleMACS system for isolation of limb blastema cells has been previously described [17]. The following method described removes the requirement for a specialized gentleMACS device.

3.1 Animal Maintenance and Usage

Axolotls (*Ambystoma mexicanum*) are maintained in 20% Holtfreter's solution at 19–22 °C on a 12-h light, 12-h dark cycle. To ensure enough cells are obtained for analysis it is recommended that the animals used be at least 1 year of age. Data presented in this protocol is generated using 4 blastema's per condition from adult 1-year-old 12 cm Tg::CAGGS:GFP animals [18] (snout to tail) and white leucistic (d/d mutant) non-fluorescent control animals for flow cytometry compensation controls. Individual blastemas can be dissociated effectively using this protocol with the solutions scaled down appropriately.

3.2 Axolotl Limb Amputation and Generation of Blastema Tissue

1. To prepare 5-day-old blastema tissue, 5 days prior to harvest, anesthetize axolotl in 0.2% Tricaine. The time taken for an axolotl to become completely anesthetized is size dependent. This can take approximately 10–20 min. An anesthetized axolotl will not react to a skin pinch performed with a pair of surgical forceps (*see* **Note 3**).
2. Place the anesthetized axolotl onto the surgical stage lying on its side.
3. Secure the forelimb with a pair of surgical forceps. Using a pair of surgical scissors amputate 1 mm below the wrist joint (for a distal amputation) or 1 mm below the elbow joint (for a proximal amputation) (*see* **Note 4**).
4. Place the axolotl into a fresh tank of 20% Holtfreter's solution to regain consciousness and recover.

3.3 Blastema/Spleen Isolation and Tissue Dissociation

1. Prepare all enzyme solutions in 0.7× HBSS, Mg^{++} and Ca^{++}. Keep on ice.
2. After 5-day post-amputation, anesthetize the amputated axolotl using 0.2% Tricaine (*see* **Note 3**).
3. Collect blastema tissue (approx. 1–2 mm of stump) or spleen tissue and place in a 5-mL microfuge tube.
4. Use fine scissors to mince tissues to a paste (pieces smaller than 0.2–0.5-mm-diameter ideal), ensuring consistency of mincing between samples (*see* **Note 5**).
5. Add 1–3 mL of appropriate tissue dissociation cocktail to the minced tissue in a 5-mL microfuge tube (*see* **Note 6**). *See* Subheading 2.3 for cocktail recipes (for volumes, *see* **Note 7**).
6. Flick tube several times to mix (or gently vortex for 3 s).

7. Wrap tubes with foil (foil only required if using fluorescently labeled cells) and place on a roller at 25 °C (room temperature) (*see* **Note 8**). Leave for 30 min for the first phase of digestion.
8. Triturate using a 1-mL pipette cut with scissors to create a 0.5-cm-diameter opening. Gently pipette up and down being careful not to lose any cells in the barrel of pipette (*see* **Note 9**).
9. Leave for another 15 min to digest in second digestion stage.
10. Triturate using a 1-mL pipette cut with scissors to create a 0.2-cm-diameter opening. Gently pipette up and down being careful not to lose any cells in the barrel of pipette tip.
11. Digest for another 15 min in the final stage. After the total digestion time of 1 h has been reached, if sufficient digestion has been achieved, then the enzymatic reaction can be stopped with an equal volume of 0.7× HBSS, 10 mM EDTA (to reach a final EDTA concentration of 5 mM to inhibit the enzymes and adhesion). If additional time is required for digestion, repeat **step 11** before the addition of EDTA (*see* **Note 8**).
12. Transfer the tissue suspension within each digestion tube into a fresh 5-mL tube by passing through a 70-μm cell strainer (*see* **Note 10**). Cell clumps on strainer can be rinsed by passing additional 1 mL volumes of 0.7× HBSS, 5 mL EDTA with 30 s centrifugation until the full 5 mL volume is reached.
13. Once all the cell suspension has been strained, the 5 mL volume is then mixed by inversion and then centrifuged at 200 *g* for 15 min at 4 °C *with the brakes off*, to collect cells and prevent the pelleting of lighter debris and dead cells.
14. Aspirate supernatant and re-suspend the pelleted cells with blocking buffer solution in 500 μl–1 mL, depending on starting tissue amount (*see* **Note 11**).
15. Proceed to cell counting via trypan Blue or AO/PI using a hemocytometer (*see* **Notes 12** and **13**).
16. Aliquot cells into appropriate 1.5-mL microfuge tubes for downstream analysis (flow cytometry), i.e., no more than 1×10^6 cells, top up to 1 mL with Goat Block Buffer and centrifuge 350×*g* for 5 min (brakes on) at 4 °C.
17. Aspirate liquid leaving cells in approximately 50 μl of Goat Block Buffer. Gently pipette up and down and leave on ice for 20 min before proceeding to staining.

3.4 Live-Cell Antibody Staining for Myeloid Cells from the Limb Blastema or Spleen

1. All staining reactions are conducted in 1.7-mL micro-centrifuge tubes.
2. Block cells on ice away from light for at least 20 min FACS buffer.

3. Prepare 1.7-mL micro-centrifuge tubes with a 2× dilution of primary antibody: 0.5 mg/mL in Goat Block Buffer.
4. Add 50 ul μL to each individual 1.7-mL micro-centrifuge tube containing cells in 50 μl. Mix by flicking or pipetting up and down (preferred) (*see* **Note 14**).
5. Incubate cells in primary antibody on ice away from light for at least 30 min –1 h (*see* **Note 15**).
6. Flick tube if pelleted and wash any unbound primary antibodies with 1–1.4 mL of FACS Buffer, mix thoroughly, and centrifuge at 350 *g* at 4 °C for 5 min (*see* **Note 16**).
7. Prepare a working dilution combining secondary antibodies (i.e., Streptavidin PE-Cy7 (1:500)) in Goat Block Buffer.
8. Aspirate supernatant and re-suspend pellet in 50 μL of 2nd antibody cocktail. Incubate on ice, away from light for a minimum of 10 min (*see* **Note 17**).
9. Wash any unbound secondary antibodies with 1–1.4 mL of wash solution, mix thoroughly, and centrifuge at 4 °C for 5 min.
10. Prepare a working dilution of DAPI (1:2000) in FACS Buffer.
11. Aspirate supernatant and re-suspend pellet in FACS Buffer with 1:2000 DAPI at a final concentration of 1 × 10^6 cells/ 100 μL. Transfer the stained cell suspension to a 5-mL FACS tube by passing cells through a 40-μm filter if required to prevent blockage of instrument (*see* **Note 18**).
12. Proceed to analyze the single-cell suspensions using a flow cytometer. Alternatively, cells can be sorted for applications in downstream assays using FACS (*see* **Notes 19** and **20**).

3.5 Expected Results and Conclusions

Here we describe a straightforward protocol that enables efficient isolation of cells from the regenerating axolotl limb and the spleen. We demonstrated this protocol using three alternative enzymatic formulations in parallel with a "no enzyme" purely mechanical approach. Red Blood Cells (RBCs) do not express GFP in Tg:: CAGGS:GFP axolotls. Although we used GFP+ animal tissue to confirm purity and absence of significant RBC contamination, this protocol is equally effective on standard white (d/d) axolotl tissue. An important goal of any cell dissociation protocol is to preserve viability while maintaining cellular diversity and preserving rare subsets and above all, minimizing induced changes in cellular phenotypes. Most dissociation protocols cause significant cell death and release of cellular debris. The non-enzymatic protocol (condition A) gives the best cell recovery in spleen of all the conditions tested but poor recovery of non-immune subpopulations from regenerating limb. As immune cells do not constitute structural blocks of the tissues they reside in, they are generally easier to

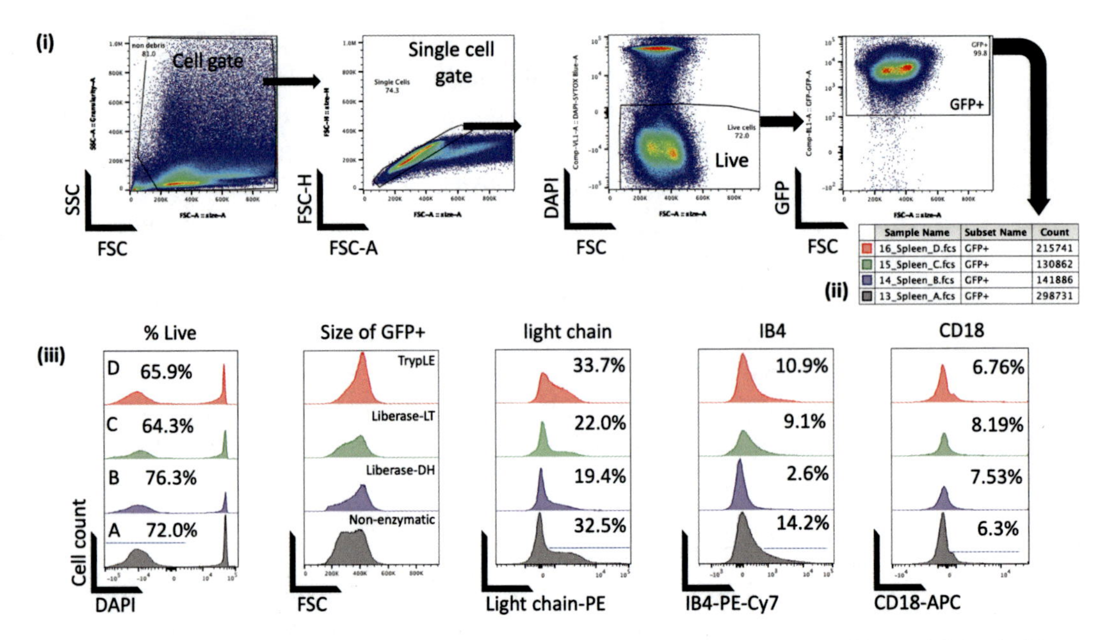

Fig. 2 Flow cytometry gating strategy on single-cell suspensions of spleen tissue. (i) Tissue debris, doublets, and non-viable DAPI+ cells are gated out. (ii) GFP+ cell counts are shown. (iii) Surface marker expression on GFP+ gate shown in light chain, IB4, and CD18 histograms for each cell isolation condition A-D. *SSC* side scatter area, *FSC-A* forward scatter area, *FSC-H* forward scatter height

release and if dissociation is not effective, then immune cells will be overrepresented in the sample. This is observed in Figs. 2 and 3. As the axolotl immune system does has some T-cells [19] not profiled in this study, but the light chain, IB4, and CD18 antibody combinations capture the majority of the axolotl immune cell populations (B cells and Myeloid cells). The current lack of robust antibodies against mesenchymal surface markers prevents direct measurement of non-immune blastema cells. Despite this limitation, their presence can be inferred from the overall cell number of non-immune cells released under each condition.

Non-enzymatic dissociation results in the brightest staining intensities as cell surface epitopes are retained. The use of Trypsin resulted in extremely low yields not suitable for downstream analysis. Unlike trypsin, TrypLE gave detectible yields good preservation of cell surface epitopes and but relative to no-enzyme and Liberase cocktails, the cell recovery was very poor.

In the spleen cell dissociations, we obtained yields of 298,731 and 215,741 live GFP+ cells from the "no enzyme" (grey histogram) and TrypLE (pink histogram) treated spleens, compared with 141,886 and 130,862 live GFP+ cells from the Liberase-DH (blue histogram) and Liberase-LT (green histogram) cocktails, respectively (Fig. 2, (ii, iii)). A distinctive CD18+ shoulder population is missing in the Liberase treatments but retained in the non-enzymatic and TrypLE treatments (Fig. 2, iii). The size profile

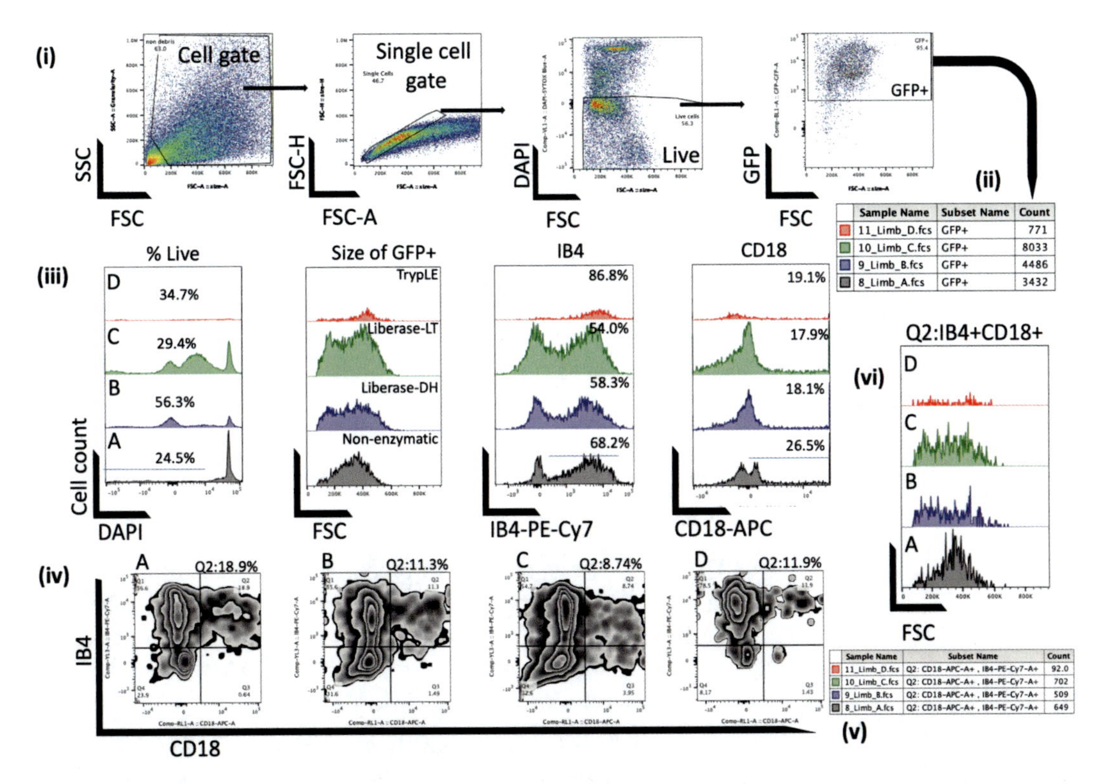

Fig. 3 Flow cytometry gating strategy on single-cell suspensions of limb blastemas obtained at 5-day post-amputation. (i) Tissue debris, doublets, and non-viable DAPI+ cells are gated out. (ii) GFP+ cells counts are shown. (iii) Surface marker expression on GFP+ gate shown in IB4 and CD18 histograms for each cell isolation condition A-D. (iv) Plots show detection of IB4+ CD18+ double-positive macrophages shown in quadrant 2 (Q2). (v) Histograms of Q2 and cell counts also shown. (vi) Cell sizes (FSC range) captured in Q2 under different enzyme conditions are shown. *SSC* side scatter area, *FSC-A* forward scatter area, *FSC-H* forward scatter height

(FSC) of the cells released in each condition varied (Fig. 2, iii). The reduced yield using Liberase cocktails relative to "no-enzyme" and TrypLE (pink histogram) may reflect suboptimal over digestion of spleen tissue with these enzymes, leading to cell losses.

In the 5-day post-amputation blastemal cell dissociations, we obtained yields of 3432 and 771 live GFP+ cells from the "no enzyme" (grey histogram) and TrypLE (pink histogram) treated blastemas, respectively, compared with 4486 and 8033 live GFP+ cells from the Liberase-DH (blue histogram) and Liberase-LT (green histogram) cocktails, respectively (Fig. 3, ii). Of all of these conditions, Liberase-LT gave the highest yield (almost double Liberase-DH) and the best profile of different cell sizes (Fig. 3, iii). However, Liberase-LT produced a greater reduction surface epitope detection in IB4 and CD18 staining (Fig. 3, iv). This results in reduced sensitivity in the flow cytometry detection of the IB4+ CD18+ macrophage population relative (Q2 population in IB4-CD18 box plot) relative to the Liberase-DH cocktail (with many of the cells in Q2 falling into Q1 or Q3 due to loss of cell surface epitopes) (Fig. 3, iii). Although reduced sensitivity is manageable in a dual staining approach, if downstream analysis is dependent on a single marker (i.e., cell sorting with one antibody), then further optimization may be required.

This protocol demonstrates the effective dissociation of regenerating axolotl tissue and the influence of alternative enzyme blends on yield and surface marker detection. This protocol can be modified with additional enzymes or incubation times to optimize target cell recovery. These examples also highlight the importance of balancing yield against changes affecting downstream phenotyping.

4 Notes

1. DAPI is light sensitive so protect stock with aluminum foil or equivalent. A working stock can be made in PBS or HBSS and added to FACS tube sample almost immediately before analysis. Dead or dying cells will take up the DAPI dye through compromised membranes allowing live/dead discrimination.
2. Commercial cell-specific antibody reagents for salamander immune cells or other target cells are extremely limited. For a list of validated commercial antibody reagents please see salamander community resource pages (i.e., "AxoBase.org "or "Slack" salamander community platform).
3. The percentage of Tricaine used to anesthetize the axolotl is dependent on the size of the animal. It is necessary to adjust this percentage to avoid any potential lethal doses. For example, animals that are 3 cm (snout–tail) in size are routinely anesthetized using 0.05% Tricaine, whereas 0.2% Tricaine is used in animals that are greater than 15 cm (snout–tail) in size [17].
4. It is recommended that surgeries are performed under a dissection microscope. Following amputation of the limb (proximal or distal), it is common to find that the bone/s have not been completely removed. To prevent any bone/cartilage protruding into the regenerating wound epithelium, it is recommended that bones be trimmed to fall in line with the plane of amputation.
5. It is important that all tissue pieces are cut very finely and there are no large chunks remaining. This step can be performed on a 35-mm glass bottom tissue culture dish for monitoring mincing. Glass minimizes cell adhesion.
6. It is recommended that DNaseI (100 μg–1 mg/mL) is added to the final solution to prevent any clumping of cells.

DNaseI can be purchased in several formats. Roche DNaseI has a typical activity of 2000 units/mg. A final concentration in a tissue dissociation cocktail of 500 μg/mL is roughly equivalent to 1000 units/mL depending on supplier. To prepare, dilute either 500 μg/mL of DNaseI powder in tissue digestion buffer or dilute a commercial DNaseI stock from 10,000 units/mL to 1000 units/mL (i.e., 100 μl of (~5 mg/mL) stock per 1 mL of tissue digestion buffer).

7. The volume of enzyme solution should be scaled according to tissue size. Typically a minimum of 1 mL of tissue dissociation buffer per 250 mg of tissue.
8. The minimum time for digestion should be determined by tissue type. Limb blastemas were found to require at least 40 min on a roller at room temperature for full digestion. Incubation times spanning greater than 1 h results in decreased cell viability. The minimum time for spleen tissue digestion is around 10–20 min on a roller at room temperature with losses in viability observable at 40 min. The suggested 1-h protocol can be modified for different tissue types by triturating every 15 min and monitoring dissociation progress.
9. Trituration can be performed with decreasing aperture in the pipette tip as digestion proceeds.
10. This protocol does not yield cells from the epidermal layer (skin). Therefore, it is expected that there will be tissue that is not digested. This will appear as small white pieces and will not pass through the 70-μm cell strainer.
11. If the cell suspension is too concentrated for counting, then dilute appropriately. If too dilute, cells can be re-centrifuged and re-suspended in a smaller volume.
12. Sample 5 μl of cell suspension and add to either 5 μl of trypan Blue or AO/PI solution and transfer 10 μl under the coverslip of a hemocytometer. If using AO/PI use a fluorescent microscope to count the number of green (AO+) live cells and red (PI+) dead cells. Perform cell counts using the formula $= n \times 10^4$/mL (counting live cells in all 25 squares) × the final dilution factor.
13. Some level of debris can be tolerated in the flow cytometry staining protocol as this will be diluted out by subsequent washing steps. If the cell solution has debris levels not appropriate for satisfactory staining or if the cells need to be used directly for single-cell RNAseq, then debris removal via a low density Percoll or Ficoll gradient should be considered. These approaches must ensure that target cells are not lost with debris and must be validated.

14. Prepare appropriate control samples (unstained, single stains, d/d:GFP 50:50 mix, and compensation beads as required).
15. Normal time for primary antibody incubation is approximately 40–60 min. We have found that this can be decreased to a minimum of 20 min without any decrease in efficiency of antibody binding capacity. Incubation times can go longer if necessary.
16. More than 1000 μL of wash solution can be used. Additional washes can be performed to improve separation of signal but may result in additional cell losses.
17. Normal time for secondary antibody incubation is approximately 30 min. We have found that this can be decreased to a minimum of 10 min without any decrease in secondary antibody binding capacity to the primary antibody. Ensure that the host species of all antibodies are known, and that each secondary antibody only targets one primary.
18. This step can be skipped, and microfuge tubes can be used directly on an Invitrogen Attune Nxt acoustically focused flow cytometer instrument.
19. In flow cytometry analyses, an initial step is to identify intact cells and remove cellular debris or enucleated cells from further analyses. Conventional fluorescence flow cytometry relies on measurement of cell size (forward scatter, FSC) and cell granularity (side scatter, SSC) to identify cells. Additionally, cellular debris can be distinguished due to its smaller size (low FSC) and higher granularity (high SSC). It is highly recommended to optimize the immunostaining protocol of each antibody in a panel to ensure target-specific staining and to optimize signal to background.
20. Compensation of fluorescence spectral overlap can be performed using UltraComp eBeads (eBioscience) or similar according to the manufacturer's instructions to minimize loss of sample.

Acknowledgments

"Research reported in this publication was supported by an Institutional Development Award (IDeA) from the National Institute of General Medical Sciences of the National Institutes of Health under grant numbers P20GM103423 and P20GM104318."

References

1. Takeichi M (1990) Cadherins: a molecular family important in selective cell-cell adhesion. Annu Rev Biochem 59:237–252. https://doi.org/10.1146/annurev.bi.59.070190.001321
2. Kumar A, Godwin JW (2010) Preparation and culture of limb blastema stem cells from regenerating larval and adult salamanders. Cold Spring Harb Protoc 2010(1):pdb.prot5367-pdb.prot5367. https://doi.org/10.1101/pdb.prot5367
3. Aronowitz JA, Lockhart RA, Hakakian CS (2015) Mechanical versus enzymatic isolation of stromal vascular fraction cells from adipose tissue. Springerplus 4:713. https://doi.org/10.1186/s40064-015-1509-2
4. Schneeberger EE, Lynch RD (1992) Structure, function, and regulation of cellular tight junctions. Am J Phys 262(6 Pt 1):L647–L661. https://doi.org/10.1152/ajplung.1992.262.6.L647
5. Frantz C, Stewart KM, Weaver VM (2010) The extracellular matrix at a glance. J Cell Sci 123 (Pt 24):4195–4200. https://doi.org/10.1242/jcs.023820
6. Halper J, Kjaer M (2014) Basic components of connective tissues and extracellular matrix: elastin, fibrillin, fibulins, fibrinogen, fibronectin, laminin, tenascins and thrombospondins. Adv Exp Med Biol 802:31–47. https://doi.org/10.1007/978-94-007-7893-1_3
7. Loganathan G, Balamurugan AN, Venugopal S (2020) Human pancreatic tissue dissociation enzymes for islet isolation: advances and clinical perspectives. Diabetes Metab Syndr 14(2): 159–166. https://doi.org/10.1016/j.dsx.2020.01.010
8. Rawlings ND, Salvesen G (2013) Handbook of proteolytic enzymes. 3rd ed, Elsevier/AP, Amsterdam
9. Autengruber A, Gereke M, Hansen G, Hennig C, Bruder D (2012) Impact of enzymatic tissue disintegration on the level of surface molecule expression and immune cell function. Eur J Microbiol Immunol (Bp) 2 (2):112–120. https://doi.org/10.1556/EuJMI.2.2012.2.3
10. Stremnitzer C, Manzano-Szalai K, Willensdorfer A, Starkl P, Pieper M, Konig P, Mildner M, Tschachler E, Reichart U, Jensen-Jarolim E (2015) Papain degrades tight junction proteins of human keratinocytes in vitro and sensitizes C57BL/6 mice via the skin independent of its enzymatic activity or TLR4 activation. J Invest Dermatol 135(7): 1790–1800. https://doi.org/10.1038/jid.2015.58
11. Price PA (1975) The essential role of Ca2+ in the activity of bovine pancreatic deoxyribonuclease. J Biol Chem 250(6):1981–1986
12. Dolmans MM, Michaux N, Camboni A, Martinez-Madrid B, Van Langendonckt A, Nottola SA, Donnez J (2006) Evaluation of Liberase, a purified enzyme blend, for the isolation of human primordial and primary ovarian follicles. Hum Reprod 21(2):413–420. https://doi.org/10.1093/humrep/dei320
13. Tsuji K, Ojima M, Otabe K, Horie M, Koga H, Sekiya I, Muneta T (2017) Effects of different cell-detaching methods on the viability and cell surface antigen expression of synovial mesenchymal stem cells. Cell Transplant 26(6): 1089–1102. https://doi.org/10.3727/096368917X694831
14. Dono K, Gotoh M, Monden M, Kanai T, Fukuzaki T, Mori T (1994) Low temperature collagenase digestion for islet isolation from 48-hour cold-preserved rat pancreas. Transplantation 57(1):22–26. https://doi.org/10.1097/00007890-199401000-00005
15. Hidvegi NC, Sales KM, Izadi D, Ong J, Kellam P, Eastwood D, Butler PE (2006) A low temperature method of isolating normal human articular chondrocytes. Osteoarthr Cartil 14(1):89–93. https://doi.org/10.1016/j.joca.2005.08.007
16. Mascotti K, McCullough J, Burger SR (2000) HPC viability measurement: trypan blue versus acridine orange and propidium iodide. Transfusion 40(6):693–696. https://doi.org/10.1046/j.1537-2995.2000.40060693.x
17. Debuque RJ, Godwin JW (2015) Methods for axolotl blood collection, intravenous injection, and efficient leukocyte isolation from peripheral blood and the regenerating limb. Methods Mol Biol 1290(Chapter 17):205–226. https://doi.org/10.1007/978-1-4939-2495-0_17
18. Sobkow L, Epperlein HH, Herklotz S, Straube WL, Tanaka EM (2006) A germline GFP transgenic axolotl and its use to track cell fate: dual origin of the fin mesenchyme during development and the fate of blood cells during regeneration. Dev Biol 290(2):386–397. https://doi.org/10.1016/j.ydbio.2005.11.037
19. Kerfourn F, Guillet F, Charlemagne J, Tournefier A (1992) T-cell-specific membrane antigens in the Mexican axolotl (urodele amphibian). Dev Immunol 2(3):237–248

Part VI

Physiological and Organismal Techniques

Chapter 26

Axolotl Metabolism: Measuring Metabolic Rate

Moshe Khurgel

Abstract

Deciphering how metabolic processes contribute to control of stem cell proliferation and differentiation is essential for understanding the mechanisms of regeneration. However, much is still unknown about axolotls' metabolism, which has not been studied in detail over their lifespan or under varied experimental conditions. We summarize the theoretical underpinnings of metabolism and respirometry, and describe a closed respirometry system to investigate metabolic energetics in axolotls as a specific aspect of metabolism. Placement of post-absorptive, fairly inactive animals in the multiple-probe respirometer for 24–48 h allows us to measure changes in concentrations of respiratory gases: oxygen (atmospheric and dissolved) and carbon dioxide, while monitoring the temperature and salinity (conductivity) of the chamber's water. Respirometry data are used to calculate oxygen intake and carbon dioxide output to estimate animal's metabolic energy dynamics during the observation periods. This method creates opportunities for study of potential fluctuations in axolotls' metabolic rate as it pertains to respiratory gases' dynamics during 24-h circadian cycle, as well as examination of changes in metabolic energy management during aging, under varied environmental temperatures, during post-amputation regeneration and many other circumstances.

Key words Oxygen, Respirometry, Metabolic energy, Stem cells, Regeneration

1 Introduction

Axolotls continue to grow in size well past their sexual maturation, and they exhibit an amazingly robust ability to regenerate damaged tissues and organs. Growth of an organism, cell turnover in tissue maintenance, as well as organ repair and regeneration in response to trauma all rely on stem cell proliferation and differentiation, which in turn require differential investments of metabolic energy and a wide range of metabolites within a complex interplay between polymer anabolism and catabolism, and epigenetic regulation of gene expression. Evidence is accumulating to support the role of "housekeeping" metabolic pathways in directing cell self-renewal- vs. lineage commitment and specification [1, 2]. Thus, the distinctive ability of axolotls to remodel anatomical structures during growth and regeneration events hints at intriguing evolutionary

Ashley W. Seifert and Joshua D. Currie (eds.), *Salamanders: Methods and Protocols*,
Methods in Molecular Biology, vol. 2562, https://doi.org/10.1007/978-1-0716-2659-7_26,

adaptations, which involve metabolic processes that enable extraordinary biosynthesis throughout most of these animals' lifespan. Yet, the metabolic aspects of axolotl biology have received far less attention than other cellular and molecular elements of development and regeneration.

It is impossible to measure an animal's entire metabolism. This is because metabolism, conceptually, is the sum of all biochemical reactions and transformations in an organism at any given point in time. As such, it is impractical to tackle the numerous metabolic pathways in any given study or any single research laboratory. Instead, we must differentiate between the overarching umbrella concept of *metabolism* and specific, narrow aspects of metabolism that might be of research interest, such as metabolic energy availability and utilization, metabolic pathway switching in relation to cell fate [3], or metabolism of specific molecules, like creatine, for example.

Respirometry, the approach to estimating the animals' metabolic rate by monitoring the exchange of respiratory gases, such as measuring removal of oxygen (O_2) from the animals' environment for respiration and consequent release of carbon dioxide (CO_2) into the same environment, is an established method for studying several aspects of metabolism, including differential utilization of nutrients and calorimetry, with a wealth of data for a large variety of animals [4, 5]. This is one of the methods we employ in our lab, as we seek to address metabolism in terms of the axolotls' management of metabolic energy during their growth, in response to environmental challenges, such as changes in water quality (temperature and/or conductivity), and during regeneration following tissue/organ damage. While the focus of this chapter is on application of respirometry to the study of certain aspects of metabolism specifically in axolotls, the presented principles are also applicable to studying similar aspects of metabolism in other urodeles, aquatic or terrestrial, with minor modifications to the experimental setup for the latter.

Oxygen is necessary for aerobic respiration, which is the pathway for the majority of energy transfer reactions from catabolism of biomolecules to ATP in animal cells. ATP-transported energy is used for a wide range of reactions in cells: synthesis of polymers, molecular transport and maintenance of cell membrane potentials, cardiac activity, locomotion, and many others. ATP synthesis is linked to animal physical activity, whereby animals' intake of O_2 increases with increase in activity, and decreases at rest [6, 7]. Since inspired oxygen is directly tied to synthesis of most ATP, and the rate of ATP usage is a reflection of the extent of metabolic reactions, oxygen intake rate can serve as an accurate indicator of the levels of underlying metabolic activities. However, care must be exercised with respect to experimental assumptions and conclusions, such as substituting O_2 *consumption* for O_2 *intake* and using the rate of O_2

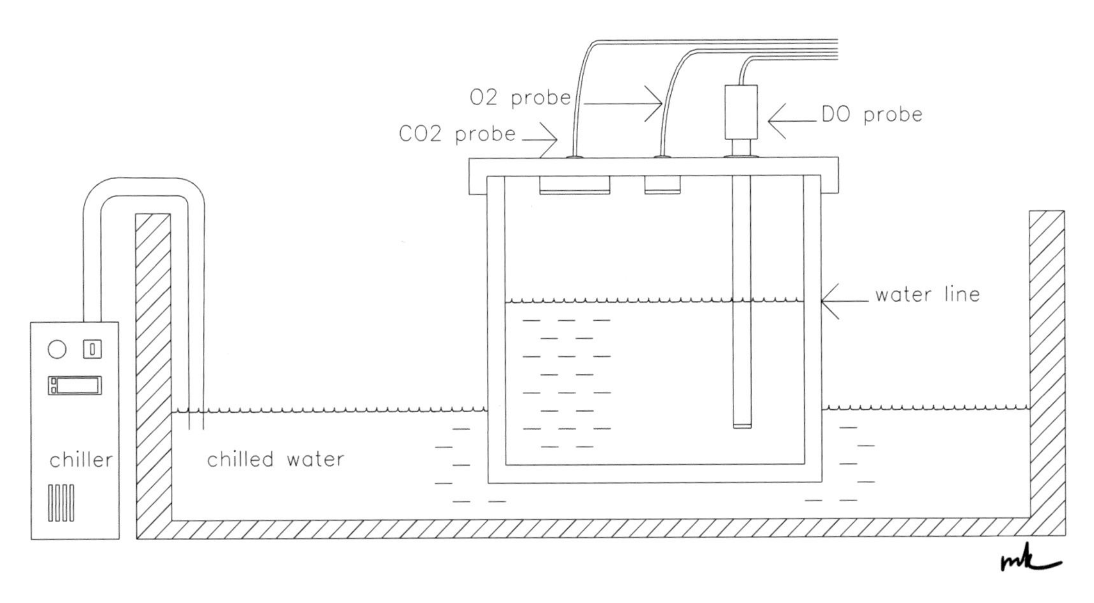

Fig. 1 A drawing of the respirometry chamber positioned in the chiller tank. The drawing conveys the relative placement of the gas probes and the relationship between water and air compartments in the chamber

intake in a given span of time as a convenient proxy for *metabolic rate*, which may entail varied rates for specific metabolic reactions. These concepts are not interchangeable [8]. Consequently, the working definition for metabolic rate that we employ with the respirometry method discussed here is of metabolic rate as the rate of respiratory oxygen intake.

Our design of the respirometry chamber is a slight modification of the two-phase respirometer [9] - a hybrid of several classic techniques to study O_2 consumption in aquatic animals [4, 10]. The chamber contains both a water compartment and an air compartment, and it is positioned in a controlled temperature water tank (Fig. 1). In addition, we employ multiple sensors for monitoring respiratory gas concentrations and related conditions, such as temperature (air and water) and water conductivity to provide a comprehensive assessment of the animals' environment. Measurements of atmospheric O_2 in the air compartment of the chamber allow us to monitor the balance between this gas in the air and its dissolved oxygen (DO) fraction in the chamber's water. Measurements of atmospheric CO_2 are used in improving our estimates of animals' metabolic rate based on O_2 intake.

Closed volume respirometry is especially well suited for studying metabolic energetics in axolotls for a number of reasons. Axolotls are fully aquatic representatives of the genus Ambystoma. This provides an advantage when seeking to detect minute changes in inspired O_2, since the concentration of O_2 is much lower in an aquatic environment than in the air due to oxygen's low solubility in water. Thus, relatively low levels of dissolved O_2 provide favorable conditions for detecting individual axolotl's removal of O_2 from a relatively small volume of water for its respiration. In addition, O_2 solubility in water depends on water temperature, where DO measurements range from 8.26 mg/L (8.26 ppm) at 25 °C to 14.60 mg/L at 0 °C for ~100% saturation of fresh water at 1 Atm. This property can be utilized for studying axolotls' respiratory metabolism under different environmental conditions. For example, water temperature of axolotls' native habitats in the Xochimilco area varies between 20 and 22 °C, with DO measurements of 3.25 mg/L in low-turbidity water, and 5.89–6.74 ml/L in higher-turbidity water [11]. The other benefit of this animal model in the study of metabolic energetics is that axolotls become fairly sedentary as adults. The relative paucity of intense locomotion supports the assumption that the contribution of anaerobic glycolysis to ATP synthesis during observation periods is minimal. This increases the predictive accuracy of O_2 measurements as reflections of energy transfer reactions to support the animals' overall metabolism.

2 Materials

2.1 Animals

This experimental setup was designed for estimating the metabolic rate in axolotls (*Ambystoma mexicanum*), which are fully aquatic animals. Animals that were used in these experiments were obtained from the Ambystoma Genetic Stock Center and subsequently reared in our laboratory's colony under standard husbandry protocols [12]. Animals may be of any strain (wild type, white, albino, etc.) and any age and size.

2.2 Respiratory Chamber

Accurate estimates of axolotl metabolic rate can be obtained by isolating an animal in a hermetically sealed chamber that contains a predetermined and constant volume of water and a volume of air, which restricts movement of respiratory gases between the chamber and the space around it. The chamber should be constructed to enable visual monitoring of the animal (unassisted or with cameras) and to convey light–dark cycle for maintenance of normal circadian rhythm.

We use optically clear polycarbonate containers, square or round (*see* **Note 1**). The dimensions of the respiratory chamber in an experiment are determined by considering the size of the animals in that experiment and the total volume of water and air that the chamber will contain when sealed, where the guiding principle is based on an inverse relationship between these volumes and signal resolution. The chamber must be large enough in its horizontal measurements to accommodate sufficient swimming space for the animals, and deep enough to accommodate the length of the respiratory gas probes, but small enough to ensure meaningful measurements of changes in the concentrations of respiratory gases. For animals that range between 13 and 14 cm total length (~7 cm SVL) with a body mass of 14–17 g, we use a 7.5-L chamber, with an 18 cm x 18 cm bottom area (*see* **Note 2**).

The containers have fitted plastic lids, which provide a liquid-proof seal. To ensure gas impermeability, we add a layer of wax seal (Parafilm) around the upper edge of the chamber prior to closing it for the experiment. Perforations are made in the lids to fit gas probes or their cables. The perforations are fitted with rubber ring gaskets or rubber stoppers to ensure tight fit for the probes or their cables, and thereafter sealed with wax.

2.3 Chiller/Heater Tanks

To enable comparative respirometry measurements in animals at different ambient temperatures, respirometry chambers can be placed into large, custom-made, or commercial (Living Stream System; http://www.frigidunits.com/products/living-stream-system) tanks, sized to fit at least one or multiple respirometry chambers. The tanks are fitted with an independent, controlled temperature water supply. Water is continuously chilled (or heated) at a desired setting by the chiller system, and circulated around the chambers to keep the chamber water at a desired temperature via temperature exchange.

2.4 Water

To extrapolate axolotl metabolism data to your other findings, the same type of water that is used for housing animals in your laboratory should be used in respirometry chambers. In many of our experiments we have used Holtfreter's solution [12] to ensure controlled water chemistry. Alternatively, treated tap water may be used. Since salinity of Holtfreter's solution will lead to a slight decrease in dissolved O_2 levels compared to fresh water, appropriate trials will need to be performed to determine optimal water choice for your experiments. Determination of the optimal volume of water/solution will also require pilot trials. We use 2 L of water for 12–15 cm animals and 7.5 L chambers described above.

2.5 Hardware and Software for Measuring Respiratory Gases and Related Variables

2.5.1 Measurement of DO

For measuring the concentration of oxygen dissolved in water, which is the fraction that axolotls directly access for respiration, as well as water temperature and atmospheric pressure, we use the following:

1. Orion Versa Star Pro benchtop meter (VSTAR90, Thermo-Fisher) as the user interface for calibration, data logging, and real-time data analysis of input from DO probes.
2. Orion RDO DO probes (087010MD, ThermoFisher).
   - Measurement range: 0–20 mg/L (0–200%).
   - Accuracy: ±0.1 mg/L up to 8 mg/L, ±0.2 mg/L from 8 to 20 mg/L.
   - Resolution: 0.01 mg/L.
   - Temperature range: 0–50 °C.
   - Temperature accuracy: ±0.3 °C.

3. Versa Star RDO/DO modules (VSTAR-RD, ThermoFisher) as obligatory interfaces between the probes and the meter.
4. Orion Navigator Pro software for collecting data. This software is not obligatory, since Versa Star meter may be used as a standalone device. However, we find that using Navigator Pro makes for a less cumbersome collection of data from multiple probes than using the meter alone.

2.5.2 Measurement of Atmospheric Oxygen

To measure the concentration of oxygen in the air fraction within the chamber (the air above the water), we use the following:

1. UV Flux 25% Oxygen Smart flow through sensor (CM-42991, CO2meter).
   - Sensing method: fluorescence quenching by O_2.
   - Measurement range: 0.1–25% O_2.
   - Accuracy: 2%.
   - Resolution: 0.10%/0.1 mbar.
   - Rate of measurement: 1 sample/s.
2. GasLab software for calibration, data logging, and real-time data analysis.

2.5.3 Measurement of Atmospheric Carbon Dioxide

To measure the concentration of carbon dioxide in the air fraction within the chamber, temperature, and % relative humidity in the air fraction within the chamber (the air above the water), we use the following:

1. K33 ELG 10,000 ppm CO2 + RH/T data logging sensor (SE-0020, CO2meter).
   - Sensing method: non-dispersive infrared.
   - Measurement range: 0–10,000 ppm.
   - Accuracy: ±30 ppm ±3% of measured value.
   - Rate of measurement: 1 sample/2 s.
2. GasLab software for calibration, data logging, and real-time data analysis.

2.5.4 Measurement of Water Conductivity

To monitor water conductivity, which influences levels of DO, we use the following:

1. Orion 4-electrode conductivity probes (013005MD, ThermoFisher).
   - $K = 0.475\ \text{cm}^{-1}$.
   - Measurement range: 1 μS to 200 μS.
2. Conductivity modules (VSTAR-CND, ThermoFisher).

3 Methods

1. Dissolved oxygen probe and conductivity probe are inserted through a rubber ring-fitted perforation in the chamber lid (*see* **Note 3**). The probes (19.0 cm length or 12 cm length, respectively) extend down the depth of the chamber (Fig. 1). Minimum required immersion depth in the chamber's water is 4 cm (*see* **Note 4**).
2. Atmospheric oxygen and carbon dioxide probes are affixed to the bottom of the lid on the interior side of the chamber, while their cables are threaded through the lid perforations fitted with rubber stoppers.
3. The chambers are filled with 2 L of freshly prepared Holtfreter's solution and sealed. Respiratory gases data are collected for 48 h to establish an animal-free baseline.
4. The housing solution is replaced with a freshly made one and experimental axolotls are placed into the chamber at least 12 h after their last meal (*see* **Note 6**). The chamber is sealed, and data recording starts 1 h after the animal's placement Respiratory gases data are collected for 48 h (*see* **Note 6**) or until DO levels decrease to 3 mg of O_2/L to prevent excessive oxidative stress (*see* **Note 7**).
5. To perform respirometry under varied ambient temperatures that are lower than the common ambient temperature in a modern climate controlled laboratory buildings (21–22 °C), respirometry chambers are positioned in large chiller tanks with water levels that are suitably high to ensure effective temperature exchange with water in the respiratory chambers (Fig. 1). The animals are acclimated to the desired temperature for at least 1 month prior to respirometry experiments. The chillers may be reset into "heaters" for studies in laboratories where animals are maintained at 16–20 °C or similar conditions [12] and there is interest in testing animals at slightly higher ambient temperatures. Maximal advised temperature for "warm" conditions is 25 °C.
6. Since O_2 use alone is not a sufficient measure of energy metabolism [8], we also employ measurements of CO_2 in the air compartment of the chamber. As axolotls release CO_2 into the water, and the levels of that gas reach saturation levels, excess CO_2 diffuses into the air above the water. We combine our measurements of O_2 removal with measurements of CO_2 generation, a second valuable approach to estimating metabolic energetics rate, to calculate the respiratory quotient (RQ). When concentrations of these two gases are obtained under resting conditions, the ratio between the two measures is indicative of the relationship between aerobic and anaerobic

fractions of the total metabolism, and, for nutrient-based experiments, the types of energy-storing substrates that were used in energy transfer reactions:

$$\mathrm{RQ} = \frac{\text{produced carbon dioxide}}{\text{removed oxygen}}.$$

The rate of oxygen use ($\dot{M}_{O2}$) for any range of time is calculated as follows:

$$\dot{M}_{O2} = Vol \cdot \Delta[\mathrm{DO}]/\Delta t \cdot wt.$$

where *Vol* is the volume of water in mL or L, Δ[DO] is the difference between the end concentration and starting concentrations of dissolved oxygen, Δt is the length of time range in min, or h and *wt* are the body mass in grams (modified from 13)

4 Notes

1. If using optically clear chambers, it is a good practice to set up the experiment in a quiet, visually isolated area in the lab, to reduce visual stimulation and behavioral changes in the animals that may result in respiratory spikes due to behavioral stimulation.
2. Avoid using containers that are too large, if using small size animals. Large volumes of air and/or water will pose a challenge for detection of changes in respiratory gases during the 24-h circadian cycle. Run trials to determine the optimal combinations of chamber size and volumes of water/air phases with respect to the size of experimental animals.
3. The sensor cap for the DO probe must be replaced every 365 days. Set up a replacement schedule, if using multiple probes with different installation dates, to ensure data accuracy and to prevent interruptions to experiments.
4. To prevent uneven concentration "pockets" of dissolved oxygen, avoid placing the probes into the corners of the chamber.
5. Feeding times, nutrient content, and amounts of food will influence respiratory gases requirements and exchange, and therefore should be carefully controlled [13].
6. With respect to duration of the respirometry experiments, an evaluation of SMR measurements in fishes showed 12 h to be insufficient, but 24 h as adequate [14].
7. Axolotls seem to be fairly tolerant of poorly oxygenated water with respect to survival, but it is unknown how low dissolved oxygen affects various aspects of these animals' metabolism.

Acknowledgments

This work was supported with faculty research grants (Bridgewater College) to the author.
Axolotls that were used in development of this methodology came from Ambystoma Genetic Stock Center (University of Kentucky), which receives financial support (P40-OD019794) from the Office of Research Infrastructure Programs (NIH). With special thanks to Laura Muzinic (AGSC) for assistance and advice on axolotls over the years.

References

1. Ryall JG, Cliff T, Dalton S, Sartorelli V (2015) Metabolic reprogramming of stem cell epigenetics. Cell Stem Cell 17:651–662. https://doi.org/10.1016/j.stem.2015.11.012
2. Ludikhuize MC, Colman MJR (2020) Metabolic regulation of stem cells and differentiation: a forkhead box O transcription factor perspective. Antioxid Redox Signal 34(13): 1004–1024. https://doi.org/10.1089/ars.2020.8126
3. Cliff TS, Dalton S (2017) Metabolic switching and cell fate decisions: implications for pluripotency, reprograming and development. Curr Opin Genet Dev 46:44–49
4. McLean JA, Tobin G (1987) Animal and human calorimetry. Cambridge University Press, Cambridge
5. Even PC, Mokhtarian A, Pele A (1994) Practical aspects of indirect calorimetry in laboratory animals. Neurosci Biobehav Rev 18(3): 435–447
6. Barton BA, Schreck CB (1987) Metabolic cost of acute physical stress in juvenile steelhead. Trans Am Fish Soc 116(2):257–263
7. Cech JJ, Brauner CJ (2011) Ventilation and animal respiration: techniques in whole animal respiratory physiology. In: Farrell AP (ed) Encyclopedia of fish physiology. Academic Press, pp 846–853
8. Salin A, Auer KA et al (2015) Variation in the link between oxygen consumption and ATP production, and its relevance for animal performance. Proc R Soc B282:20151028
9. Seifert and Chapman (2006) Respiratory allocation and standard rate of metabolism in the African lungfish, Protopterus aethiopicus. Comp Biochem Physiol A143:142–148
10. Lighton JRB (2008) Measuring metabolic rates: a manual for scientists. Oxford University Press
11. Contreras V, Martínez-Meyer E, Valiente E, Zambrano L (2009) Recent decline and potential distribution in the last remnant area of the microendemic Mexican axolotl (*Ambystoma mexicanum*). Biol Conserv 142(12): 2881–2885
12. Farkas JE, Monaghan JR (2015) Housing and maintenance of *Ambystoma mexicanum*, the Mexican axolotl. In: Kumar A, Simon A (eds) Salamanders in regeneration research: methods and protocols. Methods in molecular biology, vol 1290. Springer, New York, pp 27–46
13. Irwin LN, Talentino KA, Caruso DA (1998) Effect of fasting and thermal acclimation on metabolism of juvenile axolotls (*Ambystoma mexicanum*). Exp Biol Online 3:1
14. Chabot D, Steffensen JF, Farrell AP (2016) The determination of standard metabolic rate in fishes. J Fish Biol 88:81–121

Chapter 27

Artificial Insemination in Axolotl

Yuka Taniguchi-Sugiura and Elly M. Tanaka

Abstract

In axolotls (*Ambystoma mexicanum*), fertilization takes place internally. After courtship, the male axolotl deposits spermatophores, which the female takes up into her cloaca in order to fertilize eggs internally. The success of axolotl breedings is subject to several poorly understood factors including age, pairing, and genotype. In some cases, individuals are unable to breed naturally despite having significant scientific value. Assisted reproductive technologies represent one approach to maintaining stocks of such individuals, as well as supplementing natural breedings of laboratory stocks.

Here, we describe a protocol for artificial insemination—an assisted reproductive technology in which sperm is extracted from a male and transferred into the female cloaca, thus mimicking natural fertilization in axolotls. We believe that this simple method can be applied to other salamander species that have internal fertilization and also help restore endangered wild populations.

Key words Artificial insemination, Assisted reproductive technology

1 Introduction

The axolotl is one of the most intensively studied salamander models. Axolotls are of great scientific interest because of their amazing regenerative capacity. A variety of cutting-edge technologies, such as those listed in this book, are now available for axolotl research. To keep axolotl research going, one requirement is a continuous supply of axolotls of various genotypes. Breeding axolotls is normally performed in a laboratory by pairing animals in a tank, to encourage natural mating [1, 2]. However, natural breeding can sometimes fail because of age, animal pairing, or other reasons. To address these difficulties in obtaining fertilized eggs, we have recently begun to implement artificial insemination in axolotls. We are still working on improving the efficiency of this assisted reproductive technology, but would like to share our current protocol. Together with in vitro fertilization—fertilization in a

Ashley W. Seifert and Joshua D. Currie (eds.), *Salamanders: Methods and Protocols*,
Methods in Molecular Biology, vol. 2562, https://doi.org/10.1007/978-1-0716-2659-7_27,

dish using isolated gametes—which is well established in axolotl and other salamanders [2, 3], we are now able to obtain fertilized axolotl eggs on demand.

2 Materials

2.1 Equipment

1. Needle 30G × ½″ (e.g., BD Biosciences, 304000).
2. 1-mL Syringe.
3. Artificial leaves (e.g., Pets Menu, PM25110).
4. 1.5- or 2-mL tube.
5. P1000 pipette.
6. Wide bore filtered pipette tip (e.g., Thermo fisher, 2079GPK).
7. Filtered pipette tip (e.g., Eppendorf epTips 0030078.683).
8. Glass petri dish (e.g., Sigma-Aldrich, SLW1480/08D-10EA).
9. Forceps (e.g., Fine Science Tools, 11251-20).
10. Animal tanks, containers for eggs.
11. Fish net.
12. Temperature-controlled incubator.

2.2 Chemicals and Solutions

1. 10× MMR: 1 M NaCl, 20 mM KCl, 10 mM $MgCl_2$, 20 mM $CaCl_2$, 0.1 mM EDTA, 50 mM HEPES. Adjust pH to 7.8 with 10 M NaOH. Autoclave and store at room temperature. Recommended shelf life is 12 months.
2. 10× MMR without calcium: 1 M NaCl, 20 mM KCl, 10 mM $MgCl_2$, 50 mM HEPES. Adjust pH to 7.8 with 10 M NaOH. Autoclave and store at room temperature. Recommended shelf life is 12 months.
3. 0.1× MMR: 10 mL of 10× MMR adjusted to 1 L with MilliQ water. Autoclave and store at room temperature. Recommended shelf life is 12 months. Before use, add 10 mL of Penicillin-Streptomycin solution and store at 10 °C. Recommended shelf life is 2 weeks.
4. 0.5× MMR without calcium: 100 mL of 10× MMR without calcium, adjusted to 2 L with de-ionized water or MilliQ water.
5. 10× TBS: Weigh 24.2 g of Tris and 90 g NaCl. Add MilliQ water to 990 mL. Mix and adjust pH to 7.0 with HCl. Top up to final volume 1 L with MilliQ water. Autoclave and store at room temperature.
6. 40× Holtfreter's solution: Weigh 158.4 g of NaCl, 2.9 g KCl, 5.4 g $CaCl_2 \cdot 2H_2O$ and 11 g $MgSO_4 \cdot 7H2O$. Top up to 1 L with MilliQ water. Autoclave and store at room temperature.
7. Benzocaine (e.g., Sigma-Aldrich, E1501-100 g).

8. 10% Benzocaine: Dissolve 10 g Benzocaine in 100 mL of 100% ethanol absolute.
9. 0.03% Benzocaine: Mix 500 mL of 10× TBS and 50 mL of 40× Holtfreter's solution. Adjust to 9970 mL with de-ionized water. Add 30 mL of 10% Benzocaine and mix well.
10. Penicillin-Streptomycin solution (e.g., Sigma-Aldrich, P0781).
11. Human chorionic gonadotropin (e.g., Sigma-Aldrich, CG10).
12. Sigmacote (Sigma-Aldrich, SL2-100 mL).

3 Methods

1. Select sexually mature male and female axolotls (*see* **Note 1**).
2. Hold each animal in place using a fish net, by placing the palm of one hand over its head and trunk. Give both female and male a dose of 300–400 IU of human chorionic gonadotropin (hCG) by injecting it intramuscularly, dorso-lateral to the cloaca through the fish net (Fig. 1, *see* **Note 2**). Return to housing water (*see* **Note 3**).
3. After 20–24 h, transfer animals to fresh animal holding water (*see* **Note 4**). Put some artificial leaves into the females' tanks to allow egg deposition (*see* **Note 5**).
4. Within 2 h of the female starting to lay eggs, anesthetize first one male then the female in 0.015–0.03% benzocaine (*see* **Note 6**).
5. Once anesthetized, dip the male briefly into 0.5× MMR without calcium; then lift it out and wipe excess liquid using a paper towel (*see* **Note 7**).

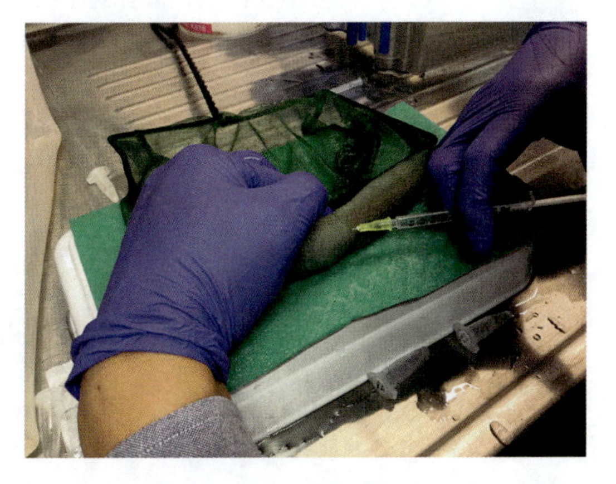

Fig. 1 Hold the animal in place by placing the palm of one hand over its head and trunk. Inject hCG intramuscularly, dorso-lateral to the cloaca, through the fish net

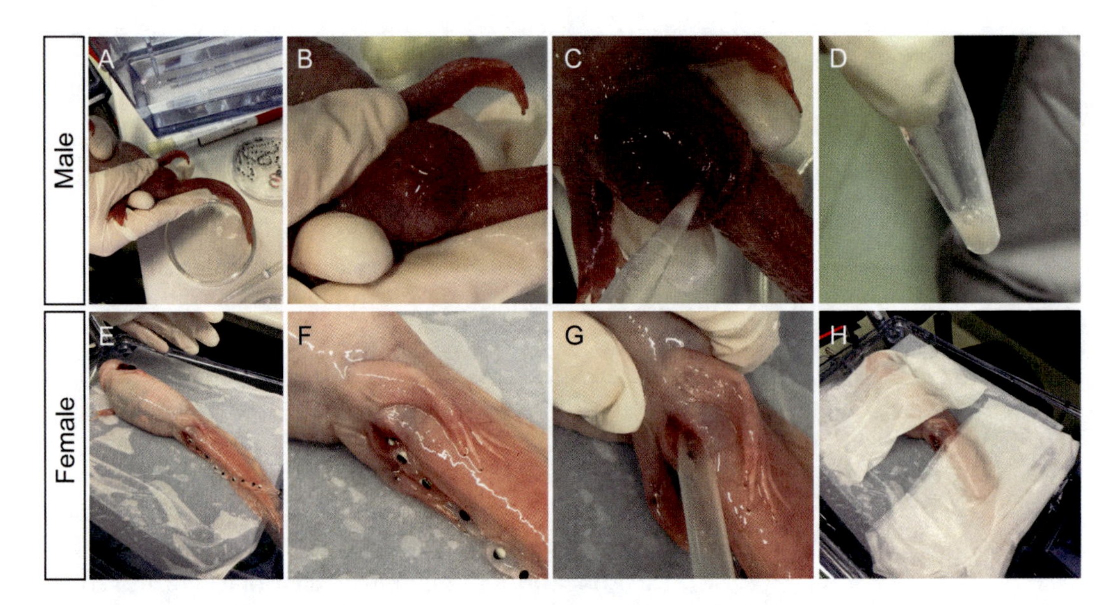

Fig. 2 Hold anesthetized male (**a**) and gently massage the male abdomen from an anterior to a posterior direction, pressing gently around the pelvis region anterior to the cloaca to release sperm from the cloaca (**b**). Collect extruded semen using a wide bore P1000 pipette (**c**), and transfer to a 1.5- or 2-mL tube on ice (**d**). Place anesthetized female on a paper towel wetted with 0.5x MMR (**e**, **f**). Insert the pipette tip ~1.5–2 cm deep into the female's cloaca before ejecting and inject 50–100 uL of semen into the cloaca using a P1000 pipette (**g**). Place the injected female on a paper towel wetted with 0.5× MMR and place wet paper towels on top to keep the female wet for 15–20 min (H)

6. Gently massage the male abdomen from an anterior to a posterior direction and press gently around the pelvis region—anterior to the cloaca—to release sperm from the cloaca (Fig. 2a and b). Collect extruded semen using a wide bore P1000 pipette (*see* **Note 8**) and transfer to a 1.5- or 2 mL-tube on ice (*see* **Notes 9** and **10**, Fig. 2c and d). The male can be returned to its tank at this point.
7. Alternatively, semen run-off from the cloaca can be collected directly onto a Petri dish. In this case, use a glass Petri dish coated with Sigmacote (*see* **Note 11**).
8. Put a small drop of the collected semen onto a Petri dish and confirm that the sperm is motile using a stereo-microscope.
9. If the collected sperm is immotile, select a different male and repeat **steps 6–8**.
10. If only spermic urine—a thin, clear liquid that contains spermatozoa—is available, spin down the liquid for 1 s at 5000 rpm, discard the watery supernatant, and use the thick part for artificial insemination.
11. Lift the anesthetized female out of its tank. Place female on a paper towel wetted with 0.5× MMR.

12. Insert the P1000 pipette tip ~1.5–2 cm deep into the female's cloaca; then inject 50–100 uL of semen into the cloaca (*see* **Note 12**, Fig. 2e–g).
13. Place the injected female on a paper towel wetted with 0.5× MMR and also place wet paper towels on top to keep the female wet for 15–20 min (Fig. 2h).
14. Return the female to the tank of tap water with artificial leaves. Place the tank in a calm, quiet room and allow the female to recover from anesthesia and to lay eggs.
15. 4–5 h after insemination, collect laid eggs into a container of fresh tap water and store them at room temperature or in a 16–18 °C incubator (*see* **Note 13**).
16. **Steps 4–14** can now be repeated once for the same female, using the same male (*see* **Notes 14** and **15**). Unlike spermatophores, ejaculated semen does not survive very long in the female's body. To obtain a large number of fertilized eggs, it is necessary to repeat the artificial insemination using freshly collected semen. Depending on the quality of the sperm and oocytes, you can expect 30–100 fertilized eggs from one round of artificial insemination.
17. The next day, discard unfertilized eggs (no cleavage divisions) and transfer fertilized eggs to fresh animal holding water. Animals can be returned to their usual tanks 4 days after hCG injection. Until then, change the housing water every day (*see* **Note 16**).
18. If the jelly coat is squeezing the egg, remove the jelly gently using forceps and keep de-jellied eggs in 0.1× MMR with pen/strep (*see* **Note 17**).

4 Notes

1. If available, select females between 1.5 and 3 years and males below 5 years. The survival rate of the embryos after artificial insemination is highly dependent on the combination of the parents, as during natural mating. Usually, we select 3 females and 3 males per genotype. In our experience, two out of three pairs generally produce fertilized offspring. Depending on the quality of sperm and oocyte, you will get 50–200 fertilized eggs from one pair of animals. Animals can be used for natural mating or artificial insemination again after a few weeks' recovery time.
2. Hold the animal in place using the palm of one hand over its head and trunk. Inject the hormone through the fish net. If this is difficult, this process can be performed under benzocaine anesthesia. We normally inject hCG between 10:00 and 11:00 am in the morning.

3. Isolate the hCG-injected animals from the water circulation system to avoid effects on other animals.
4. We normally leave animals in 19.5 °C housing water in isolated tanks that are not in a recirculating system. Place the tanks in the room at 18 °C until the following morning to keep the water temperature around 18–19 °C. The next day, transfer the animals to cold fresh animal holding water. With this condition, female starts to lay eggs around 10:00 am.
5. We normally put seven artificial leaves per 8-L tank.
6. For each mating pair, anesthetize the male first. After 5–10 min, anesthetize the female.
7. We include this step to avoid contamination of the sperm with animal holding water. We are using calcium-free medium here because calcium ions in the water may activate oocytes, and inhibit sperm motility.
8. Use a wide bore filtered pipette tip to collect sperm. We are using a commercial product but you can instead cut the tip off of a normal pipette tip to generate a larger opening.
9. If sufficient sperm is collected, it can be kept on ice and used for a second round of artificial insemination.
10. You can expect between 50 and 200 μL semen from one male.
11. With Sigmacote, sperm does not stick to the dish.
12. Do not insert the pipette tip too deep, otherwise you may damage the inside of the female's body. For this reason, we do not recommend a long-thin pipette tip (Fig. 3).
13. If you want to slow down the speed of egg development, place eggs in a 16–18 °C incubator.
14. The first artificial insemination should be performed within 2 h of the female starting to lay eggs. Usually, the first round of insemination can be performed between 11:00 and 13:00 and

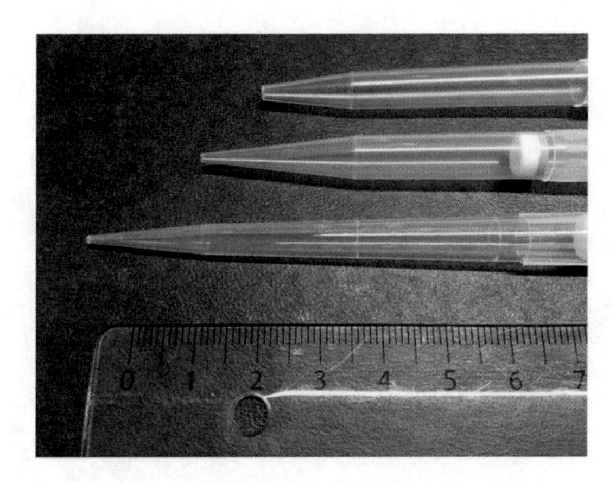

Fig. 3 Top: wide bore, Middle: normal, Bottom: long-thin P1000 filter tip

a second round between 16:00 and 18:00. However, this is highly dependent on the female, as some females only start to lay eggs more than 24 h after hCG injection.

15. To avoid contamination of different genotypes, use the same male for the second insemination.
16. Females may continue to lay eggs for several days.
17. Keep de-jellied eggs in 0.1× MMR with pen/strep may increase the survival rate of eggs.

Acknowledgments

The authors are deeply grateful to their animal team for their great axolotl care. And the authors appreciate the feedback offered by Leo Otsuki.

References

1. Khattak S et al (2014) Optimized axolotl (Ambystoma mexicanum) husbandry, breeding, metamorphosis, transgenesis and tamoxifen-mediated recombination. Nat Protoc 9(3): 529–540
2. Kumar A, Simon AS (2015) Salamanders in regeneration research: methods and protocols. In: Methods in molecular biology. Humana Press, New York, p 357
3. Mansour N, Lahnsteiner F, Patzner RA (2011) Collection of gametes from live axolotl, Ambystoma mexicanum, and standardization of in vitro fertilization. Theriogenology 75(2):354–361

Chapter 28

Screening Salamanders for Symbionts

Elli Vickers and Ryan Kerney

Abstract

Microbial symbionts are broadly categorized by their impacts on host fitness: commensals, pathogens, and mutualists. However, recent investigations into the physiological basis of these impacts have revealed nuanced microbial influences on a wide range of host developmental, immunological, and physiological processes, including regeneration. Exploring these impacts begins with knowing which microbes are present. This methodological pipeline contains both targeted assays using PCR and culturing, as well as culture-independent approaches, to survey host salamander tissues for common and unknown microbial symbionts.

Key words *Batrachochytrium*, *Oophila*, Culturing, Bacteria, Metabarcoding, Microbiota

1 Introduction

Microbial community ecology has been revolutionized by next generation sequencing. The ability to interrogate environmental samples for their un-culturable microbial diversity has provided a new window into the world of bacteria, archaea, eukaryotic protists, and fungi. These approaches rely on amplifying and sequencing conserved 16s, 18s, and internal transcribed spacer (ITS) regions of the ribosomal DNA, respectively. These metabarcode (aka amplicon) sequences are then used to identify the diversity and relative abundance of microbes in an environmental sample. This information is critical for correlating microbial community structure with experimental manipulations, host development, or natural environmental changes/circumstances. However, exploring the causative roles of host-associated microbes also requires conventional isolation, culturing, and co-culturing techniques.

Surprisingly few studies have examined the interrelation between microbial symbionts and regeneration. Recent research has discovered inhibitory effects of microbes or microbial populations in planaria [1, 2] as well as correlations between intestinal microbiota changes and regenerative competence in pre- and post-

Ashley W. Seifert and Joshua D. Currie (eds.), *Salamanders: Methods and Protocols*,
Methods in Molecular Biology, vol. 2562, https://doi.org/10.1007/978-1-0716-2659-7_28,

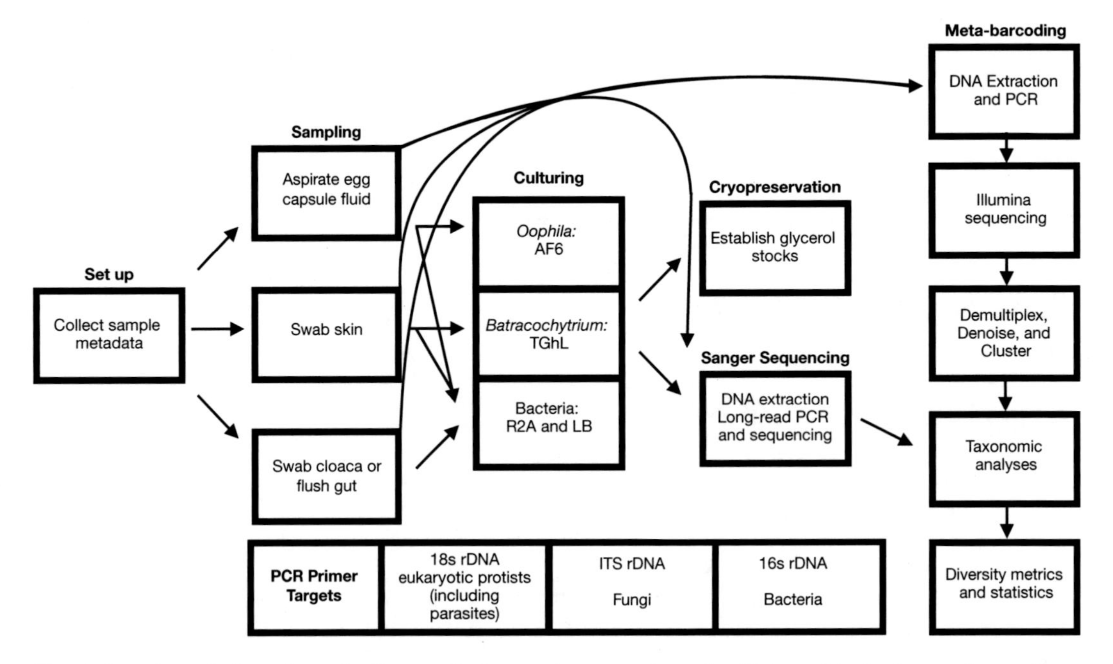

Fig. 1 Overview of steps for collecting, culturing, and sequencing microbial symbionts from salamanders. Samples can be either assayed for specific microbes through selective culturing and/or direct Sanger sequencing, or they can be screened for unknown microbes using bacterial colony sequencing and/or environmental metabarcoding

metamorphic *Ambystoma mexicanum* [3]. Changes in microbial communities during limb regeneration suggests potential regulation by the host as well as opportunistic infection of the blastema [4]. Even though various microbial communities of salamanders have been explored for their probiotic potential [5, 6] or production of beneficial metabolites [7, 8], we have found no studies to date that explore these metabolites for their ability to regulate salamander tissue regeneration.

This protocol outlines a standard pipeline for collecting, culturing, and sequencing microbial symbionts from salamanders (Fig. 1). It is based on established protocols for environmental microbiology, assays of *Batrachochytrium* sp. infection status of salamanders, screening for other common commensals and mutualists, as well as the characterization of microbial communities using metabarcode sequencing. It assumes users are familiar with the colony picking, plating, and aseptic/sterile techniques commonly used in recombinant DNA technology.

2 Materials

2.1 Sampling

2.1.1 Skin Swabs

1. Nitrile gloves [9, 10].
2. 50-mL falcon tubes with 25 mL of sterile PBS.

3. Sterile swab (e.g., MW-113 swab).
4. Sterile screw-capped 15-mL epi tubes.
5. Sterile scissors.
6. 3% sodium hypochlorite [11]: Add 3 mL of standard bleach per 100 mL of distilled water.

2.1.2 Gut/Cloaca

1. Sterile dissection tools.
2. Sterile swab.
3. 0.2% MS222 [3]: pH 6.5–7.5. Weigh 0.4 g of tricaine powder and add to 200 mL of sterile water. Initial pH will be around 4–5; buffer to pH 6.5–7.5 using NaOH.
4. Isotonic sterile saline solution [3]: Weigh 9 g of non-iodized salt (sodium chloride) and add to 1 L of sterile water. Boil to dissolve and sterilize.

2.1.3 Egg Capsule Fluid

1. 23- or 30-gauge needles and 1-mL syringes [12, 13].
2. Untreated 24-well plates.
3. Sterile phosphate buffered saline (PBS): Add 1 tablet of PBS per 200 mL of sterile water. Autoclave to dissolve and sterilize.
4. 70% ethanol.

2.2 Culturing

2.2.1 Bacterial Culturing

1. Sterile 0.05 M HEPES buffer: pH 7.15. Weigh 2.38 g of HEPES powder and add to 160 mL of sterile water. Use a stir bar to dissolve. Initial pH will be around 5; buffer to pH 7.15 with NaOH.
2. Freshwater R2A broth and plates [14, 15]: Weigh 3.15 g of R2A broth powder (0.5 g/L yeast extract, 0.5 g/L meat peptone, 0.5 g/L casamino acid, 0.5 g/L glucose, 0.5 g/L starch, 0.3 g/L dipotassium hydrogen phosphate, 0.3 g/L sodium pyruvate, 0.05 g/L magnesium sulfate) and add sterile water to a volume of 1 L. Heat and stir with stir bar until powder is dissolved. Autoclave. For agar plates, add 7.5 g of agar powder to solution. Once autoclaved, pour a thin layer into petri dishes and cool until solid. Store plates at 4 °C.
3. Miller's LB broth and plates [13, 14]: Same as above with 25 g of LB broth powder (10 g/L tryptone, 5 g/L yeast extract, 10 g/L NaCl).
4. Sterile 100% glycerol.
5. Sterile loops and glass beads.

2.2.2 Algal Culturing

1. Environmental chamber with broad spectrum fluorescent lights (e.g., Conviron).

2. AF6 media [16]: Weigh 400 mg of MES buffer and 1 mL of 2 g/L Fe-citrate and add to mL of distilled water in a 2-L flask. Add 1 mL each of the following components to the solution: 140 g/L $NaNO_3$, 22 g/L NH_4NO_3, 30 g/L $MgSO_4$ $7H_2O$, 5 g/L K_2HPO_4, 10 g/L KH_2PO_4, 10 g/L $CaCL_2$ $2H_2O$, and 2 g/L citric acid. Separately, prepare the vitamin and trace metals solutions. For the vitamin solution, dissolve 10 mg of thiamine (vit. B_1) in 950 mL of distilled water. Add 1 mL each of 2 g/L biotin (vit. H), 1 g/L cyanocobalamin (vit. B_{12}), and 1 g/L pyridoine (vit. B_6). Filter-sterilize and store in fridge. For the trace metals solution, dissolve 5 g of Na_2EDTA $2H_2O$ in 950 mL of distilled water; then individually dissolve the following metals: 0.98 g $FeCL_3$ $6H_2O$, 0.18 g $MnCl_2$ $4H_2O$, and 0.11 g $ZnSO_4$ $7H_2O$. Then, add 1 mL each of 20 g/L $CoCl_2$ $6H_2O$ and 12.5 g/L Na_2MoO_4 $2H_2O$. Filter-sterilize and store in fridge. Add 1 mL each of the vitamin and trace metals solution to the AF6 medium; then bring the final volume up to 1 L with distilled water. Adjust the pH to 6.6 and autoclave.
3. Sterile 250-mL culture flasks.

2.2.3 *Batrachochytrium* Culturing

1. TGhL broth and plates [17, 18]: Weigh 16 g tryptone, 4 g gelatin hydrolysate, and 2 g lactose and add to 1 L of sterile water. Heat and stir with stir bar until powder is dissolved. Autoclave. When cool, add 200 mg/mL of penicillin-G and 200 mg/mL streptomycin sulfate. For agar plates, add 7.5 g of agar powder to solution. Once autoclaved, pour a thin layer into petri dishes and cool until solid. Store plates at 20 °C.
2. 1% Vikron S®, 4% sodium hypochlorite, or 70% ethanol for decontamination [19]: Use sterile water to dilute to the correct concentration for each decontaminant.
3. Tissue culture flasks.
4. Sterile glass beads.
5. 10% DMSO/10% fetal calf serum.
6. Cryotubes.
7. Cryocontainer with 100% isopropanol.

2.3 DNA Extraction

2.3.1 Swabs and Tissue Samples

1. Qiagen DNeasy Blood & Tissue Kit [3, 20].

2.3.2 Egg Capsule Fluid

2. Qiagen DNeasy PowerSoil Kit.

2.4 PCR

3. PCR-grade water.
4. 2× PCR Master Mix (*see* **Note 1**).

2.5 Analytical Software

1. QIIME2 [36].
2. Microeco R package [37]: *See* **Note 4** for additional R packages.

3 Methods

3.1 Sampling

Approaches to acquiring environmental isolates have existed for over a century. This protocol reflects the techniques standardized by the amphibian disease research community. These include very specific guidelines for acquiring cutaneous microbes through sterile swabbing, as well as preferred media use for bacterial culturing. *See* **Note 5** for permit requirements and field equipment usage.

3.1.1 Collecting Sample Metadata

1. A vital part of sample collection is the recording of metadata. Examples of metadata include but are not limited to the following variables: experimental conditions, clutch size and morphology, mortality, developmental stage, colony-forming units (CFUs) of bacteria, and environmental conditions (e.g., temperature, dissolved oxygen, pH). Record relevant conditions for each experiment. *See* **Note 6** for metadata formatting.

3.1.2 Skin Swab

1. Use new nitrile gloves for handling each salamander. Rinse them briefly 2× for 30 s in sterile PBS to reduce transient microbes.
2. Use a sterile MW-113 swab to gently swab the left side dorsal/ ventral and fore-/hindlimbs 5× each (20 total). Field swabs can be stored in 75% ethanol for DNA extraction, with long-term freezing at −80 °C [38].

3.1.3 Cloaca/Gut

The amphibian intestinal microbiome can be sampled both directly and indirectly. Cloacal sampling may be an adequate non-lethal proxy for the gut microbiome [39].

1. For indirect sampling, use a sterile swab to gently roll over the external cloaca [40].
2. For direct sampling, first euthanize in 0.2% MS222, then dissect and remove intestinal tissue [3]. Intestinal contents are evacuated and rinsed 5× with 2 mL of sterile saline for DNA extraction. Centrifuge at >15,000 × g for 7 min, then remove the supernatant, and store the remaining pellet at −80 °C.

3.1.4 Egg Capsule Fluids

Exploring the function of microbial symbionts in developmental and regenerative processes requires bacterial isolation and culturing. The bacteria of the intracapsular fluid are potential sources for colonization of the adult amphibian microbiome [41] and are an under-explored resource for discovering novel metabolites [42].

1. To capture the microbes of the intracapsular fluid, begin by removing eggs from the egg jelly using gloved hands; then remove the excess jelly by gently pipetting the egg capsule through a plastic pipette with the tip cut to create a bore slightly smaller than the egg capsule.
2. In a 24-well plate, immerse each egg in 70% ethanol for 30 s, then remove the egg and wash twice in PBS, taking care to not transfer additional liquid.
3. Using a 23- or 30-gauge needle and a 1-mL syringe, aspirate as much intracapsular fluid as possible per egg without disturbing the embryo. Pool intracapsular fluid from the same conditions.
4. To store the intracapsular fluid, centrifuge at $>15,000 \times g$ for 7 min, then remove the supernatant, and store the remaining pellet at −80 °C.

3.2 Culturing

3.2.1 Bacterial Culturing

Environmental bacterial isolates are notoriously difficult to culture. Recent developments in staged bioreactors allow the isolation and clonal culturing of a wider range of bacteria than conventional techniques [43]. We use the low-nutrient freshwater media Reasoner's 2A (R2A) [14] as well as the richer Luria-Bertani (LB) media [15] which includes yeast extract, casein peptones, and peptides that aid heterotrophic growth. Environmental extracts (such as cutaneous and gut/cloacal extracts) can be swabbed onto agar plates enriched with either media or added directly as a liquid to the plates (such as egg capsule fluid or intestinal rinse). Sterile and aseptic technique should be used for all of these steps in order to eliminate the possibility of cross-contamination.

1. For fluid culturing, first dilute sample fluid 1:10 in 0.05 HEPES buffer (pH 7.15). Plate 100 μL of this dilution onto separate agar plates and spread with sterile glass beads. For swab culturing, roll the swab over the surface of an agar plate.
2. Incubate plates at a temperature that closely matches the environmental conditions of the host in order to selectively grow clonal isolates. Check plates daily until colonies have grown. Plates with bacteria can be sealed with parafilm and stored at 4 °C.
3. Screen agar colonies based on their color, form, elevation, substance, opacity, and margin [38]. If possible, colonies should be photographed before picking in order to record these characteristics to later verify 16s rDNA identification.
4. Streak-purify colonies of interest by using a sterile stick or inoculation loop to grab a single colony and streak onto a new plate.

5. To grow liquid cultures, use a sterile stick or inoculation loop to grab a single colony off of an agar plate. Dip the colony in 3–5 mL of media. Incubate shaking at 150 rpm at environmental temperature.
6. For long-term storage of liquid cultures, add 150 μL of 100% glycerol to 850 μL of liquid culture (a final concentration of 15% glycerol). Store at −80 °C.

3.2.2 Algal Culturing [44–46]

1. Inoculate a stock of algae into 100 mL of AF6 in a 250-mL culture flask. Liquid media are preferred over plates due to reports that solid media favors non-*Oophila* algae [44].
2. Grow algal culture flasks in an environmental chamber set at 18 °C on a 12-h light/12-h dark timer with 59 μmol/m^2/s (PAR) broad spectrum fluorescent light [13].
3. To prepare algal cultures for experiments, subculture a fresh 1: 10 dilution in AF6 2–3 weeks prior to the experiment.

3.2.3 Batrachochytrium Culturing

Two species of chytrid fungus, *Batrachochytrium dendrobatidis* (Bd) and *B. salamandrivorans* (Bsal), are keratin-consuming amphibian pathogens that have contributed to global declines of amphibian populations [17]. Culturing and experimentally infecting amphibians with these pathogens is a useful tool for understanding their effect on host physiology and immune responses. As these are pathogenic species, it is crucial to work with these organisms at United States Biosafety Level 2, which includes performing all culture work in biosafety cabinets [47].

1. To non-lethally isolate chytrid from amphibians in the wild, use sterile biopsy punches to collect small skin tissue samples [48].
2. Flame-sterilize an inoculating needle and use to drag the tissue samples through sterile TGhL agar containing antibiotics (200 mg/mL penicillin-G and 200 mg/mL streptomycin sulfate) to remove bacteria and yeast cells.
3. Place the decontaminated tissue samples onto fresh TGhL agar plates with antibiotics [49].
4. Keep the plates on ice until back at the laboratory, where they can be incubated at 21–22 °C and monitored for *Batrachochytrium* colony formation as well as contamination.
5. To grow liquid cultures, an actively growing *Batrachochytrium* colony or culture must be inoculated into a tissue culture flask containing TGhL broth and incubated at 17–23 °C for 3–21 days [17]. Tissue culture flasks are recommended because sporangia, the stationary reproductive phase of zoosporic fungal cells, adhere to the flask, while zoospores, the motile infective cells, remain in the liquid medium. Staring new cultures

and subculturing with zoospores results in more consistent growth than cultures containing both zoospores and sporangia.

6. Subculture liquid cultures every 4 days to prevent overgrowth of cultures and maintain uniform cell numbers for experiments [47]. To subculture into liquid medium, dilute a mature chytrid culture 1:10 in fresh TGhL, which results in 3×10^4 zoospores per mL in each passage. Draw directly from the liquid medium rather than scraping the sides of the flask when subculturing.
7. To subculture onto solid media, add 5×10^5 zoospores in liquid culture to a TGhL agar plate pre-chilled to 15 °C, using sterile glass beads to spread the culture evenly across the plate.
8. To harvest chytrid cells for long-term storage, scrape the flask and centrifuge at 1700 × g for 10 min, then decant the supernatant, and resuspend the pellet in 1 mL of 10% DMSO/10% FCS, which serves as a cryoprotectant. The pellets can then be flash frozen in cryotubes in a cryocontainer with 100% isopropanol at −80 °C. Then stored at −80 °C. To revive frozen samples, immerse in 43 °C water for up to a minute [17].
9. To infect salamanders with chytrid, topically apply 1 mL of chytrid culture with 10^5 cells/mL [17]. Infection can be confirmed using qPCR with chytrid-specific primers (Table 1). Protocols must be approved by the Institutional Animal Care and Use Committee (IACUC) and extreme caution should be used to prevent environmental release.
10. Disinfect all lab equipment and areas after working with Bd and/or Bsal. 1% Vikron S®, 4% sodium hypochlorite (dilute household bleach), and 70% ethanol have all been found to effectively kill Bsal, Bd, and *Ranavirus*. 1% Vikron S® requires 5 min of contact time to effectively disinfect, whereas the latter disinfectants only require 1 min [19].

3.3 Sanger Sequencing

3.3.1 DNA Extraction

1. For swabs and tissue samples, the DNeasy Blood and Tissue Kit by Qiagen is our preferred option for DNA extraction [3, 20].
2. For intracapsular fluid, we have found that the DNeasy PowerSoil Kit by Qiagen provides effective and easy-to-use extraction of DNA from low-abundance microbes.

3.3.2 16s PCR

1. Before staring the PCR, ensure that both the quantity and quality of the extracted DNA are sufficient using a spectrophotometer. Pure DNA has a 260/280 ratio of 1.80–2.00. The minimum DNA concentration for library preparation is 0.5 ng/mL [12].

Table 1
List of primers and primer sequences used in PCR to identify and sequence various microbes in salamander biology. Targeted amplification can detect specific microbes. Examples include pathogens (fungal: *B. salamandrivorans* and *B. dendrobatidis;* protist: *Oomycetes*); commensals (*Protoopalina* sp.); and mutualists (*O. amblystomatis*). Illumina sequencing is used for metabarcoding, of eukaryotic protists, fungi, and bacteria, which reveals greater sample diversity, including "unculturable" microbes

For identification of specific pathogenic, commensal, and mutualistic symbionts:

Target	Primer name	Primer sequence (5′–3′)	Product length	Target region
Batrachochytrium salamandrivorans [21]	STerF STerR	TGCTCCATCTCCCCCTCTTCA TGAACGCACATTGCACTCTAC	160 bp	5.8S rDNA
Batrachochytrium dendrobatidis [22, 23]	*1st PCR* Bd18SF1 Bd28SR1 *2nd PCR* Bd1a Bd2a	TTTGTACACACCGCCCGTCGC ATATGCTTAAGTTCAGCGGG CAGTGTGCCATATGTCACG CATGGTTCATATCTGTCCAG	704 bp	5.8S rDNA and ITS
Oomycetes (water molds) [24]	ITS1 ITS4	TCCGTAGGTGAACCTGCGG TCCTCCGCTTATTGATATGC	290 bp	ITS
Protoopalina sp. [25]	PK PKR	CACACCAGATATGGGTTATGC GCCCTCCAATKGATTCG	750 bp	18s rDNA
Oophila amblystomatis [26]	18s_261F 18s_604R	TCGGATCGTCTCGGTTTC CAGTGTTGCCACCGCTCG	343 bp	18s rDNA

For Sanger sequencing (*see* Note 2):

Target	Primer name	Primer sequence (5′–3′)	Product length	Target region
Fungal [27]	ITS_817F ITS_1536R	TTAGCATGGAATAATRRAA TAGGA ATTGCAATGCYCTATCCCCA	762 bp	18s rDNA
Eukaryotic protist [28]	18s_33F 18s_1768R	CCTGCCAGTAGTCATAYGCTT TGATCCTTCYGCAGGTTCACC	1735 bp	18s rDNA
Bacteria [29]	16s_27F 16s_1492R	AGAGTTTGATCMTGGCTCAG TACGGYTACCTTGTTACGAC TT	1465 bp	16s rDNA

For Illumina sequencing (*see* Note 3):

Target	Primer name	Primer sequence (5′–3′) (*without index, linker, or pad*)	Product length	Target region
Fungal [30, 31]	ITS1f ITS2	CTTGGTCATTTAGAGGAAG TAA GCTGCGTTCTTCATCGATGC	230 bp	ITS

(continued)

Table 1 (continued)

For Illumina sequencing (*see* Note 3):				
Target	**Primer name**	**Primer sequence (5′–3′) (*without index, linker, or pad*)**	**Product length**	**Target region**
Eukaryotic protist [30, 32]	Euk_1391F EukBr	GTACACACCGCCCGTC TGATCCTTCTGCAGGT TCACCTAC	260 bp	18s rDNA
Bacteria [33, 34]	16s_515F (515F) 16s_806R (806R)	GTGYCAGCMGCCGCGGTAA GGACTACNVGGGTWTCTAAT	300–350 bp	16s V4 rDNA
Illumina overhang [35]	Forward Reverse	TCGTCGGCAGCGTCAG ATGTGTATAAGAGACAG-(locus-specific primer) GTCTCGTGGGCTCGG AGATGTGTATAAGAGACAG-(locus-specific primer)	–	–

2. For PCR off of bacterial colonies, incubate a single fresh colony in 50 μL of sterile water at 95–100 °C for 5–10 min. Use 1 uL of that solution as template DNA.
3. For a single 25 μL reaction (1×), combine the following in a 1.5-mL microcentrifuge tube on ice: 12.5 μL 2× PCR Master Mix (*see* **Note 2**), 10.5 μL PCR-grade water, and 0.5 μL of each primer at 10 μM. Scale up as needed. Aliquot 24 μL per PCR tube and add 1 μL of template DNA for each reaction. Refer to Table 1 for primer sequences and template lengths. Cycling conditions vary by your choice of DNA polymerase.

3.3.3 Sanger Sequencing Analysis

The most common search engine for nucleotide sequences is NCBI BLAST. However, for 16s sequence databases, the service EzBioCloud (EzBioCloud.net) is considered the most up-to-date and comprehensive resource for reference bacterial sequence alignments [50, 51]. It is best to utilize a combination of both NCBI Ref Seq Targeted Loci, or Selected Loci BLAST (https://www.ncbi.nlm.nih.gov/refseq/targetedloci/), and EzBioCloud for identifying microbial symbionts from Sanger sequencing reads.

3.4 Amplicon Sequencing (Metabarcoding)

Standard protocols for environmental amplicon sequencing are established by the Earth Microbiome Project [30]. They include designated "universal" ribosomal DNA (rDNA) primers for eukaryotic protists, fungi, and bacteria (Table 1). A recent survey of metabarcoding papers found the Illumina MiSeq to be the most common platform, with an average amplicon read depth of 60,000 raw reads per sample (range <10,000 to >900,000) [52]. Illumina MiSeq is limited by read length (up to 100 base pairs in single reads and up to 300 base pairs for paired reads using V3 chemistry), although it has been shown that 100 base pair differences in 16s rDNA are sufficient for distinguishing bacterial communities [53, 54]. While MiSeq is most commonly adopted platform for metabarcoding, higher read depth per PCR library is achievable through the newer NovaSeq platform [52].

There are a variety of commercial services that will prepare and sequence amplicon libraries from extracted DNA samples (e.g., Mr DNA—MrDNALab.com; Integrated Microbiome Resource—https://imr.bio/index.html; Genewiz; CD Genomics; and institution-specific sequencing core facilities). These vendors employ a range of approaches, and standard protocols vary by MiSeq chemistry (versions 2 or 3), the number of pooled samples, and the regions amplified. Given this variability and the rapid advance of these technologies, users are referred to their sequencing facility of choice for library preparation and sequencing protocols. Specific sample preparation protocols can be found through the Earth Microbiome Project for smaller sample size projects [1, 28, 32, 55]; Illumina product material [35]; or the Schloss lab [56].

Amplification of host-associated bacterial 16s rDNA can result in a large number of off-target mitochondrial genome amplifications. These can be reduced using a PNA clamp which binds to mitochondria-specific sequences, which limits vertebrate mitochondrial 16s amplification [57]. Recent work on cutaneous and intestinal salamander microbiomes does not use PNA blocking [4, 5], although we have found it effective in reducing host mtDNA signatures in egg capsule microbiomes [unpublished observation].

Sequence analysis: There are several standardized pipelines for amplicon sequencing analysis. These range from R packages [58] and command line programs [59] to the MiSeq Reporter Software v2.5.1.3 (Illumina) and web-based graphical user interfaces [60]. Our preferred program is currently QIIME2 [36], which handles raw FASTA/FASTQ reads and outputs a taxonomy assignments, basic phylogenies, and diversity statistics. The basic pipeline for analyzing metabarcoding data is as follows:

1. Demultiplex samples: Raw DNA amplicon sequence data are usually in the FASTQ format, which contains read quality information, or FASTA, which does not. Paired-end V4/V5 amplicons need to be demultiplexed based on the sequence tags in the sample metadata.
2. Denoise and cluster: QIIME2 can run DADA2 [61], which will remove noisy sequences, correct low-quality score errors, remove singletons, filter chimeric sequences, join pair-end

reads, and dereplicate the sequences. The output produced is amplicon sequence variants (ASVs), which represent the sequenced products of individual PCR products that have been clustered based on a probabilistic model to reduce false identification of variants from sequencing artifacts [62].

3. Taxonomic analyses: There are several options for determining the taxonomic identity of ASVs [63]. The largest and most up-to-date validated reference sequence database is SILVA for bacteria, archaea, and non-fungal eukaryotes [64]; the UNITE database is the best resource for validated fungal ITS sequences [65]. The SILVA database needs to be trained for an individual dataset of ASVs based on the target region that was sequenced [66]. QIIME2 will create a custom "trained" feature classifier using a multinomial naïve Bayesian classifier [66]. The taxonomic identities are then assigned using a probabilistic model through machine learning [67, 68] insert the citation to reference 68 here. The output will be ranked hierarchical taxonomic frequencies down to the genus level for each metabarcode library.
4. Diversity metrics: There are a large number of comparative statistics available for environmental microbiology. The most common community-level comparisons are divided between alpha and beta diversity metrics. Alpha diversity measures *community composition* using a range of algorithms. These can include consideration of taxa abundance (e.g., Fisher's index, Pielou's evenness) or just species present (e.g., Brillouin's index), with or without consideration of phylogenetic diversity (e.g., Faith's phylogenetic diversity index) [69]. Beta diversity compares *species richness* between samples and is similarly divided between phylogenetic (e.g., unifrac) and non-phylogenetic metrics (e.g., Bray-Curtis) [70]. The independent variable(s) for any of these comparisons is derived from the sample metadata (e.g., amputation conditions, regenerative status, etc.).
5. Additional analyses are too numerous to mention here. Common approaches include determining differences in longitudinal community composition [71], explaining the variance in community composition with independent variables [72], or predicting the metagenomic functional interactions from the metabarcoding data (e.g., PICRUSt) [73].

4 Notes

1. 2× PCR Master Mix simplifies the PCR process and minimizes error, but individual PCR components can be used as well. For a single 25 μL reaction (1×), this includes the following: 2.5 μL

10× *Taq* buffer, 0.5 μL 10 mM dNTPs, 0.5 μL each of 10 μM primer, 0.125 μL *Taq* DNA polymerase, and 1 μL template DNA. Add PCR-grade water up to 25 μL.

2. 16s_27F and 16s_1492R are twofold degenerate primers for Sanger sequencing of most bacteria, especially enteric bacteria. There are also twofold (27f-CM AGAGTTTGATCMTGGCTCAG, where M is A or C), fourfold (27f-YM (AGAGTTTGATYMTGGCTCAG, where Y is C or T), and sevenfold degenerate primers that capture a broader range of bacteria, although this comes with a small loss of amplification efficiency The sevenfold degenerate primer 27f-YM+3 is four parts 27f-YM (fourfold degenerate primer) and one part each of the 27f primers specific to *Bifidobacteriaceae* (27f-Bif, AGGGTTCGA TTCTGGCTCAG), *Borrelia* (AGAGTTTGATCCTGGCTTAG), and *Chlamydiales* (AGAATTTGATCTTGGTTCAG) [29].

3. PCRs for Illumina MiSeq library construction also require sample-specific index sequences (also called barcode sequences), which enable multiplexing and Illumina adapter sequences. For running libraries on Illumina MiSeq and HiSeq, 5–10% PhiX will need to be added to the run [30]. Additionally, Euk_1391F is a universal primer, and EukBr favors eukaryotic DNA but can mismatch with prokaryotes. For samples associated with hosts, there is a blocking primer designed to reduce amplification of vertebrate host DNA [30].

4. There are a large number of additional R packages for microbial ecology. A current comprehensive list is available [58], which is updated regularly by the European COST action network on Statistical and Machine Learning Techniques in Human Microbiome Studies.

5. Field sampling of amphibians requires regional permits and is often regulated out of concern for pathogen spread and population impacts. Please consult CITES (cites.org/eng) and regional wildlife authorities. Nitrile gloves are preferable for killing chytrid zoospores [8, 10], but rinsing bare hands with 70% ethanol and drying is an acceptable alternative [59]. All field equipment should be sanitized between uses and sites following established protocols. These include either complete drying (>48 h) between sites and/or treatment 3% bleach spray which can kill chytrid zoospores and *Ranavirus*, an amphibian and reptilian virus [19, 38]. Designating gear for specific sampling sites is the preferable way to reduce possible pathogen spread.

6. Metadata values must be specifically formatted in a table depending on the analytical software used later on. The Earth

Microbiome Project website has a metadata guide for formatting these values [30]. Additionally, Keemei is an open-source website for validating bioinformatic metadata spreadsheets for subsequent analysis in QIIME2 [74].

Conclusion

This guide is intended to give users tools for the identification of both specific symbionts as well as surveys of culturable and unculturable microbes from environmental samples. We hope that these protocols will instigate more research into the intersection of symbiotic microbiota with salamander biology. Investigations of microbial diversity have already revolutionized studies of salamander ecology and conservation [e.g., 6]. However, this remains a ripe frontier for salamander regeneration and developmental research [75].

Acknowledgments

This work was supported by the Gordon and Betty Moore Foundation and the Gettysburg College Cross-Disciplinary Science Institute. The authors would like to thank lab members River Larson-Pollock, Matt Urbano, and Yan Zhou from Gettysburg College for making this work possible and Hui Yang and Solange Duhamel from the University of Arizona for insights into these methods.

References

1. Arnold CP, Merryman MS, Harris-Arnold A, McKinney SA, Seidel CW, Loethen S, Proctor KN, Guo L, Sánchez Alvarado A (2016) Pathogenic shifts in endogenous microbiota impede tissue regeneration via distinct activation of TAK1/MKK/p38. eLife 5:e16793
2. Lee FJ, Williams KB, Levin M, Wolfe BE (2018) The bacterial metabolite indole inhibits regeneration of the planarian flatworm *Dugesia japonica*. Science 10:135–148
3. Demircan T, Ovezmyradov G, Yıldırım B, Keskin İ, İlhan AE, Fesçioğlu EC, Öztürk G, Yıldırım S (2018) Experimentally induced metamorphosis in highly regenerative axolotl (*Ambystoma mexicanum*) under constant diet restructures microbiota. Sci Rep 8:10974
4. Demircan T, İlhan AE, Ovezmyradov G, Öztürk G, Yıldırım S (2019) Longitudinal 16S rRNA data derived from limb regenerative tissue samples of axolotl *Ambystoma mexicanum*. Sci Data 6:1–7
5. Ellison S, Rovito S, Parra-Olea G, Vásquez-Almazán C, Flechas SV, Bi K, Vredenburg VT (2019) The influence of habitat and phylogeny on the skin microbiome of amphibians in Guatemala and Mexico. Microb Ecol 78:257–267
6. Rebollar EA, Antwis RE, Becker MH, Belden LK, Bletz MC, Brucker RM, Harrison XA, Hughey MC, Kueneman JG, Loudon AH, McKenzie V, Medina D, Minbiole KPC, Rollins-Smith LA, Walke JB, Weiss S, Woodhams DC, Harris RN (2016) Using "omics" and integrated multi-omics approaches to guide probiotic selection to mitigate chytridiomycosis and other emerging infectious diseases. Front Microbiol 7:68
7. Brucker RM, Harris RD, Schwantes CR, Gallaher TN, Flaherty DC, Lam BA, Minbiole KPC (2008) Amphibian chemical defense: antifungal metabolites of the microsymbiont *Janthinobacterium lividum* on the salamander

Plethodon cinereus. J Chem Ecol 24: 1422–1429

8. Avguštin JA, Bertok DZ, Kostanjšek R, Avguštin G (2012) Isolation and characterization of a novel violacein-like pigment producing psychotropic bacterial species *Janthinobacterium svalbardensis* sp. nov. Antonie Van Leeuwenhoek 103:763–769
9. Thomas V, Van Rooij P, Meerpoel C, Stegen G, Wauters J, Vanhaecke L, Martel A, Pasmans F (2020) Instant killing of pathogenic chytrid fungi by disposable nitrile gloves prevents disease transmission between amphibians. PLoS One 15:e0241048
10. Mendez D, Webb R, Berger L, Speare R (2008) Survival of the amphibian chytrid fungus *Batrachochytrium dendrobatidis* on bare hands and gloves: hygiene implications for amphibian handling. Dis Aquat Org 82: 97–104
11. Phillott AD, Speare R, Hines HB, Skerratt LF, Meyer E, McDonald KR, Cashins SD, Mendez D, Berger L (2010) Minimizing exposure of amphibians to pathogens during field studies. Dis Aquat Org 92:175–185
12. Jurga E, Graham L, Bishop CD (2020) *Oophila* is monophyletic within a three-taxon eukaryotic microbiome in egg masses of the salamander *Ambystoma maculatum*. Symbiosis 81: 187–199
13. Kerney R, Leavitt J, Hill E, Zhang H, Kim E, Burns J (2019) Co-cultures of *Oophila amblystomatis* between *Ambystoma maculatum* and *Ambystoma gracile* hosts show host-symbiont fidelity. Symbiosis 78:73–85
14. Reasoner DJ, Geldreich EE (1985) A new medium for the enumeration and subculture of bacteria from potable water. Appl Environ Microbiol 49:1–7
15. Bertani G (1951) Studies on lysogenesis I. J Bacteriol 62:293–300
16. Wantanabe MM, Kawachi M, Hiroki M, Kasai F (2000) NIES collection list of strains. National Institute for Environmental Studies, Japan
17. Boyle D, Hyatt A, Daszak P, Berger L, Longcore J, Porter D, Hengstberger SG, Olsen V (2003) Cryo-archiving of *Batrachochytrium dendrobatidis* and other chytridiomycetes. Dis Aquat Org 56:59–64
18. Blooi M, Pasmans F, Rouffaer L, Haesebrouck F, Vercammen F, Martel A (2015) Successful treatment of *Batrachochytrium salamandrivorans* infections in salamanders requires synergy between voriconazole, polymyxin E and temperature. Sci Rep 5: 11788
19. Rooji PV, Pasmans F, Coen Y, Martel A (2017) Efficacy of chemical disinfectants for the containment of the salamander chytrid fungus *Batrachochytrium salamandrivorans*. PLoS One 12:e0186269
20. Prunier J, Kaufman B, Grolet O, Picard D, Pompanon F, Joly P (2012) Skin swabbing as a new efficient DNA sampling technique in amphibians, and 14 new microsatellite markets in the alpine newt (*Ichthyosaura alpestris*). Mol Ecol Resour 12:524–531
21. Martel A, Spitzen-van der Sluijs A, Blooi M, Bert W, Ducatelle R, Fisher MC, Woeltjes A, Bosman W, Chiers K, Bossuyt F, Pasmans F (2013) *Batrachochytrium salamandrivorans* sp. nov. causes lethal chytridiomycosis in amphibians. PNAS 110:15325–15329
22. Kosch TA, Summers K (2012) Techniques for minimizing the effects of PCR inhibitors in the chytridiomycosis assay. Mol Ecol Resour 13: 230–236
23. Shin J, Bataille A, Kosch TA, Waldman B (2014) Swabbing often fails to detect amphibian chytridiomycosis under conditions of low infection load. PLoS One 9:e111091
24. Eissa AE, Abdelsalam M, Tharwat N, Zaki M (2013) Detection of *Saprolegnia parasitica* in eggs of angelfish *Pterophyllum scalare* (Cuvier-Valenciennes) with a history of decreased hatchability. Int J Vet Sci Med 1:7–14
25. Kostka M, Hampl V, Cepicka I, Flegr J (2004) Phylogenetic position of *Protoopalina intestinalis* based on SSU rRNA gene sequence. Mol Phylogenet Evol 33:220–224
26. Kerney R, Kim E, Hangarter RP, Heiss AA, Bishop CD, Hall BK (2011) Intracellular invasion of green algae in a salamander host. PNAS 108:6497–6502
27. Borneman J, Hartin RJ (2000) PCR primers that amplify fungal rRNA genes from environmental samples. Appl Environ Microbiol 66: 4356–4360
28. Kim E, Simpson AG, Graham LE (2006) Evolutionary relationships of apusomonads inferred from taxon-rich analyses of 6 nuclear encoded genes. Mol Biol Evol 23:2455–2466
29. Frank JA, Reich CI, Sharma S, Weisbaum JS, Wilson BA, Olsen GJ (2008) Critical evaluation of two primers commonly used for amplification of bacterial 16S rRNA genes. Appl Environ Microbiol 74:2461–2470
30. Thompson LR, Sanders JG, McDonald D et al (2017) A communal catalogue reveals Earth's multiscale microbial diversity. Nature 551: 457–463
31. Smith DP, Peay KG, Ackermann G, … Thompson L, Walters WA, White TJ, Earth

Microbiome Project Consortium (2018) EMP ITS Illumina amplicon protocol. Protocols.io https://doi.org/10.17504/protocols.io.pa7dihn

32. Amaral-Zettler LA, Bauer M, Berg-Lyons D, ... Thompson L, Vestheim H, Walters WA (2018) EMP I8S Illumina amplicon protocol. Protocols.io https://doi.org/10.17504/protocols.io.nuvdew6
33. Apprill A, McNally S, Parsons R, Weber L (2015) Minor revision to V4 region SSU rRNA 806R gene primer greatly increases detection of SAR11 bacterioplankton. Aquat Microb Ecol 75:129–137
34. Parada AE, Needham DM, Fuhrman JA (2016) Every base matters: assessing small subunit rRNA primers for marine microbiomes with mock communities, time series and global field samples: primers for marine microbiome studies. Environ Microbiol 18:1403–1414
35. Illumina (2013) 16S metagenomic sequencing library preparation: preparing 16S ribosomal RNA gene amplicons for the Illumina MiSeq system. Available at: https://support.illumina.com/documents/documentation/chemistry_documentation/16s/16s-metagenomic-library-prep-guide-15044223-b.pdf
36. Bolyen E, Rideout JR, Dillon MR et al (2019) Reproducible, interactive, scalable and extensible microbiome data science using QIIME 2. Nat Biotechnol 37:852–857
37. Liu C, Cui Y, Li X, Yao M (2021) microeco: an R package for data mining in microbial community ecology. FEMS Microbiol Ecol 97: fiaa255
38. Muletz-Wolz CR, DiRenzo GV, Yarwood SA, Campbell Grant EH, Fleischer RC, Lips KR (2017b) Antifungal bacteria on woodland salamander skin exhibit high taxonomic diversity and geographic variability. Appl Environ Microbiol 83:e00186–e00117
39. Zhou J, Nelson TM, Rodriguez Lopez C, Sarma RR, Zhou SJ, Rollins LA (2020) A comparison of nonlethal sampling methods for amphibian gut microbiome analyses. Mol Ecol Resour 20:844–855
40. Müller AS, Lenhardt PP, Theissinger K (2013) Pros and cons of external swabbing of amphibians for genetic analyses. Eur J Wildl Res 59: 609–612
41. Warne RW, Kirschman L, Zeglin L (2017) Manipulation of gut microbiota reveals shifting community structure shaped by host developmental windows in amphibian larvae. Integr Comp Biol 57:786–794
42. Nyholm SV (2020) In the beginning: egg–microbe interactions and consequences for animal hosts. Philos Trans R Soc Lond B Biol Sci 375: 20190593
43. Browne HP, Forster SC, Anonye BO, Kumar N, Neville BA, Stares MD, Goulding D, Lawley TD (2016) Culturing of 'unculturable' human microbiota reveals novel taxa and extensive sporulation. Nature 533: 543–546
44. Kim E, Lin Y, Kerney R, Blumenberg L, Bishop CD (2014) Phylogenetic analysis of algal symbionts associated with four North American amphibian egg masses. PLoS One 9:e1088915
45. Correia N, Pereira H, Silva JT, Santos T, Soares M, Sousa CB, Schüler LM, Costa M, Varela J, Pareira L, Silva J (2020) Isolation, identification, and biotechnological applications of a novel, robust, free-living *Chlorococcum (Oophila) amblystomatis* strain isolated from a local pond. Appl Sci 10:3040
46. Rodríguez-Gil JL, Brain R, Baxter L, Ruffell S, McConkey B, Solomon K, Hanson M (2014) Optimization of culturing conditions for toxicity testing with the alga *Oophila* sp. (Chlorophyceae), an amphibian endosymbiont. Environ Toxicol Chem 33:2566–2575
47. Robinson KA, Pereira KE, Bletz MC, Carter ED, Gray MJ, Piovia-Scott J, Romansic JM, Woodhams DC, Fritz-Laylin L (2020) Isolation and maintenance of *Batrachochytrium salamandrivorans* cultures. Dis Aquat Org 140: 1–11
48. Cook KJ, Voyles J, Kenney H, Pope KL, Piovia-Scott J (2018) Non-lethal isolation of the fungal pathogen *Batrachochytrium dendrobatidis* from amphibians. Dis Aquat Org 129: 159–164
49. Longcore J (2000) Culture techniques for amphibian chytrids: recognizing, isolating, and culturing *Batrachochytrium dendrobatidis* from amphibians. Proceedings of Getting the Jump! On Amphibian Diseases Conference, Cairns, Australia, p. 52–54
50. Yoon SH, Ha SM, Kwon S, Lim J, Kim Y, Seo H, Chun J (2017) Introducing EzBioCloud: a taxonomically united database of 16S rRNA and whole genome assemblies. Int J Syst Evol Microbiol 67:1613–1617
51. Dunitz MI, Lang JM, Jospin G, Darling AE, Eisen JA, Coil DA (2015) Swabs to genomes: a comprehensive workflow. PeerJ 3:e960
52. Singer GAC, Fahner NA, Barnes JG, McCarthy A, Hajibabaei M (2019) Comprehensive biodiversity analysis via ultra-deep

patterned flow cell technology: a case study of eDNA metabarcoding seawater. Sci Rep 9: 5991

53. Caporaso JG, Lauber CL, Walters WA, Berg-Lyons D, Lozupone CA, Turnbaugh PJ, Fierer N, Knight R (2010) Global patterns of 16S rRNA diversity at a depth of millions of sequences per sample. PNAS 108:4516–4522

54. Liu Z, Lozupone C, Hamady M, Bushman FD, Knight R (2007) Short pyrosequencing reads suffice for accurate microbial community analysis. Nucleic Acids Res 35:e120

55. Caporaso JG, Ackermann G, Apprill A ... Turnbaugh PJ, Walters WA, Weber L, Earth Microbiome Project Consortium (2018). EMP 16S Illumina amplicon protocol. Protocols.io https://doi.org/10.17504/protocols.io.nuudeww

56. Kozich JJ, Westcott SL, Baxter NT, Highlander SK, Schloss PD (2013) Development of a dual-index sequencing strategy and curation pipeline for analyzing amplicon sequence data on the MiSeq Illumina sequencing platform. Appl Environ Microbiol 79:5112–5120

57. Víquez-R L, Fleischer R, Wilhelm K, Tschapka M, Sommer S (2020) Jumping the green wall: the use of PNA-DNA clamps to enhance microbiome sampling depth in wildlife microbiome research. Ecol Evol 10: 11779–11786

58. Sudarshan (2018) List of R tools for microbiome data analysis. Available at: https://microsud.github.io/Tools-Microbiome-Analysis/

59. Schloss PD, Westcott SL, Ryabin T, Hall JR, Hartmann M, Hollister EB, Lesniewski RA, Oakley BB, Parks DH, Robinson CJ, Sahl JW, Stres B, Thallinger GG, Horn DJV, Weber CF (2009) Introducing Mothur: open-source, platform-independent, community-supported software for describing and comparing microbial communities. Appl Environ Microbiol 75: 7537–7541

60. Dhariwal A, Chong J, Habib S, King IL, Agellon LB, Xia J (2017) MicrobiomeAnalyst: a web-based tool for comprehensive statistical, visual and meta-analysis of microbiome data. Nucleic Acids Res 45:W180–W188

61. Callahan BJ, McMurdie PJ, Rosen MJ, Han AW, Johnson AJA, Holmes SP (2016) DADA2: high-resolution sample inference from Illumina amplicon data. Nat Methods 13:581–583

62. Callahan BJ, McMurdie PJ, Holmes SP (2017) Exact sequence variants should replace operational taxonomic units in marker-gene data analysis. ISME J 11:2639–2643

63. Balvočiūtė M, Huson DH (2017) SILVA, RDP, Greengenes, NCBI and OTT: how do these taxonomies compare? BMC Genomics 18:114

64. Yilmaz P, Parfrey LW, Yarza P, Gerken J, Pruesse E, Quast C, Schweer T, Peplies J, Ludwig W, Glöckner FO (2014) The SILVA and "All-species Living Tree Project (LTP)" taxonomic frameworks. Nucleic Acids Res 42: D643–D648

65. Nilsson RH, Larsson KH, Taylor AFS, Bengtsson-Palme J, Jeppesen TS, Schigel D, Kennedy P, Picard K, Glöckner FO, Tedersoo L, Saar I, Kõljalg U, Abarenkov K (2019) The UNITE database for molecular identification of fungi: handling dark taxa and parallel taxonomic classifications. Nucleic Acids Res 47:D259–D264

66. Werner JJ, Koren O, Hugenholtz P, DeSantis TZ, Walters WA, Caporaso JG, Angenent L, Knight R, Ley R (2012) Impact of training sets on classification of high-throughput bacterial 16s rRNA gene surveys. ISME J 6:94–103

67. Liu K, Wong T (2013) Naïve Bayesian classifiers with multinomial models for rRNA taxonomic assignment. IEEE/ACM Trans Comput Biol Bioinform 10:1334–1339

68. Pedregosa F, Varoquaux G, Gramfort A, Michel V, Thirion B, Grisel O, Blondel M, Prettenhofer P, Weiss R, Dubourg V, Vanderplas J, Passos A, Cournapeau D, Brucher M, Perrot M, Duchesnay E (2011) Scikit-learn: machine learning in Python. J Mach Learn Res 12:2825–2830

69. Walters KE, Martiny JBH (2020) Alpha-, beta-, and gamma-diversity of bacteria varies across habitats. PLoS One 15:e0233872

70. Jost L, Chao A, Chazdon RL (2010) Compositional similarity and β (beta) diversity. In: Biological diversity: frontiers in measurement and assessment. Oxford University Press, pp 66–84

71. Bokulich NA, Dillon MR, Zhang Y, Rideout JR, Bolyen E, Li H, Albert PS, Caporaso GJ (2018) q2-longitudinal: longitudinal and paired-sample analyses of microbiome data. mSystems 3:e00219–e00218

72. Anderson MJ (2001) A new method for non-parametric multivariate analysis of variance: non-parametric MANOVA for ecology. Austral Ecol 26:32–46

73. Langille MGI, Zaneveld J, Caporaso JG, McDonald D, Knights D, Reyes JA, Clemente JC, Burkepile DE, Vega Thurber RL, Knight R, Beiko RG, Huttenhower C (2013) Predictive functional profiling of microbial communities

using 16S rRNA marker gene sequences. Nat Biotechnol 31:814–821

74. Rideout JR, Chase JH, Bolyen E, Ackermann G, González A, Knight R, Caporaso JG (2016) Keemei: cloud-based validation of tabular bioinformatics file formats in Google Sheets. GigaScience 5:27

75. Kerney R (2021) Developing inside a layer of germs—A potential role for multiciliated surface cells in vertebrate embryos. Diversity 13:527. https://doi.org/10.3390/d13110527

Chapter 29

Assessing Leukocyte Profiles of Salamanders and Other Amphibians: A Herpetologists' Guide

Andrew K. Davis and John C. Maerz

Abstract

Assessing numbers of leukocytes in salamanders and other amphibians can be useful metrics for understanding health or stress levels of individuals in a population. In this chapter we describe the procedures for obtaining blood samples from amphibians, preparing blood films for microscopy, counting, and identifying cells. We also provide reference values for amphibian leukocytes for use in interpreting leukocyte data. From our assessment of the published and unpublished literature, "non-stressed" salamanders would have a leukocyte profile where 60–70% of cells are lymphocytes, 17–30% are neutrophils, 1–4% are eosinophils, 4–12% are basophils, and 2–6% are monocytes. In *Ambystoma* spp., the eosinophil abundance can be notably higher (30% of all white blood cells), for reasons unknown. Finally, the neutrophil–lymphocyte ratio of most non-stressed salamanders tends to be between 0.3 and 0.4 (sometimes less), while the ratios of stressed salamanders tend to be over 1.0.

Key words Leukocyte profiles, Salamander, White blood cells, Neutrophil–lymphocyte ratios, Stress indices, Microscopy, Blood smears, Physiology

1 Introduction

For the herpetologist studying salamanders (or other amphibians), it is useful to have some way(s) of ascertaining the physiological status of the animals under study. Doing so can yield insights into how individuals or populations are faring under different natural and anthropogenic pressures, for example. It can also be of value to know if the animals under study are "stressed" by the pressures they are exposed to, or in other words, are they currently experiencing a physiological stress reaction. While there are a variety of ways and procedures for obtaining this information, with each varying in level of training or sophistication, herpetologists are increasingly turning to the evaluation of "leukocyte profiles" as part of their research projects [e.g., 1–9]. This approach involves obtaining small blood samples from animals to make blood smears (sometimes called films), which are examined under a compound

Ashley W. Seifert and Joshua D. Currie (eds.), *Salamanders: Methods and Protocols*,
Methods in Molecular Biology, vol. 2562, https://doi.org/10.1007/978-1-0716-2659-7_29,

microscope to count the number of white blood cells (leukocytes). The abundance of leukocytes in salamanders, and other vertebrates, if measured and interpreted properly, can provide information about the health status and stress levels of animals in a population (below). This approach is also fairly inexpensive, requiring only a microscope and some glass slides. Moreover, the technique can be less sensitive to handling and sample collection time, alleviating some of the challenges associating with direct measures of stress hormones, and has the advantage of measuring relevant physiological responses to stress rather than simply concentrations of circulating hormones. For the herpetologist then, this procedure can be an extremely valuable part of their research projects, especially if blood is being taken anyway as part of other procedures.

There are five major types of mature leukocytes in amphibians which serve as one of the major elements of the vertebrate immune system and these cell types are typical of those found in most other vertebrate taxa. While each white blood cell has a different "role" in the immune response, determining the exact responsibilities of each cell is still an active line of research in vertebrate immunology even within human medicine [10–12]. For details of the different cell type-specific roles we refer the reader to several veterinary textbooks that we have identified in Table 1 which have at least some information devoted to amphibian biology. Nevertheless, since the cells in amphibians are really not that different from those found in mammals, even reference texts devoted solely to mammals can be useful. We at least outline below the general roles, if known, for the different cells, which are helpful to know when interpreting the data obtained when one conducts a leukocyte assessment.

Neutrophils are phagocytic cells that are typically one of the first to respond to infections [13, 14]. Thus, higher than normal neutrophils can be a sign of infection. Similarly, monocytes (also called macrophages) are also phagocytic and can respond to infections [13, 14], though these cells are usually fewer in number than neutrophils. Therefore, increased neutrophil numbers and, to a lesser extent, monocyte numbers can indicate infection. Lymphocytes are involved in immune memory with different categories of lymphocytes performing different memory functions (B lymphocytes, T lymphocytes, etc.). However, these different categories cannot be distinguished on a simple stained blood film [15]. Eosinophils are thought to play a role in defense against multicellular parasites and good evidence for this actually comes from research on amphibians [16–18]. Other work, however, suggests that eosinophils do not always respond similarly to parasite infections across amphibian species [19], which highlights how this knowledge is still evolving. In amphibians, eosinophils also appear to have a role during metamorphosis; acute elevations in this cell during metamorphosis have been documented in ambystomid

Table 1
Partial list of hematology information for herpetofauna or other vertebrates

Source title	References	Notes
Atlas of adult *Xenopus laevis* hematology	[47]	This is an article in a comparative physiology journal rather than a standalone book, but it is an excellent source for photographs and descriptions of anuran red and white blood cells.
Hematology techniques and concepts for veterinary technicians	[48]	Although this book is intended for veterinarians and therefore mammalian hematologists, the concepts for mammals (and the blood cell morphology) are similar for amphibians. Additionally, there are good descriptions of techniques such as making and staining blood smears, and microscopy.
Veterinary hematology and clinical chemistry	[15]	This veterinary textbook has a chapter devoted to amphibian hematology that covers the basic concepts and gives brief descriptions of blood cells and their function.
Amphibian medicine and captive husbandry	[34]	This is also a book with a chapter on amphibian hematology that covers the basics of cell identification and function.
The Wildlife Leukocytes Webpage (www.wildlifehematology.uga.edu	[42]	A website by one of the authors that contains photographs of amphibian blood cells and listings of published leukocyte data from amphibians. These data can aid researchers in interpretation of hematological data from their study species
Clinical pathology of amphibians: a review	[49]	An open-access article in the journal, Veterinary Clinical Pathology, written for the veterinary professional, but with detailed instructions and descriptive photos of techniques

salamanders [20] and in ranid frogs [21]. The role of basophils is still under investigation, though they are not known to become elevated during trauma, stress, or illness [15]. Photos and descriptions of these cells are provided below under Microscopy and Blood cell identification tips.

For the researcher studying salamanders, knowing how many of each of these cells are present in the blood of each study animal can provide important information about what the animals are currently dealing with (if anything), from an immunological standpoint. In general, higher than normal abundance of leukocytes is consistent with a general inflammatory response (i.e., response to an infection, or injury). Consider the following examples using amphibians. Evaluation of multiple internal parasites of red-backed salamanders (*Plethodon cinereus*) revealed how parasitism leads to increased leukocytes in circulation [8]. Infection with a seemingly unremarkable and miniscule blood-borne rickettsial

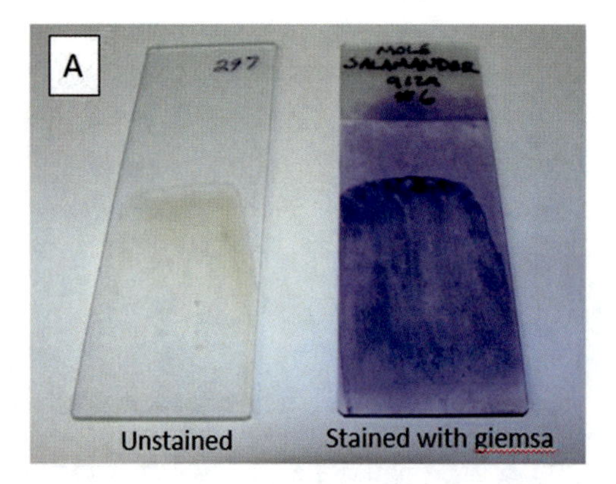

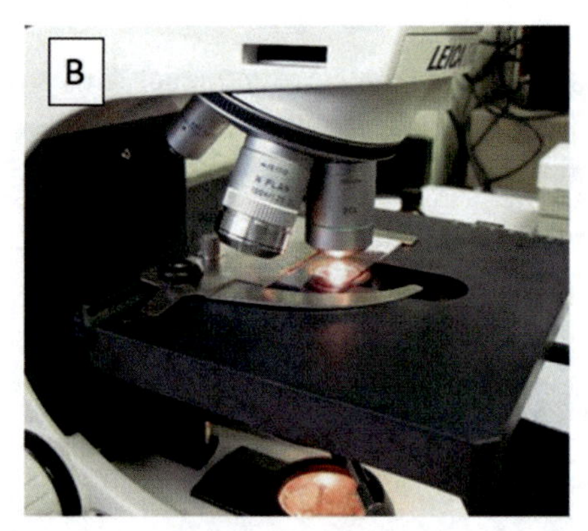

Fig. 1 (**a**) Blood smears of amphibians for assessing leukocyte profiles. Left is an air-dried, unstained smear. Right is a smear that has been stained with giemsa. (**b**) Standard light microscope used for reading amphibian blood smears, with a giemsa-stained smear on the stage. Typically, the highest magnification used is 1000× oil immersion (i.e., 100× objective × 10× eyepiece)

bacteria in red-backed salamanders was also associated with greater abundance of leukocytes [22]. Similarly, infections with lung trematodes were associated with high leukocyte concentrations among bullfrogs [23].

Another use of leukocyte data that have had a long history in other taxa but are now being revived in the world of herpetology is estimating physiological stress levels [reviewed in ref. 24]. When animals become stressed their level of plasma stress hormones (corticosterone in amphibians and reptiles) increases, which then causes a chain reaction of events in the body, which are all designed to aid the animal in dealing with the stressor [reviewed in ref. 25]. One of these events is a temporary alteration in the numbers of two of the five white blood cell types in the circulating bloodstream. Specifically, elevations in stress hormones cause the abundance of neutrophils to increase in circulation and the numbers of lymphocytes to decrease. From reviewing a considerable amount of current and historical literature, it was determined that this specific leukocyte alteration is typical of all animal taxa, including amphibians [24]. The reason for this alteration is thought to be a mechanism for redistributing the different types of cells to where they would most be needed during the stress event [26, 27]. For the amphibian researcher though, this effect provides a convenient way to indirectly estimate levels of stress by counting numbers of white blood cells from standard blood films under a light microscope (Fig. 1). Because stress alters these cell numbers in opposite directions, the ratio of the two (neutrophil–lymphocyte or "N/L" ratio) can be used as a "hematological stress" index.

It is important to point out that this measure of stress in animals has some key differences over the more conventional approach of directly measuring the concentration of corticosterone in blood, and this should be considered during interpretation of leukocyte data. In other words, the two approaches are not necessarily interchangeable [reviewed in ref. 28]. In fact, most studies that have evaluated both corticosterone concentration and N/L ratios in animals have shown that the two are rarely correlated during routine collections of animals. This is thought to be because of the temporal nature of the two responses to the stressor in question [reviewed in ref. 28]. Briefly, the level of corticosterone in animals increases quickly following a stressor (minutes), but then typically wanes after a few hours. Meanwhile, the leukocyte reaction has a longer lag time before increasing (*see* below in Animal Handling) and then remains high for many hours to days, even after hormone levels return to baseline. Thus, in any given animal population, those animals sampled would each have probably had different recent histories of stressors and would be each experiencing different temporal stages of the stress reaction (or none at all). For more details of this temporal difference, plus discussion of other approaches to measuring short- vs long-term stress in animals, the reader is encouraged to consult a recent review by Gormally and Romero [29]. Because responses are rapid, direct measurement of corticosterone is challenging, particular for researchers working in the field where trapping and handling animals results in rapid hormone increases. Therefore, one advantage of using N/L ratios from leukocyte profiles is it is relatively insensitive to the immediate effects of handling.

2 Materials

2.1 Blood Collection

1. Standard glass microscope slides, with frosted edges for labeling.
2. Slide case for storing and organizing prepared slides.
3. Standard dissection tools, including scissors.
4. Sterile, 26-gauge needles (one for each animal to be sampled).
5. Heparinized microcapillary tubes (these come in packs of 200).
6. Stain for slides. There are a variety of types of stain kits available, and each requires different numbers of steps. Staining the slides soon after preparation usually produces the best results, so fewer, simpler steps is an advantage. In the authors' experience, the most user-friendly and quickest product is Camco Quik-Stain II (Cambridge Diagnostic Products, Inc.™, Fort Lauderdale, FL, USA).
7. Coplin slide staining jar. These come in glass or plastic. Plastic is preferable if taking it into the field.
8. Container of rinse water. This can be a simple Tupperware container, or any container that is OK to get stain on.

2.2 Microscopy

1. Standard light microscope, with at least 3 objective lenses, and 1 that is 1000× oil immersion. A unit with a trinocular head is ideal, for attachment of a camera, and for generating real-time images of microscope fields on a connected desktop.
2. Immersion oil (for use at the 1000× magnification only).
3. Clickers for enumerating cells (optional).

3 Methods

3.1 Blood Collection

3.1.1 Terminal Blood Collection

If the salamander is to be euthanized, one of the most effective ways to obtain a usable blood sample is to obtain it directly from the heart after euthanasia. In salamanders, the heart is located just posterior to the head and is most accessible in the ventral side. Expose the heart using a scalpel and nick the ventricle so blood wells from the nick. Siphon the blood using a microcapillary tube [30]. Sampling directly from the heart also ensures that the blood is pure and not contaminated with other body fluids such as lymph or even water.

3.1.2 Non-lethal Blood Collection

If the animals under study are *not* to be euthanized, there are alternative procedures for obtaining non-lethal samples, though the amount of blood obtained is usually less, they may only be effective for particular species and larger animals, and, these procedures typically require more skill and practice (Davis, *pers. obs.*). For aquatic salamanders with gills, it is possible to obtain just enough blood for a small smear (~5 μl) by clipping a frond of the gill, and then siphoning the welled blood with a microcapillary tube [31]. For larger species of terrestrial salamanders (e.g., *Ambystoma* spp.), some researchers have been able to obtain non-destructive samples by puncturing the caudal vein with a 21-guage needle and siphoning the welling blood with a microcapillary tube [32]. For more information on blood sampling amphibians, or slide preparation, a listing of sources with good information is provided in Table 1.

3.2 Slide Making and Preparation

Once blood has been obtained, either in a syringe or a capillary tube (this one is preferred by the authors), it needs to be immediately used to create a blood smear. There are numerous instructional videos on the internet on how to create blood smears, since they are routine tools in human and veterinary medicine. The most common technique to use is the "two-slide wedge" procedure, where a drop of blood is placed on one end of a microscope slide, and another slide is used to push the drop so that it spreads evenly along the base slide. A lesser-known procedure, that is often used by the authors, is to place a drop of blood on the center of one slide, then place another slide on top to "sandwich" the blood drop, and spread it out radially. Then with a smooth motion draw the top slide off to spread the blood drop across the base slide. The goal of either method is to obtain a thin monolayer of cells that are evenly distributed on the slide. Hesitation or rough spreading can lead to clumps of blood on the smear that are too thick to view individual cells. The most appropriate method may depend on researcher preference, though a thorough investigation of techniques for making smears (for iguanas) did show that the two-slide wedge method produces superior slides, i.e., less cell damage and a more even cell distribution [33].

The final step is to stain the slide, which facilitates identification of blood cells under light microscopy. With the Camco Quik Stain, the blood smear is dipped in a coplin jar containing the stain for 15 s, and then removed and swirled in a second container of rinse water. Other stain products may require longer immersion. Slides are then placed to air dry, and once dry, should be stored in dry slide cassettes such that slides do not touch or rub on adjacent slides. Slide cassettes should be stored in dry areas and avoid exposure to high temperatures, humidity, or light. Slides that are stored in a humid environment can become colonized by bacteria. *See* Fig. 1 for images of stained and unstained slides.

3.3 Microscopy

Stained blood films should be examined with a standard light microscope under oil immersion at 1000× magnification (Fig. 1). Within each field of view, all visible leukocytes of each type should be counted. Recall that the five types of amphibian leukocytes are neutrophils, lymphocytes, eosinophils, basophils, and monocytes [14, 34], and these are shown in Fig. 2. Tips for cell identification are given below. For all blood films, the slide should be moved on the microscope stage in a standard zig-zag pattern, and all leukocytes counted in each field of view until at least 100 cells are counted. If the WBCs are sparse on the slide, one can usually cease counting when 150 fields of view have been examined (the latter is the authors' own rule of thumb). Since the objective is to obtain data on the relative proportion of each cell type, then obviously, the more cells counted, the better to obtain these values. Conversely, undercounting leukocytes will increase the uncertainty in measurements, leading to higher variances that might obscure differences. This is especially important to consider if more than one observer is examining the slides. With these cell estimates in hand, the percentage of each cell type can be calculated for each individual amphibian and the neutrophil–lymphocyte ratio obtained based on the percentages of these two cell types.

3.4 Blood Cell Identification Tips

Of the entire leukocyte profiling process, perhaps the element with the steepest learning curve is the actual identification and enumeration of cell abundance under the microscope. For the average herpetologist, who perhaps is not a student of veterinary medicine,

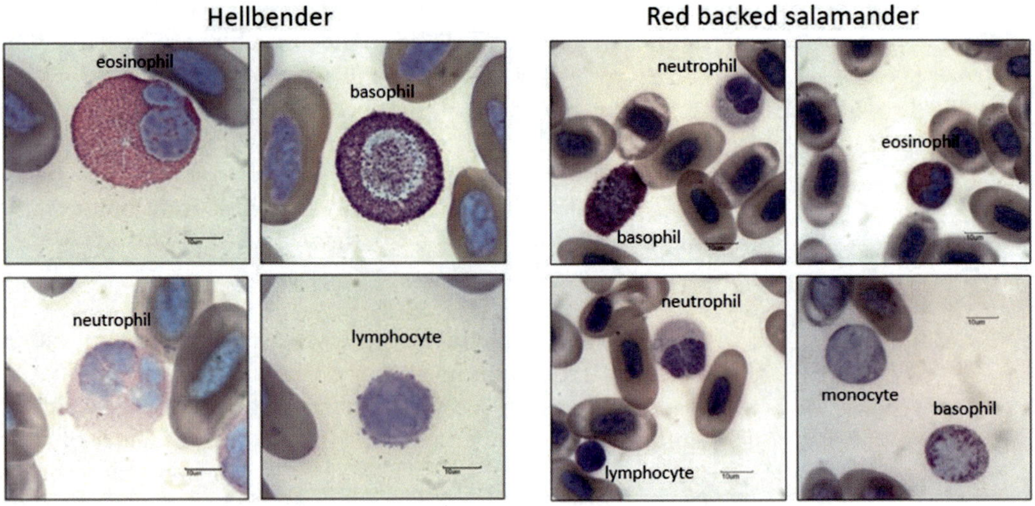

Fig. 2 Photomicrographs of example white blood cells from two representative salamanders: *Cryptobranchus alleganiensis* and *Plethodon cinereus*. Shown are eosinophils, basophils, lymphocytes, neutrophils, and monocytes. All images = 1000×. Note the general similarities in appearance of cells from each species (especially color), though the overall sizes may differ. Hellbenders in fact have some of the largest cells of any living animal species

this can be daunting at first, but after some practice (maybe 2–3 days of practice) this part can be accomplished with accuracy and consistency. In the sources listed in Table 1, one can find descriptions of all five cell types of amphibians. In the following section, we provide our own cell identification tips, garnered from experience, for aiding new users of this technique. And, note that these tips are based on our experience in assessing *amphibian* leukocytes. Refer also to Fig. 2 of this article for photographs of cells from two representative salamander species.

Lymphocytes.– These cells are usually the most numerous of all WBCs. The cells tend to be smaller than, or nearly the same size as, erythrocytes. They are usually blue in appearance. Importantly, the nucleus takes up the majority of the cell, such that there is only a thin band of cytoplasm around the edge. Note that this is the surest way to distinguish between lymphocytes and monocytes, which have some similarities, at least in staining color.

Neutrophils.– These cells tend to be rounded, and usually larger than erythrocytes. They can have small, wavy, cytoplasmic extensions. They are conspicuous by their pinkish cytoplasm and multi-lobed, darker nucleus. The number of lobes of the nucleus is actually a diagnostic feature—immature neutrophils tend to have no lobes, and older cells have many. In some cases, these cells can be faded if the stain did not take properly, so care must be taken not to miss them. In animals facing a severe illness these cells can take on a darker appearance.

Eosinophils.– These cells have very conspicuous orange-staining, round granules, which usually obscure all else within the cell. The cells can be round is some species, but tend to be oblong in *Ambystoma*. Also in the genus *Ambystoma*, they make up an unusually large proportion of the WBCs, where 30% of the leukocytes are eosinophils [7].

Basophils.– These cells are nearly always purple after staining—in fact they are usually the only purple cells on the entire smear. The purple comes from the cytoplasmic granules, which make up most of the cytoplasm. These cells are usually round, but they can take on oblong shapes. Note the morphological difference between the two species in Fig. 2, but that in both, the cell is purple.

Monocytes.– These cells are usually much larger than erythrocytes and other WBCs. They are usually round, but can have cytoplasmic pseudopodia. The cytoplasm is bluish-gray. Importantly, the ratio of nucleus to cytoplasm is close to 50–50 (this is a key distinction between this and lymphocytes). The nucleus appearance can be mottled. The cytoplasm often has vacuoles and material from phagocytosis.

Thrombocytes.– This is one more cell type that can be seen in an amphibian blood smear, though these are not leukocytes. Thrombocytes are the amphibian equivalent to mammalian platelets. These cells can be mistaken for lymphocytes; they are usually smaller than erythrocytes, and can be rounded and bluish-color. However, they usually have some faint pink granules in the cytoplasm, and they tend to be found in groups of cells stuck together. Note that some older studies of amphibians have counted these cells alongside the leukocyte numbers [e.g., 35], though they may not have any diagnostic value.

4 Notes

4.1 Animal Handling

1. The status of the animals immediately prior to blood collection must be considered by herpetologists, because of the known impact of stress to leukocyte profiles. For example, if the goal is to determine how habitat alterations affect leukocyte profiles of salamanders, then the researcher needs to know that the data obtained are not artifacts of the collection procedure, but are in fact reflective of the habitat differences. The only way to really accomplish this is to obtain blood samples "soon" after capture (*see* below). Long-distance transport or storage of animals overnight prior to blood sampling can lead to stress-related changes in abundance of leukocytes, which would swamp any variation due to habitat (to use this same example).
2. How soon should the blood collection be performed? The temporal nature of the stress reaction on *amphibian* leukocytes

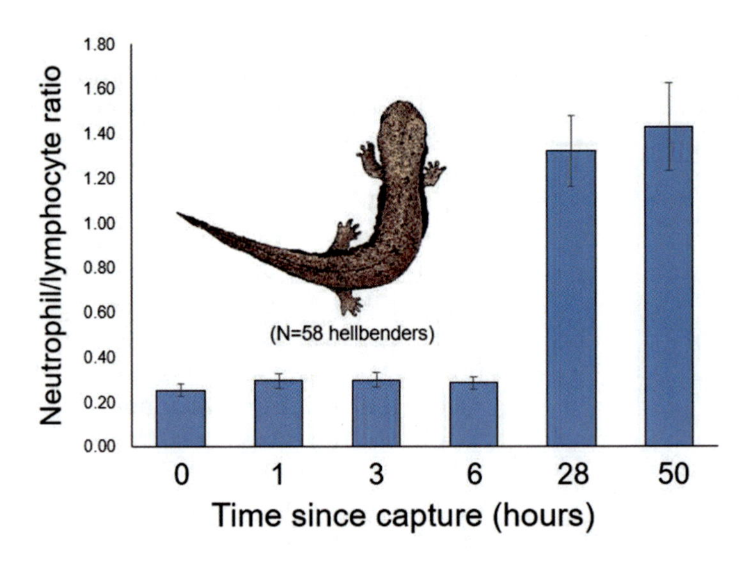

Fig. 3 Neutrophils to lymphocyte ratios for 58 hellbenders sampled serially after capture, showing the temporal nature of the leukocyte response to acute stress in amphibians. Hellbenders were collected as part of an ongoing research project in the state of Virginia by Bill Hopkins. Leukocyte profiles assessed by the author (Davis). Bars show the mean N/L ratio at each time point. Whiskers around means indicate standard errors. Note that N/L ratios after capture were close to 0.3, which is typical for non-stressed amphibians

is something that has been thoroughly researched both by prior workers and the authors [36–40]. From this body of work, it is clear that (1) the capture, handling, and transport of amphibians *do* alter the leukocyte profile, but (2) the timing of this can take *many hours or sometimes days* in amphibians. A useful illustration of this temporal effect is shown in Fig. 3. Here, the author (Davis) evaluated leukocytes of eastern hellbenders that were captured as part of an ongoing study and sampled serially over 2 days. Note that the average N/L ratio of the animals did not rise for at least 6 h after capture (and after 3 prior blood draws!), but then was dramatically higher after 28 h [41]. Thus, herpetologists should expect that leukocyte profiles will change over time in the animals they capture, though the time lag of the response is such that the animals may not need to be handled quickly to obtain baseline data. In fact, we recommend that the stress of capture, handling, and/or transport is negligible in cases where animals are sampled within 1–3 h of capture.

4.2 Slide Preparation

1. We add a cautionary note about slide preparation for researchers with little experience in this area. While these procedures may sound straightforward, we have found that there can be numerous (seemingly minor) methodological issues that can cause slides to be completely unusable, and these *must* be

sorted out prior to beginning a field or lab project. Finding out later that all of the 100 or so slides that have just been made for a project are unusable is not a desirable outcome for any researcher, and we point out that this has actually happened more than once in our lab! Thus, we encourage researchers to conduct some preliminary tests of their blood collection and slide preparation procedures, and most importantly, examine the slides under a microscope, to see if they are of sufficient quality for research (i.e., that leukocytes are present and not ruptured, and that they stained well enough to be identified). We have found that problems with collection of usable slides can occur at any point in the process. Sometimes the issue is related to the technique used for making the smear, or it can be related to the staining step, or even the storage of slides later (if slides are stored in a humid environment, they can become colonies by bacteria). Thus, we encourage trying a variety of techniques to ensure the best result, which is a "usable" blood smear.

2. Once the blood smear has been made the slide needs to air dry. Importantly, in veterinary medicine, a common practice is to then dip the smear in methanol to "fix" it, or render it permanent. However, in the authors' experience, this practice is not necessarily suitable for amphibian blood, as this can actually cause more problems during the staining step (below). Specifically, sometimes methanol-fixed slides fail to take up the stain.
3. Camco Quik-Stain II contains methanol, which fixes the slides as part of the staining process. A benefit of this approach is that it can be done relatively easily in the lab or the field.
4. In some veterinary or biomedical circles, blood smears are capped with a cover slip. This is unnecessary in the authors' experience, and can sometimes interfere with cell observation at 1000× magnification.

4.3 Microscopy

1. While searching for leukocytes, observers will also notice nucleated thrombocytes, which are NOT leukocytes, but are the amphibian analog to mammalian platelets [14]. These are sometimes confused with small lymphocytes. These are not part of a routine leukocyte profile.

4.4 Interpreting Cell Data

1. When groups of aquatic salamanders are collected from ponds and sampled for leukocytes, the data obtained are difficult to interpret unless the investigator knows what the "normal" proportion of each cell type is in the species in question. For the vast majority of amphibian species, there is no "reference" data of normal cell numbers to consult for this, like there is with domestic mammalian or bird species. In response to this problem, one of us (Davis) has spent considerable time

compiling the information that is available on amphibian blood cell profiles, in an effort to at least reveal if there are consistencies within taxonomic groups. If so, that would mean that the entire taxonomic group can at least serve as a reference, in lieu of having data from the species in question. The results of this effort have been made public on a website so that other researchers can consult it [www.wildlifehematology.uga.edu, 42], which is an online compilation of published and unpublished records of leukocyte data of amphibians, birds, and reptiles. The lists for each taxon were compiled by searching the scientific literature for records where species or groups of species were examined and their white blood cell profiles reported. For each species studied, the average *percentage* of each white blood cell type is listed, no matter how many individuals were examined. In some cases, over 300 animals were examined, while in others less than five were. Also included in the tables are certain unpublished records from the author's own laboratory investigations of amphibians. Currently, there are 52 records of 29 amphibian species, representing more than 1700 individuals.

Sometimes there can be large discrepancies in the relative number of certain cell types reported, particularly for the neutrophils and lymphocytes. An example of this is the multiple records from American bullfrogs (*Rana catesbeiana*). Coppo et al. [43] report that the percentage of neutrophils and lymphocytes in this species is 26.8 and 60.9, respectively. However, Cathers et al. [44] report values of 62.9 and 22.0 (nearly opposite cell values). Prior work showed average lymphocyte and neutrophil percentages in adult bullfrogs are 67.4 and 8.8 [21]. These differences are no doubt the result of different handling or husbandry procedures being used in each study, and the influence of stress in the animals on their leukocyte distributions. If one examines these bullfrog papers more closely, this effect is clear. In the Coppo et al. study, all frogs were transported 24 h before sampling, which would increase stress levels. In the Cathers et al. study, all frogs were captive (and presumably acclimated to captivity) and anesthetized before sampling. In the Davis study of bullfrogs, which had the lowest percentage of neutrophils, the highest percentage of lymphocytes, and the lowest overall neutrophil–lymphocyte ratio, all frogs were captured in the wild and sampled immediately in the field. This scenario illustrates the magnitude of the stress effect on leukocytes and is exactly why one must always take into account the methods used in each published study before taking the reported cell values into consideration.

2. For the purposes of this chapter, we derived a simple reference table for the herpetologist to consult, which would provide a point of comparison for their records (Table 2). For this table

Table 2
Summary of "reference" leukocyte profiles of anurans, urodeles (without *Ambystoma*), and *Ambystoma* salamanders, based on compilation of published data, listed at www.wildlifehematology.uga.edu. *N* is the number of records included in the calculation

Group	Cell Type	*N*	Mean	lower CI	Upper CI	Minimum	Maximum
Anurans	Lymphocytes	22	62.9	57.7	68.2	30.1	83.3
	Neutrophils	22	18.9	15.9	21.8	7.0	27.3
	Eosinophils	22	6.5	4.4	8.5	1.1	17.0
	Basophils	22	8.2	4.4	12.0	0.0	40.5
	Monocytes	22	3.6	1.2	6.0	0.0	22.7
Urodeles	Lymphocytes	10	66.7	60.4	73.0	55.6	88.0
	Neutrophils	10	23.8	17.7	30.0	7.0	37.6
	Eosinophils	10	2.7	1.4	4.0	0.9	6.2
	Basophils	10	2.6	0.7	4.5	0.0	8.8
	Monocytes	10	4.1	1.9	6.3	0.8	9.3
Ambystoma	Lymphocytes	8	46.6	39.3	53.9	31.7	59.0
	Neutrophils	8	14.3	10.7	17.8	5.7	19.6
	Eosinophils	8	29.6	19.6	39.6	19.6	51.2
	Basophils	8	9.2	2.7	15.7	0.0	24.2
	Monocytes	8	0.4	0.1	0.7	0.0	1.0

we used the aforementioned published records of leukocyte profiles of amphibians, which amounted to 40 records representing 25 species of amphibians, both anurans and urodeles. We did not include records where the N/L ratio was greater than 1, as these animals were likely stressed. In both the anurans and urodeles, average percentages of each cell type appear similar, with the majority of cells being lymphocytes, followed by neutrophils, and a low percentage of the other three types. Neutrophil–lymphocyte ratios of all amphibians ranged from 0.08 to 0.88. Of the 40 records of amphibian species, the majority reported ratios between 0.30 and 0.39. Further, despite the different percentages of leukocytes among *Ambystoma* salamanders, the neutrophil–lymphocyte ratios of all amphibian groups were similar.

The white blood cell values shown in Table 2 could be viewed as typical for most amphibians. Two pieces of evidence can be provided to justify this. First, these averages are derived from many collections of amphibians, spanning the Hylidae, Ranidae, Bufonidae, and Plethodontidae families. Second, when the author

Davis has collected data on non-breeding and otherwise unstressed amphibians, regardless of the species, the leukocyte profiles obtained are usually close to these means. As an example, this author had collected 11 *Eurycea wilderae* salamanders as part of a research study. These animals came from a relatively high-quality stream free of anthropogenic disturbance, and, blood was obtained within minutes of capture. Their average leukocyte profile was 68.7% lymphocytes, 22.3% neutrophils, 1.5% eosinophils, 5.0% basophils, and 2.4% monocytes, and the group-wide average neutrophil–lymphocyte ratio was 0.32. Note that these values are all very close to the means for other urodeles reported in Table 2.

3. One final point to consider when interpreting leukocyte data from amphibians, and especially the N/L ratio, is that statistically significant increases in N/L ratios in a particular group of animals may not necessarily mean that this group is "stressed," especially if the elevation is minor. In other words, minor differences in N/L ratios can occur but may not be biologically important differences. To know if an animal is truly undergoing a physiological stress reaction (complete with leukocyte shifts) one needs to first understand how much the N/L ratio changes after known stressors are applied. For example, among *Ambystoma* salamanders, in which typical N/L ratios are close to 0.3, capture and injection with saline resulted in a 4-fold increase in N/L ratios over baseline after 24 h [40], so that the "stressed" ratios were usually over 1.0. In another species of *Ambystoma*, capture and application of a different stressor (chasing for 30 s) also led to a 4-fold increase in N/L ratios after 24 h [45], going from 0.3 to over 1.0. For anurans, the same scenario has been found; capture and restraint of cururu toads (*Rhinella icterica*) for 24 h in moistened cloth bags also resulted in a 4-fold increase in N/L ratios [46]. Thus, one should interpret salamander N/L ratios that are closer to, or over 1.0, as evidence that the animals in question are truly "stressed." To further illustrate this, note the N/L ratios of hellbenders shown in Fig. 3, and how they began at or near 0.3 (unstressed), and then when the leukocyte reaction had occurred by the following day, they were over 1.0.

If finding high N/L ratios in an amphibian population is an indication of stress, then the opposite holds as well. A recent evaluation of leukocyte profiles of red-backed salamanders from Mountain Lake, Virginia, which is in the core range of this species and in ideal montane habitat, revealed these salamanders have one of the lowest average N/L ratio (0.05) yet reported for any salamander population or species [8], which was interpreted as evidence that this population is remarkably stress free! For any population of animals, this would be the ideal scenario.

References

1. Forson D, Storfer A (2006) Atrazine increases ranavirus susceptibility in the tiger salamander, *Ambystoma tigrinum*. Ecol Appl 16(6): 2325–2332
2. Gervasi SS, Foufopoulos J (2008) Costs of plasticity: responses to desiccation decrease post-metamorphic immune function in a pond-breeding amphibian. Funct Ecol 22(1): 100–108
3. Cabagna MC et al (2005) Hematological parameters of health status in the common toad Bufo arenarum in agroecosystems of Santa Fe Province, *Argintina*. Appl Herpetol 2: 373–380
4. Raffel TR et al (2006) Negative effects of changing temperature on amphibian immunity under field conditions. Funct Ecol 20:819–828
5. Rutherford PL, McRuer DL, Forbes MR (2005) Condition and immune traits of frogs from Ontario baitshops: risks of practice not ameliorated by sale of healthy frogs. Herpetol Rev 36(2):129–133
6. Shutler D, Smith TG, Robinson SR (2009) Relationships between leukocytes and hepatozon spp. in green frogs, Rana clamitans. J Wildl Dis 45(1):67–72
7. Davis AK, Durso AM (2009) Leukocyte differentials of northern cricket frogs (Acris c. crepitans) with a compilation of published values from other amphibians. Herpetologica 65(3):260–267
8. Davis AK, Golladay C (2019) A survey of leukocyte profiles of red-backed salamanders from Mountain Lake, Virginia, and associations with host parasite types. Comp Clin Pathol 28: 1743–1750
9. Barriga-Vallejo C et al (2015) Assessing population health of the Toluca axolotl Ambystoma rivulare (Taylor, 1940) from Mexico, using leukocyte profiles. Herpetol Conserv Biol 10(2):592–601
10. Kim HJ, Jung YJ (2020) The emerging role of eosinophils as multifunctional leukocytes in health and disease. Immune Netw 20(3):14
11. Carnevale S et al (2020) The complexity of neutrophils in health and disease: focus on cancer. Semin Immunol 48:13
12. Demir A et al (2020) Ratio of monocytes to lymphocytes in peripheral blood in children diagnosed with active tuberculosis. J Pediatric Res 7(2):97–101
13. Jain NC (1986) Schalm's veterinary hematology, vol 1221. Lea & Febiger, Philadelphia
14. Thrall MA (2004) Hematology of amphibians. In: Thrall MA, Baker DC, Lassen ED (eds) Veterinary hematology and clinical chemistry: text and clinical case presentations. Lippincott Williams & Wilkins, Philadelphia, PA
15. Thrall MA et al (2006) Veterinary hematology and clinical chemistry. Blackwell Publishing, Ames, IA, p 518
16. Kiesecker JM (2002) Synergism between trematode infection and pesticide exposure: a link to amphibian deformities in nature? Proc Natl Acad Sci 99(15):9900–9904
17. Rohr JR et al (2008) Agrochemicals increase trematode infections in a declining amphibian species. Nature 455(7217):1235–1U50
18. Shutler D et al (2015) Nematode parasites and leukocyte profiles of Northern Leopard Frogs, Rana pipiens: location, location, location. Can J Zool 93(1):41–49
19. Koprivnikar J et al (2019) Endocrine and immune responses of larval amphibians to trematode exposure. Parasitol Res 118(1): 275–288
20. Ussing AP, Rosenkilde P (1995) Effect of induced metamorphosis on the immune system of the axolotl, *Ambystoma mexicanum*. Gen Comp Endocrinol 97(3):308–319
21. Davis AK (2009) Metamorphosis-related changes in leukocyte profiles of larval bullfrogs (Rana catesbeiana). Comp Clin Pathol 18(2): 181–186
22. Davis AK et al (2009) New findings from an old pathogen: intraerythrocytic bacteria (Family Anaplasmatacea) in red-backed salamanders Plethodon cinereus. Ecohealth 6(2):219–228
23. Marcogliese DJ et al (2009) Combined effects of agricultural activity and parasites on biomarkers in the bullfrog, *Rana catasbeiana*. Aquat Toxicol 91(2):126–134
24. Davis AK, Maney DL, Maerz JC (2008) The use of leukocyte profiles to measure stress in vertebrates: a review for ecologists. Funct Ecol 22:760–772
25. Wingfield JC, Romero LM (2001) Adrenocortical responses to stress and their modulation in free-living vertebrates. In: McEwen BS, Goodman HM (eds) Handbook of physiology, section 7: the endocrine system, Vol. IV: coping with the environment: neural and endocrine mechanisms. Oxford University Press, New York, pp 211–234
26. Dhabhar FS (2002) A hassle a day may keep the doctor away: stress and the augmentation of immune function. Integr Comp Biol 42(3): 556–564

27. Dhabhar FS (2009) Enhancing versus suppressive effects of stress on immune function: implications for immunoprotection and immunopathology. Neuroimmunomodulation 16(5):300–317
28. Davis AK, Maney DL (2018) The use of glucocorticoid hormones or leucocyte profiles to measure stress in vertebrates: What's the difference? Methods Ecol Evol 9(6):1556–1568
29. Gormally BMG, Romero LM (2020) What are you actually measuring? A review of techniques that integrate the stress response on distinct time-scales. Funct Ecol 34(10):2030–2044
30. Reynolds AE, Pickard BL (1973) Normal blood values in some plethodontid salamanders. Herpetologica 29:184–188
31. Rivera M, Davis AK (2013) Evaluating a method for non-destructively obtaining small volumes of blood from gilled amphibians. Herpetol Rev 44(3):428–430
32. Homan RN et al (2003) Impact os varying habitat quality on the physiological stress of spotted salamanders (Ambystoma maculatum). Anim Conserv 6:11–18
33. Perpiñán D et al (2006) Comparison of three different techniques to produce blood smears from green iguanas, Iguana iguana. J Herpetol Med Surg 16:99–101
34. Wright KM (2001) Amphibian hematology. In: Wright KM, Whitaker BR (eds) Amphibian medicine and captive husbandry. Krieger Publishing Company, Malabar, FL, pp 129–146
35. Friedmann GB (1970) Differential white blood cell counts for the urodele Taricha granulosa. Can J Zool 48:271–274
36. Bennett MF, Alspaugh JK (1964) Some changes in the blood of frogs following administration of hydrocortisone. Virginia J Sci 15: 76–79
37. Bennett MF et al (1972) Changes in the blood of newts, Notophthalmus viridescens, following administration of hydrocortisone. J Comp Physiol A 80:233–237
38. Bennett MF, Newell NC (1965) A further study of the effects of hydrocortisone on the blood of frogs. Virginia J Sci 16:128–130
39. Davis AK, Maerz JC (2008) Comparison of hematological stress indicators in recently captured and captive paedomorphic mole salamanders, *Ambystoma talpoideum*. Copeia 2008(3): 613–617
40. Davis AK, Maerz JC (2010) Effects of exogenous corticosterone on circulating leukocytes of a salamander (Ambystoma talpoideum) with unusually abundant eosinophils. Int J Zool 2010:735937. https://doi.org/10.1155/2010/735937
41. DuRant SE et al (2015) Evidence of ectoparasite-induced endocrine disruption in an imperiled giant salamander, the eastern hellbender (Cryptobranchus alleganiensis). J Exp Biol 218:2297–2304
42. Davis AK (2009) The wildlife leukocytes website: the ecologis's source for information about leukocytes of wildlife species. http://wildlifehematology.uga.edu
43. Coppo JA et al (2005) Blood and urine physiological values in captive bullfrog, Rana catesbeiana (Anura: Ranidae). Analecta Veterinaria 25(1):15–17
44. Cathers T et al (1997) Serum chemistry and hematology values for anesthetized American bullfrogs (Rana catesbeiana). J Zoo Wildl Med 28(2):171–174
45. Davis AK, Maerz JC (2011) Assessing stress levels of captive-reared amphibians with hematological data: implications for conservation initiatives. J Herpetol 45(1):40–44
46. de Assis VR et al (2015) Effects of acute restraint stress, prolonged captivity stress and transdermal corticosterone application on immunocompetence and plasma levels of corticosterone on the cururu toad (Rhinella icterica). PLoS One 10(4):21
47. Hadji-Azimi I, Coosemans V, Canicatti C (1987) Atlas of adult Xenopus laevis laevis hematology. Dev Comp Immunol 11:807–874
48. Voight GL (2000) Hematology techniques and concepts for veterinary technicians. Iowa State University Press, Ames, IA, p 139
49. Forzan MJ et al (2017) Clinical pathology of amphibians: a review. Vet Clin Pathol 46(1): 11–33

Part VII

Epilogue

Chapter 30

Best Practices to Promote Data Utility and Reuse by the Non-Traditional Model Organism Community

Garrett S. Dunlap and Nicholas D. Leigh

Abstract

The dramatic increase in accessibility to sequencing technologies has opened new avenues into studying different processes, cells, and animal models. In the amphibian models used for regeneration research, these new datasets have uncovered a variety of information about what genes define the regenerating limb as well as how genes and cells change over the course of regeneration. The accumulation of data from these studies undoubtedly increases our understanding of regeneration. Throughout these studies, it is important to consider how data can be made most useful not only for the primary study but also for reuse within the scientific community. This chapter will focus on best practices for data collection and handling as well as principles to promote access and reuse of big datasets. However, the deposition and thorough description of data of all sizes generated for a publication (e.g., images, fcs files, etc.) can also be done following this generic workflow. The aim is to lower hurdles for reuse, access, and re-evaluation of data which will in turn increase the utility of these datasets and accelerate scientific progress.

Key words Data collection, Data sharing, Data repositories, Next-generation sequencing, Big data, Data reuse

1 Introduction

In recent years, there has been a surge in data-driven science [1]. This has dramatically changed the methods and techniques used, how we share findings, and the utility of datasets. An important consideration going forward is that large datasets (e.g., Next-Generation Sequencing (NGS)- and imaging-based data) can be reused and integrated into other datasets. In addition, these datasets are considerably more expensive to generate and analyze than more traditional methods, making it important that quality data are obtained and shared. There is also an exciting push for more wet lab scientists to conduct bioinformatic analyses, and it is important that these individuals also learn how to effectively manage and share data. In addition, the multidisciplinary approaches used to generate many datasets requires that more than one person be involved in

Ashley W. Seifert and Joshua D. Currie (eds.), *Salamanders: Methods and Protocols*,
Methods in Molecular Biology, vol. 2562, https://doi.org/10.1007/978-1-0716-2659-7_30,

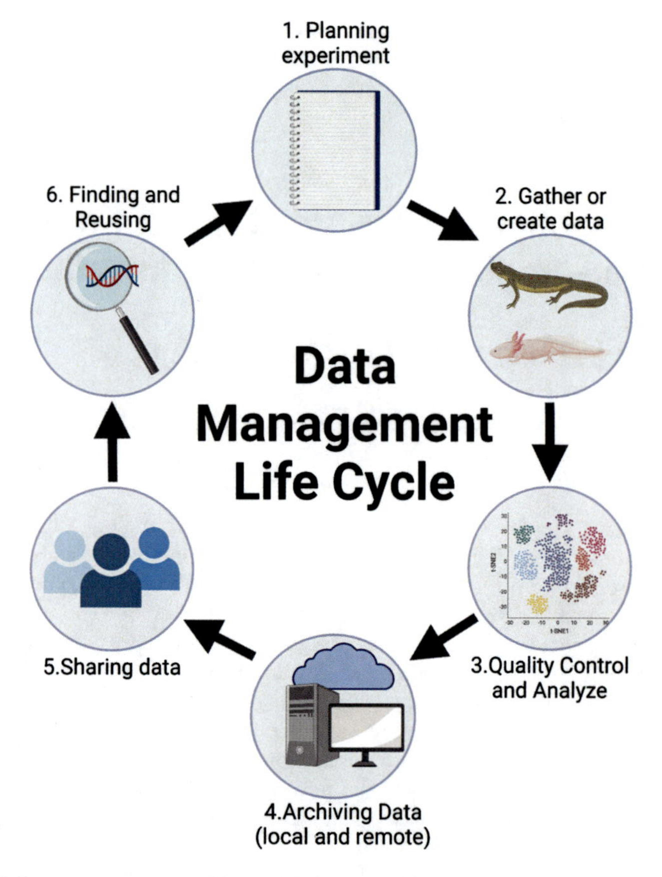

Fig. 1 Data management life cycle. Proper planning paired with appropriate acquisition, recording, handling, and sharing of data is essential to preserve data and promote its reuse. Data shared with the scientific community can reenter this cycle continuously, aiding the original lab and the community

the experimental pipeline. It is therefore crucial to adopt standard approaches that facilitate communication between individuals working on the project and to consider how the data will be distributed in its final form to facilitate reuse.

To ensure a high utility of datasets, we review best practices for the generation, deposition, sharing, and promotion of reuse of these datasets in the non-traditional model organism research community (Fig. 1). The movement toward sharing these large datasets also allows for one to re-evaluate how more traditional datasets (e.g., fcs files, in situ hybridization images, etc.) are shared and apply similar principles regardless of the scale of the data. Simply indicating the corresponding author can be contacted to obtain data is problematic as relying on personal communications may cause a variety of problems (e.g., missed emails, changed email address, language barriers, etc.). Fortunately, an increase in data sharing platforms provides a means to share all data and methods associated with your work. Many of these data repositories allow for a digital object identifier (DOI) to be associated with your dataset, and while they can be used in conjunction with a publication can also be shared as a standalone dataset outside of the standard publishing system. As for protocol repositories, these provide a forum for asking questions about protocols and allow for protocols to be "living" documents that can be updated as protocols are improved or questions are asked to clarify steps. The movement of journals and funders toward ensuring quality data (and metadata) sharing principles serves as an incentive to establish standard operating procedures within the lab as a means to guarantee proper handling and sharing of datasets.

This chapter will focus on the best practices for sharing big (i.e., >Gigabytes) datasets, with a focus on the steps we can take to make these data as useful as possible to the research community. It is clear that open data practices increase the reach of a work [2, 3] and that poor data practices can at the least lead to misinterpretations and in the worst case retractions [4]. The salamander regeneration field has seen a massive influx of big data, including next-generation sequencing [5–10] and more recently images and videos [11–16]. Therefore, decreasing the hurdles to data access and re-analysis will benefit the community. The principles described here can be applied to any dataset generated within the lab, regardless of dataset size. The limiting factor is the willingness and time of the researcher to carefully curate and share datasets with the intention of making reuse as easy as possible for themselves, their lab groups, and the community as a whole.

2 Materials

The availability and reuse of your dataset are solely dependent on the willingness of the lab members and head to faithfully record the experimental procedure, share adequate data (including metadata), and make available the processing performed on the dataset. The only materials required are a computer, access to the internet, and adequate storage for the datasets.

2.1 Computer and Internet Access

1. Any computer that is capable of accessing the internet is suitable.
2. High-performance computing (HPC) clusters are often available via universities and will provide more computing power than a personal computer. They are also typically maintained by knowledgeable professionals and have many ready-to-use bioinformatics tools and pipelines available (*see* **Note 1**).
3. High-speed internet is important to allow for rapid uploading and downloading of large datasets.

2.2 Storage

1. Purchase a large standalone hard drive and identify a cloud-based storage location to back up files from the computer. Preferentially, both of these options can be set up to automatically back up all files. Always make sure to keep raw data in three locations (this can and should also be backed up online, *see* below). In regard to hard drives, it is important to consider options for storage that best fit the budget and usage needs. Some drives (e.g., Solid State Drives (SSD)) have limited storage but can read and write data quickly, are not vulnerable to shock and sudden movement, and are thus good options for laptops. Others drives, such as Serial Advanced Technology Attachment (SATA), are cheaper, can have more storage capacity, but read and write data more slowly, and are susceptible to shock. This makes SATA drives good options for a desktop but not laptop computers.
2. Carefully consider the folder/directory structure [17] to keep separate raw and processed data to ensure that raw data are not inadvertently deleted or changed (Fig. 2).
3. Make use of storage resources provided by your university (e.g., HPC clusters likely have storage associated). If these are not provided, there are a variety of free-of-charge repositories that can host large datasets and have options to keep data private prior to publication (*see* **Note 2**).

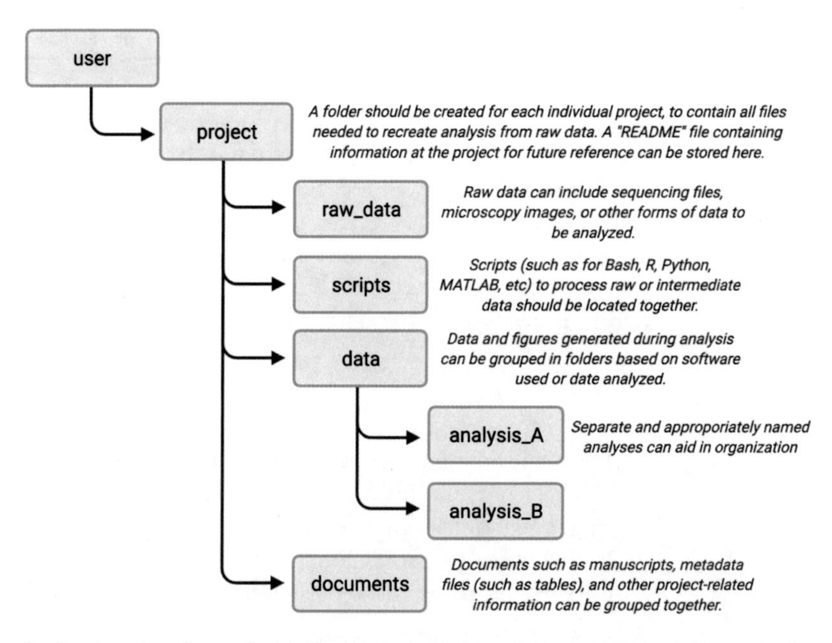

Fig. 2 Example directory tree for project management. Organized storage of an entire project can aid in the utility of the dataset. It is particularly important to preserve raw data. A consensus directory tree structure in the lab can also aid other and future lab members when revisiting your work

3 Methods

3.1 Planning Experiments

1. Determine the experimental setup, including adequate controls, the goal of the experiment, and the technical and financial considerations at hand (*see* **Note 3**).
2. Plan for how the data will be acquired, analyzed, managed, and shared (*see* **Note 4**).

3.2 Conducting Experiments

3. Perform experiment as planned above, noting any deviations (*see* **Note 5**).
4. Collect data throughout the process of sample preparation and data collection, ensuring detailed notes are recorded. These will serve as metadata.
5. Quality control should be performed whenever possible throughout data collection and processing. These checks can serve as important metadata and may serve as a means to understand unexpected variation in your dataset.
6. Organize data making sure files have consistent, informative, and intuitive naming schemes [17]. Also, collate metadata and ensure that all data are raw and uncorrupted.
7. Save raw data in secure locations and in open formats. Adhere to the 3-2-1 backup rule, keeping at least three (3) copies of your data, two (2) of which are on different storage media and one (1) copy at an offsite location. Open data formats include non-proprietary file formats (e.g., html, fastq, etc.) Generate and include a README file that is associated with your project (*see* **Note 6**).

3.3 Analyzing Data

7. Revisit your analysis plan from Subheading 3.1 and consider that if many months passed between planning, conducting, and collecting data that the standard operating procedures for how to analyze data may have changed drastically. In addition, for sequencing-based studies it is critical to carefully consider what assembly is used for alignment (*see* **Note 7**).
8. Use a version-controlled system, such as an electronic lab notebook or preferably a Git-based version control system, to track changes, save the pipeline, or access/revert to previous analysis pipelines if necessary (*see* **Note 8**).
9. Thoroughly annotate all of the steps taken in data analysis (*see* **Note 9**).
10. Ask someone else to review and reproduce your analysis using only the raw data and information (e.g., annotated code) that will be provided with the published dataset.

3.4 Archiving and Sharing Data

11. Upload data, metadata, and thorough descriptions of the protocols to a freely accessible repository (*see* **Note 10**).
12. Generate a linked repository (such as a GitHub page) with information about the dataset (*see* **Note 11**).
13. Optional, but recommended: Upload data to an interactive data repository (*see* **Note 12**).
14. Continue to think about ways to improve the access of your data and spend time to provide relevant updates should mistakes be found in the uploaded data or descriptive repository (*see* **Note 13**).

4 Notes

1. Computing resources beyond what a personal computer can provide can also be readily accessed via Amazon Web Services or Google Cloud Platform. There are also a number of portals that allow for data analysis by researchers with limited experience of data analysis in the cloud (e.g., Terra (https://app.terra.bio/), Galaxy [18], etc.).
2. Examples of storage resources include government funded sites for specific data types like the NCBI Sequence Read Archive (SRA) or European Nucleotide Archive (ENA) designed for sequencing-based data. Non-profit organizations for generalist datasets like Zenodo, FigShare, or Open Science Framework may work depending on the size and scope of the project. There are also a variety of specialist repositories that may better fit your dataset (https://www.nature.com/sdata/policies/repositories). Many of these allow for a digital object identifier (DOI) to be associated with your dataset, making it permanently available and citable. Further, data can be uploaded and stored privately until an author-specified release date (i.e., when paper is published). An example of the simple steps that are required for data upload to FigShare can be found here: https://help.figshare.com/article/how-to-upload-and-publish-your-data.
3. There are a variety of technologies, chemistries, and platforms capable of generating data and choice of methodology can have a major impact on the data generated and its utility. Further, different methods can have different success rates and this may impact group size determination. All of this should be carefully considered, and it is advised to approach a core facility, company, or colleague with expertise in designing experiments using your methodology of choice to ensure your experiment is well designed.

4. The multidisciplinary nature of many datasets requires that more than one person be involved in collecting the data. Before performing the experiment consider how these data will be used by your lab, but also how you are going to distribute and make the data useful for other researchers in the scientific community.
5. It is critical to collect metadata associated with your experiment and make this information available when you publish your dataset. This includes, but is not limited to, manipulations performed, size and age of animal(s), sex, species, housing temperature, transgenic strain and origin, instrument setting (e.g., electroporator and microscope), water origins/parameters, and time point(s) collected. In addition, keeping detailed records of the experiment procedure may help explain variations observed between experiments or labs. If possible, keep information about the animal origins (e.g., ordered from a stock center or bred in house) and as much lineage information as possible. Additional examples of metadata that can be captured throughout the duration of a next-generation sequencing experiment can be found in Ref. 19.
6. Many funding agencies now require data management plans and your lab likely has one in place. Generating a cryptographic hash (e.g., MD5) of the data can also ensure that it has not been corrupted prior to reuse and should be distributed with the data. Now is a good time to generate a README file which will reside in the project directory. This should include information on the content and structure of the project. It is important to provide enough information so that others know at a minimum the basics about the project and the included dataset (s) and if it may be useful for them.
7. In order to increase the utility of datasets it is important to use consensus genomes, transcriptomes, and annotations. If the work is an improvement of an already existing resource, it is useful to provide information about interconversion between the old annotations/version and the new. This effort to conform to standard use instead of reannotation and re-construction will greatly facilitate the use of the dataset. Further, it is useful to share all files that were used to generate genome/transcriptome references (e.g., gff3, gtf files, etc.).
8. Ensure that details of the processing pipeline are recorded, including versions of any software used, and that code is deposited. For advanced users, it may be helpful to share saved virtual environments, containers, or a pipeline or workflow that is capable of re-running all the code. This can greatly facilitate the re-analysis and reuse of the dataset.

9. Annotation should include who wrote the code, its purpose, and in general should be commented thoroughly enough to be able to be understood by another user. Annotation will also be useful for the original user.
10. Data upload can be an exceedingly time consuming and complicated step, but taking the time to do it properly at the onset can save an immense amount of time later on and dramatically increase reuse and utility of the dataset. Choosing to deposit the data instead of indicating that the corresponding author can be contacted to obtain data also eliminates some of the potential pitfalls of relying on corresponding authors and thus personal communications (e.g., missed emails, changed email address, language barriers, etc.). Examples of repositories are shown in *see* **Note 2**.
11. Since there are no page limits, using a repository allows one to be thorough and informative about the data and how to re-process it. This also provides a means to provide contact information outside of an official publication and a forum to interact with others to promote reuse of your data (e.g., protocols.io). The communication between authors and interested colleagues can help improve the protocol for clarity or improve the protocol as more scientists contribute their ideas and optimizations. Further, the ability to fork protocols (i.e., copy and modify with attribution to the original authors) on sites, like protocols.io, allows for one to quickly customize or adapt protocols from others.
12. Consider adding the data and associated metadata to a browsable, centralized repository. Examples for single-cell RNA sequencing data include the EMBL-EBI Single Cell Expression Atlas (https://www.ebi.ac.uk/gxa/sc/home) and the Broad Institute Single Cell Portal (https://singlecell.broadinstitute.org/single_cell). If a relevant centralized repository does not exist, it can be extremely beneficial to set up a browsable database on a personal/lab website; exemplary examples include https://ambystoma.uky.edu/ and https://www.axolotl-omics.org/. Keep in mind that lab websites may not be permanent and typically only contain data from one lab, so if a centralized browsable repository becomes available it can be beneficial to also provide an option to access the data there.
13. If data are shared among the community, soliciting feedback from individuals who have accessed the data can ensure that the file structures and pipelines are understandable to a broader audience. It may also be useful to integrate datasets from newer publications from your or another lab, making these integrated datasets available to the community.

Acknowledgments

This work was generously supported by the Knut and Alice Wallenberg Foundation (NDL) and the Swedish Research Council (NDL, Registration # 2020-01486). Figures were created with BioRender.com.

References

1. Marx V (2013) Biology: the big challenges of big data. Nature 498:255–260
2. Piwowar HA, Vision TJ (2013) Data reuse and the open data citation advantage. PeerJ 1:e175
3. Colavizza G, Hrynaszkiewicz I, Staden I, Whitaker K, McGillivray B (2020) The citation advantage of linking publications to research data. PLoS One 15:e0230416
4. Munafò MR, Nosek BA, Bishop DVM, Button KS, Chambers CD, du Sert NP et al (2017) A manifesto for reproducible science. Nat Hum Behav 1:0021
5. Elewa A, Wang H, Talavera-López C, Joven A, Brito G, Kumar A et al (2017) Reading and editing the Pleurodeles waltl genome reveals novel features of tetrapod regeneration. Nat Commun 8:2286
6. Nowoshilow S, Schloissnig S, Fei JF, Dahl A, Pang AWC, Pippel M et al (2018) The axolotl genome and the evolution of key tissue formation regulators. Nature 554(7690):50–55. https://doi.org/10.1038/nature25458
7. Bryant DM, Johnson K, DiTommaso T, Tickle T, Couger MB, Payzin-Dogru D et al (2017) A tissue-mapped axolotl de novo transcriptome enables identification of limb regeneration factors. Cell Rep 18:762–776
8. Matsunami M, Suzuki M, Haramoto Y, Fukui A, Inoue T, Yamaguchi K et al (2019) A comprehensive reference transcriptome resource for the Iberian ribbed newt Pleurodeles waltl, an emerging model for developmental and regeneration biology. DNA Res 26:217–229
9. Gerber T, Murawala P, Knapp D, Masselink W, Schuez M, Hermann S et al (2018) Single-cell analysis uncovers convergence of cell identities during axolotl limb regeneration. Science 362(6413):eaaq0681. https://doi.org/10.1126/science.aaq0681
10. Leigh ND, Dunlap GS, Johnson K, Mariano R, Oshiro R, Wong AY et al (2018) Transcriptomic landscape of the blastema niche in regenerating adult axolotl limbs at single-cell resolution. Nat Commun 9:5153
11. Duerr TJ, Comellas E, Jeon EK, Farkas JE, Joetzjer M, Garnier J et al (2020) 3D visualization of macromolecule synthesis. elife 9: e60354. https://doi.org/10.7554/eLife.60354
12. Pende M, Vadiwala K, Schmidbaur H, Stockinger AW, Murawala P, Saghafi S et al (2020) A versatile depigmentation, clearing, and labeling method for exploring nervous system diversity. Sci Adv 6:eaba0365
13. Subiran Adrados C, Yu Q, Bolaños Castro LA, Rodriguez Cabrera LA, Yun MH (2020) Salamander-Eci: an optical clearing protocol for the three-dimensional exploration of regeneration. Dev Dyn 250(6):902–915. https://doi.org/10.1002/dvdy.264
14. Masselink W, Reumann D, Murawala P, Pasierbek P, Taniguchi Y, Bonnay F et al (2019) Broad applicability of a streamlined ethyl cinnamate-based clearing procedure. Development 146(3):dev166884. https://doi.org/10.1242/dev.166884
15. Pinheiro T, Mayor I, Edwards S, Joven A, Kantzer CG, Kirkham M et al (2020) CUBIC-f: an optimized clearing method for cell tracing and evaluation of neurite density in the salamander brain. J Neurosci Methods 348:109002
16. Currie JD, Kawaguchi A, Traspas RM, Schuez M, Chara O, Tanaka EM (2016) Live imaging of axolotl digit regeneration reveals spatiotemporal choreography of diverse connective tissue progenitor pools. Dev Cell 39: 411–423
17. Directory Structure. [cited 21 Jan 2022]. Available: https://datamanagement.hms.harvard.edu/collect/directory-structure
18. Afgan E, Baker D, Batut B, van den Beek M, Bouvier D, Cech M et al (2018) The Galaxy platform for accessible, reproducible and collaborative biomedical analyses: 2018 update. Nucleic Acids Res 46:W537–W544
19. Füllgrabe A, George N, Green M, Nejad P, Aronow B, Fexova SK et al (2020) Guidelines for reporting single-cell RNA-seq experiments. Nat Biotechnol 38(12):1384–1386

Chapter 31

Now that We Got There, What Next?

Elly M. Tanaka

Abstract

As seen in the protocols in this book, the opportunities to pursue work at the cellular and molecular work in salamanders have considerably broadened over the last years. The availability of genomic information and genome editing, and the possibility to image tissues live and other methods enhance the spectrum of biological questions accessible to all researchers. Here I provide a personal perspective on what I consider exciting future questions open for investigation.

Key words Salamander, Future directions, Genomics, Evolution, Cell biology, Regeneration

Riding the wave of remarkable methodological developments in molecular biology, salamanders, such as *Ambystoma mexicanum* and *Pleurodeles waltl*, have come into full force as easily breedable animal models with an impressive arsenal of molecular genetic tools to dissect cellular and molecular mechanisms of fascinating, complex, biological phenomena, such as regeneration. Prior to this time, enthusiastic students of regeneration universally suffered when trying to identify molecules responsible for regeneration due to the lack of an annotated genome and the lack of a means to functionally test molecules in vivo. Less than a decade ago, the prospect of functionally testing gene function by knock-down was basically limited to morpholino electroporation which suffered from caveats in issues of penetration. Now this has been rectified, with full access to a sequenced, highly contiguous genome sequence [1], and multiple bulk and scRNASeq profiling of regeneration [2–10], coupled with CRISPR-mediated mutation and gene modification as well as electroporation and viral approaches [11–18]. These developments have opened up a new era in axolotl research with bright prospects to allow creative forces to fulfill their dreams to identify and analyze the cellular interactions and the molecular components that constitute the regeneration program and other fascinating questions, such as the evolution of

Ashley W. Seifert and Joshua D. Currie (eds.), *Salamanders: Methods and Protocols*,
Methods in Molecular Biology, vol. 2562, https://doi.org/10.1007/978-1-0716-2659-7_31,

developmental form and function. This chapter aims to provide some perspectives on future directions that are now open for investigation, as well as further development of the system that could continue to help hone our understanding of regeneration.

1 Genome Dynamics During Regeneration and Over Evolutionary Timescales

Salamanders were long known to have a giant genome riddled with repetitive sequences which made sequence assembly difficult but with long-read sequencing, the placement of genes among the sea of repeats in the axolotl genome has been resolved with a chromosome level assembly [1]. This, coupled with genome/chromatin profiling methods such as ATACSeq to identify open chromatin regions, promises to help identify enhancers that are responsible for expression of genes in specific cell types, or that fire during regeneration [19]. Naturally, the bioinformatic analysis of such regions for motif enrichment will help uncover the genetic cascade of gene regulatory events that occurs at the onset of regeneration and in different regions of the regenerate. A key unanswered question is how the limb developmental program is held as a latent yet activatable form in mature limb cells. Additional profiling, as well as functional tools such as CRISPRi and CRISPRa, to target chromatin modifications to defined loci will be necessary to understand which and how loci transit from a silent to an active state. Recent work in several regeneration organisms has implicated Fosl1, Jun, and AP-1 complexes as important initial inducers of a pro-regenerative state [20–24]. As Fos/Jun are factors very often implemented as early immediate genes in metazoan stress responses, how these factors couple to the regeneration program will be an important future question.

From an evolutionary perspective, how regeneration capability evolved is a fascinating, unavoidable topic that seems more approachable now with the genome assembly and associated methods. Interesting approaches would involve not only comparing transcriptional programs among regenerating and non-regenerating animals but also trying to understand how genome structure evolved surrounding important genes involved in regeneration. In that sense, those pesky repeats probably deserve a closer look. Over the last years, it has become clear that a major force in evolution of novel gene regulation is the movement of transposons that shuffle enhancer elements into new areas [25]. It has been proposed that salamanders harbor a giant genome in part due to weak mechanisms to eliminate transposons from its genome [26]. An intriguing question is whether the tendency to maintain transposable elements facilitated the sculpting of the genome to a robust, pro-regenerative state. While the repetitive sequences in the axolotl genome assembly has so far been annotated with standard pipelines, a much more in-depth examination of repeats, the relation to each other and their location will be necessary, as well as rapid enhancer function assays. Another fascinating question is whether the presence of many repeats has led to the salamander genome adopting a distinctive organization of euchromatin versus heterochromatin that facilitates the silencing and activation of genes in a way that facilitated regeneration. Analysis of long-distance contacts to probe large-scale organization, as well as even biochemical approaches, may help clarify such possibilities.

Finally, with the recent sequencing of not only the giant axolotl genome but also two lungfish genomes, there are now much better possibilities for understanding which genome features arose at the base of the tetrapod lineage and which were already present in a more ancient common ancestor with fish [27, 28]. Therefore, the existence of the axolotl genome has great significance for the understanding of tetrapod history. A deep comparison of genomes from gar, lungfishes, frogs, and salamanders promises to be a rich treasure trove for understanding evolutionary history of vertebrates. For example, the comparison of Major Histocompatibility Complex of axolotl with *Xenopus* and human revealed that the axolotl locus appears to have gene organization with surprising features that may be conserved with the human locus [1]. These results implore the investigation of the lungfish MHC locus.

2 Cellular and Molecular Dynamics of Regeneration

A transformative dimension of recent developments is the ease of making F0 and F1 transgenic and genetically modified animals not only for interfering or augmenting gene expression but also for marking cells with fluorescent reporters with high cell-type fidelity. Currie et al. demonstrated the fundamental importance of performing live imaging of cells during regeneration, which allowed them to define the source zone of regeneration, to resolve the relation between temporal waves of cell migration and cell fate, and to elucidate clonal cell lineage [29]. With the ability to use CRISPR technology to knock-in gene sequences into defined loci, fluorescent proteins can be expressed in specific cells based on cell-type-specific expression which allowed the tagging of the Pax7 gene to understand the contribution of satellite cells to limb regeneration [13]. There are still many opportunities to perform live imaging on cell cohorts that generate or regulate the blastema to uncover new insights into the process. For example, the intimate interaction between wound epidermis and nerve, and how this results in apical ectodermal cap formation are key issues [30–32]. Such approaches combined with the molecular knowledge coming from single cell sequencing can allow interrogation of regeneration at unprecedented single cell resolution. In addition, sophisticated constructs reporting cell cycle or retinoic acid transcriptional activity have been made to analyze dynamic events in regeneration, and future generations of sophisticated genetical reporters for signaling states or metabolic states are likely to yield important new information for regeneration [33–35]. Finally, global imaging approaches coupled to unbiased image analysis and machine learning approaches could help us glean cellular relationships that have not yet been unearthed.

The use of gene editing to mutate genes and test their role in regeneration is just coming to the fore. This method yields penetrant enough modification so that F0 animals can be directly analyzed for phenotypes [11, 12]. Interestingly, several mutants have shown mild or no phenotype with a severe phenotype during regeneration, demonstrating the utility of trying "straight" gene modification as a first pass. Of course, it will become necessary to study the role of genes that might also have a function earlier in development and the further development of means to modulate gene function inducibly in the limb is an important future direction. A beautiful notable recent example was described in which an optogenetically inducible construct that induces a multi-headed myosin was electroporated into the blastema and used to study filopodia in blastema cells [36]. Still open for development is floxing of genes for cre-induced gene disruption, and also use of the several viral systems to somatically express Cas9 and gRNAs for gene disruption. Currently, a pol III promoter that works efficiently in axolotl cells has not been identified.

Regeneration has gratifyingly entered the era in which potential molecular players can be identified based on expression, and then the function of a given gene can be studied in extensive detail. For deepening the analysis of single players, keeping with the most modern methods will continue to be important, but in addition, a move toward genome-wide or systems-type approaches to integrate multiple pathways and understand the dynamic network of molecular signaling at the heart of regeneration can be an important future goal. With the availability of easy transduction methods such as electroporation, and four viral systems, coupled with primary culture, cell lines, and in vivo transduction, as well as multiplexed in situ hybridization, there is no reason why genome-wide or multidimensional perturbation methods could not be used to uncover unexpected players and molecular relationships operating in regeneration.

In that regard, the complexity of the immune system, characterizing its cell types, and understanding which cell types and signals could promote regeneration and which types may hold back regeneration is an area of great interest and a great place to apply such ideas. Work in salamander and frog regeneration has identified a number of immune cell types, and cytokines that work either for or against regeneration but more precise mechanisms could still be teased out, and the larger question of whether differences in immune system functions could underlie in part differences in regenerative capacity among vertebrates is still a large frontier [37–40]. Clever transgenics to modulate immune cell function in a time- and spatial-dependent manner and monitoring the cells that build the regenerate seems like an important and fruitful direction forward. The recent single cell characterization of immune cell types in axolotl regenerating tissues as well as the amazing finding in zebrafish that regulatory T cells adapt their expression of organ-specific regenerative factors to tissue context to promote regeneration are exciting new developments in this field that will help support future research [8, 41].

In addition, the late stages of limb regeneration are scarcely explored but promise to yield fascinating insights, such as how to match regenerate size with body size. For example, a recent study shows that limb regeneration occurs by forming a "tiny-limb" that expands in size to match host size. The host nerve plays a crucial role in this late matching process and understanding the mechanisms in this matching process will likely yield generalizable insights into organ size control [42].

3 Commonality and Divergence Among Salamanders as Meaningful for Regeneration

Ambystoma mexicanum and *Pleurodeles waltl* are salamanders that diverged approximately 125 million years ago. By studying muscle dedifferentiation as well as satellite cells, it has become clear that while both species regenerate, there are significant divergences in the process. Whereas *Pleurodeles* muscle fibers show dedifferentiation, axolotl muscle does not [43]. On the satellite cell side, axolotl muscle progenitor cells are solely driven by the Pax-family member Pax7 (missing Pax3), whereas *Pleurodeles waltl* harbors an intact Pax3 that contributes to muscle progenitor cell regulation [5, 44]. These differences show the spectrum of possibilities that exist in regeneration spaces—regeneration can be executed with significant variation, and it is likely that other variances will be discovered by studying these species in parallel. How such differences may have evolved will be fascinating to pursue, and with chromosome level assemblies of axolotl available and *Pleurodeles* in preparation, it will be interesting to see whether we will be able to align these genomes, and make a search for meaningful changes that resulted in differences in the propensity to dedifferentiate cells or not.

Given that sequencing, CRISPR and transgenesis methods are broadly applicable, the implementation of additional salamander species that have variant regeneration capabilities or that display other non-regeneration traits is now accessible opening up a broader view of salamander-related traits [45–47]. In addition, the probing of phylogenetically close relatives, which seemed previously daunting, is now in reach for comparative studies [48, 49].

With such cross-species comparisons, an ambitious question is naturally whether the comparison to quite distant species that do not regenerate, such as frog or mammals can be made meaningfully, by comparing cell types, their gene expression states, or the features of the genome. A recent impressive functional comparison of microRNAs and their downstream targets after spinal cord lesion in axolotl versus rat provides a harbinger of what is to come [50]. The histological, functional, and molecular comparison of wound healing and bone regeneration between salamanders and mammals promises also interesting insights [51, 52].

4 Beyond Regeneration

This chapter has largely focused on the axolotl in the context of regeneration research, the author's traditional field of research. But the model also has a long history in electrophysiology and behavior. With the expanded tools now available, the possibility is open to explore the axolotl and other salamanders for other aspects of physiology such as aging, and hormone-induced transitions, and their lack thereof, for example, paedomorphy. The accessibility of animals at all life stages, the continuous growth of the animals and the relatively long lifespan provide some unique contexts for linking cellular and molecular features to organismal properties. Another area in which the salamander has already played an important role is in neuroscience, as a great electrophysiology model due to its large cells and eggs, but also in the study of motor control systems [53, 54]. Now with the availability of genetic activity sensors, the ease of making salamander transgenics and hardiness to imaging, the salamander promises to be an excellent organism for systems-type neuroscience approaches that integrate live imaging of cell populations with behavioral tasks. Given its position in the phylogenetic tree such studies could yield insight into organization and function of basic vertebrate neuronal circuits. These are just a few ideas and perspectives for the potential of this truly remarkable class of animals that have zoomed into the modern molecular age.

References

1. Schloissnig S, Kawaguchi A, Nowoshilow S, Falcon F, Otsuki L, Tardivo P, Timoshevskaya N, Keinath MC, Smith JJ, Voss SR et al (2021) The giant axolotl genome uncovers the evolution, scaling, and transcriptional control of complex gene loci. Proc Natl Acad Sci U S A 118(5): e2017176118
2. Stewart R, Rascon CA, Tian S, Nie J, Barry C, Chu LF, Ardalani H, Wagner RJ, Probasco

MD, Bolin JM et al (2013) Comparative RNA-seq analysis in the unsequenced axolotl: the oncogene burst highlights early gene expression in the blastema. PLoS Comput Biol 9:e1002936

3. Bryant DM, Johnson K, DiTommaso T, Tickle T, Couger MB, Payzin-Dogru D, Lee TJ, Leigh ND, Kuo TH, Davis FG et al (2017) A tissue-mapped axolotl de novo transcriptome enables identification of limb regeneration factors. Cell Rep 18:762–776
4. Jiang P, Nelson JD, Leng N, Collins M, Swanson S, Dewey CN, Thomson JA, Stewart R (2017) Analysis of embryonic development in the unsequenced axolotl: waves of transcriptomic upheaval and stability. Dev Biol 426: 143–154
5. Nowoshilow S, Schloissnig S, Fei JF, Dahl A, Pang AWC, Pippel M, Winkler S, Hastie AR, Young G, Roscito JG et al (2018) The axolotl genome and the evolution of key tissue formation regulators. Nature 554:50–55
6. Leigh ND, Dunlap GS, Johnson K, Mariano R, Oshiro R, Wong AY, Bryant DM, Miller BM, Ratner A, Chen A et al (2018) Transcriptomic landscape of the blastema niche in regenerating adult axolotl limbs at single-cell resolution. Nat Commun 9:5153
7. Gerber T, Murawala P, Knapp D, Masselink W, Schuez M, Hermann S, Gac-Santel M, Nowoshilow S, Kageyama J, Khattak S et al (2018) Single-cell analysis uncovers convergence of cell identities during axolotl limb regeneration. Science 362:eaaq0681
8. Rodgers AK, Smith JJ, Voss SR (2020) Identification of immune and non-immune cells in regenerating axolotl limbs by single-cell sequencing. Exp Cell Res 394:112149
9. Qin T, Fan CM, Wang TZ, Sun H, Zhao YY, Yan RJ, Yang L, Shen WL, Lin JX, Bunpetch V et al (2021) Single-cell RNA-seq reveals novel mitochondria-related musculoskeletal cell populations during adult axolotl limb regeneration process. Cell Death Differ 28: 1110–1125
10. Li H, Wei X, Zhou L, Zhang W, Wang C, Guo Y, Li D, Chen J, Liu T, Zhang Y et al (2021) Dynamic cell transition and immune response landscapes of axolotl limb regeneration revealed by single-cell analysis. Protein Cell 12:57–66
11. Fei JF, Schuez M, Tazaki A, Taniguchi Y, Roensch K, Tanaka EM (2014) CRISPR-mediated genomic deletion of Sox2 in the axolotl shows a requirement in spinal cord neural stem cell amplification during tail regeneration. Stem Cell Reports 3:444–459
12. Flowers GP, Timberlake AT, Mclean KC, Monaghan JR, Crews CM (2014) Highly efficient targeted mutagenesis in axolotl using Cas9 RNA-guided nuclease. Development 141(10): 2165–2171
13. Fei JF, Schuez M, Knapp D, Taniguchi Y, Drechsel DN, Tanaka EM (2017) Efficient gene knockin in axolotl and its use to test the role of satellite cells in limb regeneration. Proc Natl Acad Sci U S A 114:12501–12506
14. Echeverri K, Tanaka EM (2003) Electroporation as a tool to study in vivo spinal cord regeneration. Dev Dyn 226:418–425
15. Roy S, Gardiner DM, Bryant SV (2000) Vaccinia as a tool for functional analysis in regenerating limbs: ectopic expression of Shh. Dev Biol 218:199–205
16. Whited JL, Tsai SL, Beier KT, White JN, Piekarski N, Hanken J, Cepko CL, Tabin CJ (2013) Pseudotyped retroviruses for infecting axolotl in vivo and in vitro. Development 140: 1137–1146
17. Khattak S, Sandoval-Guzman T, Stanke N, Protze S, Tanaka EM, Lindemann D (2013) Foamy virus for efficient gene transfer in regeneration studies. BMC Dev Biol 13:17
18. Oliveira CR, Lemaitre R, Murawala P, Tazaki A, Drechsel DN, Tanaka EM (2018) Pseudotyped baculovirus is an effective gene expression tool for studying molecular function during axolotl limb regeneration. Dev Biol 433:262–275
19. Wei X, Li H, Guo Y, Zhao X, Liu Y, Zou X, Zhou L, Yuan Y, Qin Y, Mao C et al (2021) An ATAC-seq dataset uncovers the regulatory landscape during axolotl limb regeneration. Front Cell Dev Biol 9:651145
20. Sabin KZ, Jiang P, Gearhart MD, Stewart R, Echeverri K (2019) AP-1(cFos/JunB)/miR-200a regulate the pro-regenerative glial cell response during axolotl spinal cord regeneration. Commun Biol 2:91
21. Beisaw A, Kuenne C, Guenther S, Dallmann J, Wu CC, Bentsen M, Looso M, Stainier DYR (2020) AP-1 contributes to chromatin accessibility to promote sarcomere disassembly and cardiomyocyte protrusion during zebrafish heart regeneration. Circ Res 126:1760–1778
22. Lee HJ, Hou Y, Chen Y, Dailey ZZ, Riddihough A, Jang HS, Wang T, Johnson SL (2020) Regenerating zebrafish fin epigenome is characterized by stable lineage-specific DNA methylation and dynamic chromatin accessibility. Genome Biol 21:52
23. Wang W, Hu CK, Zeng A, Alegre D, Hu D, Gotting K, Ortega Granillo A, Wang Y, Robb S, Schnittker R et al (2020) Changes in

regeneration-responsive enhancers shape regenerative capacities in vertebrates. Science 369(6508):eaaz3090

24. Wu HY, Zhou YM, Liao ZQ, Zhong JW, Liu YB, Zhao H, Liang CQ, Huang RJ, Park KS, Feng SS et al (2021) Fosl1 is vital to heart regeneration upon apex resection in adult Xenopus tropicalis. NPJ Regen Med 6:36
25. Feschotte C (2008) Transposable elements and the evolution of regulatory networks. Nat Rev Genet 9:397–405
26. Sun C, Lopez Arriaza JR, Mueller RL (2012) Slow DNA loss in the gigantic genomes of salamanders. Genome Biol Evol 4:1340–1348
27. Meyer A, Schloissnig S, Franchini P, Du K, Woltering JM, Irisarri I, Wong WY, Nowoshilow S, Kneitz S, Kawaguchi A et al (2021) Giant lungfish genome elucidates the conquest of land by vertebrates. Nature 590: 284–289
28. Wang K, Wang J, Zhu C, Yang L, Ren Y, Ruan J, Fan G, Hu J, Xu W, Bi X et al (2021) African lungfish genome sheds light on the vertebrate water-to-land transition. Cell 184(1362–1376):e1318
29. Currie JD, Kawaguchi A, Traspas RM, Schuez M, Chara O, Tanaka EM (2016) Live imaging of axolotl digit regeneration reveals spatiotemporal choreography of diverse connective tissue progenitor pools. Dev Cell 39: 411–423
30. Kumar A, Nevill G, Brockes JP, Forge A (2010) A comparative study of gland cells implicated in the nerve dependence of salamander limb regeneration. J Anat 217:16–25
31. Farkas JE, Freitas PD, Bryant DM, Whited JL, Monaghan JR (2016) Neuregulin-1 signaling is essential for nerve-dependent axolotl limb regeneration. Development 143:2724–2731
32. Tsai SL, Baselga-Garriga C, Melton DA (2020) Midkine is a dual regulator of wound epidermis development and inflammation during the initiation of limb regeneration. Elife 9:e50765
33. Monaghan JR, Maden M (2012) Visualization of retinoic acid signaling in transgenic axolotls during limb development and regeneration. Dev Biol 368:63–75
34. Duerr TJ, Jeon EK, Wells KM, Villanueva A, Seifert AW, McCusker DD, Monaghan JR (2021) A constitutively expressed fluorescence ubiquitin cell cycle indicator (FUCCI) in axolotls for studying tissue regeneration. BioRxv. https://doi.org/10.1101/2021.03.30.437716
35. Cura Costa E, Otsuki L, Rodrigo Albors A, Tanaka EM, Chara O (2021) Spatiotemporal control of cell cycle acceleration during axolotl spinal cord regeneration. Elife 10:e55665
36. Zhang Z, Denans N, Liu Y, Zhulyn O, Rosenblatt HD, Wernig M, Barna M (2021) Optogenetic manipulation of cellular communication using engineered myosin motors. Nat Cell Biol 23:198–208
37. Fukazawa T, Naora Y, Kunieda T, Kubo T (2009) Suppression of the immune response potentiates tadpole tail regeneration during the refractory period. Development 136: 2323–2327
38. Godwin JW, Pinto AR, Rosenthal NA (2013) Macrophages are required for adult salamander limb regeneration. Proc Natl Acad Sci U S A 110:9415–9420
39. Tsujioka H, Kunieda T, Katou Y, Shirahige K, Fukazawa T, Kubo T (2017) Interleukin-11 induces and maintains progenitors of different cell lineages during Xenopus tadpole tail regeneration. Nat Commun 8:495
40. Debuque RJ, Nowoshilow S, Chan KE, Rosenthal NA, Godwin JW (2021) Distinct toll-like receptor signaling in the salamander response to tissue damage. Dev Dyn 251(6): 988–1003
41. Hui SP, Sheng DZ, Sugimoto K, Gonzalez-Rajal A, Nakagawa S, Hesselson D, Kikuchi K (2017) Zebrafish regulatory T cells mediate organ-specific regenerative programs. Dev Cell 43:659–672, e655
42. Wells-Enright KM, Kelley K, Baumel M, Vieira WA, McCusker CD (2021) Neurotrophic control of size regulation during axolotl limb regeneration. BioRxv. https://doi.org/10.1101/2021.04.27.441633
43. Sandoval-Guzman T, Wang H, Khattak S, Schuez M, Roensch K, Nacu E, Tazaki A, Joven A, Tanaka EM, Simon A (2014) Fundamental differences in dedifferentiation and stem cell recruitment during skeletal muscle regeneration in two salamander species. Cell Stem Cell 14:174–187
44. Elewa A, Wang H, Talavera-Lopez C, Joven A, Brito G, Kumar A, Hameed LS, Penrad-Mobayed M, Yao Z, Zamani N et al (2017) Reading and editing the Pleurodeles waltl genome reveals novel features of tetrapod regeneration. Nat Commun 8:2286
45. Arenas Gomez CM, Gomez Molina A, Zapata JD, Delgado JP (2017) Limb regeneration in a direct-developing terrestrial salamander, Bolitoglossa ramosi (Caudata: Plethodontidae): limb regeneration in plethodontid salamanders. Regeneration (Oxf) 4:227–235
46. Sun C, Mueller RL (2014) Hellbender genome sequences shed light on genomic expansion at

the base of crown salamanders. Genome Biol Evol 6:1818–1829

47. Palacios-Martinez J, Caballero-Perez J, Espinal-Centeno A, Marquez-Chavoya G, Lomeli H, Salas-Vidal E, Schnabel D, Chimal-Monroy J, Cruz-Ramirez A (2020) Multi-organ transcriptomic landscape of Ambystoma velasci metamorphosis. Dev Biol 466:22–35
48. Dwaraka VB, Smith JJ, Woodcock MR, Voss SR (2019) Comparative transcriptomics of limb regeneration: identification of conserved expression changes among three species of Ambystoma. Genomics 111:1216–1225
49. Tracy KE, Kiemnec-Tyburczy KM, DeWoody JA, Parra-Olea G, Zamudio KR (2015) Positive selection drives the evolution of a major histocompatibility complex gene in an endangered Mexican salamander species complex. Immunogenetics 67:323–335
50. Diaz Quiroz JF, Tsai E, Coyle M, Sehm T, Echeverri K (2014) Precise control of miR-125b levels is required to create a regeneration-permissive environment after spinal cord injury: a cross-species comparison between salamander and rat. Dis Model Mech 7:601–611
51. Cook AB, Seifert AW (2016) Beryllium nitrate inhibits fibroblast migration to disrupt epimorphic regeneration. Development 143: 3491–3505
52. Hutchison C, Pilote M, Roy S (2007) The axolotl limb: a model for bone development, regeneration and fracture healing. Bone 40: 45–56
53. Rozenblit F, Gollisch T (2020) What the salamander eye has been telling the vision scientist's brain. Semin Cell Dev Biol 106:61–71
54. Ryczko D, Simon A, Ijspeert AJ (2020) Walking with salamanders: from molecules to biorobotics. Trends Neurosci 43:916–930

Index

A

B

C

D

E

Ashley W. Seifert and Joshua D. Currie (eds.), *Salamanders: Methods and Protocols*,
Methods in Molecular Biology, vol. 2562, https://doi.org/10.1007/978-1-0716-2659-7,

T

W

Z

Printed in the United States
by Baker & Taylor Publisher Services